Praise for *Total Load Theory*

From the Medical Community

"In *Total Load Theory*, Patricia Lemer distills dec
and human empathy into a work of rare scope ar
both science and storytelling, she reframes the cc
tal, societal, and physiological stressors into a col
the power to transform lives. Lemer's expansion

to encompass the full spectrum of neurodevelopmental and psychiatric ...
visionary and urgently needed. Every chapter is meticulously constructed, moving seam-
lessly from foundational science, spanning polyvagal theory, immunology, toxicology, and
the microbiome, to pragmatic, evidence-based interventions that restore function and
resilience. What sets this book apart is its capacity to speak to both the clinician seeking
depth and the caregiver seeking hope, bridging the gap between technical mastery and
deep human healing. *Total Load Theory* is not just a book, it is a landmark contribution
to the fields of developmental neuroscience and integrative health, and a guide for anyone
committed to dismantling the barriers to optimal human potential."

> **—Christian Bogner, MD, FACOG, CFMNP, and cofounder,
> Researched Elements LLC & Autism is Biomedical, Inc.**

"A well-organized and passionately written book offering a multidimensional exploration
of the potential and multifactorial causes of neurodevelopmental disorders. Patricia S.
Lemer presents a comprehensive synthesis of both Eastern and Western healing modalities, equipping families and caregivers with practical tools and informed perspectives.
With deep compassion and clarity, she helps readers navigate the complexities of our
increasingly toxic, fast-paced, and electromagnetic world, empowering them to make
informed choices for healing and prevention."

> **—Dr. Rob Brown, diagnostic radiologist and fellow of the American
> College of Radiology, vice president for Scientific Research and
> Clinical Affairs, the Environmental Health Trust**

"*Total Load Theory* brings together Lemer's decades of dedication to supporting individuals and families. It offers a wide-ranging look at how genetics, environment, nutrition,
sensory development, communication, and life skills all play a role in health and learning.
This comprehensive resource provides practical strategies and fresh perspectives for those
seeking to better understand and support neurodiverse individuals."

> **—Emily Gutierrez, DNP, CPNP, APRN, PMHS, CCN, IFMCP**

"Children with autism often face more than developmental challenges—they carry a silent
weight of environmental toxins. This book explores the total body burden of heavy metals,
chemicals, pollutants and other toxins, revealing how they impact the brain, immune system, and behavior. With clear explanations and actionable strategies, Patty offers parents
hope for reducing exposures and supporting healthier outcomes for people struggling
with neurodevelopmental issues. This is a must-read for any parent looking to help their
child thrive."

> **—Darin Ingels, ND, FAAEM, FMAPS**

"Patty has been on this journey with me for over 25 years working with patients with special needs. The ideas and strategies that Patty outlines for identifying and addressing our toxic load are the very basis for helping individuals with ADD/HD, autism, anxiety/OCD, and neurodevelopmental delays because this population is particularly vulnerable. This book is a must-read for parents who are trying to raise healthy children in this toxic world."

—Anju Usman Singh, MD, FAAFP, FMAPS, medical director and owner of True Health Medical Center and Pure Compounding Pharmacy

From Organizations

"In *Total Load Theory*, Patricia Lemer masterfully expands her groundbreaking framework to address not just autism but a spectrum of neurodevelopmental disorders—from ADHD and learning disabilities to sensory processing issues and mental health challenges. Lemer illuminates how an endless barrage of environmental insults can trigger genetic and epigenetic changes that undermine and overwhelm our bodies' capacity to methylate and detoxify the massive toxic load we face daily. For years, Patty has been an invaluable interviewer for my organization, the Autism Research Coalition. Her knowledge is deep and wide. She is a beacon of truth and sanity for desperate parents, and this book cements her legacy as a visionary guide to healing and prevention."

—Enrique Basurto, Founder Autism Research Coalition

"An insightful book that expands the scope of Total Load Theory beyond autism to address a wide range of neurodevelopmental and mental health challenges. Clearly written and informative, it offers practical strategies to identify and reduce stressors. It provides a compelling case for clinicians, educators, families, and individuals to approach complex conditions through the lens of total load management. Important reading for anyone seeking a holistic, transformative approach to improving quality of life for those affected by neurodevelopmental disorders and mental health concerns."

—Stephen M. Edelson, PhD, chief science officer, Autism Research Institute

"Patricia Lemer, author of the best-selling book *Outsmarting Autism, Updated and Expanded* believes, 'We have to look at *TOTAL LOAD*!'—a term she coined. And yes, I think she is right. In this groundbreaking new book, Lemer dives deep into the root causes, offering a comprehensive roadmap for transforming health for those with learning and behavioral disorders. In twenty-three guided chapters, she covers everything—diagnosis, genetics, environmental factors (from pollution to water to vaccines), societal factors, and detoxification—and provides prioritized therapies.

You'll need this one, and you'll probably want a few extra copies to share."

—Mary S. Holland, Esq., CEO of Children's Health Defense

"Patricia Lemer's *Total Load Theory* is the ultimate resource for parents and practitioners looking to understand not only why children are so sick today but also how these conditions can be fully reversed. This book gives tremendous hope!"

—Beth Lambert, founder of Documenting Hope and author of *A Compromised Generation* and *Brain Under Attack*

From Practitioners

"Patricia Lemer's *Total Load Theory* is an extraordinary resource for parents navigating autism, ADHD, learning challenges, and mental health struggles. Building on the brilliance of *Outsmarting Autism*, this new book goes deeper—shining light on the hidden stressors that contribute to a child's struggles and offering practical, holistic strategies to reduce the total load. It's insightful, thorough, and deeply rooted in decades of hands-on experience.

I've purchased *Outsmarting Autism* for every single client I've ever worked with—because it's that essential. Parents tell me it's the most dog-eared, underlined, and trusted book on their shelf. And I'll be doing the same with *Total Load Theory*. Patty continues to set the gold standard—and this book is another generous gift to families everywhere."

—Len Arcuri, cofounder of All In Parent Coaching, host of Autism Parenting Secrets Podcast

"Patricia Lemer coined the phrase 'autism spectrum' to describe the range of neurological and developmental impairment in children and was first to introduce the concept of 'total load' to assessment and treatment. Now she has produced the ultimate resource for parents, clinicians, and policymakers dedicated to recovery of compromised children affected by the most devastating epidemic of the modern age. Read it, learn it, share it!"

—Hannah Vance Bradford, MAc, MBA, health researcher and parent coach

"After five decades of research in the field of health and wellness, Patty Lemer's new book, *Total Load Theory* is truly a phenomenon. Step by step, this remarkable navigational tool forensically uncovers, and unravels, the etiologies of so many chronic illnesses. Before you throw your hard-earned money at yet another 'therapy of the day,' pick-up Patty's new book. Whether you're a healthcare professional, or just beginning your quest toward greater health, *Total Load Theory* will empower you to become an even more astute, and savvy, health and wellness consumer—for you and your family."

—Mary Coyle, DIHom

"*Total Load Theory* is an encyclopedic resource for parents and clinicians to understand how the overwhelming world around us influences the inner world. Patty Lemer breaks down how environmental, emotional, and nutritional inputs shape cellular responses, nervous system reactivity, and behavioral expression. Regardless of a particular diagnosis, we are all impacted by the physiological consequences of the world we live in."

—BG Mancini, AP, PCP, MSOM

"*Total Load Theory* is both inspiring and empowering. Patricia Lemer uncovers the hidden stressors behind today's epidemic of childhood health and developmental challenges and offers a clear, actionable path to healing. This book is essential for professionals committed to whole-person care and for families seeking hope, answers, and lasting solutions for their children."

—Svetlana Masgutova, PhD, founder of the MNRI® / Masgutova Method® and Indie Medal award-winning author of *Autism: Maximizing Brain Potential with MNRI Reflex Integration*

"Patricia Lemer's Total Load Theory is not only a remarkable book—it's a true paradigm shifter. In today's world of one-size-fits-all medicine, where the 'answers' from health practitioners are too often pharmaceuticals and other quick fixes, this book stands apart. I cannot recommend this book more highly. It will be one of my go-to resources, a book I consult often. And while Patricia is a well-known children's advocate, Total Load Theory is also a 'must read' for adults—even those with no children!"

—**Julia Schopick, patient advocate and author of Honest Medicine and The Power of Honest Medicine**

From Optometry

"Patricia Lemer understands the profound role of vision and vision development on a child's behavior and success, and how visual considerations are often masked or overlooked. She is an authentic soldier in this epidemic scourge of childhood disorders, and at her core, she is a great advocate for kids' health!"

—**Hans F. Lessmann, OD, developmental optometrist at Vision Development Institute**

"Patricia Lemer's new book crucially expands on the concept of 'total load theory.' It represents the culmination of Patty's long and impressive career in counseling and advocacy for children who are labeled but not well understood. She makes very useful distinctions between genetics and environmental load factors as stressors and as opposed to societal load factors. Whereas the former includes susceptibility to toxic elements in food and in the air, the latter features stressors that can provide an unbearable load in the physical, educational, language, and emotional domains. This book is a wonderful addition to the library of any parent or professional looking to reduce load factors from prior conception and throughout the lifespan."

—**Leonard J. Press, OD, FAAO, FOVDR**

"Patricia Lemer has done it again! In this, the fourth and latest of a series of well researched and comprehensive books, she gives actionable steps to remediate and care for those clearly overburdened in our society. Ms. Lemer has been committed to addressing vision concerns due to its importance for so many areas of life. She gives a clear overview of vision and the visual skills needed for moving, learning, thinking, and engaging with our world in such a way that parents and professionals can easily understand and appreciate the necessity of addressing deficits in visual skill development. She outlines the ways to identify vision difficulties and gives the resources to help, including the appropriate vision evaluation and treatment using the optometric power tools of lenses, prisms, colored light filters, and vision therapy."

—**Randy Schulman, MS, OD, FCOVD**

From Parents

"*Total Load Theory* reveals a profound truth: It's not just one factor that overwhelms a child, but the cumulative burden of many stressors, that creates what Patricia Spear Lemer calls the 'Total Load.' In this groundbreaking book, Lemer brings clarity, compassion, and decades of expertise to one of today's most urgent challenges: understanding and supporting children with autism and other complex neurological conditions. Rather than viewing autism as a fixed label, she reframes it as the body and brain's response to an overload of stressors—an insight that opens new pathways to healing. Drawing on her extensive clinical experience, Lemer provides a roadmap for identifying root causes and relieving the total load. Her approaches empower parents and practitioners to move beyond symptom management toward integrative strategies that foster resilience and recovery. More than just a theory, *Total Load Theory* is a beacon of hope. With warmth, wisdom, and an unwavering belief in every child's potential, Lemer shows families that healing is not only possible—it's within reach."
—Judy Ryder Duffy, founder of the "Power Moms: Autism" Facebook groups

"Patricia Lemer is the most caring, knowledgeable, and well-researched person I know on the subject of diagnosing and prioritizing therapies for those with learning and behavioral disorders. Her third book sits prominently at eye level on my shelf for easy and frequent reference. It is the only book that I consistently recommend to families seeking treatment alternatives. Now, adding to her literary accomplishments comes *Total Load Theory* targeting a broader audience with the latest steps for maximizing function. She did *ALL* the legwork so you can focus on parenting. This is the only book you need to read. Period."
—Filomena LaForgia, founder of Filanthropists.com

"Patty Lemer is beyond a maverick in the autism and special needs communities. Soon after my daughter was born and diagnosed with hypotonia (a useless label), and a neurologist claimed she would never speak, I attended a talk featuring Patty. She blew my mind and changed the course of my daughter's life and mine. Today, Isabella never stops talking and leads a fulfilling life at a gluten-free cafe and pottery studio and as an equestrian. Patty also impacted the direction of my career, guiding me as registered dietitian nutritionist to return to my original passion about the impact of food on mood."
—Laura Lagano, functional nutritionist

"If there was one book I would recommend to parents of children with autism, ADHD, or other developmental delays, it would be this one! Patricia painstakingly leaves no stone unturned, when researching to find the root cause of symptoms. After having applied many of the things I learned from reading Patricia Lemer's books, my son is now a thriving, happy, and confident highschooler, free of many of the symptoms he once had!"
—Dr. Carrie Fick, optometrist

"At just four years old, my twin son was diagnosed with autism. I was fortunate to discover the work of Patricia Lemer, whose intelligent and holistic approach to uncovering the root causes of autism proved nothing short of transformative. Through her guidance, my son has made remarkable progress—he now engages positively at school and in social settings, free from the anxiety and dysregulation that once held him back. Lemer's work has been a true blessing to our family, and I am deeply grateful for her wisdom and dedication."
—Aleta St. James, bestselling author and spiritual energy healer

International Praise

"Patricia Lemer has spent her career helping children and people with disabilities through the most comprehensive possible methods. She is THE EXPERT who I totally trust for my work with autistic children and families in Vietnam. We loved her book *Outsmarting Autism* so much that we translated it into Vietnamese. *Total Load Theory* has been my main guidance for working on the core challenges of children with autism and ADHD. I have no doubt that this new book will transform many more lives."

—Hoa Le, director of GANH XIEC NHA JU, Vietnam

"This book offers a refreshing and empowering perspective on autism and other childhood developmental disorders. By introducing the Total Load Theory, it helps parents and professionals alike understand the complex interplay of genetics, environment, and lifestyle in shaping a child's health. I am in awe of Patricia Lemer. At 80 years old, she has poured her heart, her decades of wisdom, and her life's work into this incredible book. As a parent, I am so grateful for the way she asks the tough 'why' questions about childhood developmental disorders—and then goes further to offer clear, practical answers and treatment options. This book feels like a gift and a legacy of knowledge that will guide families for years to come. Patricia's dedication and passion shine through every page, and her work gives parents like me both hope and direction on how to truly help our children thrive."

—Ilana Gerschlowitz, author of *Saving My Sons: A Journey with Autism*, director of The Star Academy (Center for Autism), South Africa

"Patricia Lemer has done it again. Her books are eloquently written, covering a wide gamut of treatments and therapies for autism and related disorders. What is unique is the manner in which she chronicles the sequence in which to introduce the various therapies. Never content to rest on her laurels, Patricia is constantly updating her resources every few years as new research and new treatments come available. If you are a parent or a practitioner who desires to learn more about autism treatments, this is the one book you should buy. I highly recommend her books to all my patients, friends and colleagues."

—Sandra Morton Weizman, DMH, DHHP, Calgary, Alberta, Canada

TOTAL LOAD THEORY

TRANSFORMING LIVES IN AUTISM, ADHD, LD, SPD, AND MENTAL HEALTH

PATRICIA S. LEMER
Foreword by James Maskell

Skyhorse Publishing

Children's
Health Defense

Visit our website at www.skyhorsepublishing.com.
Please follow our publisher Tony Lyons on Instagram @tonylyonsisuncertain.

10 9 8 7 6 5 4 3 2 1

Library of Congress Cataloging-in-Publication Data is available on file.

Paperback ISBN: 978-1-64821-203-1
eBook ISBN: 978-1-64821-204-8

Cover design by David Ter-Avanesyan

Printed in the United States of America

This book is dedicated to all the remarkable mothers, fathers, grandparents, siblings, teachers, and others who have supported "special" kids and taught me empathy and patience.

CONTENTS

FOREWORD

By James Maskell

I have long believed that the transformation of our healthcare system would not be driven by policy alone, nor by top-down reform, but by families—ordinary people with extraordinary courage—whose lives have been disrupted by the modern epidemic of chronic illness. In this moment of historic change, with leaders like Robert F. Kennedy Jr. at Health and Human Services and Dr. Marty Makary at the FDA, we are witnessing the emergence of a new paradigm. One where complex chronic illness is no longer a mystery, but a solvable puzzle—with children at the center of the story.

That story is not easy. It is full of heartbreak, sleepless nights, dashed expectations, and unimaginable strength. But it's also the story of how families, doctors, and thinkers across the country are beginning to see the full picture of how we got here. And that picture is best understood through the lens of total load.

Patricia Lemer's book arrives at a crucial inflection point. The term "total load" refers to the cumulative burden of stressors—environmental, chemical, emotional, infectious, dietary, and more—that overwhelms an individual's biology and leads to dysfunction. It's a concept that is simultaneously simple and revolutionary.

Simple because it mirrors common sense: The more weight we carry, the harder it is to move. Revolutionary because modern medicine has become so reductionist, so myopically focused on single causes and single interventions, that it has failed to see the obvious—that a sick person is not broken, but burdened.

And in America, that burden is heavier than anywhere else.

We live in the most toxic culture on Earth. We are exposed to more synthetic chemicals, artificial foods, sensory overload, screen time, EMFs, antibiotics, and stress than any generation before us. Our food is depleted, our soil poisoned, our air increasingly unbreathable, and our water often laced with pharmaceuticals. But it's not just the physical environment—it's the social and emotional terrain as well. Broken families. Isolated communities. Fractured education systems. Medical models that prioritize suppression over investigation. Children are absorbing it all.

And they are responding.

Autism. ADHD. OCD. PANS/PANDAS. Anxiety. Sensory processing disorders. Learning disabilities. Allergies. Autoimmunity. These are not random misfortunes. They are patterned responses to overwhelming load. And in those patterns lies the answer. Children are showing us where the system is broken. They are the canaries in the coal mine.

Total Load Theory is more than a book. It's a map. A map for families to navigate the minefield of modern childhood. A map for practitioners to see beneath the surface symptoms and uncover root causes. A map for policymakers to finally understand the hidden costs of our industrial way of life. And a map for all of us—because if we can build a society where the most vulnerable can thrive, we all benefit.

Patricia Lemer is uniquely qualified to write this book. Her decades of service—first through Developmental Delay Resources, then Epidemic Answers, and now through this masterwork—have laid the foundation for a new kind of medicine. One rooted not in the pharmaceutical pipeline, but in the lived experience of parents, caregivers, and clinicians who have been on the front lines. I first connected with Patty through the Functional Forum, when she helped build one of the early meetup groups in Pittsburgh. It's been incredible to witness her evolution, and to now write this foreword is both an honor and a full-circle moment.

This book is comprehensive, courageous, and clear. It expands the scope of *Outsmarting Autism* to encompass not just developmental disorders, but the full spectrum of modern pediatric chronic illness—including trauma, nervous system dysregulation, dentistry, vision therapy, reflex integration, and more. It is as much a practical guide as it is a philosophical reckoning.

As a health economist, I often think about incentives and systems. Our current system incentivizes sickness. Chronic disease is a business model, not a medical failure. But what happens when enough families opt out? When they demand a different path? When they begin to heal, and then turn to help others do the same? We get a movement. We get momentum. We get a new era.

This book captures that moment—and propels it forward.

If you are a parent, clinician, policymaker, or concerned citizen, this book will open your eyes. It will challenge you to see beyond diagnoses and into dynamics. It will introduce you to the pioneers who are forging new roads, and the brave families walking them. It will ground you in science, but inspire you with story. Most of all, it will help you understand that healing is not only possible—it is happening.

We are entering the next chapter of medicine. One that values connection over control. One that respects the intelligence of the body and the wisdom of lived experience. One that honors the child as the teacher.

Patricia's work reminds us that, while the path is complex, the compass is clear: lighten the load. Support the terrain. Restore what's been lost. Reconnect what's been broken. And listen—really listen—to what the children are telling us.

Because they are not just sick.

They are signaling.

And now, finally, we're starting to listen.

—James Maskell
Author, *The Evolution of Medicine* and *The Community Cure*
Founder, Functional Forum & TruNeura
El Dorado Hills, CA

ACKNOWLEDGMENTS

This book is the culmination of a lifetime asking *why* questions that had few answers. Fortunately, experts are now able to answer those endless questions, having discovered reasons for physical and mental distress. As I turn 80 next year, my gratitude is boundless.

I dedicate this book to the families who have taught me so much about biology, courage, patience, perseverance, and empathy. Some of you have trusted me with counseling three generations. I am so grateful for the road I have traveled in these crazy, constantly changing times.

My deepest thanks to health visionary leader James Maskell for his beautiful foreword. It is proof that the way to get something you need and want is to ask a busy person! We have known each other since he started his Functional Forum in 2014, and I jumped on board as the organizer of the Pittsburgh chapter. *Total Load Theory* resonates with everything James believes in.

I am so grateful to the following not-for-profit organizations for their leadership and the friendships they have fostered:

*The Autism Research Institute (ARI) and its founder, Bernie Rimland, and Medical Director, Liz Mumper. Bernie encouraged Kelly Dorfman and me to found Developmental Delay Resources in 1993. What an experience it was to run it for 20 years!

*The Optometric Extension Program (OEP) and the Optometric Development and Vision Rehabilitation Association (ODVRA), to whom my last book was dedicated. You gave me a professional family and "vision." OEP honored me as the first non-optometrist to deliver Regional Clinical Seminars, published my first book, several journal articles, and awarded me for my work. You appointed me as a bridge between optometry and the public by encouraging me to teach consumers about vison and optometrists about working with schools and other professionals. OVD invited me to speak twice to their members, ten years apart, on the biomedical and other sensory aspects of disabilities. I cannot thank you enough, especially Bob Williams, Len Press, Lynn Hellerstein, Randy Schulman, Mary VanHoy, and the late Irwin Suchoff.

*Documenting Hope (formerly Epidemic Answers) and Executive Director Beth Lambert, and Media Director Maria Rickert Hong for giving DDR and its board of directors a new home and for promoting the concept of Total Load Theory.

*American Academy of Environmental Medicine (AAEM), whose founders originated the term "Total Load Theory" as related to medicine. Attending their conference in 2023 was the origin of this book. Thank you to Barry and Linda Smeltzer for introducing me to the work of Neil Nathan and Peter McCollough, to Darin Ingels, who edited much of this book, and Jessica Tran for being my guides.

Autism is Biomedical (AiB) and its cofounders, Christian Bogner and Alex Zaharakis, the next generation of fathers giving back their knowledge, who have welcomed me into their fold as a senior advisor.

I am also grateful to the following people:

*Physicians extraordinaire who have led me to resources, edited my work, and taught me so much: Letran Hoang, Larry Palevsky, Jared Skowron, and Anju Usman Singh.

*Peter Sullivan for your beautiful updated chart on environmental stressors.

*Others who contributed time and knowledge: Len Arcuri, Teresa Badillo, Hannah Bradford, Preston Brooks, Mary Coyle, Anne Dachel, Emily Gutierrez, Ryan Hedrick, BG Mancini, Julia Schopick, Lauren Stone, Kristan Weisdack, Sandra Weizman, Deb Wilson, I value and thank you.

*My "Dream Team" at Skyhorse Publishing, editor Zoey O'Toole, Lou Conte, Hector Carosso, and Brian Peterson. You are simply *the best*! Thank you, Tony Lyons, for believing in this book and making it the "audition" for Skyhorse entering the AI world.

*The extraordinary advocates and warriors at Children's Health Defense. Thank you RFK Jr. for spearheading this amazing organization and Mary Holland for leading it with such grace.

*Amr Mohammed for the unbelievable honor of being the first book *ever* with a searchable function.

*Friends Filomena Laforgia, Janine Burnham Ruth, Mary Rentschler, Joanne Spence, Kristi Wees, and the boyfriend, daughter, and granddaughter for whom I have been too busy to be there.

*And those who have kept my body, mind, and house in tip-top shape while I work: Kathy, Sarah, and Michael.

My humble gratitude and thanks.

—Patty

PREFACE

Total Load Theory: Transforming Lives in Autism, ADHD, LD, SPD, and Mental Health is the most complete book addressing the epidemic of illness among young people today. Not only does it address the numerous environmental and genetic factors that contribute to neurodevelopmental disorders, it also provides practical solutions for the many physical and mental symptoms that accompany them. It is the first book *ever* to offer an AI search engine, allowing you to get the answers to your important questions immediately.

More than ten years ago, when the original *Outsmarting Autism* was published, I was excited because it included simply *everything* anyone could possibly need to know about the possible etiologies and treatments for those with autism diagnoses. And it did . . . until we learned more about genetics, environmental triggers, detoxification, the gut, and the immune system. As the first wave of kids with autism transitioned into adulthood, advanced laboratory testing, new technology, novel treatments, and exciting employment options became available, forcing me to update what I had written.

Now, even the updated and expanded edition of *Outsmarting Autism* is outdated. When it went out of print in 2024, I thought of putting my 50 years of writing to rest. However, my followers had other ideas. They said, "No, the information in this book is too valuable to be retired." They encouraged me to update once again and expand the audience beyond autism as the same approach can help relieve symptoms of countless other conditions with similar physical roots.

The timing was perfect! Metabolic psychiatrists—like Christopher Palmer, Georgia Eide, and Robert Lustig—recently hopped onto the autism bandwagon and have begun teaching that "mental illness" starts in the body and affects the mind. My lifelong dream that mental health practitioners become open to the possibility that our lifestyle choices—especially around diet, sleep, and exercise—affect both our physical and mental health was coming true!

A "spectrum" of disorders has been grouped under the heading "autism," and many adults are embracing that diagnosis to explain their quirky behaviors and sensory differences. Thanks to pioneers like Vander Kolk and Stephen Porges, we learned about trauma, the nervous system, and the vagus nerve. Now we were at the real roots of illness!

Then we learned that just because someone does not speak, it doesn't mean they have nothing to say. A whole generation of nonspeaking autistics had begun to communicate; how tragic that we had "dumbed them down" throughout their schooling! "Assuming competence" has resulted in a new version of *Awakenings* as their personalities have emerged through letterboards.

Robert F. Kennedy Jr. has been appointed as Secretary of Health and Human Services and is mandating taking toxins out of our air, food, and water. He has sparked interest in the potential risks of over-vaccinating and has put experienced vaccine scientists in positions of authority.

Please don't let this book's size scare you. Yes, it's *big*! But I have made it extremely easy to navigate. Not only does it have a remarkable index, it is also the first book of its kind with a search feature using artificial intelligence (AI). **Just visit ai.patricialemer.com, set up an account, and try it out.** I think you will be intrigued by this add-on, which also includes a search of over 100 episodes of interviews from my bimonthly podcast, *The Autism Detective*. I am over the moon to be pioneering this feature!

How to Use This Book

I hope you will read about Total Load in sequence, but if you decide not to, at least read Chapters 1–6 before jumping ahead. Understanding the concept of Total Load Theory and the many stressors that comprise it is so important!

Beyond that, here are a few guidelines, depending on your background and purpose for reading:

- *For parents or grandparents of a child diagnosed within the past two years:* Please start at the beginning and read chapter by chapter. Every child is unique, and your child deserves an individualized approach to intervention. Which stressors does your family have? Chapter 6 on the importance of feeling safe is vital to getting well. Finding a healthcare practitioner who knows a variety of approaches and is willing to delve into your family history with pointed questioning and testing is necessary to act as a "case manager" and guide you. You may think you can do this yourself, but support is essential.

    Be careful not to fall into the trap of taking therapies "off the shelf." What your school system offers and what insurance pays for may or may not be all your child needs. Palliative care and mere management of symptoms is not acceptable for the newly diagnosed. Go step by step with the goal of possible healing in mind. These disorders may lessen or recede altogether! We don't know how much progress we can make until we get started and see what happens.

Take care of yourself while you are caring for your family members. Follow this strategy as if you are on an airplane and the flight attendant says to put on your own oxygen mask first—you will need it. As you apply your new knowledge, everyone in the family will become healthier.

- *For families whose member(s) were diagnosed over two years ago and have been doing interventions for a while:* The key to true restoration of your family's health is in *your* hands, not in the school, clinic, or therapist's office. Everything you do at home has the potential to make the efforts of your school and outside therapies work better.

- *For veterans who have been in the trenches for many years* and whose children are adolescents transitioning from school or adults for whom living arrangements and employment are considerations: since my previous book was published, we have learned more about "turning off" or mitigating the effect of problematic genes through supplementation and detoxification, the importance of certain nutrients to eliminate specific symptoms, and the preferred order of interventions. Revisit some of the therapies you may have sampled in the past; approaches are much more sophisticated than they used to be. Technology and social networking have changed the world of disabilities. If you are not using them, now is the time to start.

- *For adults who have never thought about the physical roots of their symptoms before:* the ideas in this book may feel like they are out of left field, but they have all helped countless individuals to feel and function better—many of them with symptoms or issues very similar to yours. If you keep an open mind, you may just find something in these pages that changes your life for the better.

- *For educators and licensed or certified specialists who provide services in clinics, schools, or privately:* put new tools in your tool chest and approach this book with an open mind. So many brilliant therapists have packaged multidisciplinary programs to address motor, language, social skills, and cognition by focusing on the foundations. These materials are perfect for schools, home therapy, and summer programs.

- *For information junkies who are just curious as to why so many kids are sick and delayed:* the information in this book will surprise and amaze you! Our media is not giving you the whole story. Be prepared to have your questions answered. Use the AI tool to ask them. Discover step by step what you can do to protect your own health and that of those you love.

Total Load Theory: Transforming Lives in Autism, ADHD, LD, SPD and Mental Health is meant to challenge some of your beliefs. Be available to this exciting new information, some of which goes against mainstream philosophy. If you have questions, please feel free to contact me through my website at PatriciaLemer. com.

Whether you are a parent, grandparent, therapist, doctor, educator, or an interested consumer, I believe you will find my fourth book invaluable. Thank you for reading it!

—Patricia S. Lemer
Fall 2025

CHAPTER 1

TOTAL LOAD THEORY: A NEW PARADIGM FOR ENDING TODAY'S EPIDEMIC OF CHRONIC ILLNESS

About one in six American children, more than 17%, aged 3–17 years are diagnosed with a developmental disability. And this is not just an American phenomenon according to the Global Burden of Disease Study back in 2019.[1] Today's kids are not only delayed, they are also sick, anxious, and depressed. And you know at least one, or you would not have picked up this book.[2]

Included in this group are the one in thirty-one with autism (3.2%),[3] one in nine with attention deficit/hyperactivity disorder (ADHD) (9.5%),[4] one in five with a learning disability (LD) (7.9%),[5] and one in six with anxiety disorders (8%),[6] among others. According to the Pew Research Center, that adds up to over seven million children in the United States;[7] UNICEF reports almost 240 million children worldwide.[8]

Most of these statistics are more than five years old and still rising.[9] Shocking isn't it? Is this an "epidemic," defined by the CDC as "the occurrence of more cases of disease than expected in a given area or among a specific group of people over a particular period of time"?[10] Some healthcare professionals think so, and I agree.

For over four decades I was a diagnostician whose career was based on assisting parents in determining why their children were having difficulty learning, behaving badly, or not talking. My intake history usually revealed a myriad of health issues that accompanied and often preceded these developmental, behavioral, and symptom-based diagnoses. Allergies, asthma, reflux, constipation, seizures, "failure to thrive," sleep issues, ear infections, vaccine reactions, and others were common in these kids' health histories. Was there a relationship between these health problems and developmental disorders?

So many concurrent conditions were too important to ignore. Growing up in the 1950s, and during college and graduate school in the 1960s, I knew almost no one with developmental disorders. The only "mental illness" we learned about in my psychology, neurology, education, and counseling courses was schizophrenia. Depression combined with too much anxiety was called having a "nervous breakdown." There was no talk of autism, attention deficits, or obsessive-compulsive disorder.

I needed answers. How was it possible that so many kids were struggling? What was behind the numbers? I watched in alarm as, over the years, my clients presented with more and more serious lags in cognition, speech and language, reading skills, and socialization.

New diagnoses with acronyms such as ADD, ASD, CDD, SPD, OCD, and PDD began emerging as psychiatrists, psychologists, and neurologists tried hard to name and blame each disorder by describing symptoms. I resisted using these terms, which I viewed as merely an alphabet soup that didn't get to the root of the problem.

What Is Total Load Theory?

Total Load Theory is a concept from engineering that explains why, as a heavy truck travels over a bridge, the structure suddenly collapses. Who or what is to blame? The truck driver? The trucking company? The engineer who designed the bridge? The weather? The ship captain whose tanker bumped into the moorings endless times? None of these is the single cause, even though each stressor contributed to the outcome. An accumulation of numerous stressors caused the bridge's collapse.

Have you ever had a bad day at work, gotten caught in traffic on the way home, yelled at your kids for something insignificant, and, as you fell into bed, felt a cold coming on? You experienced the same phenomenon as that bridge: an excess of "**Total Load**," the cumulative effect of multiple individual assaults on the body.

Hans Selye (1907–1982), a Viennese-born physician who spent most of his career in Canada, is considered the founder of the stress theory. The word "stress" was used in physics to refer to the interaction between a force and a resistance or counterforce, but it was Selye who first incorporated this term into the medical vocabulary to describe the "nonspecific response of the body to any demand." He named the body's response to accumulated stressors that eventually led to disease the *general adaptation syndrome*. It later became known as *Selye's syndrome*. Selye was nominated for the Nobel Prize in 1949 and wrote his best-known book, *The Stress of Life*,[11] in 1956.[12]

Like I do, Selye rejected the study of specific disease signs and symptoms and instead focused on patients' bodily responses to stress. His concept of stress is now accepted universally in many fields of medicine to explain disease.

In the 1970s, I learned about a group of doctors, mostly allergists, including the late William Rea and Doris Rapp, who founded what is now known as the American Academy of Environmental Medicine (AAEM). They were the first to recognize and apply "Total Load" to the accumulation of underlying stressors as the cause of their patients' health woes.

In the 1980s, I read a paper, "ADD: Acronym for Any Dysfunction or Difficulty."[13] The author's need to know *why* kids were struggling resonated with me. Their parents seemed to be "good enough," a concept introduced by pediatrician Donald Winnicott back in 1953,[14] and they were growing—but not thriving. What was going on?

In the 1990s, my colleagues and I started applying this engineering theory to the multitude of developmental, psychological, and behavioral disorders of childhood we were seeing in our practices. We posited that, just like a bridge, every person has a unique combination of stressors and a unique load limit.

Envision a wellness threshold, below which individuals are "healthy" and above which they are "sick." As load factors accumulate and move a person closer and closer to their personal line, a single assault may be enough to take them over the top: the proverbial "straw that breaks the camel's back."

In 2005, microbiologist Christopher P. Wild coined the name of a new field: *exposomics*, defined as "the measure of all the exposures of an individual in a lifetime and how those exposures relate to health." Each individual has a unique combination of stressors that mount up and can gradually overload their "bridge" to its breaking point. Every stressor adds to the body's burden or Total Load. An individual's cumulative exposure, called the *exposome*, begins before birth, and includes insults from our environment, diet, and lifestyle. These insults interact with our genetics and impact our health.[15]

As physical and environmental stressors mount, they cause nervous, immunological, digestive, neurological, psychological, respiratory, and other systems to weaken. Issues in these biological systems coexist with developmental, sensory, motor, language, social-emotional, cognitive, and attentional symptoms as an individual's body struggles and approaches its personal limit, and their relationship is very complex. The larger the load of stressors, the more severe the attention, behavior, and cognitive difficulties.

Even though the relationship between stress and disease is now well established, scientists are only beginning to understand the complex synergy among stressors and the extent to which they affect causation. They increasingly recognize, however, that many of today's conditions of both child- and adulthood are a result of high Total Load.

The Total Load Theory of childhood disorders encompasses not only ADHD and autism, but many other psychiatric conditions, such as obsessive-compulsive disorder, panic attacks, and even bipolar disorder, as well. It considers many possible etiologies for physical and "mental" illness, as various stressors combine with individual biological factors to produce an infinite number of outcomes. Whatever diagnosis we end up with is the end product of many systems of the body being stressed to their limits.

Diagnosis

Diagnosis depends on whether the purpose is for insurance companies to allow or deny reimbursement for treatment (medical) or to qualify for school services (educational). Unlike other chronic conditions, such as diabetes, these childhood disorders have no laboratory test that nails the label.

For a school-age child, a medical diagnosis does not guarantee educational services. Each school system has its own procedures for determining placement and eligibility for services. Small districts may just review medical and other reports and approve a student for services. In large, complex school districts, the process is more complicated because special education law now allows for over a dozen categories of disability. The multidisciplinary team could decide that a young student diagnosed with autism by his doctor is "speech-language impaired," "developmentally delayed," and/or "other health impaired." It's even possible for a highly functioning student to have the school system deny services altogether.

A specific diagnosis is applied *only* when an individual exhibits certain specific behavioral symptoms *and* an expert gives a subjective opinion that those behaviors and other symptoms fit a certain category. Often, a child's complex of symptoms satisfies more than one category, and the child receives more than one diagnosis; the disabilities are then known as *comorbid* or *co-occurring*. An interesting result of designating comorbid conditions is the emergence of yet additional acronyms, such as *AuDHD* for someone who has *both* autism and ADHD, which strikes me as ridiculous. Do you know anyone with autism who *doesn't* have ADHD? Maybe a handful.

The Diagnostic and Statistical Manual (DSM)

The tool of choice for medical purposes is the *Diagnostic and Statistical Manual of Mental Disorders* (DSM), published under various titles since 1952 by the American Psychiatric Association. The DSM is updated every ten to twelve years and has experienced seven revisions. The newest version, *Diagnostic and Statistical Manual of Mental Disorders—Fifth Edition Text Revision* (*DSM-5-TR*), was released in March 2022[16] and is over 1,000 pages, up 150 pages from the previous tome. Many parents are shocked to learn that the diagnostic determination for these disorders is "psychiatric."

The DSM lists hundreds of "mental" disorders that traditional psychiatry believes are **not caused** by the physiological effects of one or more medical conditions. Remember this exclusion. It is critically important, because it is on this point that Total Load Theory disagrees. Each and every psychiatric condition listed in the DSM and considered in this book can have one or more biological medical conditions as a root cause.

Here is a chart demonstrating possible root causes for the disorders that follow:

	Food	Nutritional Deficiencies	Mold	Bacteria/ Viruses	Reflexes	Vision	Heavy Metals	Vagus Nerve	Sensory
ADHD	X	X	X	X	X	X	X	X	X
ASD	X	X	X	X	X	X	X	X	X
OCD	X	X	X	X	X	X	X	X	X
ANXIETY	X	X	X	X	X	X	X	X	X
BIPOLAR	X	X	X	X	X	X	X	X	X
SLD	X	X	X	X	X	X	X	X	X
SPD	X	X	X	X	X	X	X	X	X

That's correct! *Everything* is caused by *everything*, in an infinite number of combinations!

Attention Deficit/Hyperactivity Disorder (ADHD)

The DSM-5 classifies ADHD as a neurodevelopmental disorder. To be diagnosed with ADHD, a patient must have a persistent pattern of inattention, disorganization, and/or hyperactivity-impulsivity that negatively affects their development and functioning. Inattention and disorganization result in the apparent inability to stay on task, pay attention, and/or organize the materials needed for tasks at levels that are inconsistent with age or developmental level. Hyperactivity-impulsivity entails overactivity, fidgeting, intruding into other people's activities, and/or an inability to wait or stay seated. The number of symptoms required for a diagnosis depends on the type of ADHD and the patient's age.

Autism Spectrum Disorder (ASD)

ASD is characterized by impairments in social interaction and communication in the presence of repetitive and stereotyped behavior, interests, and activities. This includes deficits in social-emotional reciprocity, social interaction, and in developing and maintaining relationships. Common behaviors include strong adherence to routines, ritualized patterns, highly restricted or fixated interests, and hyper- or hypo-reactivity to sensory input.

The DSM-5-TR has changed a little in its language regarding the presence of criteria, replacing the word "persistent" with "all" to describe the intensity of behaviors. Revision from the DSM-IV to the DSM-5, however, eliminated several discrete diagnoses and clumped them as mild (*Level 1*), moderate (*Level 2*), and severe (*Level 3*). Individuals with previous diagnoses of autism, pervasive developmental disorder (PDD), and pervasive developmental disorder, not otherwise specified (PDD-NOS), became simply "autistic." Childhood disintegrative disorder (CDD), Rett syndrome, and Asperger syndrome were not only no longer considered to be forms of autism, they were eliminated from the DSM-5 altogether.

With this landmark revision more than ten years ago, many people who had previously been diagnosed with one of the disorders that no longer exist—and researchers who focused on them—freaked out. They feared that removing the Asperger and other diagnoses from the DSM would affect long-term research on these populations and they could lose their supports and services.

Members of more than 20 organizations banded together and wrote a letter to the Lancet Commission objecting to the term "profound autism." Thus, the first shot was fired in "The Autism Wars." The DSM-5 reduced all categories of autism down to a single term, "autism spectrum disorder." Many speculated that renaming and excluding some individuals was an attempt to bring down the numbers and thus further deny an epidemic.[17]

The spectrum is not linear and could be presented in a variety of ways, with considerable overlap. Many agree that "when you've seen one person with autism, you have seen one individual with autism." You simply cannot generalize.

Let's consider how the concept of a "spectrum" arose. While individuals with a specific diagnosis show specific symptoms, the range of cognitive, social, language, and attentional strengths and deficits is extremely broad. Historically, "autism" only applied to those who were unable to function in mainstream society, including those who are now considered Level 3 and some of Level 2. Until recently, few people realized that individuals whose symptoms were *so* different were members of the same diagnostic category. How could someone with a PhD who teaches at a university and a nonspeaking, not toilet-trained, self-injurious individual both have "autism"? My opinion is that they should not be classified together.

It is these individuals at the more severe end of the spectrum that Bobby Kennedy spoke so passionately about in April 2025, when he declared that "autism destroys families" and that these individuals will "never hold a job, play baseball, pay taxes, get married, write a poem, or use a toilet unassisted." They are those children for whom the Profound Autism Alliance was founded in 2023 by the Ursitti family with the mission "to improve their health and connection through inclusive research and focused advocacy that will result in meaningful services and supports." It is to these families, whom I have worked with for 30 years, that this book is dedicated.

Specific Learning Disorders (SLD) or Learning Disabilities (LD)

Specific learning disorders are neurodevelopmental disorders characterized by persistent impairment in at least one of three major academic areas: reading (speed, accuracy, and comprehension), written expression (spelling, grammar, organization, and punctuation), and/or mathematics (concepts, calculation, and reasoning). Each has its own name (*dyslexia*, *dysgraphia*, and *dyscalculia*, respectively) and criteria for inclusion; the level of skills must be substantially below what is expected for age, as well as cause problems in school and everyday activities or at work. In addition to naming the type of disorder, the severity is designated as *mild*, *moderate*, or *severe*. Learning disorders are typically diagnosed in elementary school but may not be recognized until adulthood, when academic, work, and day-to-day demands are greater.

Obsessive-Compulsive and Related Disorder (OCD)

Earlier editions of the DSM classified OCD as an anxiety disorder, but the latest edition allots OCD its own distinct category. OCD is characterized by repetitive and persistent thoughts, often accompanied by repetitive behaviors the individual feels compelled to perform in order to relieve the obsessive thoughts. To qualify for this diagnosis, these obsessions and compulsions must take more than an hour per day, or considerably impact the individual's ability to work, socialize, or otherwise function. An individual is specified to have either *good*, *fair*, or *poor* understanding of whether their obsessive thoughts are "unrealistic" or "untrue."

If a patient with an OCD diagnosis has a history of strep infection followed by tics or OCD, a doctor should immediately suspect and test for *pediatric autoimmune neuropsychiatric disorder associated with streptococcus* (PANDAS) or *pediatric acute-onset neuropsychiatric syndrome* (PANS). Read about testing for these in chapter 5 and what goes wrong and how to fix it in chapter 8.

"Mental Illness" or "Mental Health Disorders"

NAMI, the National Alliance on Mental Illness, the nation's largest grassroots mental health organization, classifies about a dozen disorders, including autism,

ADHD, and OCD, under the umbrella of mental illness. Other well-known diagnoses are depression, psychosis, schizophrenia, and eating disorders. Two deserve elaboration for the purposes of this book:

Anxiety Disorder

Anxiety disorders are characterized by persistent, excessive fear or worry in situations that are not threatening. People typically experience both emotional symptoms, such as restlessness and dread, as well as physical symptoms, such as sweating and diarrhea. Anxiety disorders can manifest themselves chronically in everyday life socially, as extreme shyness for example, as sudden panic or terror, phobias, or as selective mutism. Anxiety disorders can affect sleep, appetite, relationships, and a myriad of other areas.

Bipolar Disorder

This "mental illness" is characterized by dramatic shifts from high to low mood, energy, and ability to think clearly—the poles of which are known as mania and depression—which differ from the typical ups and downs most people experience. Symptoms and severity vary, from mild to so severe that the sufferer cannot get out of bed when depressed or needs restraining or hospitalization when manic.[18]

Disorders Not Included in the DSM

Nonverbal learning disability (NLD/NLVD) is defined as a set of strengths in verbal memory and vocabulary, accompanied by visual-spatial, fine motor, and social difficulties that include decoding body language and understanding inference and humor.[19]

Sensory processing disorder (SPD) is manifested by hyper- or hyporeactivity to sensory input. Read more about this disorder in Chapter 15.

A new diagnosis, *pathological demand avoidance disorder*, or *pathological drive for autonomy*, both abbreviated *PDA*, can also be known as rational demand avoidance (RDA), extreme demand avoidance (EDA), or demand avoidance phenomena (DAP). (No, I am not kidding you.) While many believe this to be a subset of autism spectrum disorders,[20] I believe it is a potentially protective sensory processing disorder of varying degrees.

Despite prodigious, continuous efforts, neither NLD, SPD, nor PDA is included as a standalone condition in the DSM-5. Why? Probably because experts have not agreed upon definitions or they view them as comorbid conditions of specific learning disabilities, autism spectrum disorder, oppositional defiant disorder, or conduct disorders.

Neurodiversity and Person-First Language

Today, the neurodiversity movement is a social justice crusade that seeks civil rights, equality, respect, and full societal inclusion for anyone whose neurology varies from "normal" in any way. In the past 30 years, it has morphed from an early focus on "awareness" to an aggressive move for "acceptance."

A discussion of the complex subject of neurodiversity is included here to clarify the book's intention of embracing differences while improving and even transforming lives, without judgment. We are all neurodiverse. Acceptance of everyone's unique differences is the *only* option.

Origins

The term ***neurodiversity*** is attributed to Judy Singer, an Australian student of sociology and anthropology who, in the late 1990s, coined the word to describe her mother's "singular oddness, unusual body posture, harsh unregulated voice, egocentricity, and inability to sense what others were feeling, or how their minds worked."[21]

As the World Wide Web became increasingly accessible during the late nineties, several leaders in the autism world found each other and began communicating online. Temple Grandin, Stephen Shore, Donna Williams, Valerie Paradiz, and others bonded over their high levels of achievement despite often being written off as "hopeless" when they were preverbal young children.

The Language of Neurodiversity

Neurodiversity is a way of saying that human brains are all different from each other; neurodiverse brains, just like fingerprints, are unique. The word derives from *neuro*, referring to the nervous system, and *diversity*, referring to variations or differences. Neurodiversity is a biological fact in humans; no two brains are the same.

The ***Neurodiversity paradigm*** is a specific perspective on neurodiversity that holds these differences are natural and valuable to society. Because humans are neurodiverse, everyone's nervous system processes their sensory world differently, resulting in different experiences. Some people have shared differences in their experiences and reactions that distinguish them from the majority; their differences are similar enough to each other that a single term like "autism" or "learning disabled" can describe them.

Neurotypical (NT) means having a style of brain functioning or thinking that results in "typical" behavior that falls within the dominant standards in a society. NT people do not have brain-based disabilities.

When someone's significant differences have to do with how information from the senses is processed by the brain, they are described as **neurodivergent (ND)**. The brains of ND people function significantly differently cognitively, sexually, or otherwise from the dominant societal standards of "normal." Individuals diagnosed with autism, dyslexia, or many of the other DSM-5 diagnoses, whose brains function differently from dominant societal norms, fall in this category.

The neurodiversity paradigm rejects pathologizing any form of neurodivergence. It supports the idea that there is no such thing as a "normal" brain, and that those with brain-based conditions like autism, intellectual disabilities, learning disabilities, or debilitating mental health conditions like anxiety should be accepted and included in society just like neurotypical people. No one can argue with that!

But the paradigm also includes the contradictory idea that neurodiverse conditions affecting the way some people learn, move, communicate, and experience the world should be considered developmental disabilities that qualify people for the accommodations they need to maximize their potential.

I agree that disabilities are a natural part of human diversity. And who wouldn't agree with the paradigm's premise that neurodiverse people should be allowed to exist and that society should work to make sure that everyone gets the accommodations they need to reach their full potential?

The neurodiversity movement got a huge boost in 2015 because of the bestseller, *Neurotribes: The Legacy of Autism and the Future of Neurodiversity*, by the late Steve Silberman (1957–2024). Silberman, a multiply neurodiverse science writer argued that "conditions like autism, dyslexia, and attention deficit/hyperactivity disorder (ADHD) should be regarded as naturally occurring cognitive variations with distinctive strengths that have contributed to the evolution of technology and culture, rather than mere checklists of deficits and dysfunctions."[22] Not since Kanner blamed autism on "refrigerator mothers" has so much controversy arisen. While "geeky" techies in Silicon Valley rejoiced in being understood, parents of offspring who require 24/7 supervision were horrified; for them, neurodiversity was not a gift!

Person-First Language, Yes or No?

Person-first language (PFL) is the idea that when referring to people with disabilities one should emphasize the person first and the disability second: Susie is a "person with a disability," rather than Susie is a "disabled person." Ever since I started working with children with cerebral palsy and Down syndrome (person first) in the 1960s, I have used person-first language. Most advocates I know prefer this word order.

Recently, however, some ND adults have objected to this attempt at respect and asked to be referred to by nouns like "Autists" (with or without a capital A). Jim Sinclair, one of the early members of the neurodiversity movement defends this stance:

> Saying "person with autism" suggests that the autism can be separated from the person. But this is not the case. I can be separated from things that are not part of me, and I am still the same person. . . . But autism is part of me. Autism is hard-wired into the ways my brain works. I am autistic because I cannot be separated from how my brain works.

He continues, "We talk about left-handed people, not 'people with left-handed-ness,' and about athletic or musical people, not about 'people with athleticism' or 'people with musicality.'"[23]

Jane Strauss, a mother of five diagnosed in her fifties, concurs: "I am no more a 'person with autism' than I am a 'person with femaleness' or a 'person with Jewishness' or a 'person with cleverness' or a 'person with photographic skill.' I am an Autistic, Jewish, clever, woman photographer."[24]

As I have learned more about neurodiversity, I came to understand their reasoning. Henceforth, I will always put the person first by asking and respecting what each individual wants to be called. In this book, I use both styles—not because language doesn't matter, but because I respect individuals on both sides of the issue.

Another person on the forefront of advocating for and helping the public understand neurodiversity is John Elder Robison. He teaches that neurodiversity is a culture, akin to Deafness. Many of the problems faced by autistic adults in modern society, he says, are not just their own individual challenges, but also societal failures. He feels that we must make societal changes that integrate autistic people into the modern working world.[25]

A Social Versus Medical Model of Neurodiversity

Many in the neurodiversity movement, like Robison, divide thinking into two camps: the Medical Model and the Social Model.

- *The Medical Model* pathologizes neurodivergent individuals and recommends extinguishing "undesirable" behaviors with the use of medications, counseling, and behavior-modifying therapies so that a neurodivergent person can "fit in" and conform to an inflexible "normal" society.

- *The Social Model* blames society, claiming it is overly structured and lacks sufficient flexibility to accommodate differences and meet the sensory and other needs of the neurodivergent.[26]

Clearly, society *is* the problem when it does not accommodate people with disabilities. That's why we have laws prohibiting discrimination. But I believe both models miss the mark. I strongly and passionately support the premise that *all* behavior results from meaningful attempts to meet sensory and other needs. If the behavior seems odd—like the irrational rages associated with PANDAS— underlying causes should be investigated and mitigated whenever possible. Simultaneously, everyone should work toward a more flexible and inclusive society that accommodates individual differences.

Brain inflammation, immune and digestive system dysregulation, compromised detoxification, vitamin deficiencies, and autoimmunity are not "diversity." They are medical conditions that can cause all types of behavior. *Everyone* with one or more of the DSM diagnoses is physically sick. You wouldn't qualify for the diagnosis if you weren't struggling. When you address the above biological dysfunctions, the brain gets better along with the rest of the body and struggling decreases.

Are you or your child one of these people? Wouldn't you like to feel better and depend less on the world's willingness to adjust to your ills? If so, keep reading. I can help you.

We all have the same goals: healthy, happy, productive citizens. Let's be respectful of each other, our sameness and our differences.

What a Diagnosis Means

A new diagnosis of *anything* can be overwhelming and confusing to a family! What everyone must understand is that it merely means that an individual's symptoms match a specific cluster of behaviors. Individuals with the same diagnosis may have similar symptoms; however, they do not necessarily require the same treatments. Prescribing treatment requires identifying a *cause*.

Consider the following example:

Symptom: A pounding pain in the right front temporal lobe
Diagnosis: Headache

Possible Causes	Treatment for That Cause
Tension and stress	Aspirin, bed rest
Brain tumor	Surgery
MSG poisoning	Alka-Seltzer Gold
Nagging spouse	Divorce

If a person with a headache has not ingested Chinese food, bottled salad dressing, or another food containing MSG and is unmarried, causes can be narrowed down to stress or a brain tumor. While this example might seem frivolous, it clearly demonstrates the difficulty inherent in determining cause and appropriate treatment.

Are Childhood Disabilities Overdiagnosed?

Many cynics and some scientists believe that the recent dramatic rise in childhood illness, especially autism, is not real. In their extreme, these activists loudly deny any epidemic of childhood illness and refute any notion of environmental causes.

Eminent, Yale-educated autism physician Sid Baker has stated,

> Explanations about how autism was always there, and not increasing . . . are nonsense. It's surprising that some people still believe this, considering that the data is now very solid discounting the idea that better diagnosis, diagnostic substitution, and genetics can fully account for the rise.[27]

Clearly the public is far more aware of autism and other diagnoses today than in the past. So many people are learning about the characteristics of these disorders that they are even diagnosing themselves and family members!

Though we are indeed doing a far better job of identifying those whose symptoms fit the criteria, it's not clear whether brilliant "geeks" in history—such as Einstein, Thomas Jefferson, or even Steve Jobs—or those diagnosed in their forties or later, should have the *same* diagnosis as young children with severe delays and behaviors, or even a diagnosis at all. Perhaps we should join Steve Silberman and call them neurodiverse without pathologizing them.

Cynthia Nevison makes an excellent argument that a combination of increased awareness, better identification, and more inclusive diagnostic criteria is insufficient to explain the numbers occurring worldwide.[28] In addition to her compelling data, it simply is not credible that so many children with a seriously debilitating condition would go unnoticed for so long, that well-educated parents would remain silent for so long, and that professionals would suddenly wake up—and all at the same time!

I find it difficult to deny the reality of the epidemic of childhood illness. Autism and ADHD may be slightly overdiagnosed, but parents don't go to the trouble of getting a diagnosis if a child isn't struggling. The worldwide epidemic of childhood illness is real. Still in doubt? I recommend Mark Blaxill and Dan Olmstead's rigorous historical analysis of psychiatric disorders in childhood,

Denial: How Refusing to Face the Facts about Our Autism Epidemic Hurts Children, Families and Our Future.[29]

Prevalence

Numbers are rising! Autism is the country's fastest-growing developmental disability according to the Autism Society of America, with a 10–17% annual growth rate.[30] Look at this graph:

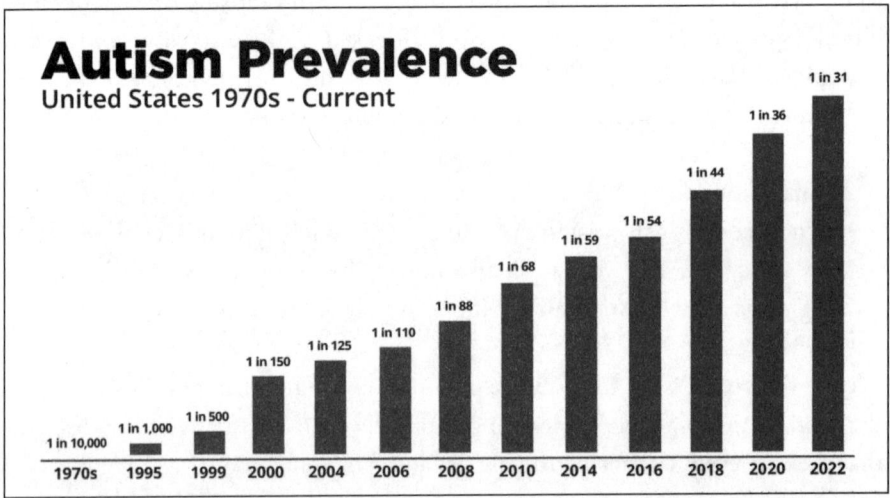

Autism Prevalence
United States 1970s - Current

| 1970s | 1995 | 1999 | 2000 | 2004 | 2006 | 2008 | 2010 | 2014 | 2016 | 2018 | 2020 | 2022 |
| 1 in 10,000 | 1 in 1,000 | 1 in 500 | 1 in 150 | 1 in 125 | 1 in 110 | 1 in 88 | 1 in 68 | 1 in 59 | 1 in 54 | 1 in 44 | 1 in 36 | 1 in 31 |

One would think such a dramatic rise in a developmental disorder would be cause for alarm, but the US government largely ignored autism until a little less than twenty years ago. The CDC is still unwilling to call it an "epidemic." In fact, they had the audacity to report inaccurately that "U.S. autism rates appeared to be stabilizing" between 2014 and 2016.[31]

At the end of 2006, legislators came out of their slumber and finally recognized the need to focus some attention on autism by passing the Combating Autism Act. Now known as Autism CARES (for Collaboration, Accountability, Research, Education and Support), this Act has undergone several reauthorizations to expand services, most recently in 2024. The latest Act ensures support for research and prevalence tracking through September 2029, at the cost of billions of dollars to the taxpayer.[32] But there's so much the Act has left undone.

Disability Law

Discussing childhood disorders is impossible without some understanding of two very important federal laws: the Individuals with Disabilities Education Act (IDEA) and the Americans with Disabilities Act (ADA). Entire books have been written about the interpretation of these laws; an overview will have to suffice here.

Education for All Handicapped Children Act (EHA) and Individuals with Disabilities Education Act (IDEA)

When I began my career in the late 1960s, no laws existed to mandate services. In 1975, Congress passed Public Law 94–142 or the Education for All Handicapped Children Act (EHA), which assured a free and appropriate public education (FAPE) to school-aged children with over a dozen handicapping conditions, including visual and hearing impairments, physical disabilities, and cognitive deficits, but not autism. Public schools are required to design and implement an Individualized Educational Program (IEP) tailored to each child's specific needs.

For about 30 years, I was part of teams that determined eligibility for services at all levels of education from preschool through college. We focused almost solely on those who were deaf and/or blind or had cognitive and physical impairments or learning disabilities. No autism.

EHA was reenacted in 1990 as the Individuals with Disabilities Education Act (IDEA) and is up for re-authorization every five years.[33] This revision brought with it a couple of important changes: it extended services to developmentally delayed children under age five, making it possible for young children to receive help even before they begin school (known as "early intervention"). And it added autism as a new category of disability. The timing is of particular interest here, as autism did not show up on the radar screen in 1975 but was prevalent enough in 1990 to be written into disability law. Furthermore, it expanded the category "Other Health Impaired (OHI)" in IDEA to include attention deficit disorder.

The Americans with Disabilities Act (ADA)

In 1990 (a banner year for disabilities) the US government passed the Americans with Disabilities Act, extending services beyond secondary school into college and the workplace.[34] Now individuals with all types of disabilities, including autism, could receive special attention beyond high school and age 21.

Post-secondary placements were now required to establish guidelines to accept, hire, and accommodate individuals with disabilities when they graduated from high school and sought further education and jobs. Community colleges, vocational schools, residential colleges, and universities struggled with what to do. They built ramps for wheelchairs—that was a no-brainer. But what types of accommodations were necessary and appropriate for those with invisible issues such as anxiety disorders?

Employers faced the same dilemmas. The law said they must make "reasonable accommodations" to help workers who have any handicap do their job. Such accommodations could include shifting job responsibilities, modifying equipment, or adjusting work schedules. The ADA also guarantees equal employment opportunity for people with disabilities and protects disabled workers against job discrimination.

To learn about how these laws impact students as they age out of schools and enter the workplace, read Chapter 17.

IDEA and ADA enumerate the rights of individuals with a variety of handicapping conditions, starting from birth and continuing through school attendance, college, and in the workplace. States have some leeway about how to interpret these federal laws. The scope and breadth of each state's services depends upon both their budgets and numbers of individuals with disabilities. Some offer excellent services with copious hours of "related services" such as occupational therapy, speech-language therapy, and even one-on-one aides. Others place all students in mainstream classes, arguing that inclusion is better for everyone, and barely support teachers and pupils with any related services. Only when parents legally challenge whether services are appropriate or students have been placed in the least restrictive environment (LRE) in which they can be successful (requirements of IDEA) are changes made.

Some lawyers specialize in disability law. Many have made their reputations by taking school systems and employers to court over violations of their clients' disability rights. For more than 20 years, I testified as an expert at hundreds of due process hearings, most resulting in my clients obtaining additional services for their children.

Old Thinking Versus New Thinking About Treating Neurodevelopmental Disorders

Parents sometimes think that getting the right diagnosis will be the hard part, but they usually find that the next step can be anything but easy. The relationship between symptoms and the best interventions is not always linear, but rather a synchronistic matrix with multiple variables. Every family must tenaciously search for the right combination of "magic bullets" that will provide relief, and then save the money to cover the costs.

What interventions can jump-start language, reduce the need for self-stimulatory or self-soothing behaviors, or increase social interactions? And who is going to pay? Three choices: an insurance company, taxpayers, or the family; most often money comes from all three pockets. Nobody wants to touch this hot potato!

Old Thinking: Manage Childhood Conditions by Eliminating Symptoms

Historically, medical, psychological, and educational approaches focused on *managing* behaviors with medication, special education, counseling, and applied behavior analysis (ABA). Services are often provided through departments of "behavioral health." The ultimate goal is to replace hyperactivity, attentional difficulties, anxiety, depression, mood swings, agitation, aggression, self-injurious behavior, insomnia, perseveration, and impulsivity with relatedness, eye contact, self-control, attention span, and confidence.

Medications

The goal of medication, as well as all other treatments, is to maximize an individual's functioning.[35] True, medications sometimes alleviate behavioral and attentional symptoms, but often with undesirable side effects. Pharmacologic intervention typically is viewed as only one component of a multi-modal treatment plan, and drugs are usually prescribed one at a time so that effectiveness and side effects can be accurately determined. Unfortunately, some clinicians recommend drug "cocktails" that allegedly balance out one another's side effects.

Since they were introduced over 60 years ago, medications have become more specialized and powerful. Medications used for "mental illnesses," including, but not limited to, anxiety, bipolar disorder, depression, OCD, schizophrenia, are

- *Antipsychotics.* Examples: Abilify, Risperdal, and Zyprexia.
- *Anticonvulsants/mood stabilizers.* Examples: Depakote, Klonopin, Topomax, and Valium.
- *Antidepressants.* Examples: Celexa, Luvox and Prozac.
- *Antihistamines.* Reduces allergic reactions. Example: Benadryl.
- *Antifungals.* Used to treat yeast infections. Examples: Diflucan and Nystatin.
- *Antipurinergics.* Used in ASD to improved language and socialization. Example: Suramin
- *Stimulants.* Available in short- and long-acting formulas for ADHD to target focus. Examples: Adderall, Concerta, Ritalin, and Strattera.

Many of these medications are used off-label. Minor to severe health-related side effects are inevitable with nearly all drugs, so use of all medications must be monitored by a physician.[36] In addition, behavioral issues can crop up as secondary drug effects. It is imperative that all healthcare providers evaluate a patient's medications when taking a health history and become familiar with the side effects of individual drugs by reading the latest *Physicians' Desk Reference*. They must then decide whether sleep disturbances, constipation, seizures, and potentially fatal heart problems are acceptable trade-offs for increased attention and decreased aggression.

Since 1964, psychiatrist Peter Breggin, MD, has been outspoken about the dangers of a traditional psychiatric approach. He has published dozens of scientific, peer-reviewed articles and about 20 professional books, including the first and only medical book devoted wholly to the dangers of psychiatric drugs to both body and brain.[37]

From 1967–2009, the Autism Research Institute (ARI) and its beloved founder, Bernard Rimland, PhD, collected ratings from over 27,000 parents on the behavioral effects of over 40 medications on individuals with autism. For the majority of drugs, parents reported improved behavior less than 50% of the time.

James B. Adams, PhD, Professor at Arizona State University (ASU) and Director of the Autism/Asperger's Research Program, picked up where ARI left off. He compiles research on the efficacy of medications, as well as nutrients and diets, on the functioning of those with autism.[38]

Adams's recent findings are consistent with those found over the years by ARI: parents report that drugs can sometimes improve their children's behavior, but far less well than biomedical interventions and often with undesirable side effects. See Chapters 7–11 for how diets, vitamins, minerals, and other interventions that address underlying imbalances can be helpful.

Behavior Management and Special Education

While behavior management and special education provide external methods of monitoring and handling behavior, they may fail to allow the child's own sensory systems to learn from experience how to modulate and integrate information and to develop internal controls. Counseling programs help parents cope with chronic issues such as fecal smearing, picky eating, and sleep problems, but do not address possible causes.

The benefits of these symptom-alleviating treatments are short term at best. At worst, benefits may not outweigh side effects. Also of great concern is the huge long-term financial burden placed on both the healthcare and the educational systems, which were not designed to manage so many children.

New Thinking: Ameliorate these Conditions by Determining Underlying Causes and Eliminating Them

Total Load Theory is a new paradigm. New-thinking **professionals** believe that the variety of childhood conditions we're seeing is the outward manifestation of an accumulation of stressors on the body—starting with a dysregulated nervous system and including exposure to toxins, malfunctioning digestive systems, poorly functioning detoxification pathways, and immune system failures—all of which I address throughout this book. They want to *heal* children, not just prescribe palliative treatments.

New-thinking **parents** *know* that their kids' sensitivities, digestive issues, allergies, vaccine reactions, and behavioral and learning issues are related, and that these relationships deserve closer attention. They recognize that traditional medical treatments and educational management techniques can certainly have palliative effects, but they are looking for better outcomes than simply helping children compensate.

New thinkers understand and address underlying biological issues. They subscribe to what veteran physician Leo Galland, MD, calls "patient-centered diagnosis." They look for antecedents, genetic and environmental risk factors that predispose a person to an illness, and triggers, those straws that break the camel's back. They depend upon in-depth interviews, thorough physical examinations, laboratory tests, and often their intuition and experience to determine the kaleidoscope of possibilities that makes each patient unique.[39]

Shebani Sethi, MD, is credited with introducing the term "Metabolic Psychiatry." Mental health professionals all over the world who were frustrated by the poor results they were getting with pharmaceuticals and counseling started looking at diet, nutrition, and lifestyle factors in their clients. Metabolic psychiatry is applying a multidisciplinary Total Load approach to so-called mental illness.

I had the privilege of hearing highly respected veteran psychiatrists, neurologists, nutritionists, biochemists, psychologists, environmentalists, and other professionals share their experiences about the latest scientific research in October 2024 at the Integrative Medicine & Mental Health Conference near Washington, DC. I was blown away that, *finally*, these professions were exploring biochemical, psychological, lifestyle, and other factors in what we have previously called mental health disorders. All of them spoke about the *end* of the concept of "mental" illness and embracing the idea that *all* illness starts in the body and affects the brain. Mental disorders are metabolic disorders of the brain. A dream come true for me!

Veteran pediatric neurologist Martha Herbert, MD, and renegade physician Michael J. Goldberg, MD, agree. In her book *The Autism Revolution,*[40] Herbert targets foundational systems like immunity, gut function, and detoxification, because autism is neither a "disease" nor a "brain disorder," but rather a "disorder that affects the brain." In *The Myth of Autism,*[41] Goldberg says, "Call it whatever you want, but don't call it autism." He calls it the *neuro-immune dysfunction syndromes (NIDS)*.

Chiropractic neurologist, Anthony Ebel, DC, represents the new generation of doctors who refuse to treat behavioral and emotional symptoms. For him, nervous system dysfunction and dysregulation underlie chronic illness in our kids, and focusing on neurology *must* precede working on the gut and the immune system. He uses a tree analogy, where the nervous system is the roots and the gut and immune system are represented by the trunk. The soil is all the environmental factors affecting the tree's growth and development. Learn more about Dr. Ebel throughout this book and access his network of drug-free practitioners at PXDocs.com.

World-renowned physician, Neil Nathan, MD, ties all these ideas together in his most recent book, *The Sensitive Patient's Healing Guide*.[42] His explanation of what happens to the nervous system and protocol for calming it *before* trying to heal other systems is now recognized by many new thinkers as key to long-term healing and eventual health. This book subscribes to his approach which targets the limbic system, vagus nerve, and an inflammatory and immune condition called *mast cell activation*. Unless these areas are addressed first, the brain does not feel safe; lack of safety prevents healing.

Because new thinkers understand individual variability, they believe that subscribing to a generic treatment plan is just plain stupid. One size doesn't fit all. So, they evaluate each individual's unique history, identify possible causes, and set priorities. As interventions address causes one by one, stressors fall away, freeing up new energy in the body for development and learning. Each patient requires a different path because of biological uniqueness.

Where Can I Turn for Help?

Treatments and therapies for many disorders have become big business in a thriving marketplace of interventions delivered by talented and experienced practitioners, most of whom have something to offer. New approaches are emerging every day, claiming to be the missing link. I recommend you seek therapy, advice, and support soon after diagnosis. Do not "wait and see."

But the looming question is "from whom?" Today's parents face almost *too* many options. Let's start with doctors:

Pediatrician—Most families suspect "something is wrong" well before getting a formal diagnosis. Many beg their pediatricians to make a referral for early intervention.

Since 2007, the American Academy of Pediatrics (AAP) has strongly urged their member physicians to routinely screen babies for autism at 18 and 24 months and not to wait for a diagnosis before referring them for services.[43] Unfortunately, those screenings often have significant drawbacks: they happen at the same time as a round of vaccines and often result in dead-end therapies.

In the past, doctors collaborated with their patients as partners and took parents' concerns about the CDC vaccine schedule seriously. More recently however, pediatricians began "firing" vaccine non-compliant families from their practices, resulting in families being forced to seek alternative healthcare providers.

Once a pediatrician agrees something is amiss, they refer the family to a medical specialist, most likely a neurologist or developmental pediatrician, who in turn usually sends them for palliative approaches that only target symptoms, such as speech and language therapy for toddlers who are not talking or physical therapy for those whose motor skills lag.

Neurologist—This referral is a "name it and blame it" attempt. A child who isn't talking, may be aphasic. A child who isn't learning may be dyslexic. Once in a blue moon, a child's issues come from a brain tumor or a seizure disorder that can be addressed by a neurologist; in my over 40 years working with families, I rarely saw any real help come from a neurologist.

Developmental pediatrician—These children's doctors are specialists for kids with a myriad of genetic, language, motor, and behavioral disorders. Some developmental pediatricians may be "old thinkers" who suggest a trial of medication, behavioral management, and special education that provides intense language, motor, and social-emotional therapies. Lately, however, more and more are suggesting treatments that focus on physical symptoms, like the immunological, dermatological, digestive, sensory, neurological, respiratory, cognitive, psychological, and developmental markers that preceded the diagnosis.

Functional medicine doctor—A new breed of physician has emerged! These well-trained doctors may be MDs, DOs, naturopaths (NDs), chiropractors (DCs), or doctors of Traditional Chinese Medicine. What they have in common, is that they have post-graduate training from the Institute of Functional Medicine (IFM). Functional medicine is a multidisciplinary approach that addresses the unique physical, mental, and emotional needs of all patients. It combines lifestyle, nutrition, exercise, environmental, structural, cognitive, emotional, and pharmaceutical therapies that take a comprehensive, "whole person" approach to treat the root causes of disease and restore healthy function.

This approach was conceived in 1991 by Jeffrey Bland, PhD, a biochemist. For the past decade functional medicine has been championed by visionary James Maskell, the founder of the Evolution of Medicine (GoEvomed.com) and the author of the foreword to this book. Today, thousands of healthcare practitioners are trained in functional medicine. Go to IFM.org to access a list of certified professionals worldwide, as well as courses in all aspects of functional medicine.

Chiropractic neurologist—These doctors of chiropractic medicine have taken 300 hours of post-doctorate education and training in functional neurology and passed rigorous tests. Many are also certified functional medicine doctors. This hybrid profession marries the biomechanical aspect of chiropractic care with the latest techniques of assessment and rehabilitation of the central nervous system. To find a chiropractic neurologist near you, go to ACNB.org.

MAPS doctor—MAPS stands for Medical Academy of Pediatric Special Needs. It is a group of functional medicine doctors offering continuing education to many types of medical professionals. MAPS doctors are specialists who have completed a rigorous fellowship program focusing on the unique gastrointestinal, metabolic, immunological, nutritional, neurological, genetic, and other issues of patients with autism and related disorders. For more information, go to MedMAPS.org.

Health coach—Health coaching is a hybrid profession to which practitioners come from all walks of life. Their commonality? A passion for sharing what they know about lifestyle changes, non-pharmaceutical options, energy techniques, and more. Health coaches are usually "certified" by one of a dozen or so agencies with online courses, such as the Institute for Integrative Nutrition and the Nutritional Therapy Association. As an emerging profession, health coaches are generally unlicensed, and almost anyone can hang up a shingle and call themselves a "health coach." They could be a parent who has become an expert on reversing chronic childhood illness, such as asthma, allergies, or even autism. Or they could be a counselor who specializes in nutrition. The right one can be invaluable to getting you started while waiting to see a doctor.

Friends and family—Well-meaning relatives and parents encourage each other to try therapies they have heard about or personally found helpful. This approach shows concern, but, unfortunately, can waste both time and money. Simply taking therapies off the rack like a dress in a department store is like going into the wilderness without a map and compass; you need a professional guide.

Conferences—Informative conferences take place at airport hotels and online. In addition to learning from national experts in powerful educational sessions, parents learn from and network with each other. At the hundreds I have attended, I am always amazed at the incessant buzz about the latest treatment option that resonates throughout the halls as parents share their experiences, successes, and failures.

The exhibit halls are exciting and offer promising options, which are sometimes not as rewarding as they look. Space in these marketplaces sells out quickly to vendors with payment plans intended to make their wares accessible to everyone. Few companies are pedaling snake oil; careful screening of exhibitors assures products have efficacy. But buyer beware! The salient question is, will it work for my child?

The Internet—The Internet offers a global society, and everyone has a Facebook page and/or a website. Private Facebook groups offer support from a variety of people on a variety of subjects, although recent censorship has limited their usefulness. Webinars replace in-person conferences, allowing attendance anytime, anywhere. The discovery of new, exciting therapies with the potential to change lives spreads faster than California wildfires. How do you know where to start?

This book—Reading this book is a perfect beginning, whether you are a parent with a newly diagnosed child, one who has "tried everything" for a teenager or young adult, or an independent adult looking for ways to optimize your own health. Use it with confidence along with your hand-picked healthcare team.

Why do we need yet another book, written by neither a physician nor a parent? Because discoveries are occurring very rapidly, and much of the information in this book is not available from the resources listed above. Because we need a step-by-step prioritized model that includes not only biological, but also sensory, motor, language, social-emotional, academic, and vocational options across the lifespan. Because we need an up-to-date, practical, easy-to-read guide that is appropriate for parents, professionals, and laypeople, complete with resources on where to get help in each area.

Total Load Theory:
A Definitive Guide to End the Epidemic of Illness

Veteran Harvard psychiatrist Christopher Palmer, MD, quotes Albert Einstein and Leopold Infeld to describe how a new theory develops:

> Creating a new theory is not like destroying an old barn and erecting a skyscraper in its place. It is rather like climbing a mountain, gaining new and wider views, discovering unexpected connections between our starting point and its rich environment. But the point from which we started still exists and can be seen, although it appears smaller and forms a tiny part of our broad view gained by the mastery of the obstacles on our adventurous way up.[44]

In *The Four Pillars of Healing*, Leo Galland likens bringing a body back to health to restoring a fine painting.[45] In both cases, you must know the history. Anyone with a diagnosis deserves at least as careful a look as a painting. In order to reveal someone's true abilities and allow them to emerge, one must know their history. Galland views disease as the appearance of symptoms as the result of an accumulation of load factors the body has experienced in its environment. Peel back and treat each layer, one by one, until the person is well.

The following chapters address eliminating important load factors. Go to the Table of Contents to see what each includes.

Take-Home Points

Total Load Theory is a new paradigm for understanding childhood disorders. To quote Christopher Palmer's talk at last year's Integrative Medicine for Mental Health conference, "The mental health field is on the verge of a paradigm shift in which we begin looking for the root causes of mental disorders. The reality is all of those diagnoses are simply symptoms. I believe that we can do better."

New thinking prioritizes addressing a dysregulated nervous system before focusing on a distressed gut, immune system, genetic susceptibility, and environmental stressors. The ultimate goal is to remove nervous system sensitivities, the largest load factors, first and foremost. Then energy can be directed toward healing the gut and strengthening the immune system. Once the body moves out of fight-or-flight mode—and is putting less energy into digestion, circulation, respiration, and detoxification—cognitive, language, and physical development and learning can take place.

In other words, at first, other biological issues must take a back seat to calming the nervous system. Next, language, social-emotional, and academic development take a back seat to other biological issues. After underlying physical issues have been addressed, a remedial team can develop the specific hierarchy of adjunct therapies.

Patience is essential. While it may sound crazy to wait to address delayed talking or socialization, even for a short time, this plan will waste less time and money in the long term. Allowing the body to use its own wisdom to heal could possibly result in bypassing the need for some therapies altogether.

As Christopher Palmer says, "for any new theory to be taken seriously, it must incorporate what we already know to be true. It can't just replace it; it must tie together our existing knowledge and experience into a broader understanding—one that will widen our perspective and offer new insights."[46] Total Load Theory does just that, and this book will teach you how to apply it. Let's get started.

CHAPTER 2

GENETIC AND ENVIRONMENTAL LOAD FACTORS AS STRESSORS

What is contributing to this epidemic of childhood illness? That's the question scientists have been trying to answer for several decades. Most experts agree it does not have a single cause.

The multitude of possible causes can all be captured under a single umbrella: *stress*! From where? Our genetic makeup interacting with toxins in our air, food, water, and drugs. While we continuously welcome new technology and inventions that make our world more convenient, some of these representations of progress are hurting us.

This chapter enumerates dozens of genetic and environmental load factors that burden individuals with developmental, behavioral, and learning problems. The next chapter looks at yet more areas of stress: lifestyle choices, structural and sensory issues, dental status, educational demands, and emotional factors, such as trauma, fears, and expectations. These all come crashing together in an infinite number of combinations to form an individual's unique, personal Total Load.

Genetics, Genomics, and Epigenetics

The science of genes, those inherited combinations of DNA that determine characteristics like eye color that make up our uniqueness, is called *genetics*. Biologists have been fascinated by genes since Mendel experimented with them in the late 1800s.

Genetics 101

The human body has about 20,000 genes made up of billions of strands of DNA.[1] Genes come in pairs, one from each parent. Sometimes, an error occurs during early cell division, causing a true genetic condition, such as Down, fragile X, Rett, or Angelman syndrome. This error results in a baby being born with physical, intellectual, and/or emotional challenges.

Today, some prospective parents undergo genetic screenings prior to pregnancy for as many as 500 genetic conditions. Others screen genetics once the pregnancy is confirmed to see if they are carrying a baby with a serious genetic disorder. These controversial screenings have resulted in an estimated over 90 percent termination of pregnancies with positive results.

Genomics

All the DNA in the cell is called the genome. The study of very small changes in genomes is called genomics. Subtle variations in DNA can create health differences by changing the instructions for everything the proteins that govern what our cells do.

Mutations can occur in genes due to the insertion, deletion, duplication, or substitution of DNA. Researchers are studying variations in genes, called *SNPs,* for *single nucleotide polymorphisms,* to understand the health issues they can cause, including the ones covered in this book. Though everyone has some SNPs in their genes, scientists have been able to identify hundreds of SNPs and combinations of SNPs that are of particular interest in those with neurodevelopmental delays.

Why is all of this important? Because several categories of SNPs are affected. While all areas are important, detoxification is of particular interest, as that is how the body "takes out the garbage." When toxic "garbage" accumulates, people get sick. Go to the index to locate information on how SNPs in specific genes can downregulate immune function, cause inflammation, interfere with neurotransmitters, and even affect our emotions.

Since the Human Genome Project (HGP) was completed in 2003, billions of research dollars have been spent on countless studies searching for "the gene" for autism, ADHD, schizophrenia, and other "mental health" disorders.

But so far, researchers have made little headway toward finding single genes that cause any of these disorders. In fact, according to the Centers for Disease Control and Prevention (CDC), genetics alone account for about 10 percent of disorders.[2] It shouldn't be surprising that no single gene has been implicated in the majority of autism cases, because so few families choose to knowingly give birth to a child with a genetic disorder.

Biologic Stressors

Oxidative Stress

Stress that occurs at a cellular level is called *oxidative stress,* which, as one of the most significant biological stressors to the body and mind, has far-reaching effects in neurodevelopmental disorders. The brain is particularly susceptible to oxidative stress because of its high oxygen consumption and high fat content.[3]

Through a chemical process called *oxidation*, toxic substances known as *oxidants* overwhelm a cell's natural defenses, gradually disrupting function, and eventually contributing to DNA damage or even cell death.

Oxidants enter the body either from the environment or as byproducts of the body's normal metabolism. Because most cells in the body are vulnerable to oxidation, limiting oxidants through diet and using natural products are extremely important interventions included in the upcoming chapters.

The body fights off oxidants with compounds called *antioxidants*, which include vitamins and minerals from food and special enzymes that the body makes. When antioxidants outnumber oxidants, health is good. When oxidants prevail, cells suffer oxidative stress. In many individuals with neurodevelopmental disorders, high levels of toxins overpower low levels of antioxidants, allowing their bodies to continuously accumulate more toxins, thus causing chronic oxidative stress.[4]

Immunological Stress

Of all the possible load factors associated with neurodevelopmental disorders, dysfunction of the immune system is the one that has received the most attention. I once heard Jane El-Dahr, MD, a professor of pediatrics at Tulane Medical Center, liken the immune system to the United States Department of Defense (DOD). Just as the DOD acts to keep the country safe from invaders, the body's immune system acts to defend it against harmful substances. The "ideal" immune system recognizes all foreign organisms efficiently, rapidly destroys invaders, prevents further infection, and never causes damage to itself.

The immune system of those with neurodevelopmental disorders can be over- or underactive. Symptoms of immune system stress include seasonal allergies as well as reactions to common foods, vaccine reactions, frequent infections, and more. Go to Chapter 9 for more on the immune system and what can go wrong.

Toxins in Our Food, Water, and Air

Toxins

An epidemic of childhood illness due to genetics alone is biologically impossible as genes don't change that quickly. Genetics are said to "load the gun." However, something has to pull the trigger. More and more, scientists are pointing a finger at toxins in the environment as the "trigger" that activates the loaded genetic "gun." By *environment* I go broader than the water, the soil, and the air. The toxic soup we live in includes a great deal more: everything tangible and intangible outside our body that affects our biological, physical, and psychological status.

We now live in a toxic world. The world has changed in so many ways since my own childhood in the 1950s. At no point in history have there been such rapid dramatic changes in the physical environment. The 21st century has brought genetically altered and preserved foods, polluted air, electromagnetic fields, tainted and treated water, and depleted soil.

Everyone carries a water bottle filled with filtered or spring water to avoid drinking toxic tap water. Farmers use poisonous herbicides to prevent insects from damaging their crops. My favorite drink, root beer, is no longer made from sassafras root, but rather from chemicals. Breakfast cereals now come in bright colors.

Exposure to toxins, called *xenobiotics*, starts prenatally and continues throughout the lifespan. Compared to previous generations, today's children are exposed to dramatically more toxic chemicals (including metals), drugs, pesticides, plastics, and food additives that are foreign and poisonous to living systems just through living and breathing in a toxic world. Pound for pound, young children eat, drink, and breathe more than adults, making their exposure to toxins disproportionately high. The dose that is considered toxic depends upon a person's weight. Experts continually revise "safe" thresholds of exposure downward as our knowledge about neurotoxic chemicals increases.

The earlier the exposure to environmental toxins, the more significant the damage. International environmental health researcher Phillippe Grandjean argues convincingly that toxic chemicals can affect intellectual ability, cause ADHD, autism, cerebral palsy, and other potentially lifelong disorders.[5]

Billions of pounds of toxic chemicals are released into the air and water, as well as onto the land, each year. The Toxic Substances Control Act (TSCA) of 1976 grandfathered some 62,000 chemicals already on the market, despite the lack of safety data to support this policy. More than 80,000 chemicals are registered in the United States today; an estimated *2,000 new ones are introduced annually* for use in such everyday items as foods, personal care products, prescription drugs, household cleaners, and lawn care products.[6] Scientists looking at cancer "hot spots" have discovered they are often downstream from chemical plants.

Toxins disrupt every system of the body, causing them to become inflamed, distressed, and dysfunctional. They can also alter and suppress immune system function. Total Load Theory posits that genetically vulnerable immune systems eventually become exhausted from continuous efforts to ward off repeated assaults from the environment.

Toxins that primarily affect the nervous system are called *neurotoxins* because they damage the brain's ability to develop and function normally, resulting in the development of many neurodegenerative and neurodevelopmental conditions. A review of the top 20 chemicals reported on in the 2000 Toxics Release Inventory reveals that nearly half are known or suspected neurotoxins.[7]

Looking at the role environmental insults play in illness is not new. Rachel Carson shocked the world with her 1962 classic, *Silent Spring*.[8] This frightening exposé about how pesticides and other poisons have disrupted the bodies of all creatures on earth was a wake-up call to everyone. Theo Colburn disrupted our sleep again in 1997 with *Our Stolen Future*,[9] which confirmed that a generation later things were getting worse, not better.

Attorney Robert F. Kennedy Jr. (RFK Jr.), an outspoken environmental champion, and founder of Children's Health Defense, has been fighting the destruction of our health with toxic chemicals like mercury and protecting our children for almost 50 years. It will be fascinating to watch what happens with him in charge of the Department of Health and Human Services and the agencies that should be monitoring toxins in our environment.

The figure below depicts environmental stressors in four general categories: Thanks go to Peter Sullivan from Clear Light Ventures for this concept.

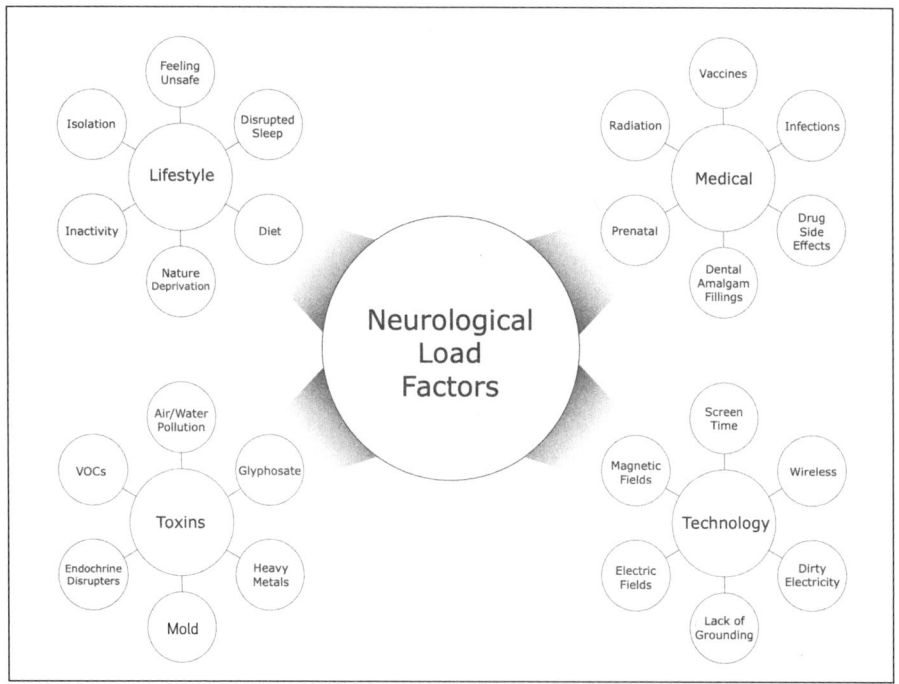

Toxins in our Food

For more than 30 years, one of America's premier environmental activists, David Steinman, Director of the Chemical Toxin Working Group, and Founder of the Healthy Living Foundation, has been warning us about the increasing number of dangerous toxins lurking in our foods. First, *Diet for a Poisoned Planet: How to Choose Safe Foods for Your Family* in 1990[10] and, now, *Raising Healthy Kids: Protecting Your Children from Hidden Chemical Toxins*[11] almost convince you to become your own farmer.

Genetically Modified Foods

Farmers have been engineering crops and domestic animals to improve the way they grow, thrive, look, or taste for years. We've been crossbreeding plants and animals for centuries to make hybrid fruits like tangelos (a cross between a tangerine and a type of grapefruit called a pomelo) and specialized breeds of livestock, like a mule, which is a cross between a male donkey and a female horse. Artificial selection like this can preserve the most desirable traits of each species. These techniques have been limited to naturally occurring traits until recently.

The problem is that new technology has enabled us to directly change genetic information to make crops like corn, soybeans, cotton, and various fruits and vegetables resistant to weed killers. The resulting species are *genetically modified organisms*, or GMO. When farmers repeatedly spray large quantities of herbicides on the food we later eat, we get a dose of weed killer in every bite.

The more GMOs our food contains, the more weed killer we ingest. GMOs are ubiquitous in kids' favorites, such as chicken nuggets, hot dogs, frozen meals, chips, and highly sweetened cereals and drinks. The herbicide of choice is glyphosate, first produced and marketed by chemical giant Monsanto, now Bayer, in a product called Roundup.[12]

Glyphosate

The "queen" of glyphosate education is MIT research scientist Stephanie Seneff, PhD. Since 2010, her research has focused on the effects of drugs, toxic chemicals, and diet on human health and disease. Her book *Toxic Legacy: How the Weedkiller Glyphosate is Destroying Our Health and the Environment* is a scary overview of the dangers of this ubiquitous herbicide.[13]

Glyphosate is everywhere. Over 80% of genetically modified (GMO) crops grown worldwide are sprayed with glyphosate. These crops are known as glyphosate-tolerant or "Roundup Ready" crops.

According to the Detox Project, a watchdog organization for toxic chemicals, as these crops are processed by General Mills, Kellogg's, Campbell's, and other food companies, the glyphosate ends up in our breakfast cereals, cookies, crackers, soups, snacks, and more. David Steinman, introduced above, tested the "garbage" marketed as "food" from these major brands and found not only glyphosate, but also the poison acrylamide, an established carcinogen, and reproductive toxin, associated with obesity, diabetes, anxiety disorders, and ADHD.

When animals eat crops sprayed with glyphosate and we eat processed foods made from these crops, glyphosate enters the animals' and our blood, urine, and brains, as well as the planet's air, water, and streams. Research shows that it even crosses the blood-brain barrier.[14]

Fish are no better; the largest, and most popular, like tuna contain PCBs, another industrial chemical that disrupts the thyroid. How about meat? Unless grassfed, most animal foods contain other endocrine-disrupting hormones. And dairy? The same.

Zen Honeycutt, the founding executive director of Moms Across America, and a leader in the Clean Food Movement, has taken on glyphosate and the poisons in public school lunches, and she is *unstoppable*! In her book by that name,[15] she details not only the dangers of glyphosate in food, but also how it is connected to decline in fertility, endocrine disruption, depression, and even school violence.

But despite the horrific state of America's health and the poisons in our foods, she is optimistic. Farmers are using amazing new ways of removing glyphosate, transforming sick soil using nontoxic oils, and even chasing bugs away with certain sound frequencies! Grassroots organizations are bringing back real kitchens to our schools, and regenerative organic farming is taking hold. Want to join the movement of unstoppable moms? Go to MomsAcrossAmerica.org.

While the food giant insisted that glyphosate was safe, in 2018 a team of lawyers won a historic $290 million landmark case against Monsanto, claiming the product likely caused the cancer in their client. Researchers believe that glyphosate's damage to health can occur slowly and insidiously, taking months and even years to show effects. It is thus extremely difficult to show a cause-and-effect relationship.

The location where glyphosate attacks our bodies and destroys our health is, not surprisingly, the gut. There, glyphosate disrupts and even kills the workings of the shikimate pathway, a crucial mechanism found in our beneficial gut microbes. We need these bugs to help our bodies synthesize amino acids and enzymes critical to the detoxification of toxic chemicals and other xenobiotics.[16]

As a result, people exposed to glyphosate by eating foods treated with it become doubly at risk; not only are they vulnerable to the damaging effects of glyphosate itself, but also to other toxins in the environment, which they are unable to get rid of.[17] Thus begins a cascade of further damage that starts with the disruption of the functioning of beneficial bacteria in the digestive tract. A condition known as "leaky gut" can result from permeability issues in the intestinal wall. Read more about this condition, especially problematic in many neurodevelopmental disorders, in Chapter 6.

Way back in May 2000, the Greater Boston Physicians for Social Responsibility released the 140-page booklet *In Harm's Way: Toxic Threats to Child Development*, now available as a free download from GBPSR.org. This comprehensive report details how overexposure to toxic substances has caused millions of American children to exhibit learning, behavioral, and developmental disorders, including ADHD and autism.[18] Even though it is a little dated, this booklet still contains valuable information about toxic exposures and their outcomes.

What can we safely eat? That's a real problem, unless you are vigilant! Buy organic, local, and in season. Read labels, and make sure your detox pathways work. More on that later.

Excitotoxins

A group of neurotoxins called *excitotoxins* are additives put into almost all processed foods to enhance taste. Excitotoxins and their chemical reactions are an important part of Total Load. Continuous stimulation of children's unprotected and immature neurons is implicated in seizures, inflammation, oxidative stress, endocrine disruption, hyperactivity, sleep deprivation, and head banging, all present in neurodevelopmental disorders.

Some of these ubiquitous poisons have familiar names such as aspartame (Equal™ and NutraSweet™), hydrolyzed vegetable protein, and monosodium glutamate (MSG). Some are disguised by terms such as "natural flavoring," "spices," "yeast extract," "textured soy protein," "xanthan gum," "whey protein isolate," and "carrageenan." They are added to soups, gravies, diet sodas, ice cream, candy, cigarettes, cheese products, chewing gum, gelatin, and infant formulas. They also lurk in some "secret" places, such as vaccines, in the form of processed free glutamic acid, described as a "stabilizer," an ingredient used to keep the antigen intact.

One of the first people to warn us about excitotoxins was Russell Blaylock, MD, a board-certified neurosurgeon who became interested in the subject when his father was diagnosed with Parkinson's. After he started investigating the side effects of aspartame and MSG and learned that enormous amounts were being consumed by those later diagnosed with neurological problems, he became intrigued. His now classic book *Excitotoxins: The Taste That Kills*[19] is a must read for anyone interested in what is happening to our food and how it is wreaking havoc with our health.

Excitotoxins do not have to be artificial; some are found in nature—such as free glutamate, aspartate and cysteine, all of which are amino acids. MSG is actually a modified form of glutamic acid to which sodium has been added. Biochemist Katie Reid, PhD, discovered that glutamates, a class of additives, all of which are excitotoxins, were hurting her young daughter, who began showing signs of autism at 11 months. By researching brain science, food manufacturing processes, and applying her background in protein chemistry, she eliminated the little girl's autism symptoms by removing all foods with added excitotoxins.

Reid's nonprofit Unblind My Mind (UMM) has extremely useful information, as well as a downloadable list of ingredients that contain free glutamate on its website, UnblindMyMind.org. Learn more about The REID Program (Reduced Excitatory Inflammatory Diet) in Chapter 7.

Toxins in the Water

Fluoride, a chemical byproduct of aluminum, steel, and other manufacturing—including computer screens, fluorescent light bulbs, herbicides, and, yes, toothpaste—is a known neurotoxin. It is more poisonous than lead and slightly less poisonous than arsenic.

Fluoride from air, water, and food bioaccumulates over the lifetime. It damages the immune, musculoskeletal, respiratory, circulatory, and digestive systems; thyroid, liver, and brain function; and can cause hyperactivity in children.[20] A 2024 monograph from the National Toxicology Program of the US government reports "a large body of evidence on associations between fluoride exposure and IQ" and "some evidence that fluoride exposure is associated with other neurodevelopmental and cognitive development in children."[21] A recent article from *JAMA* confirms that prenatal fluoride exposure increases risk.[22]

Then, you might ask, "Why is it added to drinking water in most American cities?" Allegedly to prevent cavities. Up until the early 1930s, both the American Dental Association and the US Public Health Service agreed that fluoride was dangerous. Their own research showed that it could cause dental problems and damage the central nervous system.

However, in 1931, Andrew Mellon, Secretary of the Treasury and founder of the Aluminum Company of America (Alcoa), convinced the US Public Health Service, then under jurisdiction of the Treasury Department, that adding

fluoride to drinking water would prevent tooth decay. Two Alcoa-funded studies showing the safety of fluoride suddenly overrode many scientific papers
demonstrating its negative effects. Without further testing, fluoride was added
to public drinking water.

In the 1950s, doctors discovered that swallowing even a small amount of flu-
oride-containing toothpaste can cause colic or gastritis in a young child.[23] When
older patients stopped drinking fluoridated water, their migraines, stiffness, and
gastric distress all disappeared.

In the 1970s, experiments proved fluoride was also a carcinogen, not only
causing liver and bone cancer, but also responsible for symptoms resembling
Alzheimer's disease and attention deficit disorder. Yet, it was still permitted to be
added to drinking water.

In the early 1990s, a toxicologist named William Marcus, PhD, working for
the Environmental Protection Agency (EPA) tried unsuccessfully to blow the
whistle on fluoride. His statements that not only did fluoride *not* prevent tooth
decay, but it also *caused* cancer and a myriad of other health concerns, resulted
in his being abruptly fired.[24] And now, finally, RFK Jr. has demanded that this
charade cease. As the Secretary of Health and Human Services under President
Trump, he promises to make it illegal for fluoride to be in public drinking water.

Toxins in the Air

Hazardous particles in the air come from everywhere: construction and demoli-
tion, mining operations, agriculture, highways, and chemical processes such as
burning carbon-based fuels. They can be gases; metals such as asbestos, cadmium,
mercury and chromium; or complex chemical compounds.

Building materials are major sources of airborne neurotoxins from off-gassing.
"Sick building syndrome"—discomfort, "allergies," or illness that result from living or working in a building with airborne contaminants or inadequate ventila-
tion—is not just limited to office buildings. Some people reside in "sick" homes.[25]
Siding, cabinets, carpeting, glue, insulation, and other materials can off-gas for
days, months, and even years.

Fine particulate matter poses a particular threat to developing bodies and
brains.[26] These differ in size; some are as large as one-tenth the diameter of a
strand of hair, but many are so small they can only be seen with an electron micro-
scope. The differences in size determine how particles affect us.

Environmental tobacco smoke is a dynamic, complex mixture of more than
7,000 chemicals.[27] Passive tobacco exposure is unavoidable, and remains a significant air pollution problem despite extraordinary measures to educate the public.

This air pollutant can be contributing to the epidemic of childhood asthma, allergies, and thus Total Load.[28]

There simply is no safe threshold of breathing in fine particles. Everyone, especially young bodies and brains, are vulnerable to the health effects of air pollution. Maternal exposure to fine and ultrafine particulate matter can directly and indirectly cause adverse birth outcomes, as well as negatively impact children's respiratory systems, immune status, brain development, and cardiometabolic health.[29]

Noise Pollution

Environmental noise pollution from highway, rail, and air traffic and other sources is a form of air pollution and is more severe and widespread than ever. Environmental stress from sounds that are persistently loud and annoying can affect both our physical and mental health, leading to hearing loss and sleep disruption as well as behavioral and learning issues.

Muffling the noise with headphones can help. However, the current craze of using wireless earbuds to listen to preferred sounds may come with hidden risks. Earbuds, like all wireless products, emit electromagnetic fields. While these emissions are low, as compared to that of other devices, long-term use could be problematic.[30]

Toxic Metals

The following metals are the most problematic:

Mercury, arguably the second most toxic metal on the planet, next to plutonium. Exposure to mercury comes from dental amalgams, excessive fish consumption, vaccines, power plants, light bulbs, cleansers, cosmetics, RhoGAM (before 2001), and other sources.[31] According to Klinghardt, mercury is problematic on its own and becomes even more so as it interacts with other toxic substances. For instance, when it enters the brain as a gas and binds to the neurons of brain cells, it can make the brain more susceptible to microwave radiation. When hiding in the gut lining, it provides a breeding ground for microbes such as the herpesviruses.[32]

Mercury is highly suspect in neurodevelopmental disorders. It exists in three forms: methylmercury, ethylmercury, and inorganic mercury. Simple laboratory tests cannot provide an adequate representation of an individual's mercury level. Read in Chapter 5 about how doctors use a special technique called mercury speciation analysis to separate and measure the different types of mercury in their patients.

An Rh-positive baby born to an Rh-negative mother with a mouthful of mercury-containing amalgams, who used her cell phone during pregnancy then received RhoGAM with the mercury-containing preservative thimerosal, was exposed to toxic levels of mercury prenatally. Even before leaving the womb, the baby's Total Load of mercury alone neared the body's threshold. Add to this infant's load a few more micrograms of mercury within hours of birth from the hepatitis B vaccine, and that baby had a high chance of being sick and delayed by age two.

Studies found that proximity to sources of mercury pollution, such as power plants, was positively related to illness. Mercury bioaccumulates, meaning that it increasingly concentrates in the flesh of organisms as it makes its way up the food chain. Eating only one serving of a predator fish a month out of a river downstream from a coal-burning power plant is potentially dangerous for pregnant women, developing fetuses and for young children.[33]

Lead from the soil, food products, and chipping paint. We have known since 1979 about the devastating effects of even low levels of lead on the mental and physical development of young children.[34]

The legal standard for lead poisoning used to be well over 10 micrograms per deciliter (mcg/dl) in most states. When research showed that levels as low as 5 or 6 mcg/dl can decrease a child's ability to read, write, and calculate,[35] the CDC officially established 5 mcg/dl as its new reference level in May 2012. Many toxicologists now believe no amount of lead is safe.[36]

Aluminum from cans, cookware, antacids, antiperspirants, and vaccines. Aluminum has no recognized biological function. Studies relate elevated aluminum to memory and learning impairments,[37] autism,[38] and cognitive impairments.[39]

Arsenic and Antimony, present in drinking water, flame-retardants, pressure-treated wood used in playgrounds, fungicides, herbicides, corrosion inhibitors, and lead and copper alloys. Arsenic is also high in some seafood and rice (and its byproducts, such as baby cereal and rice milk) grown in areas where arsenic is in the water, both in the United States and abroad. These toxins may also come from animals, such as chickens, that ingest arsenic in their feed. Both have subtle negative effects on neurodevelopment.[40]

Joseph Pizzorno, ND, thinks it's time to address our serious arsenic problem. Like lead and mercury, no level of arsenic is safe. This veteran physician believes that even chronic low-level exposure to arsenic is related to many chronic conditions in children.[41]

I know one child with autism who was found to have arsenic poisoning. He had bloody splits between his toes, large callouses on his hands, and "rainbow" skin on his back. None of the healthcare professionals who had examined him through the years recognized these physical signs of arsenic poisoning. Rather, he "only had autism." If he had not been eating a clean diet and his other toxic load factors were not so low, he probably would have died.

Cadmium, from cigarette smoke, batteries, photography, and pigments. Cadmium is an endocrine disruptor.[42] Read more about it and other endocrine-disrupting chemicals in Chapter 10.

Before leaving the topic of heavy metals, readers should hear the now-legend story of how mercury was implicated in autism. In late 1999, four families, all with children who regressed into autism, started delving into the possible sources of their children's illnesses. Sallie Bernard, Albert Enayati, Lyn Redwood, and Heidi Roger had never met, but they corresponded, comparing notes about their children. After attending a conference on biomedical approaches to autism, they came away convinced their children were suffering from mercury poisoning. After much research, they wrote a joint paper with a physician and scientist entitled, "Autism: A Unique Type of Mercury Poisoning."[43]

The authors formalized their relationship with the founding of The Coalition of SafeMinds (Sensible Action for Ending Mercury-Induced Neurological Disorders). SafeMinds is a private, nonprofit organization whose mission is "to end the autism crisis by advancing environmental research and effective treatments."

Toxins in Our Products

The chemical soup we live in also includes manmade chemicals, with unpronounceable names such as polychlorinated biphenyls (PCBs), polybrominated diphenyl ethers (PBDEs), and phthalates, which add to the body's toxic load. PCBs, which are chemicals used in coolants, plasticizers, electrical coatings, and for other purposes, have been banned for the past 25 years, yet they remain in the soil, water, and air forever. The higher the levels of PCBs in the mother's blood and breast milk, the larger the deficits in their children's intelligence, memory, and attention.[44] PBDEs, the "sons" of PCBs, are flame retardants that have been used in hundreds of everyday products since the 1970s, including children's pajamas.

American babies appear to be exposed to far higher amounts of fire retardants than babies in Europe, where some of these chemicals have already been banned.[45] Children's sleepwear, furniture, computers, TV sets, automobile parts, polyurethane foam, and stain-proofing for textiles such as Scotchgard® all contain PBDEs. These products look like PCBs chemically, behave like PCBs environmentally, and have the same toxic effects as PCBs biologically. They are in dumps everywhere, persistently contaminating the food chain and making their way into human bodies, where they act as endocrine disrupters.[46] Minute doses of PBDEs can cause deficits in sensory and motor skills, learning, memory, and hearing, as well as impair attention, learning, and behavior in laboratory animals.

Pesticides are everywhere. Even if you don't use them at home, everyone is exposed to them through nonorganic foods, landscaping, and the routine spraying of office buildings and even doctors' offices! Pest control traditionally uses toxic chemicals to kill bugs in all stages of life, from larva to adult. Most people are afraid of bugs, and assume they carry diseases. In truth, the dangers of pesticides far outweigh the dangers of most insects. Even a slight exposure to pesticides can have negative effects on children's well-being.[47]

Cleaning products are notoriously toxic. Learning all the poisonous ingredients of laundry soaps, disinfectants, oven and drain cleaners is nearly impossible. However, one clue is that if the product contains the signal words "Danger," "Warning," or "Caution," you should avoid it.[48] Dryer sheets and fabric softeners all contain a toxic brew of chemicals that can cause allergic reactions, central nervous system disorders, headaches, loss of muscle coordination, and even cancer.[49]

Art supplies from abroad often contain toxins. Crayons, paints, magic markers, and anything with color could contain mercury, asbestos, lead, and/or other poisons.[50]

Personal care products and their plastic containers contain heavy metals such as lead and mercury and phthalates, which are plasticizers used to add texture and luster to hair spray, deodorant, nail polish, lipstick, perfume, and other products. Phthalates can be absorbed through the skin, inhaled through off-gassing, and ingested by children mouthing products, so lotions and creams are especially problematic.[51]

Hundreds of studies show that phthalates can damage the liver, kidneys, lungs, and reproductive system.[52] In pregnant women, phthalates pass through the placenta to be absorbed by the fetus. Later they show up in breast milk of nursing mothers, whose babies ingest them. In exposed males, phthalates can cause testicular atrophy, leading to a reduced sperm count.[53] Plastic containers for food and drink can also be harmful to health, especially if microwaved.[54]

More than 10 years ago, the late Senator Frank R. Lautenberg (D-NJ), chair of the Senate Subcommittee on Environmental Health, declared that "America's system for regulating industrial chemicals is broken." In his efforts to protect our children from dangerous chemicals in everyday products, he introduced the Safe Chemicals Act to bring the aforementioned Toxic Substances Control Act (TSCA) up to date for the first time since 1976.[55] It did not pass.

One of TSCA's only pluses is that it has allowed states to develop their own chemical policies and restrictions. Many states have banned particularly toxic ingredients like mercury, cadmium, and BPA from particular categories of products. Some forward-thinking states have developed even more comprehensive policies that address broader classes of chemicals designed to improve public health and environmental quality.

Finally, three years after Lautenberg's death in 2013, President Obama signed into law the Frank R. Lautenberg Chemical Safety for the 21st Century Act, designed to fix America's badly outdated chemical safety law. Progress has been painfully slow because of aggressive chemical company lobbying, but small steps have been made to protect us and our children. One organization working hard toward this goal is Toxic-Free Future, by advocating for the use of safer products, chemicals, and practices through advanced research, grassroots organizing, and consumer engagement.[56]

Mycotoxins from Mold

Mold exposure is maybe the most overlooked suspect that makes people sick and dysfunctional. Unusually severe storms in recent years resulting in widespread flooding and leaks, combined with building products that seal in the moisture, mean that many water-damaged buildings have mold, often hidden from view. Neil Nathan, MD, believes that mold sickness must be ruled out before tackling almost any other cause.[57] Simply said, no one can get well from anything if living or working in a moldy building.

Molds are spore-forming organisms. Their spores travel quickly and hatch in warm places in buildings as well as in warm, moist places in the body, such as the nasal passages. They release potent neurotoxins that can damage many aspects of gastrointestinal, respiratory, hematological, immunological, and neurological function, and are extremely resistant to complete removal.

Mold causes subtle symptoms that take unsuspecting patients from specialist to specialist, who often conclude that it's "all in your head" after standard tests come back "normal."

Go to Chapter 5 to discover some exciting new ways to test for mold spores in the body, Chapter 9 to learn how mycotoxins from mold affect the immune system, and Chapter 11 to learn about remediation. Want to go even deeper? Learn from experts Ritchie Shoemaker, MD, at SurvivingMold.com or Jill Crista, ND, at DrCrista.com.

Endo- and Exotoxins from Pathogens

Bacteria and other pathogenic microbes produce secondary toxins as they undergo their biological processes. These toxins, when found within the cell are called *endotoxins*, while those secreted by the cell are called *exotoxins*.

How do these pathological bugs get into our bodies? We touch, eat, drink, and breathe them. If doctors have prescribed multiple rounds of antibiotics to kill them, the balance of good and bad bacteria in the gut can become disrupted, a condition called *dysbiosis*. An imbalanced gut is a favorable environment for colonization by opportunistic pathogens. Read more about dysbiosis in Chapter 6.

Radiation and Dirty Electricity

Radiation is generated by power lines, cell phone towers, wireless devices, security systems, and "smart" appliances. Wired appliances are gradually disappearing; we are living in an increasingly wireless world. Even though there are no wires, something connects the appliance to an electrical source. This invisible electricity is called the electromagnetic spectrum. It is real and it can be measured. How is it affecting us?

Electrosmog

The rotation of the earth, combined with the output of electromagnetic waves from today's increasingly complex appliances, creates a type of pollution called *electrosmog* that is emerging as one of today's major stressors on the body and its nervous system. Electrosmog includes electromagnetic fields (EMFs), electromagnetic radiation (EMR), and radio frequency (RF) and microwave (MW) radiation. (RF and MW are specific frequency bands within the electromagnetic spectrum.)

The frequency of the waves' oscillation is measured in Hertz, ranging from very low (the 60 Hz that powers your hair dryer) to extremely high (the gamma rays released from an atomic bomb). The frequency determines how much energy they carry. Generally speaking, the faster they oscillate, the more energy they carry and the more damage they can do to the human body.

Electrosmog includes high-frequency radiation from cellular phones, and wireless devices; intermediate frequencies from "dirty electricity," emanating from transformers, fluorescent lighting, computers, and plasma televisions; and low-frequency fields from copiers, clock radios, and electric heaters.

New words have entered our vocabulary. We have moved up to "5G" (for "fifth generation") communication, Bluetooth devices are embedded in everything from our TVs to our hearing aids, and mandatory "smart" switches have been installed on our homes' gas, electric, and water meters. "Smart" wiring that, like the other "upgrades," avoids unsightly wires and the need for human monitoring and adds convenience.

But wait! All of this convenience comes at a price. Maybe it isn't so "smart"— or safe. Our exposure to stronger and stronger radiation is far more abundant than it was just ten years ago, and the evidence that information-carrying radio waves from cell phone towers, the phones themselves, and all the appliances that utilize them are a grave risk to our health is getting stronger and stronger as well.

EMFs damage cell membranes, thus decreasing intracellular communication by disrupting microtubular connections that enable biophotons to flow between cells. The damaged cell membranes allow heavy metals to enter the cells, resulting in intracellular production of free radicals, which can significantly decrease energy production and cause great fatigue.[58] Immature children's brains and immune systems are more vulnerable to the effects of wireless radiation due to their thinner skulls and unique physiology, which allows radiation to penetrate deeper.

Bacteria, viruses, and other bugs do not like EMFs either; high-frequency waves cause them to respond as if they are being attacked. They react by stepping up their output of toxins and thus become even more virulent when exposed. A European colleague of Dr. Dietrich Klinghardt did an experiment on microbial cultures in which he compared the growth and endotoxin production of microbes shielded in special cages to those of cultures subjected to typical EMF exposure without protection. The cultures' proliferation and endotoxin production went up 600% when subjected to ambient EMFs.[59]

Devra Davis, PhD, and Magda Havas, PhD, are two of the world's authorities on the biological effects of wireless radiation. This subject is much too complex to elaborate on here. Everyone should educate themselves on this invisible threat to our health. I strongly recommend going to EHTrust.org, Dr. Davis's website, and to MagdaHavas.com to learn more.

Geopathic Stress

The negative effect of the Earth's energies on human well-being is called "geopathic stress." Underground formations, subterranean water currents, specific mineral deposits, and fault lines can emit electromagnetic radiation and create stress. Fracking or natural gas drilling releases naturally occurring radioactive elements that have been trapped within the Earth's surface for millions of years, affecting the quality of the water, air, and food.[60] Health advocates are very concerned about the health effects on people living nearby.[61]

Living and sleeping in the energetic field of natural geological stressors could be damaging to health. Vibrations emitted by these stressors do not directly cause disease; however, they are significant load factors that affect the body.[62]

Artificial Lighting

Options for artificial lighting have changed dramatically in the past decade. The incandescent bulb is now virtually extinct, replaced by compact fluorescent (CFL) and LED bulbs, which are energy efficient but neither environmentally friendly nor good for our health. Were you aware that compact fluorescent light bulbs give off high levels of EMFs?

Each CFL bulb is an environmental and health time bomb emitting radio frequencies and holding three to five milligrams of mercury.[63] The mercury is not a problem—until the bulb breaks, either in your home or in the landfill where it eventually ends up.

The audible buzz emitted by fluorescent lights and visible flickering as a bulb burns out can be problematic for many students with special needs. One student I know had her first seizure in fourth grade. When her father, an optometrist, came to the school to pick her up, he noticed that the fluorescent light was pulsing with a strobe effect. Removing her from that classroom cured her "seizure disorder."

For reading and writing, lights should not have too much blue, which can disrupt sleep. That's the primary problem with energy-efficient light-emitting diode (LED) bulbs. Wearing blue-blocking glasses can help, especially before going to bed.

Genetics + Environment = Epigenetics

Epigenetics, the biology of how environmental triggers can modify an individual's DNA by turning specific genes "on" or "off," is at the heart of Total Load. Epigenetics is so important that it made the covers of *Time* in 2010[64] and *Scientific American* in 2011.[65] In 2012, the *International Journal of Epidemiology* devoted an entire issue to it, calling epigenetics "the next big thing."[66] In 2016, *The New Yorker* ran a fascinating article called "Same but Different" describing how epigenetics alters identical twins.[67]

Epigenetics explains why some young children exposed to today's environmental chemical, viral and electromagnetic soup don't end up sick. Those that do become ill appear to have a genetic propensity, or predisposition, that manifests itself only when triggered by the "right" combination of environmental factors.[68] Any of the environmental exposures mentioned above can change gene expression, resulting in a plethora of inappropriately activated or deactivated genes.[69]

Ultimately, the more "right" the genetic material combined with the more "right" environmental insults, the more severe the disability. For an exhaustive review of the past 50 years of literature on how genetics and environmental stressors combine to cause autism and other neurodevelopmental disorders, read *The Environmental and Genetic Causes of Autism*[70] by James Lyons-Weiler, PhD.

Take-Home Points

While we cannot change our genetics, we can learn about the genes we have inherited, what our environmental exposures have done to them, and how our genetic inheritances can potentially affect our health. The best way to do that is through testing, which is discussed in Chapter 4.

Our environment is another story. We *do* have some control over the quality of the foods we eat, the air we breathe, and the water we drink. We can also make healthy decisions which limit our exposures to excitotoxins, endotoxins, mycotoxins, and other bugs that make us sick and affect our children's development, learning, and behavior.

The next two chapters enumerate additional load factors and lifestyle changes to consider. Your genes are not your destiny! Your environment is at least partially within your control. Be the captain of the ship for you and your family.

CHAPTER 3

SOCIETAL LOAD FACTORS

Chapter 2 took a deep dive into genetic and environmental load factors. Many other stressors burden the bodies and minds of our children, resulting in brain fog, lack of focus, anxiety, depression, obsessive-compulsive behaviors, and other symptoms, eventually leading to a "wake-up" mental health diagnosis. The possible combinations of accumulated load factors making up an individual's personal Total Load are infinite.

This chapter enumerates some key factors that can form the foundation for later, serious problems resulting in a neurodevelopmental diagnosis. These include additional

- **Physical stressors**, such birth trauma, dental, spinal, and other structural issues, sensory problems, and reflex dysfunction;
- **Lifestyle stressors**, such as problematic diets, sleep deprivation, dehydration, and lack of movement opportunities;
- **Medical stressors,** such as over-vaccination and overuse of antibiotics;
- **Educational stressors**, such as an inappropriate curriculum and extended screen time, and
- **Emotional stressors**, such as worries, fears, and trauma, especially related to the recent pandemic.

Today's kids spend their days sitting staring at screens, try to function on fewer than six hours of sleep, and don't drink enough water. They no longer walk to school, play unsupervised outdoors, or spend free time at recess as I did. Estranged family relationships and physically distant relatives, expectations for high achievement, and heavy-handed medicalization of sickness all add stress. No one knows how these changes add up to affect physical and mental health.

Physical Stressors

The **pregnancy and birth histories** of individuals who later demonstrate cognitive, behavioral, learning, and emotional issues often include a number of "red flags" which placed them at risk. Healthy mothers usually have healthy babies. Those born to mothers who have

- low thyroid function,
- allergies,
- chronic fatigue syndrome / fibromyalgia,
- mercury-containing dental amalgams,
- received vaccines before or during pregnancy,
- received a RhoGAM shot or antibiotics during birthing, or
- consumed excessive fish

can be negatively affected.

According to health guru Dietrich Klinghardt, MD, PhD, mothers who have had years of toxic exposure pass two-thirds of their toxic load through the placenta into their babies during pregnancy. The firstborn child is often more severely affected than subsequent ones. Then, if the baby endured birth trauma with oxygen deprivation, breech presentation, or were born by cesarean section, the risk for developmental problems is even higher.[1]

Most children demonstrate physical symptoms before they "have" ADHD, OCD or ASD. Colic, reflux, recurrent infections, formula and food allergies, feeding problems, sleep disturbances, constipation, diarrhea, fears, skipped or late developmental steps, hyperactivity, late or unusual teeth emergence, abdominal pain, eczema, vaccine reactions, and other symptoms during the first year of life are all risk factors for a later, more serious diagnosis. The end of the second year of life is a particularly vulnerable period, when development in all areas is taking place at a rapid rate.

While doctors usually recognize that these early health issues require intervention, they tend to send parents down blind alleys that treat symptoms rather than address underlying causes. These warning signs could indicate possible trouble in the structural integrity of the body. How to identify and cut these stressors off early on is covered in Chapter 6 and in depth in Chapters 12–17.

Structural Impediments

A perfectly aligned spinal cord allows normal development to proceed along a sequential path (pre-programmed by nature) without difficulty. Structure and function work hand in hand; structure determines function, and function determines structure. Changes in the body's structure changes its function.

Any imbalance among the organs, fluids, bones, and connective tissues disrupts the overall health of the growing organism. For instance, the structural

organization of the bones of the skull anchors the brain in place and determines the tension of the membranes that suspend the pituitary gland and the
hypothalamus. The ability of the hypothalamus to release and transport its
hormones is extremely vulnerable to positional changes of the pituitary. When
structure is comprised, the body must utilize much of its energy to stay struc-
turally intact, sucking whatever energy it can from the immune system and
elsewhere.

Even subtle underlying structural issues can have profound effects on early
growth and development, and later on behavior and learning. Fluctuating visual
focus, an asymmetry between the two sides of the body, or missing stages in motor
development are easy to overlook, while feeding problems, frequent reflux, vomiting, or spitting up, a crossed eye, "wiggly eyes" (nystagmus), and reduced visual or
auditory acuity are more obvious.

According to the late Dr. Viola Frymann (1921–2016), a world-renowned
osteopathic physician, birth trauma is the most common cause of developmental problems. The birth histories of children with neurodevelopmental disorders
often include one or more of the following contributors from which their skulls
and sacra may not fully recover:[2]

- Premature birth,
- Long labor and delivery,
- Breech presentation,
- Oxygen deprivation,
- Cesarean section, and
- Use of forceps, or vacuum aspiration

can all disrupt the structural integrity of the brain, spinal column, and the fluids
surrounding and inside these organs.[3] A 2006 British study of children with an
ASD diagnosis found they were twelve times more likely to have experienced
birth trauma or complications than their neurotypical siblings.[4]

Structural problems can cause symptoms immediately or take years to be evi-
dent; the larger the load of structural stressors, the sooner and more serious the
symptoms. Unless healthcare professionals address structural impediments early
in life, they can interfere with health and development, especially in the digestive
and nervous systems, interfering with the absorption of vitamins and minerals,
and the detoxification of metals, pesticides and other poisons. Remedial options
are covered in Chapter 6.

Dental Stressors

Structural abnormalities in facial development involving the head, neck, and jaw are a subset of structural issues seen in our children. As an infant grows, the jaw, along with the rest of the body, undergoes numerous structural changes. Correct jaw development is vital for breathing, eating, and speaking. This topic is so important that an entire chapter has been dedicated to it.

Facial anomalies are sometimes obvious, like a cleft palate or lip. Others, like a high-arched palate or tongue tie, are not so easy to recognize. Even slight misalignments can compromise blood and oxygen flow to the brain and impede lymphatic drainage.

Facial structure has evolved over centuries. Drawings of cavemen show wide jaws and well-spaced teeth. Today's kids have longer, thinner jaws that cannot accommodate all their teeth, often requiring expensive orthodontics. Evolutionist Paul Ehrlich calls deteriorating jaw development a "silent epidemic."[5]

Several red flags for dental issues usually present themselves very early in life. A floppy baby, who has a weak suck, cannot latch onto the breast, or falls asleep nursing may be on the way to poor tooth development. Jaw and facial abnormalities can cause oxygen deprivation, malnutrition, poor cognition, and other problems. Chapter 15 covers prevention and remediation of dental problems.

Reflex Abnormalities

Reflexes are involuntary, stereotyped movements that a person makes in response to a stimulus. We are all familiar with that tap on the knee that causes the leg to jerk upwards.

Babies are born "pre-loaded" with approximately 100 primary reflexes that provide important protections and enable survival. As the baby grows, develops, and experiences the world, these reflexes emerge, become active, and then integrate, providing a strong foundation for learning more complex movement. This process starts about week nine in utero and continues throughout the first year of life.

Only when infant reflexes emerge on time, mature, and integrate with all their variations can a baby gradually gain intentional control of both the body and its parts. Children can then strike a balance between nonvoluntary, or "reflexive," movement and behavior that provides protection under stress and intentional learning and action.

Any pre-, peri-, or postnatal load factor can interfere with reflex development, and cause infant reflexes to be too weak, too strong, late, or retained, resulting in involuntary and inappropriate responses to sensory stimuli. Babies born to mothers confined to bed rest for long periods during their pregnancies are also at risk, as the inactivity might cause them to miss important stages of reflex integration with harmful effects on their motor development.

Retained infant reflexes can interfere with attention, language, cognitive, motor, sensory, and social-emotional development. Other undesirable outcomes include, but are not limited to, poor lateralization, weak muscle tone, misarticulations, vestibular dysfunction, vision and auditory problems. Teachers and parents might notice learning and social-emotional issues such as reversals in reading and writing, poor impulse control, trouble reading body language, or unsatisfactory peer relationships, all despite good intelligence.[6]

Chapter 14 discusses methods of addressing dysfunctional reflexes.

Sensory Issues

The Emmy-winning film *Temple Grandin* depicted the life of arguably the world's most famous autistic adult and the severe sensory issues that she and so many like her experience. Grandin struggles with touch, hearing, and other sensory hyperreactivity and at an early age learned to calm herself using a branding machine on her relatives' farm.

Others with neurodevelopmental issues describe visual issues, the need for constant pressure or movement, as well as an abhorrence of certain smells, tastes, and/or sounds. Clearly, they are living in a world that is too loud, too tight, or too bright for their nervous systems to handle. This hypersensitivity makes them vulnerable to other triggers.

The source of these sensory sensitivities is as varied and complex as the individuals themselves. Whatever their source, tactile, auditory, visual, and other sensory issues are additional load factors. When piled on top of oxidative, toxicological, immunological, environmental, and physical stress, later neurodevelopmental issues are inevitable.

Ironically, kids' sensory experiences, as well as the emergence and integration of reflexes, are also being impeded by some recent practices that specialists are recommending to "keep kids safe." From day one, we are limiting children's natural sensory needs to touch and move by squishing them into molded plastic Bumbo seats, strapping them into car seats, backpacks, playpens, and walkers and limiting play activities because they are "dangerous." These modern-day straitjackets are also preventing toddlers from knowing where they are in space and using their bodies purposefully. Learn more about sensory processing disorders and their remediation in Chapters 16 and 17.

Lifestyle Stressors

Limited diets and poor nutrition are common among those with psychiatric conditions. Obviously, foods that are high calorie and low nutrition stress the body. Today's kids' diets consist of highly processed, sugar-laden foods and beverages, while consumption of fresh fruits and vegetables has waned. Not only do these "junk foods" have little-to-no nutritional value, they also contain glyphosate and other additives that can cause gastro and neurological symptoms, such as diarrhea and hyperactivity.

Diet is a special concern for those with special needs. While too much sugar and food coloring can affect their behavior and thinking, their overly sensitive bodies and brains also react to common allergens, such as wheat and other gluten-containing foods, dairy products, soy, corn, and eggs. This topic is discussed in length in Chapters 5–7.

Even many of today's fresh, frozen, and canned fruits and vegetables cannot be assumed to be safe. Many are sprayed with pesticides, some more than others. The Environmental Working Group (EWG) has declared the following "The Dirty Dozen" because they have the largest pesticide residues: strawberries, spinach, kale, collards, mustard greens, grapes, peaches, pears, nectarines, apples, bell peppers, cherries, blueberries, and green beans.[7] Unfortunately, pesticides cannot be washed off.

Special diets and nutrition are covered in depth in Chapters 6 and 7.

Dehydration is an often-overlooked contributor to stress; it could be the hidden culprit in fuzzy thinking, impaired memory, irritability, fussiness, and other behavioral issues.[8] Human bodies are at least 50% water, and the brain is up to 85% water. Every bodily function from cellular communication to digestion depends upon good hydration.[9]

Inefficient breathing is yet another stressor. Natural, spontaneous, relaxed, deep breathing is essential for both mind and body. It supplies oxygen to blood, cells, and brain, thus reducing tension and providing energy that allows us to pay attention to the environment.

Inefficient breathing is at the foundation of many illnesses. Symptoms include

- yawning and/or sighing;
- rocking;
- poor blowing skills, such as with candles on a cake, bubbles, or a musical instrument;
- restless sleep;

- hyperventilation; and
- holding of the breath when expending effort.

According to breathing guru James Nestor, "breathholding is associated almost entirely with disease, especially panic and anxiety disorders."[10] When we habitually hold our breath or breathe in a shallow way due to stress, we are more likely to experience exhaustion. We literally "lack inspiration."

The inability to integrate movement and breathing stresses muscles and nervous systems and reduces attention.[11] The resulting tension can lead to mismatches in how the brain interprets the world.[12]

Insomnia is a hallmark of stress. Today, virtually *everyone* is sleep deprived. Kids are awake at night texting friends on their phones and fall asleep on the bus or at school during the day, while their parents try to function on five to six hours per night.

Consistent, uninterrupted, restorative sleep is essential to good health. From the cellular level up, sleep replenishes all systems of the body—endocrine, immune, and metabolic. During darkness, the pineal gland, if not calcified by fluoride, produces melatonin and the brain and body produce serotonin, hormones necessary for mood stabilization, coping, attention, and memory. The less melatonin and serotonin available, the less restorative sleep we get, and the less able we are to deal with the stressors of life.

Inefficient breathing during sleep is also common and rarely recognized in young children. Sleep-disordered breathing is called obstructive sleep apnea. It is often caused by enlarged tonsils and adenoids that block airways as the muscles in the throat relax.

Although the body can survive for a month or more without food, death can occur in a week without sleep. Seizures can occur after only 24 sleepless hours. Just three nights without restorative sleep can produce a state known as *sleep deprivation psychosis*, in which rational thinking is impossible.[13]

When deprived of sleep, the body and brain begin a slow deterioration that impacts all areas of health and function. Sleep deprivation impairs sugar metabolism, immune function, and motor skills and increases stress hormones.

The symptoms of sleep deprivation are almost identical to those of ADHD: irritability, moodiness, distractibility, poor focus and attention, depression, and learning difficulties. Non-medicated adults showed symptoms of ADHD when subjects were deprived of sleep in one study.[14] In fact, undiagnosed sleep disorders can be misdiagnosed and masquerade as ADHD.[15]

Remember the DSM guideline that medical conditions must be ruled out to make a diagnosis? Here is another one that is frequently missed. Bottom line: it's a good idea to evaluate the quality of sleep with a sleep study before slapping on a neurodevelopmental diagnosis.

Lack of exercise is a real stressor to the body and mind. Exercise is now something one does at a gym or on a playing field, not as a part of daily routines, such as walking to school or joining a pickup game of ball. Our busy, over-scheduled days allow little time for kids to engage in unstructured play. Think about this: isn't free play, without adults hovering over kids, where social skills begin to develop?

Children must make playdates instead of picking up games spontaneously and building forts out of cushions at home or stones in the woods. Parents are enrolling even the youngest and least coordinated kids in group sports and lessons that replace free play. By requiring children's bodies and minds to conform to motor schemes and rules of specific games, few are learning how to control their bodies spontaneously through trial and error. Instead, they are acquiring "splinter" athletic skills and playing games by rote.

Medical Stressors

Even before a baby is born, parents carefully choose a pediatrician to monitor growth and development. This physician's philosophy and training will affect how a family approaches wellness and sickness.

About 30 years ago, I wrote an article entitled, "Do You Really Need a Pediatrician?" As a new mother, I had only Dr. Spock and my own mother and doctor to help me know whether my baby was doing okay. Today's parents have dozens of books and the Internet to learn about babies' growth and development. Why would they need a pediatrician? Dr. Robert Mendelsohn stated two generations ago, in his classic book *How to Raise Healthy Children in Spite of Your Doctor,* that "The purpose of the pediatrician is to indoctrinate your child into a lifelong dependency on drugs."[16]

Today's answer is "to keep them up to date on their shots." Yes, "well child" visits at preset intervals are about staying current on vaccines. Did you know that many of today's babies are given up to three dozen vaccines before they turn two?[17]

Nobody knows what so many vaccines do to growing bodies and minds because no study has looked at the cumulative effect of the full CDC schedule. This constant assault on the immune system is felt to be a load factor by many of today's medical professionals.

Added to this barrage of pathogens with their additives, the number and power of frequent rounds of antibiotics is also of concern. As these potent drugs kill off both good and bad bacteria, they are damaging our guts. Read more about these issues in Chapter 8.

As mentioned in Chapter 1, the "standard of care" approach to treating symptoms of the neurodevelopmental issues that are the subject of this book usually includes prescribed medications. All medications have side effects that may require additional medications to alleviate. Before you know it, a child is taking a cocktail of several meds, without any knowledge of what is causing the original problem. Parents must educate themselves about medication side effects, especially those that disrupt appetite or sleep.

Some other medically approved practices are also probably interfering with development. For instance, recommending that babies sleep only on their backs to avoid sudden infant death syndrome (SIDS) prevents them from learning to use their hands to bear weight and coordinating their hands with their eyes for crawling.

Children with limited time on their stomachs often experience problems with later learning tasks that require crossing the midline, such as writing. The panacea "tummy time" probably helps a little but is not very well tolerated by many babies and is not equivalent to hours of sleep on the belly.

Finally, the last medical stressor comes from ongoing stress on the endocrine system, which causes stress hormones like cortisol to rise and affect thyroid hormone availability. Read more about these hormones in Chapter 6.

Social Stressors

Lack of family meals. Food is not just a nutritional issue; it's not just about *what* we eat, but also *where, when and how.* Fast food is everywhere, vending machines are ubiquitous, and snacks at sports events often constitute dinner. Few families eat together three times a week, let alone three times a day.

Regular family meals provide much more than companionship. Eating together builds vocabulary, and teaches listening, good manners, and family history. The ritual of family meals can be transforming.

Too much screen time. Computers, tablets and phones are today's entertainment and babysitters. Toddlers, handed phones by busy parents, can independently find their favorite games and shows. Our excitement about computer technology is accelerating as fast as the technology itself.

While most applaud the extraordinary opportunities our electronic wonders offer, some are concerned that for our youngest children three-dimensional sensory experiences are being replaced with two-dimensional activities. Instead of feeling blocks, puzzles, dolls, and game pieces, they are manipulating virtual blocks, dragging puzzle pieces, clicking on games, and moving make-believe dolls and flat letters and numbers by touching a screen or by pointing and clicking.

One psychologist who very early on expressed concern about what effect screens have on the developing minds of all children is Jane Healy, PhD. In her book *Failure to Connect: How Computers Affect Our Children's Minds—and What We Can Do About It,*[18] she argues convincingly for limiting screen time of all types until age seven, at least.

Social psychologist Jonathan Haidt describes how the "play-based childhood" started to decline in the 1980s and has been replaced by a "phone-based childhood." Instead of playing ball against the steps, today's kids are watching movies and playing games, often violent, on a computer or phone. They are spending endless hours texting and chatting, even in the middle of the night. Onscreen images of characters in action have replaced hands-on play that lays the foundation for higher thoughts, creativity, inspiration, imagination, and even intuition.[19]

In his 2024 best seller *The Anxious Generation*[20] Haidt demonstrates how this "great rewiring of childhood" has interfered with children's social and neurological development, covering everything from sleep deprivation to attention fragmentation, addiction, loneliness, social contagion, social comparison, and perfectionism.

Too much screen time extends to school as well, where screens are replacing people. Artificial intelligence (AI), still in its infancy, has entered the building. Students can tap into ChatGPT for help writing research papers, instead of going to the library. Kids still need to look at and interact with humans and stimulate their brains socially, not by staring at a screen.

The latest research confirms that too much screen time leads to significant mental health issues, including anxiety, depression, focus challenges, and impulsive conditions.[21] Furthermore, excessive screen time has brought us new names for new maladies.

Integrative psychiatrist Victoria Dunckley has coined the term *electronic screen syndrome* (ESS) to describe what might mimic a psychiatric disorder with similar symptoms to mood, anxiety, and behavioral diagnoses. She defines ESS as a disorder of dysregulation that shifts the nervous system into a fight-or-flight mode, affecting various biological systems.[22] (See Chapter 6.) Journalist Richard Louv, author of ten books, says alienation from nature and insufficient experiences out-of-doors can lead to *nature deficit disorder* (NDD).[23]

Safety concerns prohibit venturing out without being tethered to home by a mobile phone. Many parents track their kids using sophisticated software, knowing their exact whereabouts at all times. What does this constant vigilance do to a child's ability to develop confidence and self-esteem?

Lack of public understanding. Uneducated adults often misinterpret behaviors such as stimming, temper tantrums, and self-talk as signs of bad parenting. Dealing with problematic behaviors can be stressful at home; dealing with ignorant folks during a meltdown in the grocery store is doubly stressful.

Betsy Hicks Russ solved this problem with a vest her son Joey wears that says simply, "Autistic. Be Kind." On the other hand, while kindness to those with neurodevelopmental conditions is essential, the condition is not an excuse for bad behavior.

Family stressors. Families have changed; the majority no longer follow the traditional model with Mom at home and Dad at the office. When a child has two parents, most likely both work. They may or may not be the same sex, race, culture, or religion. Divorce and estrangements are common. Often kids must learn how to cope with different styles of parenting as custody arrangements may require them to move from parent to parent, home to daycare, or even city to city.

Many aspects of the family system related to having a child with special needs are immeasurably stressful. Some families have two or more children requiring extra care. Take Kim Rossi, whose three autistic daughters with limited verbal skills are now grown but will probably need monitoring and care for the rest of their lives. About 80 percent of such marriages end up in divorce, because of the stress of managing so many relationships.[24]

Sibling relationships can be trying. Most are quite tolerant and protective of their brother or sister with autism or OCD. However, sometimes siblings feel resentful of the "special" child taking up so much of the parents' time, and they may lash out. Feeling secure and safe is far more difficult. With distance often separating close relations, families lack the traditional security of knowing and spending frequent time with their grandparents, aunts, uncles, and cousins. A few precious days with them during holidays and school vacations in a whirlwind of ice cream, pizza, movies, and other Disneyland treats is not the same as developing day-to-day relationships.

Educational Stressors

For many kids, school is an unrelenting source of stress. Despite the fact that they may not have had the experience of sitting through a 15-minute meal, they are expected to sit through 45 minutes of circle time in kindergarten and a 60-minute lesson in first grade. Many stare out the window for visual relief or wiggle and squirm to keep alert.

Here are some specific aspects of the educational system that are stressing out our kids.

Developmentally inappropriate curricula as a result of the No Child Left Behind Act have hastened the teaching of reading and writing to a generation of children who cannot sleep through the night, tie their shoes, or speak in complete sentences. What was once taught in first grade is now part of the kindergarten curriculum. Many kindergarteners still have poor control over their own bodies, let alone pencils and scissors.

Pushing kids ahead academically without first establishing a strong foundation of motor, sensory-motor, and language skills forces them to learn by rote, at best. I believe that many "learning disabilities" could be prevented by putting a brake on teaching academics prematurely. Let's bring back old-fashioned skills like handwriting, keyboarding, and doing mathematics mentally, instead of using a calculator. These brain-based exercises all build mental "muscle."

Decreased time for the creative arts is common in schools today. Music and art are becoming extinct as we focus on academics. After-school lessons in playing an instrument and painting a still life are not the same as providing art materials, keyboards, or drums and allowing freedom of expression.

Classes vary in size and abilities. Today's teachers must contend with up to 30 children of varying achievement levels, often without assistance. The common solution of hiring an aide, or "shadow," for those with serious special needs can help unburden the teacher. However, sometimes aides act as police, keeping students "on task," further squelching their instincts to move and touch.

As schools are cutting back on their budgets, classes can also span two or more grades. If a system has 40 first graders and 20 second graders, they might make a one-two combination. Placing higher-level first graders with lower-functioning second graders is a questionable practice.

The ***increase in safety concerns*** in our schools is daunting. School shootings have become a stark reality, requiring necessary safety precautions.

Individual kids are at risk every day. Some are wanderers, others climb, jump, and otherwise put themselves in harm's way. Everyone must be so vigilant that schools are hiring one-on-one aides to watch individual children, and some parents have resorted to tagging their kids with electronic devices.

Decreased opportunities for movement are common, with some districts eliminating recess and physical education altogether. Kids need to move! Movement is food for their brains.

Emotional Stressors

Novel and potentially painful situations like visiting the dentist or eye doctor are common, and mostly manageable, emotional stressors for neurotypical kids. But for young children with dysregulated nervous systems, even the unexpected roar of a lawn mower or an ambulance siren can ruin their day. Unfortunately, adults cannot anticipate every possible scary or overstimulating experience.

Children's fears used to be simpler. Today's fears are bigger, more abstract, and scarier. We no longer have fire drills; we now have active shooter drills. Add fear of viruses that can kill you or your grandparents and nuclear war that can kill us all, and the world becomes quite a terrifying place. With constant reinforcement from well-meaning people repeating, "Be safe," simple fears can turn into phobias that become consuming.

The extended *COVID-19 lockdowns*, mask-wearing, social distancing, separations from friends and family, and more caused untold fear and distress that didn't end when schools reopened and masks were tossed in the garbage. Emotional responses to these seemingly necessary precautions are now being recognized as contributory to a new pandemic of loneliness, depression, anxiety, suicide risk, and mental illness.

Lockdowns certainly caused stress. One study reported that usage of mental health facilities increased by 18% in regions with a lockdown compared to a 1% decline in regions without a lockdown.[25] Findings of a British study revealed that infected participants had an increased risk of developing various mental health disorders, including psychotic, mood, anxiety, alcohol use, and sleep disorders, and the risk was higher still for individuals who were hospitalized due to COVID-19.[26]

Societal modifications and mandates profoundly affected students already diagnosed with special needs. Changes in routine were confusing, remote learning was difficult, if not impossible, and limited access to related services, such as speech-language and occupational therapy, combined to cause loss of skills, both academic and social.[27]

Being bullied is a well-founded fear of those with special needs. They are particularly easy targets for bullies because of the frequent combination of naiveté, poor communication skills, and loyalty to the rules. They rarely tattle on the bully because of their additional fears of ridicule, embarrassment, or retaliation. Unlike their typical peers, they may not even recognize bullying and instead think that a bully is a friend. They can thus be lured into compromising positions or places and convinced to do dangerous things.

Every Child Is Unique

Remember, those with special needs vary markedly despite similar behavioral symptoms. It is worth repeating that healthcare professionals, educators, and therapists must treat each patient as a unique individual with a unique health and developmental history. Determining appropriate treatments and their sequence requires knowing the multiple unique causes of symptoms in each patient. Laboratory tests, discussed in Chapter 5, can frequently assist in pinpointing individual needs.

In her groundbreaking book, *The Autism Revolution*,[28] esteemed pediatric neurologist, Martha Herbert, MD, says, "the child's brain and body have been battered so much by the world that they have lost the resources to protect themselves." Any stressor—biological, environmental, lifestyle, physical, medical, educational, or emotional—can be a load factor for anyone, battering the brain and body to the point of exhaustion.

About ten years ago, Dr. Herbert and I developed informal checklists for both adults and children that we named the "Everyday Epigenetic Evaluator" or E³ for short. I have since revised these slightly to reflect societal changes.

Before you read the next chapter, take the test for yourself and give it to each of your immediate family members. Total possible points are 100. The fewer the points, the fewer the stressors. Score over 20? Consider making some changes!

Everyday Epigenetics Evaluator

Everyday Epigenetics Evaluator for Adults©—E3A
Copyright Martha Herbert MD and Patricia Lemer, M. Ed., 2015
Revised 2024

DIET (25)

1. ____ Eat mostly carbs for breakfast (2)
2. ____ Eat canned food > 3x / week (3)
3. ____ Eat/drink dairy products daily (2)
4. ____ Buy nonorganic vegetables (2)
5. ____ Eat < 5 servings fruit and vegetables/day (3)
6. ____ Drink soda (1)
7. ____ Drink diet soda (2)
8. ____ Eat out > 3x / week (2)
9. ____ Eat at fast food restaurants > 2x / week (2)

10. ____ Eat > 15 grams sugar/day (3)
11. ____ Eat sushi and/or tuna fish (1)
12. ____ Microwave food (2)

EMFS (7)

13. ____ Known cell phone tower within 1 mile of home (1)
14. ____ TV and/or computer in bedroom (1)
15. ____ Talk with cell phone to ear or with ear buds > 1 hr / day (2)
16. ____ Use home alarm system (2)
17. ____ WiFi on at night (1)

SLEEP (9)

18. ____ Get fewer than 8 hours sleep > 3 nights / week (3)
19. ____ Use nonorganic mattress (2)
20. ____ Get < 30 minutes of exercise per day (2)
21. ____ Use sleep number bed (2)

SOCIAL/EMOTIONAL (6)

22. ____ Estrangements in family (2)
23. ____ Stress in marriage (2)
24. ____ Financial worries (2)

MEDICAL/DENTAL (31)

25. ____ Have "seasonal" allergies (2)
26. ____ Carry an EpiPen (3)
27. ____ Poop < 1x / day (2)
28. ____ Get annual COVID and/or flu shots (2)
29. ____ Antibiotics in past 5 years (2)
30. ____ Take Tylenol (2)
31. ____ Low thyroid (3)
32. ____ Known low vitamin D level (3)
33. ____ Mercury amalgams (4)
34. ____ Root canal(s) or dental implants (2)
35. ____ Wear glasses with progressive lenses or monovision contacts (2)
36. ____ Take > 2 prescription meds (2)
37. ____ Major surgeries to remove organs (2)

TOXINS (18)
38. ____ Work exposure to toxins (2)
39. ____ Play or live near a golf course (1)
40. ____ Use a lawn service (1)
41. ____ Fluoride in toothpaste (1)
42. ____ Use dryer sheets (1)
43. ____ Use perfume, lipstick, dye hair (1)
44. ____ Use anti-bacterial soap and/or hand sanitizer regularly (1)
45. ____ Spray for bugs regularly (1)
46. ____ Have been exposed to mold (2)
47. ____ Smoke now or in the past (2)
48. ____ Use deodorant containing aluminum (1)
49. ____ Recently renovated/painted home (1)
50. ____ Live near freeway (1)
51. ____ Use plug-in air fresheners (1)
52. ____ Work or live in space without windows that open (1)

HYDRATION (4)
53. ____ Drink < 32 ounces of H2O / day (2)
54. ____ Drink unfiltered tap water (2)

Everyday Epigenetics Evaluator for Children©—E3C
Copyright Martha Herbert MD and Patricia Lemer, M. Ed., 2015
Revised 2024

PRENATAL (13)
1. ____ Conceived w/ fertility assistance (2)
2. ____ Birth mother with > 5 amalgams (2)
3. ____ Mother with chronic fatigue, low thyroid, or fibromyalgia (3)
4. ____ > 3 sonograms during pregnancy (1)
5. ____ Mother vaccinated pregnant (2)
6. ____ Mother used drugs, smoked, or drank alcohol (2)
7. ____ Born by C-section (1)

DIET & HYDRATION (24)
8. ____ Formula fed as infant (1)
9. ____ Picky eater (3)
10. ____ Eats bread or crackers daily (2)

11. ____ Eats/drinks dairy products daily (2)
12. ____ Eats < 3 servings fruit and vegetables / day (3)
13. ____ Drinks soda or diet soda (2)
14. ____ Eats at fast food restaurants > 2x / week (2)
15. ____ Eats > 15 grams sugar / day (3)
16. ____ Eats microwaved food (1)
17. ____ Eats foods with artificial ingredients (1)
18. ____ Drinks < 32 ounces of H2O / day (2)
19. ____ Drinks unfiltered water (2)

MEDICAL/DENTAL (26)

20. ____ History of ear infections (1)
21. ____ Hep B shot at birth (2)
22. ____ One or more vaccine reactions (2)
23. ____ Has "seasonal" allergies (2)
24. ____ Carries an EpiPen (2)
25. ____ Poops < 1x / day (1)
26. ____ Gets annual flu and/or COVID shot (2)
27. ____ Took antibiotics in past year (1)
28. ____ Takes Tylenol (2)
29. ____ Known low vitamin D level (2)
30. ____ Mercury amalgams (3)
31. ____ Root canal(s) (1)
32. ____ Wears glasses (1)
33. ____ Needs orthodontics (1)
34. ____ Walked early; didn't crawl (2)
35. ____ Shallow or mouth breathing (1)

TOXINS (8)

36. ____ Fluoride treatments or toothpaste (1)
37. ____ Swims in chlorinated pool (1)
38. ____ Uses chemical-laden dryer sheets in laundry (1)
39. ____ Uses lotions and shampoo with phthalates (1)
40. ____ Use antibacterial soap and/or hand sanitizer (1)
41. ____ Home sprayed for termites (2)
42. ____ Use bug spray and/or sunscreen (1)

SLEEP (8)

43. _____ Sleeps fewer than 10 hours > 3 nights per week (3)
44. _____ Uses flame-retardant mattress (2)
45. _____ Gets < 30 minutes of exercise per day (2)
46. _____ Attends school or lives in space without windows that open (1)

EMFS (5)

47. _____ Known cell phone tower within one mile of home (1)
48. _____ TV and/or computer in bedroom (1)
49. _____ Uses tablet/laptop > 1 hr / day (1)
50. _____ House has alarm system or smart meters (2)

SOCIAL/EMOTIONAL (11)

51. _____ Stressful/abusive family situation (2)
52. _____ Major sudden loss or trauma (2)
53. _____ Adopted or in foster care (2)
54. _____ Sees one parent infrequently (1)
55. _____ Bullied at school (1)
56. _____ Has > 3 siblings (1)
57. _____ Has < 3 friends (2)

ACADEMIC (5)

58. _____ In lower grade than age (2)
59. _____ Has IEP or accommodations (2)
60. _____ Receives tutoring (1)

Take-Home Points

In summary, physical, lifestyle, emotional, and educational stressors are sometimes difficult to identify because many have become normal parts of today's fast-paced life. However, everyone agrees they exist and can profoundly affect individuals of all ages. They go hand in hand with all the other stressors in the previous chapter. Focusing on stressors that we have control over can greatly reduce the Total Load. Then the joint efforts of professionals tackling toxins, viruses, metals, and genomics, can reap even bigger benefits.

CHAPTER 4

REDUCING AND ELIMINATING LOAD FACTORS THROUGH LIFESTYLE CHANGES

As soon as a neurodevelopmental disorder is suspected, the very first step is to reduce as many stressors as possible. Regardless of the severity, lessening everyday load factors *before* embarking upon expensive and time-consuming therapies can immeasurably improve future outcomes and quality of life, as well as save considerable time, effort, and expense.

The previous chapters enumerate the common stressors that accumulate as load factors in the bodies and minds of our loved ones, eventually affecting everyone's health, learning, and behavior. Making lifestyle changes can improve the health and well-being of each family member—diagnosed or not. Diet and nutrition are important major stress factors included only minimally in this chapter; they deserve a chapter of their own, which follows.

Remember, these are general ideas, as each person is different. A few of the suggestions in this chapter may not yet be applicable for families of children who are still nonverbal, families with significant global challenges, or even some generally higher-functioning adults with isolated severe challenges. Go slow, and keep some ideas on the shelf for a later date.

Reducing Genetic Stressors

Understanding how genomic changes affect function has given rise to a new and exciting field, nutrigenomics. Following genomic testing, described in Chapter 2, healthcare professionals can apply their knowledge of a patient's combination of SNPs by using nutrition to target and modify affected aspects of a patient's physiology. Personalized action plans change how hormones are secreted, how nutrients are balanced, and more.

IntellxxDNA, a company mentioned in the previous chapter, offers testing and analysis of over 40 specific, clinically significant genetic variants (SNPs) in a "Spotlight on Anxiety & Depression Report," targeted at specific vulnerabilities that research has shown to contribute to these symptoms. By looking at such areas as an individual's nutrient absorption, tendency to inflammation, ability to detoxify, and even their hormonal stress response, the IntellxxDNA comprehensive report provides a metabolic understanding of mental health.

An expert clinician can then create a specific treatment plan that addresses a full spectrum of factors affecting an individual's mental health. You and your doctor can discuss special dietary and nutritional strategies tailored to your ability to absorb and transport specific vitamins and minerals, choose supplements that can remove unwanted chemicals and/or boost wanted ones, and make lifestyle modifications. These interventions may be able to eliminate medications that have undesirable side effects.

Remember from Chapter 1 that for each and every diagnosis listed, the DSM stipulates that a physical reason for a cluster of symptoms must be ruled out. Psychiatrists, neurologists, and therapists are just learning about the power of nutrigenomics to turn off genes that are interfering with function and thus improve both health and behaviors in their patients. Every success makes it clear that many, if not all, disorders previously thought to be "mental" have physiological roots.

Reducing and Eliminating Toxins in Our Food, Air, Water, and Products

Eliminating environmental stressors starts by identifying sources of exposure to any and all toxins, whether they be in the air, water, food, or products or are related to jobs, family, or other relationships.

Reducing Toxins in our Food

Buy organic, especially the "Dirty Dozen": strawberries, spinach, kale, collard and mustard greens, grapes, peaches, pears, nectarines, apples, bell peppers, cherries, blueberries, and green beans.[1] That way, you are limiting the pesticides and glyphosate you and your family are ingesting. Eat in season. Grow a vegetable garden; go to local farmers' markets.

Minimize eatables that come in jars, boxes, bags, or cans. Canned foods leach BPA, a component of the container's plastic lining.[2]

Cook! In his book, *Cooked: A Natural History of Transformation,* food activist Michael Pollan states that cooking is the single most important thing that a family can do to improve health and general well-being.[3] In American households today, the amount of time spent cooking is half of what it was in the 1960s. In fact, Americans spend less time cooking than any other nation![4]

Cooking food distinguishes humans from other animals. In *Catching Fire: How Cooking Made Us Human*, Harvard anthropologist Richard Wrangham develops the "cooking hypothesis."[5] Cooking is the evolutionary prerequisite that allowed us to transcend animals by using fire in the place of bodily energy to break down complex carbohydrates and make proteins more digestible. Eating cooked food releases more bodily energy for other functions, such as thinking. Obviously, raw food enthusiasts would disagree. This is another area where you should listen to your own body.

Limit sugar, wheat, and dairy products, as well as kid and fast food. Ask for your parents' and grandparents' favorite recipes and cookbooks. Find new recipes using vegetables, grains, and fruits your children will eat. Take some cooking classes and learn how to use unfamiliar nutrient-rich foods such as quinoa, amaranth, seaweeds, and parsnips.

Toss the microwave. Use gas or electric stoves. Never, ever microwave food in plastic containers.

Eat together at least once a day—and in the car doesn't count! Choose one meal per day and give it the respect it deserves. Sit together without stress for at least 10 minutes. Make meals an enjoyable experience. Turn off the TV and computer; do not answer the telephone. Let kids take their time to accept or not eat what is put in front of them. Don't force feed, prod, plead, or bribe.

Share growing and cooking with children, who are more apt to eat something they grew or made.

Reducing Toxins in the Air

Invest in a good air filter. Buy a good air filter for your home. However, spending money on air filters and purifiers is a waste unless tobacco smoke and off-gassing from toxic new products are eliminated beforehand.

Go outside. Since outdoor air is significantly healthier than indoor air, the best possible antidote to breathing toxic air is to spend more time outdoors. Second best is opening the windows, even in winter, to bring the outside in.

Monitor tobacco exposure. If any family members are still smokers, insist they smoke well away from the house. Boycott restaurants and bars that permit smoking. Minimize contact with adults wearing clothes saturated with smoke, which transfers to others through touch and movement.

Choose building and remodeling materials carefully. Buy recycled or formaldehyde-free products that do not off-gas for renovations. Choose cork, bamboo, and other natural flooring instead of carpet treated to be stain resistant. Ensure that school buildings have no asbestos or lead paint. Use no-VOC paints and wood that is not treated with formaldehyde, especially when preparing a baby's room. Make sure renovation doesn't occur while people are in residence.

Practice safe pest control. The best way to control pests is to keep them outside. Since shoes track in outside dirt as well as pesticides, imitate the Japanese and leave shoes at the door. Try using herbs and other natural products, such as bay leaves for cockroaches, black pepper to deter ants, or lavender, cedar oil, and camphor to repel rodents. Choose integrated pest-management systems instead of toxic chemicals, even when treating head lice. Protect pets with natural bug repellants, such as diatomaceous earth—a fine, white powder made from ground-up, fossilized remains of diatoms, a sea algae.

Use nontoxic cleaning products. Wash clothes with perfume-free laundering products. Minimize dry cleaning, which also uses toxins. If clothes must be dry cleaned, air them out to let the chemicals dissipate before bringing them indoors. Buy "green," toxin-free dishwasher soaps, bathroom cleaners, detergents, facial tissue, diapers, baby wipes, paper products, and other essentials that are good both for you and your family and for the environment. Or, better still, use common household products like vinegar, baking soda, and lemon juice to make your own cleaning products. To learn more, read "Everything You Need to Know about Green Cleaning."[6]

Reduce Toxins in Our Products

Replace toxic with nontoxic. Every time you run out of something, like shampoo, toilet tissue, dishwasher detergent, or toothpaste, and need to replace it, think nontoxic. Upgrade gradually, and within a year or so your home will be "green."

- Throw out all those old containers for leftovers and replace them with a set of multi-size glass ones with tight lids. That will avoid the use of plastic wrap and aluminum foil as well.
- Read numbers carefully on plastic products. Avoid #3, #6, and #7, which can leach harmful chemicals into food and drinks. These numbers occur in some bags, wraps, toys, Styrofoam cups, to-go containers, metal can liners, and clear plastic bottles.
- Swap "nonstick" and aluminum pots and pans for cast iron, stainless steel, or ceramic.
- Read labels on art supplies and toys. Use only lead- and asbestos-free crayons, water-based glues and paste, and magic markers denoted "nontoxic." Seek out toys made of natural materials like wood. Look for terms like

"nontoxic" and "eco-friendly" on plastic toys. Avoid using chalk (which contains talc, the ingredient removed from baby powder because it causes cancer), oil-based paints (which contain toxic metals), and spray adhesives (which contain other poisons). Look for products with the Art and Creative Materials Institute (ACMI) Nontoxic Seals.[7] They have been evaluating the safely of art supplies for over 80 years.

- Use 100% natural clothing and bedding. Buy organic whenever possible. Non-organic cotton is sprayed with pesticides. Look for organic cotton, linen, silk, and wool clothing. Avoid flame retardants in mattresses and sleepwear. A good way to avoid this problem is to buy a futon, which can be made of 100% organic cotton because it is not officially a "mattress," and 100% cotton outfits that are not officially "pajamas." Replace electric and synthetic blankets with natural bedding. Three safe mattress brands are Vivetique, White Lotus, and Naturepedic. Online resources for these brands are Lifekind.com and Greensleep.com.

Buy fluoride-free toothpaste. Fluoride is a neurotoxin. No worries about cavities if you limit sugar. Try a new option from Pure Haven that contains an all-natural mineral called hydroxyapatite, which is scientifically proven to help remineralize and whiten teeth. This ingredient is a great replacement for fluoride since its unique structure recreates a new, thicker enamel and deposits calcium and phosphate ions to the teeth. Visit PureHaven.com/7869.

Choose natural cosmetics and shampoo and conditioner without additives.

Some trusted companies that are committed to safe personal care products are Dr. Bronner's, Aubrey Organics, Hello, Desert Essence, and Field Day.

Reduce Exposure to EMFs and Dirty Electricity

- *Buy a Gauss meter* and take measurements of EMFs at home and school. Stay out of "hot rooms" such as computer labs in schools.
- *Purchase filters for electrical outlets.* Remove all electrical and wireless appliances from everyone's bedrooms, including plasma TVs, treadmills, computers, cordless phones, routers, cell phones, and their chargers. Use a battery or wind-up alarm clock.
- *Turn off Wi-Fi at night.* Make sure you know what is on the other side of the wall from the head of all beds. If a TV is on that wall, move it or the bed.
- *Never use laptops on the lap or cell phones too close to the ear.* Attach an external keyboard to laptops to reduce the amount of radiation sent to the hands.

- *Reconsider buying an electric vehicle.* The jury isn't in on this one. If you are making a short commute to work, it might be okay. No one knows how long trips in an all-electric car are affecting the nervous system. We are participating in an experiment. Stay tuned.
- *Hire a Building Biology Inspector* (BuildingBiologyInstitute.org). Learn whether your home is affected by geopathic stress or has dirty electricity by having it inspected. Even better, take one of their courses or seminars on indoor air quality, electromagnetic radiation, healthy building design, and more. To learn even more about how you can minimize your exposure to dangerous radiation, read *Disconnect: A Scientist's Solutions for Safer Technology, and How to Protect your Family.*[8] *The Non-Tinfoil Guide to EMFs,*[9] *Zapped: Why Your Cell Phone Shouldn't Be Your Alarm Clock and 1,268 Ways to Outsmart the Hazards of Electronic Pollution,*[10] *Electromagnetic Fields: A Consumer's Guide to the Issues and How to Protect Ourselves,*[11] and *BioGeometry: Back to a Future for Mankind.*[12]

If all of this sounds totally overwhelming, get help from The EMF Safety SuperStore. Browse their amazing catalog: LessEMF.com

Increasing Sleep, Hydration, Deep Breathing, Exercise, and Good Lighting

In Chapter 3, you read about the devastating effects of poor sleep quality, dehydration, inefficient breathing, and not enough movement. While some fixes require professional intervention, many simple lifestyle changes can make a major difference.

Establish a sleep hygiene program. Just as you have grooming hygiene—brushing teeth, showering, washing hair, etcetera—good quality sleep requires the right routines and environment. Set a schedule, beginning at least 30 minutes before lights out:

- Establish bedtime—before 8 p.m. for preschoolers and 11 p.m. at the latest for school-age.
- Banish TV, computer, or video games for at least an hour before bedtime.
- Draw a 10–15 minute warm bath, followed by a deep towel massage. Add Epsom salts for detox and calming. Speak quietly and soothingly. Put on pajamas and get straight into a warm bed.
- Read or tell a story and turn out the lights.

Follow the same routine every single night, especially with young kids. For their circadian rhythms to work consistently, weekends should be no different than other nights.

According to experts, school-aged children need 9–12 hours of sleep a night, teenagers, 8–10 hours, and adults, 7 or more.[13] Put infants and toddlers on their tummies when awake or sleeping in the presence of an adult. Make sure your sleeping area is safe from EMFs and all electrosmog. If your home is exposed to cell phone towers, televisions in neighboring apartments, or geopathic stress, consider buying special netting, sheeting or other mitigators from Safe Living Technologies (SLT.co) to deflect the EMFs.

Bedrooms are for sleeping. Make sure the bedroom is not a home to electronic devices, including computers. If your loved one snores, or after taking all these precautions still has disordered sleep, ask your doctor to order a sleep study to check for sleep apnea. Diagnosing this underlying problem and fixing its cause early is imperative to normal physical and emotional growth and development. If sleep apnea is present, consult with a sleep dentist immediately. Go to Chapter 15 to learn how they are changing lives with various methods.

Falling asleep, staying asleep, and waking up can, at least in part, be traced to sensory regulatory issues. Read about these in Chapter 16. A weighted comforter or soothing music can be calming and promote restful sleep. Try developmental music therapist Joe Romano's CD *Dreaming in The Land of Can*. It can play throughout the night to promote deep, restful, and relaxing sleep.

If none of these work, try the Dreampad™ from Unyte.com. The Dreampad is a pillow that uses bone conduction to deliver ambient sounds and calming music that only the user can hear via transducers imbedded in the pillow. Via an app on your phone, you can choose which type of music you prefer, as well as the length of time for it to play.

Stay well hydrated. Drink good quality water frequently. Drinking water throughout the day can head off many problems before they show up. What kind of water should kids drink? Only clean, fresh, mineral-rich, natural spring, filtered, non-fluoridated, unadulterated water. Not coffee, not tea, not juice, not flavored water, and certainly not soda. The body identifies these water-based products as food, not water.

Invest in good water filters. Most practitioners recommend using a reverse osmosis water filter that uses pressure and a fine membrane to remove all impurities, such as bacteria, pesticides, and other particulates. Some add electrolytes and a special mineral formulation called "M water" that increases absorption and utilization of nutrients, oxygen availability to the cells, and the body's ability to detoxify.[14]

A whole house filter is unnecessary unless you have a well. Put filters on the kitchen sink and in showers and tubs. That is the water that goes in and on your body.

Breathe deeply. Recognizing the importance of efficient breathing, therapists from many disciplines add breathing exercises to their remediation programs to improve movement, language, vision, cognition, learning, and behavior. Remind kids throughout the day to take long, deep breaths, especially when they are upset. Introduce them to yoga and martial arts that emphasize breathing to reduce stress by calming the vagus nerve (see Chapter 6).

Move and play. In his bestselling book, *The Last Child in the Woods: Saving Our Children from Nature-Deficit Disorder*, Richard Louv describes the disconnection between today's children and nature.[15] Let kids go outside to play, even in bad weather. Do your exercise outside (not on a treadmill), walk, play ball. Teach your kids to bike and swim—both essential life skills and great exercise.

Outdoor play has many advantages over playing computer games and watching TV. Playing outdoors antidotes sedentary activity. Developmental specialists recognize that young students' bodies need movement as much as a nutritious breakfast. Let kids play freely, make up games with found objects, walk on walls, ride their bikes around the neighborhood (with a helmet, of course), and engage in freestyle motor activities instead of just group sports.

Using the outdoors as a classroom for science and social studies is a natural for homeschoolers. Have the curriculum follow the seasonal changes by locating skunk cabbage in early spring and watching the leaves turn in the fall.

Change your lighting. Stock up on incandescent bulbs or switch to LEDs, which stands for light emitting diode. LED lighting products produce light approximately 90% more efficiently than incandescent light bulbs. LED lighting is not only more efficient, but it also is more versatile, and lasts longer. Learn more about the advantages of LEDs at EnergyStar.gov.

Remediating Mold

Inspect for and remediate mold. Mold illness is underlying many "mystery" ailments. Ubiquitous mold from invisible leaks could be lurking anywhere in the home from roof to basement. Look in the tray under the refrigerator, in the dishwasher, laundry room, in the bathrooms for wobbly toilets, and in the paneled basement. If you have *ever* had a roof leak, you need an inspection.

Schools can be moldy too. One practitioner I know says that many of her patients are teachers with mold illness. Our educational institutions don't want to hear that they are poisoning their students, but if your child comes home with headaches or complains of nausea, instead of popping a pill or thinking it's psychological stress, consider mold.

Remediation companies are available everywhere. To avoid conflicts of interest, choose different vendors for inspection and remediation.

Altering Medical Interventions

Choose medical interventions wisely. Our bodies are prepared to fight illness and infection. In fact, some old-fashioned doctors think that makes them stronger than giving them a pill, cream, injection, or syrup to do the job.

Of course it is inconvenient to have a sick child. Everyone's routines are disrupted, lessons are missed, and work must wait. But, sometimes, it's the wiser decision in the long run to wait out a fever, cough, or the flu. Learn about alternatives to antibiotics, which can disrupt the gut flora, and traditional ways to bring down a fever. Make sure you have a copy of *Smart Medicine for a Healthier Child*[16] on your nightstand. It is a wealth of information on choices with alphabetical lists of both conditions and interventions—from pharmaceuticals to herbs, acupuncture, and more—so you can make an educated and informed choice. But don't be foolish; recognize that a doctor is needed for a fever over 103.

Reducing and Eliminating Social, Educational, and Emotional Stressors

Eliminating and Reducing Social Stressors

Establish day- and night-time routines; think seasonally. Introduce time concepts to kids early. Use words like "next week," "tomorrow," "in five minutes." Keep the same sequence of events for meals, homework, and bedtime rituals. Kids like the safety and security of routines. They do not push limits if they know you are serious. Be stalwart in sticking to them, except in very special circumstances, such as birthdays or holidays.

Build seasonal routines and traditions like my daughter has with my granddaughter. They go strawberry picking in May and apple picking in September. Every year, they visit a pumpkin patch, celebrate the Fourth of July with fireworks, and build a gingerbread house. They attend local annual festivals and volunteer at social service organizations so my granddaughter can understand how to help those who are less fortunate.

Read and tell stories. Share books and relate what went on during your day, especially as a bedtime ritual. Reading with children has the same advantages as the family dinner: vocabulary building, visual imaging, waiting, and listening.

While listening to stories, children's minds create their own pictures, which lead to ideas that lead to action.

Making up imaginary stories is great fun for kids. In her book, *See It, Say It, Do It,* [17] Denver-area optometrist Lynn Hellerstein recommends taking kids on imaginary hot air balloon rides over their day each night before bed. Looking down upon people, places, and things, they conjure up colors, sounds, textures, temperature, and emotions.

Find alternatives to screen that encourage creativity. Monitor TV time and computer games, especially close to bedtime. Kids need free time to use their imaginations and find something to do when nothing grabs their imagination. Encourage them to dig for worms in the garden, paint, make crafts, sing songs, and play instruments, instead of staring at screens.

Be fully present when interacting with your child. Turn off the cell phone or computer and interact fully. That sends a message to your child that what he or she is doing and saying is important to you; by respecting your child, you become their model for respectful behavior.

Separate home and work duties. Our new world of virtual jobs makes this suggestion more challenging. Try to maintain strict boundaries between the two main areas of their lives. Have a home office area or room, if possible, that is off limits to family. Set strict times that you cannot be bothered. Keeping home and work physically separate can contribute to your ability to keep them emotionally separate. Otherwise, it is extremely difficult to be fully present in either place.

Keep relationships civil. I know; if you and your ex agreed with each other, you would still be married. Try hard not to speak disparagingly about the other parent. Don't make your child choose sides.

Know when to ask for and accept help. This too can be tricky. Grandparents can be a blessing or curse, especially if they live close by. Some make awesome caretakers, and some are too busy. If and when they ask to help, allow it while insisting that they stick to your dietary rules and other boundaries. Be careful about counting on your special child's siblings to do more than their share. We do not want them to feel like Cinderella.

Eliminating and Reducing Educational Stressors

Parents, school administrators and teachers can eliminate or reduce school-related stress.

Advocate. While parents cannot control most of the stressors at school, they *can* make a difference. Get involved and know what's going on. Join committees and attend school board meetings. Be vigilant about bullying and getting junk food out of the schools. Encourage healthy snacks and reinforcers free of gluten, dairy, sugar, or sugar substitutes. Include raw fruits and vegetables—no candy, trans fats, or artificially colored foods.

Have a summer cut-off for school entrance. Make sure that children are fully five years old before entering kindergarten. Give the gift of time to those with summer and fall birthdays by "red shirting" them. This may require a note from a doctor or psychologist, or even a due process hearing. Making sure our kids develop good motor control and other foundational skills before being introduced to academics is essential.

Extra time in pre-academic pursuits can also avoid lengthy, expensive testing and IEPs for those who could catch up if time permitted. Respect for kids' development will reduce the number of children in special education.

Provide a developmentally appropriate curriculum. Meet children at their developmental stages and levels. For instance, if a child with autism is chronologically eight, but has the language and social skills of a five-year-old, reading, writing, and arithmetic instruction should be at a kindergarten, not third-grade, level.

This is not dumbing kids down. It is being respectful of their need for time to develop foundational skills in the motor, sensory-motor, and language areas before asking them to tackle higher-level academics. It also does not mean that content must be babyish. Adults, like children, can build blocks, play with trains, and perform science experiments—just at a higher level. Educators must learn to adjust the content to meet the cognitive needs of each child while keeping the reading and writing demands appropriate, too.

Incorporate movement and sensory activities into the school day. Learning is not all in your head. Whatever the goal, adding movement trains muscle memory, which reinforces all the other senses. Engage the help of occupational therapists to determine the "just right" amount of touch, movement, pressure, and sound for each child. Use hands-on materials for science experiments and math problems.

Vary seating options. Young bodies should adjust well to swings and irregular surfaces. Kids love three-legged stools, balls, and beanbag chairs. Avoid hard, unforgiving plastic and wooden chairs. Consider partially inflated seat cushions if only hard chairs are available. Encourage teachers to use warm-up activities first thing in the morning before academics begin. Ensure that kids are getting adequate food for their nervous systems by including as much time for recess and physical education as for computer lab.

Use games to encourage number learning. Count money, estimate time and distance. Use the language of time freely during the day: "We will clean up in ten minutes," or "Recess is only a half-hour away." Use clocks and calendars to help kids see the passage of time. One family I know underwrote a store in their child's school so that each class could "purchase" supplies of pencils, paper, blocks, and art materials daily. Kids actually had to pay for the supplies and make change.

Provide teachers with in-service training and extra hands. Support teachers attending at least two workshops per year to learn new skills and recharge their batteries. Burnout among special education teachers is rampant. Teach them new skills that help them deal productively with difficult behaviors. Give them parent volunteers and aides whenever the pupil-teacher ratio exceeds 10:1.

Promote a varied sensory diet. Let children touch, move, or look out the window for relief. Watch for sensory overload. Wear simple clothing; avoid too many visual distractions from hanging and reflective objects. Play quiet, calming music; dim the lights, avoiding fluorescents that hum. Eliminate strong smells such as disinfectant, perfumes, and markers. Add calming scents like lavender.

Support inclusion placement in the least restrictive environment in which they can attain success. Inclusion benefits all students, who enjoy meaningful friendships; increased appreciation, acceptance, and understanding of others; and enhanced social skills. Students with special needs also gain

- peer role models,
- increased achievement of Individualized Educational Plan (IEP) goals,
- greater access to the general curriculum,
- enhanced skill acquisition and generalization,
- increased inclusion in future environments (like higher education and employment),
- higher expectations,
- increased parent participation, and
- more integration of their families into the community.

Many safeguards must be in place for inclusion to be successful. These include various types of teacher training and support. I could not find *any* research showing negative effects of inclusion when school systems implemented it appropriately. A nonprofit that has been promoting inclusion for over 30 years is Kids Together™: KidsTogether.org.

Observe behaviors and teach diagnostically. Try to determine the *cause* of a child's behavior. Is it sensory overload? Off-gassing from the carpet?

Too much gluten? A need for movement? A stressed visual system? A family in crisis? What can you do to lessen the stressor(s)?

Write realistic, measurable IEP goals. The Individualized Education Plan (IEP) is the prescription for academic services for all students with special needs. Its goals and objectives are the guidelines for teachers and related-service providers. As a team, these adults, along with the student's parents, set achievable goals and objectives each year for each student.

Ensure success by removing goals regarding compensatory behaviors, such as "increasing eye contact" and "lessening repetitive, self-stimulatory behaviors." Poor eye contact and "stimming" are a child's attempts to cope with stress. These undesirable behaviors usually disappear spontaneously when the stressors are lessened.

Bring back healthy school lunches. Every school should have the ability to cook and serve food. Talk to the director of the lunch program about switching to locally grown, no-pesticide options. The Safe School Meals Act (SSMA) places limits on heavy metals and bans certain pesticide residues and food additives, including artificial food dyes.

Eliminating and Reducing Emotional Stressors

Teach your child how not to be a victim. Our culture has adopted a law enforcement approach to bullying by treating bullying as a crime. I wonder if that has worked to lessen this common practice. Veteran anti-bullying expert and school psychologist Izzy Kalman believes that the key is not protecting people from bullies, but teaching vulnerable kids how not to be victims. When people know how not to be victims, no one can bully them.

Kalman suggests that, instead, we take a psychological approach that recognizes bullying as an inevitable facet of social dynamics, occurring in both people and animals. Instead of criminalizing the bully, he empowers the victim. His program, originally called Bullies2Buddies, equips children to use their brains so they can understand the dynamics of bullying and put a stop to it on their own. Learn more about this unique approach from The Kalman Bullying Institute at IzzyKalman.com.

Use sensory techniques to calm emotions. Joint compression and brushing are two simple interventions that can desensitize a child's overwhelmed nervous system. Parents can easily learn how to increase serotonin release in a child's body with these and other tools, thus giving an overwhelming feeling of peace. Refer to Chapter 10 for more on this important subject.

Prepare kids for new situations, people, and places. Just talking through what's coming up, even when it is routine, can make children feel secure. Scoping out new people and places and seeing what possible sensory or other traps may be lurking there can often avoid meltdowns. Use role playing and social stories for fears of the unknown. For children with sound sensitivity, practice tolerating sirens and other loud noises or provide headphones to muffle them.

Take advantage of local and online support systems. Many more parent support systems are now available. Some groups also have programs for siblings, too. Check out SiblingSupport.org for local resources.

Love children unconditionally. Many parents have told me that having a child with special needs is the best thing that ever happened to them; it helped them grow and learn more about themselves. They learn how to experience joy not only in the special moments, but also day to day. Many have the ability to see beyond the disability and embrace this whole different world that connects them to so many other parents with the same struggles, hopes, and dreams. It's imperative to love children unconditionally, but puberty is a time when this goal is even more important.

Supporting Environmentally Healthy Practices

Reduce stressors by supporting and joining national nonprofits that are working to keep us and our kids safe outside the special needs world. Here are a few:

Green America is the oldest, largest, and most diverse network of socially and environmentally responsible businesses in America. They support an environmentally sustainable society and link green businesses of all kinds to the consumers. Membership includes a subscription to *Green American* magazine, which has articles on such varied and timely subjects as "Eco-friendly Children's Clothing," "Look for Earth Friendly School Supplies," "CFLs vs LEDs: The Best Bulbs," and "Beyond Lead: Toxins in Toys." See GreenAmerica.org.

Healthy Schools Network is the leading national voice for children's environmental health. They support three core facets of environmental health at school:

- child-safe standards for school design, construction, and siting;
- child-safe policies for housekeeping and purchasing (targeting indoor air pollutants, mercury, pesticides and other toxins, and the use of safer substitutes); and
- environmental public health services for children in harm's way.

Healthy Schools/Healthy Kids Clearinghouse© offers dozens of fact sheets, guides, and peer-reviewed reports. Its first two parent guides (on indoor air and green cleaning) have been nationally distributed since 1999. HealthySchools.org.

The Safe Cosmetics Campaign is a coalition of women's, public health, labor, environmental health, and consumer-rights organizations that work with manufacturers to encourage reformulations and safer ingredients. Download their app to look up any product and learn its ingredients instantly at SafeCosmetics.org.

Keep Learning

Stay ahead of the game by reading books, listening to podcasts, webinars, and summits. Attend conferences, and subscribe to newsletters. While we cannot control *everything*, we can control quite a bit.

Books
- *Children and Environmental Toxins: What Everyone Needs to Know*[18]
- *Raising Healthy Kids: Protecting Your Children from Hidden Chemical Toxins*[19]
- *Generation Sick: The Power, Politics & Propaganda Behind America's Health Crisis*[20]
- *Healthy Child Healthy World: Creating a Cleaner, Greener, Safer Home*[21]
- *One Bite at a Time: Reduce Toxic Exposure and Eat the World you Want*[22]
- *Squeaky Green: The Method Guide for Detoxing your Home*[23]
- *The Healthy Home: Simple Truths to Protect Your Family from Hidden Household Dangers*[24]
- *What's Making Our Kids Sick?*[25]

Blogs, Newsletters, Podcasts, and Websites
- Eartheasy—solutions for sustainable living (Eartheasy.com/)
- Environmental Working Group (EWG)—empowers people to live healthier lives in a healthier environment with breakthrough research and education on everything imaginable. (EWG.org)
- Green Chi Café—Celebrate green lifestyle and culture with bestselling green living author, Annie B. Bond (AnnieBBond.com)
- Green Living Tips—Earth-friendly advice for going green and reducing costs, consumption, and environmental impact (GreenLivingTips.com)
- Holistic Moms Network—nonprofit organization connecting parents who are interested in holistic health and green living (HolisticMoms.org)

- Mothering—*the* resource for *everything* natural from pregnancy through elementary age. Join forums, ask questions, search past issues of print magazine, and find new recipes (Mothering.com)
- Nontoxic Alternatives—(NontoxicAlternatives.com)
- The Green Mama—(TheGreenMama.com)
- Wellness Mama—(WellnessMama.com)
- The Neighborhood Food Network—Connect with your neighbors to grow community gardens (NeighborhoodFoodNetwork.com)

Take-Home Points

We are gradually getting control over some of our genes, as well as our personal environments. Making some serious lifestyle changes can substantially reduce load factors, resulting in cleaner homes and schools, healthier and happier kids, closer and more accepting relationships, easier learning, and a generally better quality of life.

On a small scale we can buy positive products, hang out with positive people, and let our kids watch only positive media. We can choose to eat nutritious food, exercise or not, go to bed at a reasonable hour, and tolerate only supportive relationships. On a larger scale, we can work for and donate to campaigns for healthy air, food, and water and strive to create a healthier planet.

Make health a priority before investing in costly therapies. You will be surprised what a difference some simple lifestyle changes can make!

CHAPTER 5

TESTING FOR LOAD FACTORS

I love tests. I spent the bulk of my 40-year career as a diagnostician, evaluating children's strengths and weaknesses, and making recommendations for management. I called myself the "Diagnostic Diva."

Unfortunately, I was giving the wrong tests. My measures of intelligence and academic skills yielded only symptoms of underlying problems: difficulties with language concepts, memory, comprehension, reasoning, calculating, staying focused, and more. So, given my passion for evaluating underlying causes, I referred families to doctors, therapists, and other professionals who could run yet more tests to dig deeper.

In the 1980s and '90s, medical evaluation measures were limited. Blood tests told us whether kids had lead poisoning or vitamin and mineral deficiencies, but not which viruses were wreaking havoc with their digestion.

Developmental measures evaluated gross and fine motor function, and vision testing revealed when the two eyes were not working together. We had not even heard of the vagus nerve yet or considered trauma as a player in learning disabilities or attention deficits. Our only tools for evaluating those with anxiety, depression, and obsessive behaviors confirmed that, yes, indeed they had anxiety, depression, and obsessive behaviors.

Then, in 1995, Dr. William Shaw, a biochemist, with no previous experience with autism, identified very high levels of tartaric acid and arabinose, byproducts of yeast overgrowth, in the gut flora of two brothers he later learned were autistic. An overgrowth of yeast in the body can mimic autism, inducing behaviors like flapping and limited attention. Now we were on to something! Shaw later founded Great Plains Laboratory, now known as Mosaic Diagnostics. It became one of the first and largest medical laboratories in the world.

If you have gotten this far, you understand that behaviors are almost always symptoms of something else. What that something else is requires an investigation of a multitude of possible Total Load factors. Determining possible causes of an individual's diagnosis requires taking a complete history, including prenatal, natal, environmental, developmental, medical, and emotional concerns. This process requires many hours, patience, good listening skills, and a sharp memory. It can be greatly assisted by laboratory testing and a significant understanding of biochemistry.

Genetic, Genomic, and Nutrigenomic Tests

Deciding which tests and which labs to use can be challenging. Today's choices for uncovering what is going on deep within the cells is like entering a big box store with more options than even the most well-read parent can decipher. How do you know whether genetics, mold, mercury, or yeast is the problem?

Because one approach never fits all cases, listing a standard test protocol is impossible. You and your doctor will choose tests that are useful for diagnosis of suspected problems, not just to get extraneous information.

Start with the Pediatrician

Pediatricians usually order traditional laboratory tests, such as a complete blood count (CBC) and sometime even IgE (RAST) allergy tests. Chemist Kristi Wees, president of Empowered Medical Advocacy, recommends that, in addition to these, parents of children with behavioral and learning issues ask their doctors to add the following laboratory tests, all available through mainstream laboratories like Quest and LabCorp with a doctor's prescription:

- Comprehensive metabolic panel (CMP);
- Complete thyroid panel, including TSH, T3, reverse T3, T4, and thyroid antibodies;
- IgG, IgA, IgM immune testing, also known as quantitative immunoglobulins;
- ELISA test to measure delayed food and environmental sensitivities, including the presence of gluten and casein peptides;
- Urine organic acid test (OAT) to look for an overabundance of abnormal organic acids associated with yeast, fungal, and clostridia metabolism;
- Plasma amino acid test for levels of amino and fatty acids;
- Antibody titers for diseases associated with markedly abnormal responses to childhood vaccines, including MMR, DTaP, and polio;
- Levels of minerals, including serum copper and plasma zinc; and
- Levels of vitamins, especially vitamin D and vitamin B12/folate/methyl-malonic acid.

The results of these tests will help determine where to go next. Now it gets tricky. You can ask your pediatrician to dig deeper, but, unfortunately, few are versed in some of the tests that reveal additional possible causes. Or you can find a special-ist, often a functional medicine doctor, a MAPS-trained physician, or even a well-trained health coach, who you can work with locally or virtually. Two problems with this route: families often wait as long as a year for a popular doctor, and most of their services are not yet covered by insurance. Time is of essence in learning what is going on. A third option is to go online and order some additional tests yourself. How do you decide which route to take?

Evaluating Options

If you have chosen a pediatrician or family doctor wisely, that might be your best choice. If that doesn't work, and finances allow, a functional medicine doctor can run additional blood, stool, hair, and urine tests to unravel what is going on in the bodies and brains of their patients.

Limited resources? Use some of the home kits available online. Like with test-ing for pregnancy, we are now able to perform many tests ourselves. Are they reliable? Who interprets the results? Important questions to know the answers to before shelling out hundreds, if not thousands of dollars to get to the bottom of the behaviors.

Each healthcare professional has their favorite lab for each test. Some labs specialize in one of the tests and do not offer others. Some labs offer a full range of tests.

Most blood, stool, and urine tests pinpoint only recent exposures. Each has its own advantages and liabilities, especially if you have to travel somewhere for evaluation. Blood levels do not always yield answers because metals are generally stored in fat tissue, not blood. Excretory products can yield highly useful results about digestion, pathogens, immune status, and allergic reactions, but they can lose information as the sample degrades.

Stool tests are the most problematic. Our intestines are lengthy, and it can be days before the stool makes the long trip from beginning to end. What can be captured on a swab at the end of the road may not be representative of the com-munity at large. So many factors are at play, from time of collection to whether the sample is representative of the feces as a whole.

Measuring accurate levels of toxins, nutrients, hormones, pathogens, and even parasites has thus been elusive until recently. Fortunately, many specialized laboratories have developed hundreds of very sophisticated tests that look at both genetic and environmental factors. These identify molecular damage and pinpoint exactly what has gone awry with the immune, digestive, respiratory, and endo-crine systems, as well as the detoxification pathways. Simplified methods—such as tape-on bags for urine samples from those who are not potty trained and home test kits complete with a lance to prick a finger or a Q-tip to do a cheek swab— make sample collection easy and almost painless.

The next sections will explore several categories of tests, with suggestions for labs requiring a doctor's order as well as those that patients can access themselves. Remember that tests provide information. Some interpret that information for you and make recommendations that can result in positive health changes. But finding and working with an expert to interpret results is imperative!

Genetic Versus Genomic Testing

Return to Chapter 2 to understand the difference between *genetics* and *genomics*. Only genetic tests can provide a medical diagnosis, such as cystic fibrosis or fragile X syndrome. Genomic tests identify mutations or SNPs. Both genetic *and* genomic testing are important to get a complete picture of a child's biological profile.

Jared Skowron, ND, has found that about a third of kids in his all-virtual, mostly autism practice has a real genetic medical issue. These kids are commonly the "non-responders." Skowron feels strongly that without both genetic and genomic testing, those kids would be missing a major key to their treatment.

Remember from the previous chapter that structure and function go hand in hand. Our biological makeup is one of the most important places to apply this principle. Functional medicine doctors practice functional genomics, the science of the relationships among all the components of your biological system—your genes, your organs, and what they produce, like proteins, metabolites, etcetera— and how they work together synergistically to produce the unique being that is *you*.

Genetic Tests

Many families have purchased genetic analysis from 23andme.com or AncestryDNA.com. These tests give us interesting information about long-lost relatives and genetic heritage but are incomplete for medical purposes. However, once you receive the raw data, you can enter your results at a number of different websites or apps to get an analysis of how your SNPs affect detoxification and other health-related issues. Check out promethease.com and livewello.com.

Chromosomal microarray analysis (CMA) and *whole exome/whole genome sequencing* (WES/WGS) are more focused types of testing. CMA looks for missing or extra pieces of DNA, while WES/WGS looks for nonfunctioning genes. Both can be involved in brain chemistry and communication. These tests can show possible underlying causes for autism and other neurodevelopmental conditions that can elicit all types of behavior and learning issues. These genetic tests are usually covered by insurance when a physician orders them.

Nutrigemonics

The word nutrigenomics combines nutrition with genomics. Once a test identifies an individual's pattern of SNPs, doctors can prescribe a nutritional supplementation program to mitigate any negative influence they have. Knowing your mutations, and mitigating them, allows the body's organ systems to operate properly.

Naturopathic physicians (NDs) have taken environmental and functional medicine and healthcare into a new paradigm by interpreting information from genetic, genomic, and other laboratory results in the context of presenting symptoms. The connection between these genes and health outcomes is truly one of the most exciting discoveries of the last century.

Fair warning: frequent retesting is necessary (and expensive!) to evaluate the results of treatment, because the body constantly shifts during the healing process. All those in the nutrigenomics business make a lot of money selling expensive products and services. First they sell you the test, then they sell you interpretation of the results, then they sell you supplements that allegedly change those results from potentially negative to positive. Be an educated consumer! You really need a functional medicine expert on your team to interpret those results without bias to make educated decisions about treatment. Be aware that you cannot change your genetics. You can only change the expression of these genes through environmental control.

Nutrigenomic Testing

An early interpreter of genomics is **Ben Lynch, ND**, who got involved in the field to help his family get back their health. Author of the book *Dirty Genes*,[1] he focuses primarily on damaged methylation pathways and detoxification. His company, Seeking Health, educates both the public and health professionals on how to overcome genetic dysfunction. To learn more, go to his websites DrBenLynch. com and SeekingHealth.com.

Naturopath **Robert Miller, CTN,** founder of the NutriGenetic Research Institute, uses a computerized program to generate a Functional Genomic Analysis™ from raw genetic data. This report prioritizes treatment in the same order as the chapters in this book: correcting digestion, taming inflammation, then rebuilding the gut, while providing digestive and emotional support, before addressing detoxification and nutritional support for SNPs.

What makes this program unique is that it avoids common failures caused by moving too quickly and detoxifying in the presence of gut issues and inflammation.

Qualified healthcare professionals can purchase Miller's software and his supplements for their patients. Go to DNASupplementation.com to watch a short video on the subject, as well as learn how doctors can become certified in using this tool.

One of my favorites is the new kid on the block, **IntellxxDNA**, cofounded and led by Harvard-educated physician **Sharon Hausman-Cohen, MD.** Using IntellxxDNA's testing kit, practitioners trained by this innovative company suggest not just supplements, but a complete overview of potential intervention strategies as well, encompassing environmental, dietary, and lifestyle changes and even some medications. They are also doing research related to autism spectrum disorders[2] and various types of mental health, including anxiety and depression, and even dementia, to show that these conditions are reversible. Their blog posts are extremely informative. To read them and/or be trained to use their resources, go to IntellxxDNA.com.

Every week new companies jump on the genomics bandwagon because this truly is the future of targeted precision medicine. Some recommended by trusted friends are MaxGen Labs, DNAPal, MyHappyGenes and 3x4Genetics. Space prohibits the discussion of other good companies out there.

Microbiome Testing

The human microbiome is the collection of microorganisms that live on and in our bodies. It is an ever-changing ecological community of viruses, bacteria, fungi, and other "bugs." It is influenced by all aspects of our environment discussed in the previous chapter, from air, water, and food—and the toxins they hold—to our thoughts and experiences.

Gut testing is thus essential in understanding that cognition, emotions, and even our personality can be influenced by these critters. I once heard a doctor say that a craving for chocolate might be those bugs telling you they need some magnesium!

Recent advanced technology has allowed scientists to learn what organisms are living in all areas of our microbiome. By sampling stool, urine, blood, and saliva, we can identify not only the bugs themselves, but their byproducts as they go through their own biological processes. Without any invasive techniques, we can determine whether harmful microorganisms, such as certain strains of strep and E. coli bacteria, molds, or yeasts, are in there and causing some of the behaviors that are hallmarks of a myriad of neurodevelopmental disorders.

After taking an extended history, most functional medicine doctors and other savvy practitioners order tests to learn how a person's lifestyle and experiences have affected their microbiomes, including stool, urine, and hair analysis tests from major laboratories such as Mosaic Diagnostics and Genova. These trusted resources offer a huge array of choices for tests such as their Organic Acids Test (urine) and Comprehensive Stool Analysis. These are not available directly to the consumer but must be ordered through a medical professional. While that may be inconvenient, it means that sometimes the cost of testing can be covered by insurance.

Not surprisingly, the Internet is flooded with do-it-yourself tests to evaluate your microbiome. This is a Wild West market. Testing varies from a look at thousands to millions of the trillions of organisms, and you usually get a beautiful printout with a "score" on the diversity of your unique microbiome for slightly under $100. Sales with deep discounts happen frequently, so look for one.

None of these tests give practical information on the gut microbiome. Most clinicians do not use any of these tests. They use Doctor's Data or Genova that give the most comprehensive data on the microbiome, inflammation, digestion, pH, etcetera. The goal of many of these companies is to capture you as a lifetime customer for repeated tests and expensive probiotics and other supplements to increase your diversity and balance out the gut. Buyer beware!

That said, here are some favorites gathered from savvy consumers I know:

- **Biomesight**—Performs something called 16srRNA sequencing to identify and classify bacteria down to the genus and/or species level. Dr. Christian Bogner and his partner Alex Zaharakis use this measure to guide a Gut Balancing program. See AutismIsBiomedical.com
- **Sun Genomics**—Offers the Floré Microbiome Gut Test with a monthly subscription fee. See Flore.com
- **Tiny Health**—The first gut health test for moms and babies 0–3 years. Detect gut imbalances early on. Especially useful for babies born by C-section. See TinyHealth.com
- **Viome**—Offers Gut Intelligence™ Test, which includes a retest every 12 orders and requires a four-month commitment. See Viome.com

Testing for SIBO

SIBO is a disruptive digestive condition that occurs when bacteria that normally live in the large intestine migrate backward to the small intestine. You can read more about it in Chapter 6.

Doctors can order a simple, painless test that measures the type and amount of gases an individual's gut is producing. The test subject drinks a sugary liquid, waits about 20 minutes for it to get into the digestive tract, and then breathes into a tube every 10 minutes for the next hour or so. The presence of hydrogen indicates that a patient is experiencing diarrhea from SIBO and motility that is too fast; methane gas is detected in SIBO that causes constipation because motility is too slow.[3]

Toxic Metals Testing

Once considered unreliable, analyzing minerals and toxins excreted through the hair is now better understood. This testing method is most useful for the detection of *recent exposure* to toxic metals. The Hair Elements test from Doctor's Data (DoctorsData.com) can be used to measure levels of metals, such as lead, mercury, aluminum, antimony, and cadmium, as well as magnesium, calcium, potassium and other important minerals.

In a landmark hair analysis study completed on children with autism and their typical siblings, researchers were astounded to find that the unaffected children showed higher levels of toxic metals in their hair than their affected siblings.[4] Why? Because the neurotypical kids were able to detoxify and excrete the poisons, while those with autism still harbored the metals in their bodies.

If autism is sometimes a "unique form of mercury poisoning," as SafeMinds contends, then testing children with autism for their loads of mercury as well as for aluminum, lead, pesticides, preservatives, solvents, and other pollutants is imperative. Proving that a particular child has toxic levels of a specific poison can be extremely difficult, however, even in clear cases of overexposure, such as mercury-loading from vaccinations.

Endocrine and Hormone Testing

Thyroid and cortisol testing are essential for both treatment and monitoring progress. Testing of stress hormones, called *catecholamines,* including dopamine, epinephrine, and norepinephrine is strongly recommended. Knowing melatonin levels can also be helpful. Read more about these hormones in Chapter 12.

Finally, testing can sort out neurotransmitter imbalances that can affect mood, attention, and activity. Different labs offer different panels; one that specializes in this type of testing is NeuroScience at NeuroScienceInc.com.

Advanced Forms of Testing

New technology has made laboratory testing and analysis more and more sophisticated, and amazing new options are emerging almost weekly. Be sure to ask your doctors and friends about their experiences to find out about new ones which might be right for you.

Mercury Speciation Analysis

An environmental scientist, Christopher Shade, PhD, became interested in the unique qualities of mercury when assessing the impacts of mercury release from natural gas drilling in Southeast Asia. In 2005 he founded his Colorado-based laboratory, Quicksilver Scientific, where he developed the patented Liquid Chromatographic Mercury Speciation technology that separates and measures the different forms of mercury: methylmercury, ethylmercury, and inorganic mercury.

Methylmercury, commonly found in fish and other animal tissues, is the most highly researched form of mercury. Methylmercury is mobile and easily absorbed in the human organism. It even crosses the placental and blood-brain barriers, exposing children as early as the fetal stage, potentially causing debilitating neurological effects. Methylmercury is difficult for organisms to eliminate, so it accumulates in biological tissues.

Ethylmercury exposure comes through pharmaceuticals, including vaccines, containing the preservative thimerosal. Like methylmercury, ethylmercury can move easily into biological tissues.

Inorganic mercury occurs in sediments, soils, and some food sources, but is not very mobile, nor does it bioaccumulate to the same degree as organic mercury. It gathers in tissues when one of the more mobile forms of mercury above enters the tissue and breaks down into elemental mercury. Once inorganic mercury is absorbed into biological tissues, it becomes an immediate toxic threat and is very difficult to remove.

This information is vital for those with neurodevelopmental disorders because simple blood and hair tests alone cannot provide an adequate representation of an individual's mercury level. The Quicksilver method analyzes methylmercury, inorganic mercury, and total mercury in one simultaneous procedure that determines the ratio of methylmercury to inorganic mercury. This is critical data that facilitates understanding of possible toxic effects on bodily functions and how to remediate them. To order kits, go to QuicksilverScientific.com.

Porphyrin Testing

Porphyrins are a group of pigments that the body uses for various purposes, like making hemoglobin, which is red because of the presence of iron. Have you ever noticed that the complexions of some children appear unusually pale and pasty? That is because of insufficient iron or a defect in hemoglobin production that can interfere with their porphyrin production.

Dr. Neil Nathan is investigating the possibility that secondary porphyria, or a buildup of porphyrins, may be a player in some of our sensitive patients. Secondary porphyria can be triggered by mold, Lyme, or other toxins that interfere with hemoglobin production and build up instead of being excreted.

By measuring the presence of porphyrins in urine, laboratory tests can detect some specific incomplete, unusable porphyrins that the body has discarded. These specific porphyrins are present in urine only because of toxicity.

Nathan informally tested 20 of his patients with anxiety, depression, and panic and found that 15 of them came back as positive for high levels of porphyrins. He believes that this test could shed light on the order in which to do various treatments.[5] Doctors in the US can order tests measuring porphyrins in plasma and urine through LabCorp and Quest Laboratories.

Testing for Lyme, Infections, and Viruses

Anyone with complex illness, especially those with mood swings, depression, anxiety, OCD, or any neuropsychiatric symptoms, should be tested for Lyme. Unfortunately, until now, lab tests for Lyme and its coinfections have been unreliable, and false negatives are common. Physicians disagree about what Lyme disease is and what viruses are involved. But we are getting closer to being able to diagnose it earlier, more easily, and more precisely, even without a telltale bull's-eye rash on the body. "Lyme literate" doctors now have access to new labs with new technology, all requiring a doctor's order. Find one from the International Lyme and Associated Diseases Society (ILADS.org).

Practitioners strongly recommend the following:

- **IgeneX** has home testing kits for both blood and urine. Their labs can evaluate not only you, but the tick that bit you as well if you are able to salvage it. Their comprehensive testing evaluates Lyme and related illnesses carried by other insects as well. See IgeneX.com.
- **Infectolab-Americas** offers extensive testing for bacterial, viral, and complex immune conditions, including long-haul COVID. Their Immune Assessment test can provide a gauge of how active and functional a patient's immune system is and put the immune response in context with viral and bacterial infections. See Infectolab-Americas.com.
- **Medical Diagnostics Lab (MDL)** offers reliable testing for Lyme disease and some coinfections. They are the only lab that provides a copy of the ImmunoBlot to your doctor. Although they do not test for all tick-borne

illnesses, they do bill most insurances and have low-cost cash pricing making it affordable to people.

Testing for Food Allergies and Sensitivities

"Allergy" is defined as a specific immediate reaction to an otherwise harmless substance. These reactions, including hives, congestion, and swelling, involve a class of antibodies called immunoglobulin E (IgE). Traditional scratch testing identifies IgE triggers, such as pollen, peanuts, or strawberries, that can cause symptoms ranging from annoying to lethal.

A second type of reaction is an immunoglobulin G (IgG) response, which is sometimes called "intolerance" or "sensitivity" rather than allergy because it can be delayed and is not usually life-threatening. IgG symptoms are cumulative in nature and can appear as chronic skin problems such as eczema and thrush, gastrointestinal problems such as diarrhea and constipation, and/or behavioral reactions such as mood swings and hyperactivity.

Many healthcare professionals order the Enzyme-Linked Immunosorbent Assay, known as ELISA for those with autism spectrum disorders. An extremely helpful test, the ELISA screens the blood for both immediate (IgE) and delayed (IgG) reactions to a panel of about 100 food proteins, including gluten and casein. A basic test of the 10 most common allergies is also available. Most labs rate food sensitivities 0–4, where the highest number requires strict avoidance and the lowest is a well-tolerated food. In-between foods can be consumed once in a while, rotated, or avoided with occasional exceptions. The ELISA method is reproducible, reliable, and valid in the detection of food-specific antibodies.[6]

If either IgG or IgE testing indicates casein or gluten sensitivity, that means that chemical messages in the body have tripped an immune response. An IgG casein or gluten reaction is most likely due to poor digestion. The body's immediate response is to clear the incompletely digested particles by producing IgG antibodies.

A third type of reaction is an immunoglobulin A (IgA) response, which is indicative of inflammation in the mucous membranes, especially of the gastrointestinal lining, and leads to intestinal irritability, stomachache, or diarrhea. Healthcare professionals are seeing more of this type of immune reaction in children on the autism spectrum. IgA antibodies are often the body's reaction to a foreign substance, which could be a pathogen or a rogue food protein.

The bottom line: many people with neurodevelopmental disorders have unidentified IgE, IgG, and IgA allergic reactions.[7] Semantics permit healthcare professionals to use the word "allergy" legitimately to describe any symptoms, immediate, or delayed, occurring after exposure. Reactions to gluten or casein usually occur more than two hours after eating a food, and are most often IgG or IgA reactions, with symptoms such as sleep disturbance, bed wetting, sinus and ear infections, or crankiness. A number of immunological measures are necessary to determine an individual's immune status.

Brain Imaging Techniques

Doctors have long speculated that those with ADHD, autism, and other diagnoses have underlying brain differences that can contribute to symptoms. Sophisticated new technology has proven them correct. Differences in structure,[8] brain and head size,[9] and activity level[10] are apparent.

Since the 1970s, SPECT, for *single-photon emission computed tomography*, scans have been used to evaluate strokes, seizures, and brain tumors. In the 1980s scientists began using this technique to study ADHD, schizophrenia, depression, and other mood disorders. The research on brain SPECT is vast with over 14,000 scientific research articles on it listed on PubMed.com.

Daniel Amen, MD, is one of the most experienced doctors using SPECT. He's published more than 50 peer-reviewed scientific studies using SPECT to explain the brain's role in complex psychiatric patients. In one study he asked seven psychiatrists to evaluate 109 consecutive charts *without* brain SPECT scans and then *with* scans. In eight out of ten cases, the doctors changed their diagnosis and/or treatment plan when the scan was added. The scans revealed unexpected brain injury or toxicity in more than one in five cases.

At the time of this writing Amen has almost a dozen clinics throughout the United States, where anyone with an alphabetical list of almost 50 conditions— from ADHD, anger disorder, anxiety, and autism to suicidal thoughts, toxic metal exposure, traumatic brain disorder, and weight loss—can go to learn why their inflamed, anxious, depressed, and inattentive brains are dysfunctional. For more information go to Amen Clinics' website at AmenClinics.com.

Autonomic Response Testing and Electrodermal Screening

Autonomic Response Testing (ART). **Dietrich Klinghardt, MD, PhD**, has co-developed ART, a form of muscle testing, or kinesiology that measures autonomic nervous system stress responses in the body. Unlike most muscle testing, ART is an extremely accurate, sophisticated, rapid, and comprehensive diagnostic system that is the leading bioenergetic technique in Europe today. Autonomic response testing simultaneously diagnoses which organs are no longer coherent, and what substances restore function. It has been found to be as accurate as traditional allergy testing using a blood draw.[11]

ART is based on *biophoton theory*, the brainchild of the late German physician Fritz-Albert Popps (1938–2018). An **extremely** important concept, Biophoton theory states that our cells communicate with each other via particles of ultra-low level light called *biophotons*. In other words, the body has a highly organized system of communication among cells that is responsible for the regulation of all physiological processes.

Human bodies are essentially "beings of light." Healthy cells emit "coherent" photons. When cells lose coherence, disease and dysfunction occur.[12] This is exactly what happens when the body's electromagnetic fields are disrupted by today's 3G+ electromagnetic fields in our crazy world.

Klinghardt sometimes delays expensive laboratory testing, preferring ART to simply start his patients on a safe, gentle detoxification program of using all-natural, mild, over-the-counter agents, only adding stronger agents later if testing indicates. Every couple of months, he might use hair analysis to monitor which metals the body is excreting. Interestingly, he prefers pubic hair (if the patient is old enough) to hair from the head, as it grows more slowly and thus delivers a longer excretion history. His experience is that toxic mineral levels register low at first, and then begin rising, with lead and nickel showing up before mercury.

During ART an examiner puts pressure on an individual's outstretched arm while touching him in a variety of locations and placing possible antidotes to various toxins near the patient, using special tools (available from BioPureUS.com) that enhance the power of the energetic signal to the outstretched arm. Simultaneously, the body reveals what toxic metals, microorganisms, traumas, or other toxins are stressing the patient, the site of the stress, and which products or procedures are beneficial in relieving the stress.

If several antidotes are successful, ART has techniques to determine which will work best. The sum of all the antidotes is the patient's "prescription." For young children, an adult surrogate can act as a circuit between the examiner and the child by touching the child's body while the examiner tests his or her outstretched arm. ART is the only type of muscle testing that can establish a diagnosis and prescribe treatment simultaneously.

Klinghardt recommends using ART to fine-tune the results of laboratory testing or as a stand-alone diagnostic tool to further evaluate toxicity. With proper training, professionals from any discipline can learn to use ART as an adjunct to standard in-office laboratory and diagnostic testing. Like any technique, skillful autonomic response testing requires continuous study, practice, and discipline.

Electrodermal screening and the EAV machine are testing techniques that can give powerful feedback about the physiology, biochemistry, and pathology of an individual. While these machines are widely used in Europe for medical diagnosis, the FDA has not yet approved them as diagnostic instruments in the United States, and probably will not.

The EAV machine, short for ***Electro-Acupuncture according to Voll,*** is basically an Ohm meter, or more specifically a skin conductance meter. The FDA has an approved classification for skin conductance metering that permits its use by a variety of alternative healthcare practitioners and some biological dentists. See later in the book about EAV's use in homotoxicology and dentistry.

Developed by Reinhold Voll, MD, a German anatomy professor and acupuncturist in the late 1940s, EAV concretizes what Chinese medicine has known for centuries: the body is an electrical system, with energy pathways, called meridians. Each meridian corresponds to specific organs in the body, and certain points along each meridian that correspond to traditional acupuncture points are key to both diagnosing and treating health issues.

To test a meridian point, the patient grasps a metal cylinder, while the practitioner places a wet metal stylus on a probe point, close to the bone. The meter measures the strength of electrical impulses on various points on the skin where changes in the electrical resistance register the body's ability to conduct electricity.

Energetically healthy pathways, regardless of the subject's age, gender, weight, or other characteristic, register a reading of about 50. Inflammation anywhere along the pathway creates an increased rate of flow, registering above 65. As organs deteriorate and cellular activity slows, the meter drops.

With these parameters in mind, a healthcare professional can determine the internal state of energy of individual organs, and organ systems, such as digestion, overall immunity, and hormone health. If the professional finds a meridian with an unfavorable reading, then substances known to balance its energy can be tested. If a given remedy is favorable, the reading will move closer (either up or down) to 50.

Dr. Voll's original technology has evolved over the years. The display used to use a moving needle like the speedometer in older cars. Once computerized, the process was streamlined, enabling a practitioner to test and prescribe much more quickly. A practitioner can determine which of 20,000 different remedies are best to use in under a half hour.

EAV was originally called *electrodermal screening* (EDS), then *computerized electrodermal screening* (CEDS) and is now sometimes known as *Meridian Stress Assessment* (MSA). Various computer programs are available; some are more user-friendly than others. If you are interested in adding an EAV to the tools in your toolchest, talk to someone who uses one they like, since they are pricey. Make sure that updates and technological support are available.

Artificial Intelligence is emerging as a promising way for clinicians to differentiate among bipolar disorder, major depressive disorder, mood disorders, and other psychiatric disorders, using biomarkers in electroencephalography (EEG) data to more accurately predict treatment responses. Not surprisingly, Asian scientists are leading the investigation of this option.[13] Stay tuned!

Take-Home Points

Diagnostic testing of many kinds can be very useful for determining which treatments to prioritize and in what order. Allopathic doctors, including pediatricians often order "standard-of-care" blood tests and scans. However, these only infrequently elucidate what is going on biologically. Functional medicine doctors use additional readily available measures, as well as sophisticated blood, urine, stool, and hair analysis to delve deeper into the cells and their contents. These can be extremely helpful in gaining further important diagnostic data.

For those who want to manage testing themselves, many companies offer do-it-yourself home tests for evaluating blood, stool, and more. But I believe you really need an expert to assist you in analyzing the results.

Finally, some innovative testing techniques, such as special brain scans and energetic analysis, can uncover information about the body's electrical signals and pathways that can be beneficial in prioritizing treatment decisions. All of these techniques can speed a patient's return to health in a most efficient fashion.

CHAPTER 6

FEELING SAFE BY BALANCING AND REGULATING THE NERVOUS SYSTEM

The state of the nervous system is the deciding factor in feeling safe and secure, which is the essence of emotional health. The commonality of children and adults with ADHD, autism, anxiety, depression, schizophrenia, and the other conditions discussed in this book is a hypervigilant, hyper reactive, overprotective, and anxious nervous system. Survival is always the primary goal of every brain.

As an unwieldy Total Load of stressors overwhelms body, mind, and spirit, the brain perceives the world as a dangerous place and feelings of safety are elusive. Overcoming the physical causes of any disease can be extremely challenging without first addressing a dysregulated nervous system.

What follows is a detailed description of the various components of the nervous system and how they interact, followed by what can cause hypersensitivity and a lack of feelings of safety and security. When the above symptoms make it painful to live in the world, absolutely anything we can do to reduce their effect and bring back emotional stability is worthwhile.

This chapter offers many possible interventions, some with decades of research and support. Other treatments are new and show promise. As you read, evaluate which ones resonate with you and your loved ones.

Many in the neurodiversity movement consider themselves simply "differently wired" and are not actively seeking help to reverse their diagnoses. I fully support anyone's right to make that choice, but the techniques discussed in this book, and especially this chapter, are intended to help virtually anyone with these diagnoses *feel* better, making their lives easier, happier, and less painful.[1] And, frankly, with today's constant bombardment of environmental stressors, we can all use help with that.

The Origins of Dysregulation

An individual overall feeling of safety evolves from a lifetime of sensory experiences, starting in utero. A woman's mental health during pregnancy, delivery, and beyond, affects the earliest sensory experiences and emotional life of her unborn baby.[2]

If early experiences are positive, development proceeds in a healthy fashion. If they are negative and elicit even *mild* stress responses, the nervous system goes on alert. When small worries and occasional anxiety become compounded by more serious worries, Total Load rises to a point where the accumulation of stressors is way beyond the nervous system's ability to process. The natural manifestations of a normal stress response are replaced by more serious symptoms like hyperventilation and anxiety. The nervous system is now in *acute* stress.

Enter panic attacks, confusion, dissociation, lack of trust, insomnia, unshakeable depression, and angry outbursts, leading to difficulty with relationships and increasingly serious emotional reactions. Then difficulties with basic self-care, work, and relationships emerge. We have now entered the realm of *trauma*.

In trauma, painful flashbacks, hypervigilance, and numbness can occur as well as continuous, chronic anxiety and the behaviors that result from the body's attempt at trying to cope. Every traumatic event is stored in memory together with the emotions experienced at the time. If an acute traumatic event is physical, as in surgery, touching the point of injury can activate the emotions as well. As the physical body is treated and heals, the emotions generated by the trauma are also released, and the event is over.

Alternatively, nonspecific potentially traumatic events like unpredictable verbal abuse, especially during childhood, have no specific focal point, and the accompanying emotions have no destination. Extreme tension can be pervasive and accumulate gradually in weakened organ systems, causing diffuse distress, pain, and various other maladies.

To understand what is happening physiologically, a short course in anatomy is necessary.

Nervous System 101

The nervous system coordinates and controls the body's reactions to sensory input by sending signals from one part to another. The body and brain respond physiologically and emotionally to stress through special cells called *neurons*.

The complex nervous system is divided into two main parts: the *central nervous system* (CNS), which includes the brain and the spinal cord, and the *peripheral nervous system* (PNS), which is itself divided into two parts:

- *Somatic* (derived from the Greek word *soma*, meaning *body*) nervous system, responsible for voluntary movement and includes the *sensory* nervous system which gives feedback to the brain regarding pain, pressure, hot and cold sensation and our senses of taste, smell, and sound.

- *Autonomic* nervous system (ANS), responsible for involuntary functions, such as respiration and digestion. It is controlled by the hypothalamus (see below), a tiny organ in the brain that monitors important essential functions like breathing and monitoring heartrate.

The ANS, in turn, is divided into two opposing systems:

- *Sympathetic* (SNS), which activates a "fight-or-flight" response and a chain of other bodily functions when a perceived threat is detected. It acts like the accelerator in a car.
- *Parasympathetic* (PNS), which promotes "rest-and-digest" functions and is responsible for calming the body. It acts sort of like a car's brake.

The **Enteric Nervous System** (ENS) is a separate component, located in the lining of the gastrointestinal tract. Read about it in the next chapter on digestion.

The autonomic nervous system is constantly scanning our unpredictable environment for safety, and sending its messages to the brain. The brain then examines these sensory stimuli, compares them to past perceptions, and acts upon them. "Is it safe?" is the question that it keeps asking. If the answer is "Yes," each system keeps the body in balance. If the converse is true, and an individual is in a sympathetic mode, bodily dysfunction occurs and is accompanied by negative emotions such as fear and anxiety.

Chronic stress occurs when the body is in a sympathetic state for extended periods of time, causing both physical and emotional dysfunction. Staying in a parasympathetic mode and out of fight-or-flight most of the time is at the heart of the matter in feeling safe.[3]

The Cell Danger Response (CDR)

When a perceived threat is detected, the body goes into a sympathetic mode and activates a cell danger response (CDR). Physician-scientist, Robert Naviaux, MD, PhD, proposed the term *cell danger response* in 2013[4] as a way to describe what happens when cells of various systems detect a physical, chemical, or biological threat.

When cells sense danger, they set off an alarm that **activates inflammatory pathways** and **recruits an immune response**. Short-term inflammation can assist in healing. However, chronic and long-term inflammation is a hallmark of disease.

Inflammation can show up in every bodily system from the gut to the brain. An active immune system uses inflammation to

- *shift metabolism to a defensive state*, impairing and preventing physical repair;
- *block cellular communication* and energy production to isolate the damage; and
- *send a message to the limbic system*, our processor of emotions, to react.

Metabolism and cellular communication begin to normalize when the system detects cues of safety. You'll find more in the next and following chapters about inflammation and the CDR, its role in immunity, and how to resolve it by signaling to the cells that the threat has passed.

For now, let's move on to the limbic system, the nervous system's partner in the body's coordinated effort to help us feel safe.

The Limbic System and Mast Cell Activation

Our limbic system is a team of organs that includes the amygdala, hippocampus, and hypothalamus: the parts of the brain that govern emotions, memory, and the autonomic nervous system. When the body is in a state of homeostasis, the brain is calm and content. Under chronic stress perpetuated by the CDR's maintaining a heightened sense of threat (can be caused by trauma), the brain and body get stuck on high alert, even when the threat has passed. The CDR traps and locks cells in a defensive state known as *limbic-system impairment*, prioritizing protection over repair and connection.

The limbic system responds by activating a pathway called the *hypothalamic-pituitary-adrenal (HPA) axis*. The *hypothalamus* and *pituitary* are responsible for controlling body temperature, hunger, thirst, appetite, circadian rhythms (day/night cycling), and emotions. The *adrenals* are a pair of endocrine glands that are located on top of the kidneys. Their name denotes their location: *ad* is Latin for *near*, and *renes* is for *kidneys*.

In perceived danger, heart rate, blood pressure, and respiration increase. Digestion, immunity, and cellular repair processes are suppressed. The adrenals release the hormones cortisol, epinephrine (adrenaline), and norepinephrine in response to stress (see Chapter 12). Emotionally, a hyperactive amygdala perpetuates fear, while the rational parts of the brain, the hippocampus and prefrontal cortex, may be suppressed.

Mast Cell Activation

As the neurologically rewired limbic system becomes hyper-defensive, it triggers more inflammation and a specific immune system response called ***mast cell activation***. Mast cells are a type of white blood cell on the frontline of the body's defense against toxins, pathogens, and other invaders. They are in every tissue of the body, except the retina. Their receptors respond to everything you breathe, eat, or have injected into you. They connect your nervous system, endocrine system, and immune system. When mast cells are inflamed and overactive, with dysfunction in more than two systems, ***mast cell activation syndrome*** is diagnosed.

Dr. Neil Nathan believes that up to 80 percent of individuals with the conditions discussed in this book have mast cell activation underlying their presenting diagnoses. Furthermore, he places this neuroendocrinoimmunological condition at the foundation of other chronic conditions, including fibromyalgia, chronic fatigue, Crohn's disease, Ehrler's Danlos syndrome, postural orthostatic tachycardia syndrome (POTS), and even some cancers.[5] There's much more about mast cell activation coming up in Chapter 9.

Bottom line: Total Load can push a person into a state where they both feel unsafe and lack the problem-solving abilities to do anything about it. This combination of negative emotions and few resources to combat them eventually results in a psychiatric diagnosis in addition to the aforementioned medical conditions.

The Vagus Nerve

The bridge between the brain and the rest of the body is the tenth cranial nerve, the ***vagus nerve***. The longest cranial nerve, it relays information about internal and external safety or threat through the parasympathetic nervous system. As the vagus nerve wanders through the body, its fibers affect and are affected by many systems. It

- detects signals of inflammation and danger via sensory fibers;
- modulates immune responses and attempts to inhibit inflammation;
- stimulates the endocrine system to release the hormones oxytocin and serotonin, thus promoting social bonding, trust, and calmness; and
- reduces heart rate and supports digestion.

The vagus nerve diverges out from the brainstem, down into the abdomen, touching most major organs along the way and spreading fibers to the tongue, pharynx, vocal cords, lungs, heart, stomach, and intestines.

Until the early 1990s, the vagus nerve was considered to be a single neural pathway. Lumping together the two parts of the vagus nerve caused confusion for a long time in understanding how the nervous system functions. But in 1994, psychologist Stephen W. Porges, PhD, at the Kinsey Institute, Indiana University Bloomington, cleared this up when he discovered that the vagus nerve has two main branches.

Dorsal and Ventral Vagal

The *dorsal* branch is the more primitive part of the vagus nerve and is present in lower vertebrates, such as reptiles and amphibians. It is involved with automatic functions, such as digestion and respiration. When an individual feels threatened, this branch conserves energy by shutting down vital metabolic functions.

The *ventral* branch is more highly evolved, and present only in mammals. It is the pathway associated with social engagement and connection. When it and other cranial nerves are functioning appropriately, an individual feels safe, can trust, bond with others, and easily form relationships.

Polyvagal Theory

Porges named his two-branched model of the vagus nerve the *Polyvagal Theory* (PVT). His "biobehavioral" theory recognizes that not feeling safe is a full-body phenomenon demonstrating the interconnectedness of body reactivity, cognitive and emotional function, and social behavior.

Polyvagal Theory goes beyond the binary concept that the body's reaction to stress is either sympathetic *or* parasympathetic and proposes that there are multiple and blended states of arousal in response to real or perceived stress. It views the source of anxiety arising from the body's response to stress as an inefficient vagal system.

Polyvagal Theory posits that the brain structures regulating social behaviors are compromised in individuals with psychiatric disorders, putting them in a constant state of high anxiety. Everything sensory—lights, touch, smells, and sounds—has the potential to cause trauma. Many are monitoring every environmental sound, not just the higher-frequency sounds that make up human speech.[6]

PVT is based on three main principles.

Hierarchy

The human body shifts unconsciously among three states depending upon how safe we feel at any given moment. These states are not always distinct and sometimes overlap in a continuum, ranging from calm to mobilized actions, within both safe and threatening circumstances.

In the lowest functional level, when the body perceives absolutely overwhelming danger, the parasympathetic activates a defensive state called "freeze." In this state, a person is immobilized: just like playing dead. This is a dorsal dominant mode.

When dorsal activity is chronic, vagal tone is said to be "low," perpetuating a cell danger response and a state of chronic stress, where depression is common. As functioning improves, tone increases.

Next is a state of "fight-or-flight", when the sympathetic nervous system has been mobilized to act defensively. At the highest functional level ventral vagal is dominant, the body is "relaxed," tone is "high" and feelings of safety and sociability prevail.[7]

Neuroception

Porges has coined the term *neuroception* to describe how a group of neurons in the nervous system, called *neural circuits*, determine whether a person or situation is safe or dangerous. This subconscious process takes place through our sensory fibers, without involving the thinking parts of the brain. Neuroception has three streams of input: inside the body, outside the body (the environment), and between bodies (others' nervous systems).

Co-regulation

Polyvagal Theory calls the process that our nervous system uses to develop safe connections with other people *co-regulation*. It begins in utero when the developing baby hears its mother's voice. In healthy development, her voice becomes a sound associated with safety. Feelings of safety continue to develop as the infant pairs the voice with a parent's smiling face, and later as a toddler looking for help when distressed.

These basic experiences create a strong foundation for later successful self-regulation and health. When emotionally nurturing relationships in childhood are lacking, feelings of safety do not emerge, and opportunities for development into happy, healthy, independent, and resilient adults are reduced.[8] In addition, the reduced bonding impacts a mother's own nervous system and instead of a reciprocal, endorphin and oxytocin-rich environment, her hormones can move her more towards stress, anxiety, and less engagement.

The Integration of Systems

The CDR is modulated by the vagus nerve, which detects danger signals and communicates them to the limbic system to regulate emotional and physiological responses. Dysregulation from acute to chronic stress or trauma at any point can lead to persistent CDR activation.

When the nervous system is calm and the limbic system signals physiological safety to the brain, a resilient person can recover from life's daily stresses. When under chronic stress, however, the ventral branch of the vagus nerve becomes dysfunctional, and the dorsal branch becomes dominant. Then a person might withdraw, lapse into fight-or-flight, or shut down.

Polyvagal Theory as a Game Changer

Many consider PVT to be one of the most important links between the nervous system and behavior to be discovered in the past 100 years, and it's one of the hottest topics in mental health today. However, it is *so* complex that many practitioners find it difficult to understand and apply.

Social Worker Deb Dana has become a welcome, easily understood translator. She brings practicality to the study of how stress and its extreme manifestation, trauma, affect us and, most importantly, how we can resolve them. She has written widely, first with Dr. Porges in *The Polyvagal Theory in Therapy: Engaging the Rhythm of Regulation*[9] and more recently in her own book, *Anchored: How to Befriend Your Nervous System Using Polyvagal Theory.*[10]

Dr. Porges has recently partnered with his son Seth, a film producer and director, to write *Our Polyvagal World: How Safety and Trauma Change Us.*[11] This collaboration has brought understanding and applying PVT into the next generation.

Those wishing to learn about PVT in depth can read *The Polyvagal Theory: Neurophysiological Foundations of Emotions, Attachment, Communication, and Self-regulation.*[12] A shorter, more concise version of the theory is recently available as *The Pocket Guide to the Polyvagal Theory: The Transformative Power of Feeling Safe.*[13] For updates on PVT go to PolyvagalInstitute.org.

Reflexes

Reflexes are involuntary, stereotyped movements that a person makes in response to a stimulus. A reflexive response has three steps: sensory neurons send information about a specific sensory stimulus to the brain. The brain processes the signal, then sends a "contract!" signal back to the muscles through a different set of neurons. These infant (sometimes called *primitive*) reflexes originate in and are controlled by the brainstem, the most primitive part of the brain. Neurologically, reflexes provide infants with learning experiences that lay the foundation for all motor skills and eventually intentional motor control, which in turn are the building blocks for feeling safe, learning, and self-control.

A Historical Overview

Our understanding of reflexes goes back to the seventeenth century, when British neuroanatomist Thomas Willis, the founder of clinical neurology, described how *reflexion* in the nerves sent messages to the body's muscles.[14] Fast forward to the early twentieth century when in Russia, social scientists Pavlov,[15] Setchenov,[16] and L. S. Vygotsky,[17] noted that nerve impulses pass information from the sense organs to the muscles and glands, which respond reflexively to these stimuli. While researchers in the West believed that reflexes were active to begin with and then were "inhibited," Russian researchers spoke of reflex "integration."

Enter Svetlana Masgutova, PhD, a Russian-born psychologist, and the founder of the educational institute in Orlando, FL which bears her name. In 1989, Dr. Masgutova had just finished her PhD in psychology and had studied the work of Vygotsky and others.

One day, a tragic train accident occurred nearby, which ultimately became a watershed event in Dr. Masgutova's life. Facing the tragedy armed with her newly acquired knowledge of learning and reflexes, she went to work with the traumatized children who had survived the train wreck, engaging them in simple movement and tactile activities to enhance both physical and emotional recovery.

Dr. Masgutova has since developed a well-respected model of reflex development based on the work of Russian scientists who place reflexes in the framework of both higher and lower nervous system activity. The material in this chapter comes mostly from her 2015 book *Reflexes: Portal to Neurodevelopment and Learning.*[18]

The Developmental Dynamic of a Reflex

The body is programmed with approximately 100 reflexes. Reflex development begins around week nine in utero and accelerates throughout the first year of life, accounting for a majority of early movement patterns, such as turning the head, rolling over, and grasping.

Reflexes serve dual purposes: survival and development. They cannot do both at the same time. If the body is stressed in any way, it prioritizes survival over development.

The purpose of infant reflexes is to build the neural foundation for motor patterns that will later integrate into the nervous system to serve for both protection and development. In infancy, specific sensory stimuli elicit specific responses that we recognize as reflex movement patterns, the remarkably strong grasp of an infant around an adult's finger, for example. As this happens repeatedly, the neural pathway for grasping is strengthened, and eventually the motor pattern and its many variants become consciously controlled rather than involuntary.

Reflexes mature as their purpose is fulfilled; they integrate into the nervous system, remain available for positive protection in the presence of danger, and serve as a foundation for skillful variants of controlled intentional behavior. A fully integrated reflex eventually becomes part of an internal posture, aptitude, or skill.

Sometimes, we use the image of a movement pattern metaphorically in expressions that connote safety. For instance, integration of the various foot reflexes enables us to "be grounded," to "stand on our own two feet," and to "stand tall" under stress. Maturation and elaboration of the grasp reflex will enable readers of this chapter not only to "hang on for dear life," but also to "grasp" the exciting possibilities open to those who embrace this concept of *neurosensorimotor reflex integration* and its implications for treatment of motor, social-emotional, and cognitive disorders.

Each and every one of the 100 reflexes must complete a three-phase process from emergence (phase 1), through growth and maturation (phase 2), to integration (phase 3). Variations emerge as the cerebral cortex matures, and growth occurs in neural networks. It begins with a sensory stimulus, then a buildup of tension in the involved muscles, and ends with the release of that tension in movement.

With the emergence of reflexes in a prescribed sequence, a baby develops from helpless to ambulatory. This quite remarkable sequence of events is depicted visually on this chart:

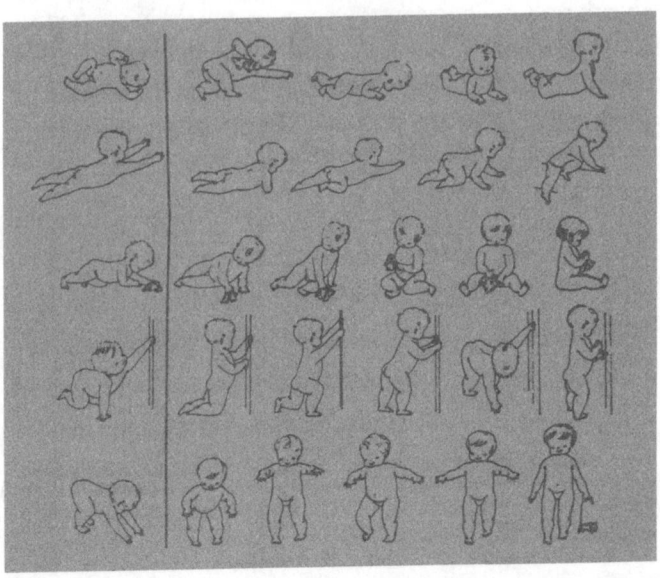

The role of fully matured reflexes ranges far beyond infants and toddlers, and the importance of reflex integration for building the control, motivation, abstract thinking, creativity, and skillful intentional behavior necessary for optimal motor, cognitive, and social skill development cannot be overemphasized.

Infant reflexes have an apparent lifespan of 6 to 24 months. As each fulfills its function and integrates, more sophisticated variations of its basic motor pattern evolve under the control of the cortex. If a baby misses steps in motor development or acquires aberrant motor patterns, such as in crawling, future movements may not be smooth or coordinated.

Reflexes and Stress

As you have read, our bodies and nervous systems are designed to tolerate the normal stressors of everyday life. However, in today's crazy world, the number of everyday stressors may be growing at a rate that outpaces our nervous systems' abilities to adapt to them.

Stressors of all types activate reflexes over and over again during our lives. That's not a problem. That's how integrated reflexes work to protect us. However, sometimes the stress is overwhelming (that is, traumatic) and the reflex circuit gets damaged. Thus, integration is not necessarily a onetime event. The same reflexive gesture may need to be integrated several times because that same gesture occurred and froze at different times of life. Time and repetition are necessary; there simply is no shortcut to integrating reflexes.

If the reflex circuit is damaged by trauma, the body remains in a hyper- or hypo-reactive state, overreacting to mild triggers or possibly underreacting to real danger. Usually, over time these triggers activate less frequently, even though the memory persists. Then, instead of a survival response, neural networks can again provide positive ways of dealing with stress.

Every stressful experience activates the reflex circuits associated with certain involuntary movements. Our response may be to gasp or throw our hands up in the air (*Moro Reflex*) or feel immobilized (*Fear Paralysis*). Even if these movements are suppressed, the pathway has been activated, signaling the brain to be on alert.

Dysfunction can exist in any part of the three steps: the sensory organs can fail to communicate with the brain; brain processing can be faulty; or the motor neurons can fail to communicate with the muscles, tendons, and ligaments. If the sensory stimulus is not recognized by the sensory apparatus and as a result is misinterpreted by the brain, or if the outgoing response is misdirected, then the reflex pattern will be inappropriate, and we cannot rely on that reflex pattern to protect us in a time of stress.

If we restrict the motor response by restraint, fear, or an external verbal command such as "hold still and pay attention," it may eventually lose its connection with the original sensory stimulation, or the sensory-motor connection may become weak, manifested by low muscle tone. Then, the correct motor response won't happen, the built-up muscle tension remains in the body, and the child is at the mercy of irregular motor reactions to sensory stimulation. Maturation and integration of the reflexes with controlled movements and skills will be slow or unreliable in the future, especially in the presence of additional learning challenges and more stress. When infant reflexes remain active and have not integrated in a timely fashion, they are considered "aberrant." Unintegrated, they provide inadequate support for more complex reflex schemes, making it harder, for instance, to do such apparently simple things like copy the words on a blackboard into a notebook.

Under threat, the ancient, early developing, survival-oriented part of the brain mobilizes for protection without involvement of the higher, more recently evolved cerebral cortex. Depending on the nature of the threat, the body must choose either freeze or fight-or-flight.

Positive and Negative Protection

To describe the role of reflexes in the functioning of both typical and neurologically challenged children, Dr. Masgutova refers to "positive" and "negative" protection.

- *Positive protection.* When the reflex systems have matured neurologically and sensory perception functions well, the brainstem recognizes stimuli and organizes protective motor responses with no disturbance to overall development.
- *Negative protection.* When a reflex fails to mature because a stressor has interfered at one of the three phases of development, a dysfunctional reflex response continues beyond a time that is necessary or useful. As the body feels unsafe, symptoms of negative protection—such as muscle tension, impulsivity, and involuntary movements—emerge. In highly dysfunctional or pathological reflex development, we might see symptoms such as flapping, or other stereotypical movements.

Furthermore, unintegrated reflexes can prevent normal sensory development as the brainstem keeps the body reacting automatically to minor sensory inputs because the immature neural circuitry does not support higher-order processing and volitional motor activity.

Retraining the Brain

Psychologists and psychiatrists have traditionally believed that anxiety, fear, depression, and other symptoms of stress come from higher levels of the brain, and thus, for years, have tried to tame negative emotions with language through talk therapy. Recent discoveries of the interactions among various organs and the brainstem, especially the dorsal and ventral branches of the vagus nerve, have the potential to revolutionize the way we treat those with "mental health" diagnoses.

Some exciting novel strategies that retrain the brain to process experiences more positively and replace undesirable emotions and behaviors are emerging. Remember that, while foundational in getting well, nervous system regulation can take place simultaneously with lifestyle changes, dietary modification, and other interventions. However, the calmer the nervous system, the faster and more efficiently the other therapies will reduce or even eliminate Total Load factors.

World-renowned physician Neil Nathan, MD, mentioned previously, has spent over 50 years treating and studying thousands of patients with every imaginable combination of illness. When his patients get "stuck," despite good compliance with his prioritized truckload of right treatments at the right time, he looks at the nervous system.

In *The Sensitive Patient's Healing Guide*,[19] Nathan delves deeply into the origins of sensitivity that prevent many sick people from getting well. The book opens with some of the methods for nervous system regulation mentioned here. Summarizing his wisdom into a few paragraphs is challenging; I am grateful for his guidance in doing so.

Restoring Limbic Balance

Nathan believes that the brains and bodies of many are stuck in a protective mode and are sending and responding to false alarms, day and night. Symptoms include multiple chemical sensitivities, chronic fatigue and anxiety, pain, headaches, sleep disturbances, and more. Everything is "on alert." Some people who were particularly afraid during the COVID pandemic found themselves experiencing a similar phenomenon after the threat had passed.

The answer to getting well is retraining the brain, and it starts with teaching the limbic system that "the coast is clear." Generally acceptable, "tried and true" methods include *daily meditation, exposure to nature, and walking barefoot*.

Jon Kabat-Zinn, PhD, the father of *mindfulness-based stress reduction*, has been helping people get well through some simple exercises since the 1970s. He has written dozens of books and articles available on his website, JohnKabat-Zinn.com.

Kabat-Zinn defines mindfulness as "paying attention on purpose in the present moment, nonjudgmentally to things as they are." His approach, available in a free download at MindfulnessForLiving.org and as an app for both Apple and Android devices, promotes focusing on being present. Practicing daily can markedly reduce stress, calm limbic hyperactivity, and lessen the brain's fear-based responses.

Gentle movement activities, like *yoga*, **and** *other mindfulness practices* can also be extremely helpful. Internationally known yoga therapist Joanne Spence, author of *Trauma-Informed Yoga: A Toolbox for Therapists,*[20] specializes in treating adults and children with depression, anxiety, ADHD, and chronic pain. Her *Trauma-Informed Yoga Card Deck* has over 50 self-guided practices to calm, balance, and restore the nervous system. She also teaches online classes. JoanneSpence.com.

Vagus Nerve Stimulation

Targeting the complex vagus nerve to reduce stress, fear, and other damaging effects of trauma is one of the hottest topics today in treating any type of neuro-developmental disorder.

Some simple activities, like exposure to temperatures below 50°F and gargling, can strengthen a sluggish vagus nerve. One study showed that when the body adjusts to extended **cold exposure,** like in a shower for about a minute, the fight-or-flight response declines and the parasympathetic response increases.[21] *Gargling, humming,* and *singing loudly* activate the vocal cords and muscles in the back of the throat, which are connected to the vagus nerve. Try adding gargling to the end of tooth brushing, humming while preparing meals, and singing at the top of your lungs in the cold shower. No one is listening!

And then there is **breathwork.** This topic is so hot that a whole book about it, *Breath: The New Science of a Lost Art,*[22] was on the New York Times Best Seller list for many weeks. Recall that respiration is regulated by the autonomic nervous system. Under stress, vagal tone decreases, respiration and heart rate increase, and the dorsal branch of the vagus nerve becomes dominant.

The good news is that breathing can be regulated by bringing attention and intention to it.[23] Two types of breathing exercises known to increase vagal tone, and reduce stress can be very powerful. Watch online videos to learn how to do them correctly.

- *Diaphragmatic Breathing (Belly Breathing)* is one of the quickest and most effective ways to get out of fight-or-flight and into safety.
- *Alternate Nostril Breathing* is easy to do almost anytime, anywhere.[24]

In both exercises, the focus is on exhaling for longer than inhaling. Consistency is the key. Daily practice is best. Meditation, yoga, and other mindfulness exercises use these techniques and are thus a two-for-one package of mindfulness and breathing.

Deb Dana's *Polyvagal Exercises for Safety and Connection*[25] and the related card deck of over *50 practices for Calm and Change* are extremely user-friendly.[26] Her newest contribution is *The Nervous System Workbook.*[27] Other easy-to-follow books are *Activate Your Vagus Nerve: Unleash Your Body's Natural Ability to Heal,*[28] *Daily Vagus Nerve Exercise,*[29] and *How to Hack Your Vagus Nerve.*[30] For even more ideas, read "19 Factors That Stimulate the Vagus Nerve" at SelfHacked.com.

Safe and Sound Program and the Rest and Restore Protocol™

Porges has developed two multidimensional interventions derived from over 40 years of neuroanatomical, neurophysiological, and behavioral research. The Safe and Sound Program (SSP) is a five-day program of hour-long acoustic stimulation sessions.

Designed to retune the nervous system, reduce stress, and enhance social engagement while promoting resilience at any age, it is based on the connection between the vagus nerve and two muscles in the middle ear—the tensor tympani and the stapedius—which control vocalization, facial expression, heart rate, and breathing. Stimulating the middle ear with filtered music, in a variety of sound frequencies, results in improved listening, lessened anxiety, and more availability, socially and emotionally.[31]

The SSP is not meant to be used in isolation; changes from the SSP accumulate and are enhanced by other therapies and learning opportunities. As patients make gains in social and communication skills, they become more available to other forms of treatment. To learn more about it, read *Safe and Sound: A Polyvagal Approach for Connection, Change, and Healing.*[32]

The Rest and Restore Protocol™ (RRP) is a new listening therapy program designed to promote deep relaxation, recovery, balance in the body and mind, and connection to self, cultivating internal awareness and self-regulation. RRP targets the healing qualities of the autonomic nervous system, helping to improve physical and mental well-being.

Early research from its developers indicates that RRP may improve sleep, digestion, anxiety, trauma-recovery, and more in patients with anxiety and other symptoms. RRP is not a stand-alone therapy; it complements existing treatment strategies for therapists to use with other methods.

The Safe and Sound Protocol (SSP) and the Rest and Restore Protocol™ (RRP) are both available from Unyte at IntegratedListening.com.

Vagus Nerve Stimulation Devices

The FDA has recently approved devices that can be implanted under the skin in the chest, worn on the body. When it fires, the device sends electrical impulses through a wire that connects to the ventral vagus nerve. These signals increase vagal tone, and thus reduce problematic physical dysfunction as well as the accompanying negative emotional and behavioral reactions to stress.

A handheld version recommended by a trusted resource is the **truvaga,** that earned the seal-of-approval from several respected medical centers, including Harvard, Johns Hopkins, and Emory. Check it out at truvaga.com.

This approach is new and promising. The Mayo Clinic is doing research on its efficacy for depression and bipolar disorder. Stay tuned.

Bodywork Treatments

Recall from Chapter 3 that any structural impediments can disrupt overall health and have both profound and subtle effects on behavior and learning. Stressors on the nervous system, derived from birth and other trauma, can weaken the brain's defenses, making it susceptible to invasion by toxic metals and microorganisms, as well as set the body up for infections.

As the body's Total Load increases, and emotional tension rises, muscles tighten to hold emotions in. Tissues of the musculoskeletal system compress the blood and lymph, decreasing their flow, adding pain to tension and stress.

A number of treatment modalities clustered under the umbrella of "bodywork" take into account the structural and skeletal systems of the body and their relationship to each other. Professionals trained in chiropractic, chiropractic neurology, osteopathy, craniosacral therapy, and massage are important members of therapeutic teams resolving distress.

In order to determine which type of structural therapy might be beneficial, an in-depth history going back to birth is essential. As with any intervention, the underlying cause of the problem rather than the diagnosis is what is important,

Not surprisingly, patients often report strong emotional release during bodywork treatment. The deeper the emotional release, the more tension diminishes and the more healing takes place.

Chiropractic

Chiropractors are physicians educated in a four-year graduate program at a chiropractic college. First, chiropractors identify interferences called *subluxations*, which can occur in any vertebral or bony area of the body, including the neck and face. They then use a variety of techniques to gently tap a segment back into place, thus restoring the structure and the function of that area.

Cranial, upper cervical, and pelvic misalignments are often present in those with neurodevelopmental and psychiatric diagnoses. Chiropractic adjustment is often an effective treatment in returning the vagus nerve to full function, thus enhancing digestion, communication, and socialization.

Chiropractic Neurology

This hybrid field marries the biomechanical aspect of chiropractic care with assessment and rehabilitation of the central nervous system. The result is a brilliant model of diagnosis and care that focuses on the two-way interaction among all of the sensory systems and the central nervous system.

Dr. Tony Ebel is a chiropractic neurologist who has devoted his practice to addressing the nervous system first and foremost in patients with psychiatric diagnoses. He is adamant that prioritizing calming the nervous system is essential to healing. Ebel uses a tree analogy: the nervous system is at the roots, the immune system is the trunk, and the symptoms are the branches and leaves. Without strong roots, the tree will fall. His *The Experience Miracles*™ Podcast on YouTube has dozens of episodes explaining his functional medicine and neurological approach.

To find a chiropractic neurologist near you, go to www.acnb.org.

Craniosacral Therapy

The late Stanley Rosenberg (1941–2023), a veteran American-born craniosacral therapist who lived in Denmark for more than 50 years, made a surprising discovery: virtually every anxious and inattentive patient in his practice had tension in the right *sternocleidomastoid muscle* (SCM), and an accompanying deformation of the skull called *plagiocephaly* or "flat back of the head." Research confirms that this finding is present in a higher percentage of those with psychiatric disorders than in typical functioning children.[33]

The SCM and trapezius muscles both originate in the bones of the cranium. Together they make up a ring of muscles in the neck, shoulders, and upper back that enable the movement of the head and neck. Tension in either of these muscles on one side pulls the body out of alignment, resulting in physical asymmetry that causes impediments in the flow of energy to the brain, ears, eyes, and other senses. Following testing, Rosenberg alleviated any dysfunction he found. Simply releasing this tension allows the blocked energy to flow, and thus increases communication and relatedness. Total Load Theory in action!

Rosenberg met Porges in 2001, became fascinated by his work, and they began collaborating. The location of the vagus nerve in the upper neck can result in it being compromised by misalignments of the C1 or C2 vertebra. Illinois chiropractor David Foss (VitalWellnessCenter.net) concurs. To address these, Rosenberg developed a special massage-like technique he named *neuro-fascial release*, which stimulates reflexes in the nerves of the loose connective tissues just under the skin over the base of the skull.

Neuro-fascial release became one of the cornerstones of Rosenberg's treatment for depression. With simple adjustments to the spine combined with lifestyle changes, practitioners report eliciting a calmer nervous system.

Read more about this procedure and other unique exercises in Rosenberg's book, *Accessing the Healing Power of the Vagus Nerve: Self-Help Exercises for Anxiety, Depression, Trauma, and Autism.*[34]

Rosenberg reminded us that an individual's nervous system can be in only one state at a time, and that an individual can shift suddenly between stress and withdrawal. He found that by improving the function of the vagus and other cranial nerves through bodywork interventions, he was able to stabilize and reduce these sudden shifts, thus enabling sustained social engagement. More at StanleyRosenberg.com.

Osteopathy

Osteopathic medicine, or *osteopathy* (from the Greek *osteo*, meaning *bone*, and *pathos*, meaning *suffering*), is a form of medical practice founded in the late 1800s. Doctors practicing osteopathy or osteopathic medicine are called *osteopathic physicians*, *osteopaths*, or *DOs*. They receive four years of post-graduate education and are licensed in all 50 states.

A unique component of osteopathic education is training in hands-on techniques designed to influence the body's structure in order to enhance the body's innate ability to heal itself and, thus, function more efficiently. This treatment is called *osteopathic manipulative therapy* (OMT).

OMT is gentle and does not use high-velocity manipulation commonly experienced in chiropractic medicine. Osteopaths, like chiropractors, strive to enhance wellness, always considering the impact that lifestyle and environment have on health. Complementary treatments often include dietary modification, nutritional supplementation, and detoxification techniques in addition to osteopathic manipulation.

Patients seeking osteopathic treatment often show injury (most probably from birth) to the back of the skull (where the first cervical vertebra attaches to the skull), which results in the neck being jammed up against the skull base or occiput. This condition, in turn, injures three groups of muscles that make up the *suboccipital triangle*. When these muscles and their fascia become contracted, they compress a space called the *jugular foramen* (literally a hole in the skull). Several nerves as well as the jugular vein, which drains 95 percent of all blood coming from the brain, pass through this area. When the hole is compressed, the amount of blood that can flow through the vein is decreased. OMT can result in marked and quantifiable changes in cerebral blood flow, thus improving brain function.[35]

A couple of osteopathic physicians have devoted some of their careers to working with patients with ADHD, autism, and psychiatric diagnoses. Mary Ann Block, DO, who attended medical school at the age of 39 to help her daughter heal, believes that structural therapies when combined with other interventions are particularly beneficial for treating those who have had chronic ear infections. She uses osteopathic manipulation to drain the fluid in the ears along with dietary modification and nutritional supplementation to enhance the immune system.[36] Her 1996 book *No More Ritalin: Treating ADHD Without Drugs* has become a classic. See BlockCenter.com.

Sherri Tenpenny, DO, whose primary interest is in vaccine education, runs a multidisciplinary health center located in Middleburg Heights, Ohio. See TenpennyIMC.com. LeTrinh Hoang, DO, is a California pediatrician who combines osteopathic medicine with homeopathy and other disciplines. Her 2015 book, *Osteopathy for Children*,[37] details how dental trauma and stress from electromagnetic radiation can be alleviated with osteopathic treatments.

Reflex Integration as Therapy for Feeling Safe

Reflex integration work often can result in profound and quick behavioral changes for deep, long-standing problems. Eve Kodiak, a now-retired developmental specialist with a varied background in early childhood music, movement and craniosacral therapy calls reflexes "the Rosetta Stone" because understanding and working with them is like decoding a language.[38] She believes that a therapeutic reflex integration program is essential for anyone whose development was compromised by too many stressors at any stage of development.

Three reflexes are extremely important in developing and restoring feelings of safety:

Fear Paralysis, also known as the startle reflex, is a freeze reflex that is present at birth. It is triggered by a loud noise, bright light, or sudden touch.

Techniques for integration: touch lightly on balls of the feet. Relaxing the feet and legs send the signal to the brain that the danger is over. Begin by holding a certain spot, which the Chinese call "Bubbling Spring," on the ball of each foot. It is below the big toe and slightly toward the second toe. Acupuncturists know it as the Kidney 1 (K-1) point, which is the beginning of the kidney meridian, and governs fear. Cover these points on both feet with the thumbs. Don't worry about whether you are on exactly the right spot—the whole general area will work just fine. Eventually the muscles relax, and calm ensues.

The Moro Reflex, also present at birth, facilitates the first "breath of life," opening the windpipe if there is a threat of suffocation. It is a fight-or-flight response to threat which immediately alerts the sympathetic nervous system of danger. Blood pressure rises, energy moves away from the core to the limbs, stress hormones are released, the skin reddens, and an emotional outburst, such as tears or anger, results. The Moro Reflex occurs in response to a vestibular (change in the relationship of the head to gravity) or proprioceptive (sudden movement in the joints) stimulus.

It should integrate sometime in the first three to four months of life.

Physically, it is composed of a series of rapid movements of the limbs away from the body, accompanied by rapid inhalation and a startle response. During the first stage of this reflex, arms and legs extend outward, and the baby gasps for breath. The second stage is exhalation, as arms and legs flex, and the third is to embrace a safe person—usually Mommy. The first stage says, "Oh no! I'm scared!" The second stage says, "I can curl up and protect myself!" The third stage says, "I can reach out for help, and someone will love me!"

Children with an active Moro Reflex are constantly reacting, running, hitting, or screaming; they seem to fight calm. Adults with active Moros yell, strike out, become hysterical, or run away. An active Moro is common in those who have experienced trauma or abuse.[39]

Techniques for integration: The classic way to integrate the Moro reflex is called the Starfish exercise. It involves alternately crossing right arm and leg over left with crossing left over right. It can be done sitting in a chair or lying comfortably on the floor, propped by pillows. The arms and legs start and end tucked close to the body, and then fling out away from the body, coming back crossed in the opposite direction. Go online to see various demonstrations. This activity can be done several times in sequence, always ending on the body with a hug.

Lack of integration of either the Moro or Fear Paralysis reflex keeps an individual in a survival mode and can cause hypersensitivity to sound, light, movement, or altered position, so that the individual is always on alert. The energy-depleting mode of constantly fighting perceived danger leaves scanty reserves for other bodily processes, such as digestion and respiration, let alone development and learning.

The Tendon Guard Reflex (TGR) is a lifelong automatic whole-body reaction to a danger message from the brainstem. Depending upon the threat level, the body chooses whether to freeze or make energy available for flight or fight.

As the TGR involves the whole body, it is of extreme importance in feeling safe. It is usually hyperactive. When integrated and matured it provides the foundation for focus and analytical thought as well as appropriate responses to real (not past or imagined) danger.

Techniques for integration: Same as for Fear Paralysis and Moro.

Some believe that single reflexes should be worked on sequentially, while others support a more global approach. Since the reflexes are all linked together in complex relationships, sometimes working on one results in the spontaneous integration of another.

YouTube has hundreds of videos showing techniques, not all of them accurate! Experiment and find some that work for you.

Svetlana Masgutova assesses a group of 30 reflexes. The more reflexes that test in the dysfunctional range, the greater the likelihood of a diagnosis covered in this book. Anyone working with those with severe challenges should have intensive training and experience. Svetlana Masgutova's remedial techniques are taught and practiced around the world as Masgutova Neurosensorimotor Reflex Integration (MNRI®). Learn more at MasgutovaMethod.com.

Reflex integration techniques are applicable to many areas of dysfunction. Psychologists, physiologists, developmental optometrists, occupational therapists, and others do reflex work as a part of their unique therapies. Recently, the American Occupational Therapy Association (AOTA) has discouraged their members from using reflex integration techniques because of a purported lack of evidence for their effectiveness. This has not discouraged those I know from including this extremely helpful tool in their repertoire.

Eve Kodiak, mentioned above, has combined her musical talents and developmental knowledge in *Rappin' on the Reflexes*, a CD and guidebook with original songs for reflex integration. More about her at EveKodiak.com.

Reading specialist Deb Em Wilson combines her knowledge of reflexes and Polyvagal Theory in her most recent work for teachers that "transforms classrooms one nervous system at a time." You can read more about her program in Chapter 20. Use the index to learn more about reflexes' role in language development, learning, vision, and more.

Dr. Nathan includes two programs that target the brain through a systems approach:

Dynamic Neural Retraining System (DNRS) is a drug-free, self-directed, neural rehabilitation program developed by Annie Hopper to overcome her own severe limbic system impairment that left her disabled and homeless. Similar to EMDR, it relies on the brain's neuroplasticity and the idea that faulty neural pathways can be retrained, as presented in the book *The Brain That Changes Itself* by Canadian psychiatrist Dr. Norman Doidge.[40]

DNRS introduces the link between the brain and various conditions through pillars of recovery reflected in the acronym IMAGINE, for intention, motivation, awareness, gains, incremental training, neurological, and emotional rehearsal. Over a minimum of six months of daily practice, this program is said to cool off threat centers in the limbic system and increase function in the prefrontal cortex, the area of rational thinking in the brain. To learn more, read Hopper's book *Wired for Healing: Remapping the Brain to Recover from Chronic and Mysterious Illnesses*[41] or go to her website at RetrainingTheBrain.com.

The Gupta Program, another brain retraining program was developed by a pioneer who treated his own illness. This one uses an app to restore feelings of safety by focusing on the overactive amygdala, the part of the limbic system that processes anxiety and fear.[42] Learn more about this self-directed program at GuptaProgram.com.

Crisis Intervention

In crisis, when all resources have been exhausted, behaviors are out of control, and a struggling loved one is in danger of hurting themselves or others, few alternatives are available. During these times, desperate family or friends resort to calling the police or going to the emergency room. Clearly, incarceration and hospitalization are not great solutions.

Fortunately, some new multidisciplinary temporary residential programs are now appearing. Based on the metabolic psychiatric model, these combine thorough laboratory testing to determine what load factors are at the root of the problem with functional medicine, nutritional intervention, mindfulness, and some of the therapeutic interventions mentioned in this chapter. Check HealingDepressionProject.com.

Take-Home Points

Feeling safe is an essential foundation for emotional health. The autonomic nervous system, the limbic system, reflexes, and the vagus nerve all play critical roles in regulating the body and brain through the cell danger response. When nervous system regulation is disrupted, the body may remain in a chronically defensive state, leading to persistent inflammation, emotional dysregulation, and eventual diagnosis of a neuropsychiatric illness.

Understanding and addressing the connections among these complex circuits is crucial to reducing the Total Load of stressors that has made someone sick. Therapies that target calming the overactive and imbalanced nervous system can lead to emotional health and healing.

Transitioning from a fight-or-flight state to feelings of safety involves a multidisciplinary response centering on downregulating stress, restoring parasympathetic tone, and resolving cellular defense mechanisms.

The vagus nerve and limbic system are central to shifting the body out of fight-or-flight and resolving the cell danger response. By enhancing vagal tone, calming limbic overactivity, the body transitions from survival mode to healing, connection, and feeling safe. Therapeutic approaches such as limbic and other brain retraining, vagus nerve stimulation, breathwork, and reflex integration can reinforce this process, supporting recovery from chronic stress, trauma, and inflammation.

CHAPTER 7

DIGESTION:
THE KEY TO ALL HEALTH

The most common topic of conversation at the many conferences I have attended may be a surprise to you. It's not temper tantrums, lack of speech, anxiety, or poor social skills. Believe it or not, what attendees talk about most—one of the most challenging issues for many taking a biomedical approach—is digestion!

Conventional doctors frequently tell patients with disorders listed in the DSM that their chronic diarrhea or constipation is because they have autism or bipolar disorder. Yes, diarrhea and constipation are hallmarks of these conditions. But these medical practitioners get the order wrong. Most individuals have autistic-like behaviors or are inattentive or anxious because their digestion is not working properly.

As we become more knowledgeable about the gut bugs with whom we live symbiotically, we may find the key to understanding behavior lies in learning more about the tiny creatures who make up the 90 percent of our cells that are "not us." That's right, current estimates are that only 10 percent of our cells are human.

Digestion 101

A short course on the normal digestive process is necessary before understanding what goes wrong in the digestion of those with psychiatric symptoms. Digestion starts in the mouth. Each bite of food mixes with saliva before being swallowed. Muscular contraction called *peristalsis* moves the partially digested food into the stomach, small intestine, and colon, where enzymes and other juices dissolve it further.

This very complex process can take hours to days, depending upon many factors. Ideally, food should begin breaking down in the mouth with saliva and chewing and not tarry too long in the intestines, as the digestive organs then reabsorb wastes. A healthy, well-fed gastrointestinal tract, assisted by an adequate supply of water, continues to break down what we ingest into usable particles, pick up wastes, and produce at least one significant bowel movement per day without any effort. As the body absorbs nutrients, the collected toxins and other waste products finish the journey and exit through the rectum as fecal matter.

The Microbiome

The entire body is a living ecological community of tiny organisms which make up an individual's microbiome. Our microbiome is everywhere on and in our body. Every little nook and cranny, such as your nose and armpits, harbors micro-organisms, and has its own population of bugs. Microbes inhabit our guts, skin, mouths, navels, genitals, hands, feet, ears, and more. Most of the organisms live in the digestive tract. Among other things, their job is assisting in the absorption of nutrients and waste elimination.

The average person's microbiome weighs more than three pounds and is comprised of about 100 trillion microbial cells that inhabit our organs, orifices, and skin (compared with about 10 trilliion human cells).[1] That's *a lot* of bugs! And they must all live in harmony for a person to be healthy. Let's look at who or what they are and how they help and/or hinder our development, functioning, and all parts of our being.

Your body's microbiome is as unique as your fingerprints. Some consider it to be another organ of the body.[2] We have known about the microbiome for about 350 years, but it took until 2007 for scientists at the National Institutes of Health to launch The Human Microbiome Project (HMP). Their goal was to identify and characterize the microorganisms found in both healthy and unhealthy humans.

The HMP currently collects samples from five areas of the body: the digestive tract, the mouth, the skin, the nose, and the vagina. The blood and male urethra will be added soon. By the end of 2017, the HMP had published over 650 scientific papers that were cited over 70,000 times.[3]

One of investigators' early surprises was the incredible diversity of organisms. Consider the human microbiome like the ocean; we will never be able to identify and name every single species living in it.[4] The diversity of the gut microbiota may be its most important characteristic; the more diverse the microbiome, the healthier a person is.

Balance is also crucial. "Good" and "bad" bacteria compete for space, with the goal of a mixture that sustains health. In general, some organisms are beneficial, especially to the digestive process and much more, whereas some can be harmful, disrupting the digestive process and overall health. The proper balance of intestinal flora is crucial to many biological processes, including our nutritional status and our abilities to detoxify, think, and feel. Bottom line: a diverse and balanced gut = a healthy person.

But wait. It's not that simple. Scientists are just beginning to understand that a few "bad" or harmful bugs can be beneficial, and that wiping out any particular strain can be almost as deleterious as having too many. That's what scientist Martin Blaser discovered when he studied *Helicobacter pylori* (*H. pylori*), the known cause of ulcers and some stomach cancers. Why did some people with this bug get sick and others did not? It was all about timing. If acquired in the first three years of life, it provided protection against disease rather than caused it.[5]

Almost every day new research is uncovering ways in which the microbiome impacts human health, especially in specific conditions. Scientists are studying the role of the microbiome in autism,[6] ADHD,[7] OCD,[8] bipolar disorder,[9] as well as in major depressive disorders, and schizophrenia.[10] Guess what they are finding? A definite relationship. Change the gut, change behavior!

Want to know who is inhabiting your microbiome? Go back to the previous chapter to find a list of tests that can tell you.

The Gut-Brain Axis

The gut and the brain are both parts of the body's autonomic nervous system, which runs and monitors the body's unconscious functions, including metabolism, endocrine production, and heart rate, in addition to digestion. The gut and the brain talk to each other through that important vagus nerve which you read about in the last chapter.[11] This channel of communication is called the *gut-brain axis* (GBA).

Doctors have long appreciated that the gut-brain axis is key to maintaining balance among unconscious functions, a process called *homeostasis*. Recently, they have broadened their scope and now recognize the importance of the microbiome as a key regulator of gut-brain function. So, it really is a triad: the microbiota-gut-brain axis. The most exciting outcome of this knowledge is the understanding that the health and function of this triad is the biological and physiological basis for many psychiatric and neurodevelopmental disorders, including, but not limited to, autism, anxiety, and schizophrenia.[12,13]

The word *vagus* comes from the Latin word meaning *rambling*, and this wandering nerve is well named. The vagus nerve diverges out from the neck to the chest and down into abdomen, spreading fibers to and influencing function of most major organs along the way, including the tongue, pharynx, vocal cords, lungs, heart, stomach, and intestines.

The interaction between our microbiome and the GBA through the vagus nerve is bidirectional—from gut microbiota to brain and from brain to gut microbiota—by means of neural, endocrine, immune, and humoral links.[14] In fact, 80–90% of the nerve fibers in the vagus nerve are dedicated to communicating the state of the viscera up to the brain.

At least 70% of the immune system is located in the digestive tract, according to Michael Gershon, MD. He has dubbed this subdivision of the autonomic nervous system that directly controls the gastrointestinal system (known scientifically as the *enteric nervous system*) "the second brain."[15]

A modern pioneer in the emerging field of *neurogastroenterology*, Gershon has greatly enhanced our understanding of the complex interactions among the gut, immune system, and brain. Those with neurodevelopmental disorders have guts damaged by toxins, immune systems fighting for life, and brains attempting to deal with it all in a rational, helpful way.

Digestion and Neurodevelopmental Disorders

Doctors are just beginning to comprehend that the many digestive complications common in individuals with psychiatric and developmental diagnoses can result from an imbalanced microbiome that's lacking in diversity. Gastroenterologists, psychiatrists, neurologists, and other doctors are beginning to recognize that these problems are not unique to those with known digestive disorders like irritable bowel syndrome (IBS), Crohn's disease, and ulcerative colitis. Digestive issues are common to many behavioral, psychological, physical, and complex neurological diseases as well.

Dysbiosis

Gut problems often begin with a condition known as dysbiosis, in which an unbalanced gut is host to high levels of harmful bacteria and low levels of beneficial bacteria.[16] An out-of-balanced gut is like a garden overrun by bugs that eat the plants, and too few bees, snails, and worms to pollinate the flowers and make good soil.

Dysbiosis contributes to health problems and behavioral issues as the overgrowth of bad microorganisms undermine an individual's resistance to infection by altering their immune response. Unwanted bugs not only wreak havoc in the gut but also force the body to deal with additional toxins produced as waste byproducts from their metabolism. And that's big trouble!

Dysbiosis in autism is well-recognized and has been studied for over ten years. In June 2014, an eminent group of clinicians, research scientists, and parents convened the 1st International Symposium on the Microbiome in Health and Disease with a Special Focus on Autism at Arkansas Children's Hospital Research Institute in Little Rock to look at dysbiosis in children with autism. This working group, headed by Richard E. Frye, MD, included Derrick F. MacFabe, MD, S. Jill James, PhD, James Adams, PhD, and Susan Swedo, MD. They sought to design a clinical trial focused on determining whether children with ASD might respond to therapy aimed at modulating and/or manipulating their microbiomes.

They uncovered some very interesting findings about children with autism:

- They have less diverse microflora;
- They lack microbes from the genus *Prevotella*, which are known to protect against pathogens;
- They harbor abundant potentially pathogenic bacteria, including clostridia.[17]

Bottom line: too little variety, not enough good stuff, and too much bad stuff.

While these doctors may not be household names, they are pioneers and heroes in discovering our relationship with our microbiome. You will read about them again and again in the upcoming chapters of this book.

Dr. Adams's latest exciting finding is the discovery of an extremely toxic substance called *P-cresol sulfite*, secreted by over 60 nasty bacteria, that affects the brains and bodies of 90 percent of his sample of children with autism.[18] Yikes! Read below how he is changing their outcomes, eliciting speech and other positive behaviors, with fecal microbiota transfers.

While the important role dysbiosis plays in mental illness is just being recognized, exciting research is being done on both sides of the world. Psychiatrists at Yale School of Medicine studied microbiota in eight different neuropsychiatric disorders,[19] while Japanese scientists looked at the guts of those with depression and anxiety.[20] Both concluded that, although more research was needed, their subjects' guts were clearly dysbiotic.

Want to observe some gut bugs in action? Watch *The Autism Enigma*, a CBS documentary on gut issues as a major contributor to autism. In this film an international group of scientists examines the gut's amazingly diverse and powerful microbial ecosystem and efforts toward healing.[21]

Leaky Gut

Leaky gut, also known technically as *intestinal hyperpermeability*, is a similar condition that often accompanies dysbiosis. The porous intestinal lining covers more than 4,000 square feet of surface area. Microvilli, finger-like hairs, increase the area of absorption in the gut even more, and aid in the absorption of nutrients and elimination of waste.

Toxins damage the microvilli, causing the intestinal lining to become porous or "leaky" and allowing partially digested food particles, proteins, and toxins to pass into the bloodstream—a process similar to ivy attaching itself to a brick house and slowly destroying the mortar.[22] Antibiotic overuse, heavy metal exposure, and GMO foods further contribute to the breakdown of the gut lining.

Yeasts and other hardy fungi and bacteria then proliferate and flood the bloodstream with more invaders. This situation triggers allergic, physical, emotional, and cognitive reactions to what the body views as foreign substances, leading to inflammation, immune dysregulation and physical disease. According to veteran nutritionist Elizabeth Lipski, PhD, a leaky gut underlies an enormous variety of illnesses and symptoms.[23] It is one of the first red flags for future autoimmune disease, which is very common in families with autism.[24]

As yeasts increase, the immune and endocrine systems weaken, the chances of infection and antibiotic usage increase, and the vicious cycle continues. The train has left the station on its trip to a neuropsychiatric diagnosis.

Though leaky gut syndrome is the nickname for this condition of increased intestinal permeability, it can affect the whole body. Symptoms occur not just in the digestive tract, but also on the skin and in the liver, thyroid, and yes, the brain. Rashes, endocrine imbalances, and even cognitive difficulties can sometimes disappear, almost magically, when the gut is healed.

SIBO

SIBO, short for *small intestinal bacterial overgrowth*, is a combination of dysbiosis and leaky gut. It occurs when bacteria that normally live in the large intestine migrate backwards in the digestive cycle to the small intestine, where they interfere. The relocation of these bacteria causes leaky small intestines, and that is yet another *big* problem! It is a common cause of many autoimmune conditions, including Crohn's, ulcerative colitis, and irritable bowel syndrome (IBS).

Migrated bacteria now not only damage the lining of the small intestines as they feed on unabsorbed food, they also emit gases that result in bloating, pain, flatulence, belching, constipation, and diarrhea. By feeding on undigested food earlier in the digestive cycle, they can snap up essential nutrients, causing deficiencies in important vitamins such as B12.

Furthermore, the byproducts of the bacteria's own metabolism, such as ammonia, can cause cognitive, neurological, and behavioral problems.[25] SIBO clearly affects mental health, especially in autism,[26] depression, and anxiety.[27] The more serious the SIBO, the more severe the symptoms.

What causes SIBO? Why would bacteria meant to live lower down in the colon migrate up to the small intestines? Yep, you guessed right. It's that pesky vagus nerve again. Recall that this meandering cranial nerve influences multiple organs and affects many unconscious bodily processes in its long journey from the brainstem to the gut and that digestion is one of its primary functions.

Some of the important jobs of the vagus nerve in the digestive process are maintaining motility and monitoring the production of enzymes and stomach acid. Vagus activation is necessary to push food forward along the intestinal tract. Bacteria moving in the wrong direction is a sign of a poorly functioning vagus nerve.[28]

SIBO doctors have had success at retraining dysfunction in the vagus nerve by stimulating it in a variety of ways. Read about this interesting intervention in Chapter 13.

In addition to the typical invasive tests that evaluate digestion, endoscopy and colonoscopy, doctors practicing functional medicine often perform a SIBO breath test. Return to Chapter 5 to refresh your knowledge about that interesting diagnostic technique.

Enzymes

Enzymes are proteins responsible for many essential biochemical reactions. A copious supply of enzymes is vitally important for proper digestion. Enzymes act as catalysts, breaking down carbohydrates, proteins, and fats into simple forms the body can absorb, burn for energy, or use to build and repair itself. When enzyme production is off, other digestive and behavioral functions can also go awry.

A damaged and unbalanced gut is not an efficient producer of the essential enzymes necessary for proper digestion. When enzymes are lacking or insufficient, symptoms such as constipation and diarrhea occur. Digestive enzymes, discussed further along at the end of this chapter, used in combination with one of the special elimination diets in the next chapter, could be the answer to normalizing digestive function.

Phenol Sulfur-Transferase and Phenols

One enzyme, *phenol sulfur-transferase* (PST), is necessary for the body to break down and remove substances called phenols. Understanding how PST works could be the key to solving many digestive problems.

Consuming high-phenolic foods isn't a problem unless PST is low, because then the phenols cannot be properly digested. While all foods contain phenolic compounds, some have higher content than others. Some foods naturally high in phenols are apples, grapes, strawberries, bananas, almonds, and vanilla.

Phenols occur not only in the abovementioned natural foods, but also in foods that contain toxic artificial dyes, colors, and flavors, as well as the preservatives BHA, BHT, and TBHQ,[29] which inhibit the production of PST and suppress its activity in the gut.[30] *Malvin*, a phenol present in foods that are naturally red, blue, or purple, such as grapes, also can also suppress PST.[31]

In a now classic study, Dr. Rosemary Waring, a British scientist who discovered the role of PST, found that as many as 90% of those with autism have limited PST activity. She believes that unprocessed phenols and toxic bacteria in the gut act as internal irritants, causing hyperactive behavior.[32] Waring notes that children with low levels of PST also have trouble digesting gluten and casein.[33] Read about these irritants further along in this chapter. Limited PST may be the root cause of their need for a special diet.

Phenols and PST deficiency are also implicated in other neuropsychiatric conditions. Researchers in Slovakia investigated their role in major depression, ADHD, and schizophrenia and found that eating phenolic foods may improve mental health.[34] Chinese[35] and Australian[36] investigators agree. Hmm . . . why is there no American research on this important subject?

The Yeast Connection

The late William Crook, MD, a country pediatrician for over 50 years, wrote extensively about the relationship among childhood ear infections, treatment with antibiotics, and a later psychiatric diagnosis.[37] As early as 1982 he noted a child named Rusty with a history of colic and ear infections in the first year of life. By age two Rusty showed both hyperactivity and autistic symptoms.

Dr. Crook believed that, like Rusty, many children with behavioral and learning problems harbor a yeast known as *Candida albicans*. Dr. Bernard Rimland of the Autism Research Institute also suspected a relationship between *Candida* and autism in 1985.[38] Children who crave sugar from candy, soft drinks, fruit juices, and baked goods are highly likely to have yeast-based problems because yeasts feed on sugar.

Symptoms of yeast overgrowth are mood swings, headaches, muscle aches, abdominal pain, itching (especially around the genitals or anus), digestive issues, excessive gas, bloating, putrid-smelling stools, irritability, depression, fuzzy thinking, crankiness, brain fog, hyperactivity, and attention problems. In babies, thrush, recurrent and persistent diaper rash, colic, recurrent ear infections with repeated or prolonged antibiotic use, and chronic allergies, including rashes, wheezing, and coughing, are all red flags for excess yeast.

In 1995, William Shaw, PhD, of Mosaic Diagnostics, proved Drs. Crook and Rimland right. Shaw, a biochemist who had no previous experience with autism, identified very high levels of tartaric acid and arabinose, byproducts of yeast overgrowth in the gut flora of brothers whom he later learned were autistic.[39]

Arabinose and tartaric acid have drastically negative effects on human metabolism. Tartaric acid inhibits and limits energy production, causing muscle weakness throughout the body. Arabinose impedes the absorption of vitamins essential for the production of digestive enzymes.

Aware that yeast metabolism is a likely source of tartaric acid and arabinose, Shaw concluded that the brothers with autism had yeast in their intestinal tracts. He began seeing a common pattern of ear infections, antibiotics, and yeast overgrowth in children who were eventually diagnosed with autism. Read more about this relationship and immune system dysregulation in Chapter 8. Today, Dr. Shaw is one of the world's authorities on yeast-based problems.

Not surprisingly, *Candida* is also implicated in bipolar disorder and schizophrenia.[40] Given Total Load Theory guidelines, this makes total sense.

Constipation and Diarrhea

Chronic constipation accompanied by serious abdominal pain and bloating or frequent diarrhea are additional hallmarks of those with neurodevelopmental disorders. Some patients go days at a time without a bowel movement; others have five or more a day. Stools are abnormal in color, consistency, and smell, as well as frequency. Impacted bowels can lead to alternating diarrhea and constipation. What appears to be diarrhea may in fact be leakage around a hardened stool.[41] Once measures are taken to remove the impaction, bowel and brain function usually improve markedly.

An examination by a functional gastrointestinal specialist is essential when these symptoms are present. The doctor performs a physical exam, which includes listening to the abdomen with a stethoscope and lightly palpating it. A hard, bloated abdomen and discomfort upon palpation is diagnostic.[42] Some doctors will try to dismiss these symptoms as "just part of autism." Don't let them.

Clinicians, researchers, and parents have been searching for answers for gastrointestinal problems in children with autism for over two decades.[43] Two GI experts specialize in autism: Arthur Krigsman, MD, with offices in both Long Island, NY, and Austin, TX (AutismGI.com), and Timothy Buie, MD, at Boston Children's Hospital (DrTimBuie.com). While they have scoped hundreds of children with ASDs, this subject has only recently attracted the attention of psychiatrists, neurologists, and others treating those with serious "mental illness."

Virtually anyone with these symptoms will improve if their gastrointestinal issues are addressed.

Gluten and Casein: The Opioid Excess Theory

Both *gluten* and *casein* are oft-missed offenders in many chronic conditions, including eczema, asthma, childhood diabetes, constipation, diarrhea, and reflux, as well as behavioral and learning problems.[44] Gluten is a naturally occurring protein in many grains—including but not limited to wheat, barley, spelt, and triticale—and in brewer's yeast, malt, and beer. It lurks in *many* processed products made from these ingredients, including cereal, chips, soup, gravy, salad dressing, cakes, crackers, and cookies. Unless a box indicates "gluten-free," a product probably contains gluten.

Casein is a protein in milk. Cow's milk has seven times as much casein as human milk, and cows also have four stomachs to our one with which to digest all that casein! The American Academy of Pediatrics recommends against introducing cow's milk into a baby's diet until after the first birthday. Why? Because cow's milk is one of the most highly allergenic foods and not a nutrient-complete food for human babies.[45]

For more than 30 years doctors have recognized that some people have problems digesting gluten-containing foods and/or cow's milk products and tested for celiac disease and lactose intolerance. Their tests often come up negative because, although presenting with similar symptoms—bloating, gas, constipation, diarrhea, joint pain, depression, brain fog, fatigue, headaches, and more—the body's reactions to these substances can be due to several different mechanisms.

Celiac disease is a genetically linked autoimmune condition and is often lifelong. In celiac, antibodies to gluten-containing foods attack the lining of the small intestine. Reactions occur within a very short time after ingesting even a tiny bit of the offender. Genetic testing, as well as a special blood test looking for tissue *transglutaminase IgA antibodies* (tTG-IgA), is necessary to diagnose celiac disease. Lactose intolerance is a lifelong condition that results from a genetic inability to produce lactase, the enzyme that breaks down lactose, the sugar in cow's milk.

Non-celiac gluten sensitivity (NCGS) and casein sensitivities, on the other hand, the kind of reaction we see most often, are often called "allergies," but they are not traditional IgE-mediated allergies. Because of this, many doctors dismiss these sensitivities. They are not imaginary.

Sensitivity reactions to gluten and casein usually occur two or more hours after eating a food trigger and can cause additional symptoms such as sleep disturbance, bed wetting, sinus and ear infections, acne, or just plain crankiness. These reactions are identified by an IgG food sensitivity blood test and managed by an elimination or rotation diet, enzymes, desensitization, and timing. The different immune reaction mechanisms and the tests used to diagnose them are discussed in Chapter 8.[46, 47]

Many people with neurodevelopmental disorders somehow manage to survive on a self-limited diet consisting almost entirely of wheat- and dairy-based food. Cereal and milk or buttered toast for breakfast, a cheese sandwich or mac and cheese for lunch, and pizza for dinner is not unusual. Amazingly, the paradoxical combination of a physical intolerance to and an emotional preference for foods containing gluten and casein is a common occurrence among those with learning and behavioral issues. Why?

In the late 1990s Karen Seroussi, mother of a 19-month-old boy with autism, began researching treatment options. She discovered that ten years prior, biochemist Karl Reichelt and his colleagues found that 90% of a sample of children with autism had abnormally high levels of certain *opioid peptides* in their urine.[48]

Opioid peptides are proteins that mimic the effects of opioid drugs, which are famous for killing pain and producing a "high." The researchers believed these opioid peptides came from an incomplete breakdown of gluten and casein. That would certainly explain why kids seem to be addicted to these food sources despite the havoc they wreak on their bodies. When Seroussi eliminated gluten- and casein-containing foods from her son's diet, he made such dramatic improvement that by age four he lost his autism diagnosis.

In 2000, Seroussi was the first person to publicly propose dietary modification as the foremost step to take in treating a child with autism and the first to use the words *autism* and *recovery* in the same sentence. Her book, *Unraveling the Mystery of Autism and Pervasive Developmental Disorder: A Mother's Story of Research & Recovery*,[49] detailing how she returned her son to health, is still one of the first many families read that offers them real hope. I cover diets in the next chapter.

Seroussi's discovery was of great interest to her then-husband, Alan Friedman, PhD, a physical chemist at Johnson & Johnson Labs. He and his colleagues compared urine samples of neurotypical children with those of children with autism. The urine of the latter contained undigested food particles, supporting the concept of a leaky gut. Furthermore, they discovered that in children with autism, the enzyme *dipeptidyl peptidase IV* (DPP IV) that helps break down gluten and casein was either reduced, inactivated, or absent via a genomic mechanism. Interestingly, mercury is known to inhibit or block this enzyme's action.[50]

The Opioid Excess Theory

What Seroussi unearthed is called the *opioid excess theory*. Many years before she put Reichelt's work into action, he and researcher Jaak Panksepp speculated that people with autism may have elevated levels of opioids because their behavior resembled that of drug addicts.[51] This finding was repeated in those with schizophrenia.[52]

The main premise of the opioid excess theory is that incomplete digestion of gluten and casein creates small peptides instead of fully broken-down amino acids. These peptides (called *gliadorphin* and *casomorphin*) pass through the damaged intestinal membrane and enter the central nervous system, where they bind to receptors in the brain and exert an opioid-like effect.[53]

The brain has receptors for opioids as well as many other different types of peptides. The purpose of the receptors is to make neurons responsive to a variety of neurotransmitters that tell the cells what to do. *Endorphins* are the endogenous version of opioids, meaning they are the chemicals produced by the body that bind to our opioid receptors.

Like the opioid drugs that mimic them, endorphins act as painkillers. While a small amount of opioid peptides can be useful, an overload is harmful, disturbing perception, behavior, mood, emotions, brain development, and immune function.[54] These partially digested gluten or casein proteins may cause behavioral symptoms of poor eye contact, irritability, or disconnection, similar to what we're seeing in today's opiate addiction epidemic. It's interesting to consider what, if any, role these undigested proteins play in causing young people to crave opioid substances.

Glyphosate

A very interesting new hypothesis about what is bothering the guts of our kids has surfaced recently. Maybe the *primary* problem is not the gluten in wheat and other grains, but rather the herbicide glyphosate used on these and so many other crops. American wheat farmers are supposedly spraying large amounts of glyphosate on their crops prior to harvest as a drying agent to ensure a more consistent yield. A 2017 study found a remarkable 0.98 correlation coefficient between the rise in autism rates in the USA and the use of glyphosate on crops.[55]

MIT researcher Stephanie Seneff, PhD, believes glyphosate is the culprit. According to Seneff's findings, fish exposed to glyphosate develop digestive problems that are reminiscent of celiac disease. Furthermore, glyphosate has a strong ability to deplete the metals iron, cobalt, molybdenum, copper, and others as well as the amino acids tryptophan, tyrosine, methionine, and selenomethionine. All of these deficiencies are associated with gluten intolerance. Lastly, Seneff believes that glyphosate is responsible for large reductions in sulfate molecules in the blood, which are vitally important for detoxification.[56]

Seneff blames glyphosate for the fact that an estimated five percent of the population in the United States now apparently suffers from gluten intolerance. Maybe that's why some gluten-sensitive Americans I know claim they can eat gluten-containing bread without side effects in France and Italy where glyphosate use is banned or restricted.

To follow Seneff's latest research, go StephanieSeneff.net.

Innovative Therapies for Improving Digestion

Recall that eminent group of autism researchers who tackled gut issues at the conference at Arkansas Children's Hospital Research Institute in Little Rock back in 2014. A year later they published an article describing several promising therapies for improving digestion and outcomes in behavioral and emotional areas.[57]

Following are a few interventions these researchers found could alter digestion positively. Anecdotal evidence from parents I know confirms improvement in the emergence of language, lessening of undesirable behaviors, and improved social skills with these innovative therapies. None of these options should be undertaken independently. Most require working with a health coach, nurse practitioner, physician's assistant, naturopathic or medical doctor.

Digestive Enzymes

We now know that heavy metals, glyphosate, bacteria, viruses, phenols, and other invaders damage the lining of the intestines, causing leaky gut. In addition, they disrupt the production of digestive enzymes, essential to breaking down all types of foods. Along with special diets, described in the next chapter, adding enzymes as supplements can often make a huge difference for those with all types of digestive and developmental problems.

At first, enzymes might cause irritation, even if the child needs them. Watch for crankiness or temporary worsening of gastrointestinal symptoms. Enzymes are something you can try yourself, especially if, as Karen de Felice, author of *Enzymes for Autism and Other Neurological Conditions,*[58] recommends, "you start low and slow."

While many manufacturers offer a variety of digestive enzymes, several really reputable companies have jumped onto the enzyme bandwagon. Kirkman, Enzymedica, Houston, and Klaire Labs are all trusted resources. Check them out online.

Probiotics and Prebiotics

Probiotics are the "good" bugs that inhabit our microbiome and assist in digestion. *Prebiotics* are substances on which the probiotics feed, thus helping probiotics proliferate. Probiotics are sometimes combined with prebiotics.

Whereas *antibiotic* translates into *against life*, *probiotic* means *for life*. Recall that antibiotics wipe out bugs indiscriminately; thus, pre- and probiotics can help restore balance by outnumbering the bad guys.

Research on treating the microbiome has sparked tremendous interest by focusing on gut balance. Because healthy guts contain more helpful bugs than pathological ones, scientists are looking at adding pre- and probiotics directly back into the gut as an attempt at maintain that balance. Both are often taken together with digestive enzymes.

Where do pre- and probiotics come from? Real food! Some natural raw prebiotic foods are asparagus, Jerusalem artichokes, dandelion greens, jicama, onions, leeks, and garlic. Anything fermented is chock full of probiotics. That's one reason the Body Ecology Diet, described in the next chapter, is so helpful; it includes making your own fermented foods, like kefir, vegetables, and coconuts. Commercial and homemade kombucha are also good choices.

Yogurt naturally contains a few probiotic strains in relatively low numbers. Make your own, instead of buying commercial brands, which often contain tons of sugar and nasty additives like carrageenan and guar gum, to which some tummies react.

A medicinal yogurt containing over 300 strains of good bacteria is Bravo Probiotic Yogurt invented by Marco Ruggiero, MD, PhD. Bravo is prepared at home from a starter kit containing a special formulation of microbes and colostrum, which can be mixed with whole milk or juice (organic, not ultra-pasteurized, of course). Fermentation for about 24 hours produces about 150 peptides.

While Bravo contains a little casein, most of the casein is metabolized during the fermentation process. If you or your child reacts to the casein in Bravo, instead of drinking the yogurt, you can deliver it through an enema, a capsule, a cream, or as a suppository, none of which need refrigeration. The rectal or topical route bypasses the gut, allowing all the healthy molecules in the Bravo to be absorbed through the rectal mucosa. Order Bravo products at Bravo-Probiotic-Yogurt.com.

Choosing a commercial probiotic. Choosing the correct probiotic formulation is like trying to find a missing piece in a 1,000-piece jigsaw puzzle. Remember, diversity of bacteria is our goal. Recent studies are focusing on specific strains that may help balance individual differences. Choosing which bugs are important missing links for each unique individual is challenging. This is where microbiome testing described in Chapter 5 can be extremely helpful.

With and without testing, combinations of various strains of *lactobacillus* and *bifidus* are proving to be the most promising. These strains are available in local drug chains as well as in apothecaries and health food stores. Consumer beware! Low quality products usually have low potency. Look for ones with over one *billion* organisms. Millions of organisms are insufficient to repopulate the guts of most individuals. Doctors see the most success with high quality, pharmaceutical-grade products, which often require refrigeration.

The new kid on the block is Flourish, an all-natural liquid probiotic fermented with molasses that keeps on multiplying in the container, even after bottling. Check it out at EntegroHealth.com.

Probiotics for mental health is an emerging field for which more research is needed. However, promising results were obtained in a small Harvard study comparing patients in the manic phase of bipolar disorder to controls. Rates of rehospitalization were cut in half in the group who took probiotics (24.2% vs 51.1%), and admissions were 74% lower in the probiotic group. Best news: an almost 90% reduction of hospitalization in the group with the highest inflammation score who took probiotics.[59]

Herbs

Plants have been used for centuries as medicine, and many work as well or better than pharmaceuticals. In fact, many prescription drugs are synthetic versions of chemicals found in wild plants long known to healers in indigenous populations all over the world.

Some herbs use the body's natural healing skills to maintain balance and reduce stressors. One of the advantages of using herbs as opposed to pharmaceuticals is that the herbs, as whole foods, often contain *other* ingredients that help the body use the active ingredient without negative side effects. In fact, Chinese herbalists often deliberately put a number of herbs together because the combination is *safer* than the individual herbs. That is the exact *opposite* of the way pharmaceuticals work.

A few that are known to heal the gut are licorice root, ginseng, ashwagandha, aloe vera, slippery elm, marshmallow, and holy basil. Consult a book or website on herbal remedies to see if any of these might work for you.

Bottom line: fix the gut so the body doesn't have to work so hard to digest or to mitigate the effects of poor digestion, which will free up the energy so important for language and social/emotional and learning skills.

Gut Balancing Program

Two dedicated fathers of sons with autism have teamed up and developed a program that is producing amazing results for anyone who wants to improve gut function. Christian Bogner, MD, a functional medicine doctor in Michigan, and Alex Zaharakis, a medical physicist in Arizona, have put together an intensive gut balancing program of at least six weeks that anyone can do with them. It is based on a Biosight genomic analysis, available through them at a discount. The analysis and custom recommendations are free of charge.

This team then recommends personalized dietary and lifestyle modifications, combined with targeted nutritional supplementation. Their approach leverages a range of all-natural powders and/or capsules meticulously formulated under their brand, Researched Elements (ResearchedElements.com).

One of these powders, Gut Guardian™, is crafted with a blend of 52 distinct plants, herbs, and compounds, including prebiotic fibers, codonopsis, and brahmi leaves, all of which have undergone rigorous clinical studies to ensure their effectiveness, potency, and collective synergistic power. With filmmaker Ryan Hetrick, Bogner and Zaharakis have produced a full-length documentary, *Restoring Balance: Autism Redefined,* about their program, with testimonials from many families that you can watch on YouTube.

Finally, these dedicated fathers have founded a 501(c)(3) not-for-profit, Autism Is Biomedical Inc., for which I am proud to serve on the Advisory Board. This entity will allow them to provide services to needy families at no cost and conduct needed research on programs for those with ASD and other biomedical conditions. See AutismIsBiomedical.com.

Ion Gut Health

In 2012 Zach Bush, MD, a brilliant physician with triple board certifications in internal medicine, endocrinology, and hospice care, discovered a family of carbon-based molecules made by bacteria. He packaged a product (originally called Restore), which has morphed into Ion Gut Health.

The active ingredient is *Terrahydrite*™, which contains fulvic and humic acids, which recent research shows can reduce glyphosate.[60] This miracle product promotes strong membrane integrity by tightening up the junctions of the bowel wall and vascular systems of the body, improving the body's ability to hydrate, and thus positively affecting the gut, blood vessels, and other organs, including the blood-brain barrier. It also restores bowel ecology with its unique bacterial content of more than 20,000 species. The minerals present in Ion Gut Health are well below dietary levels, reducing the possibility of any toxicity. Purchase it from IntelligenceofNature.com

Fecal Microbiota Transplant

Fecal microbiota transplant (FMT), sometimes called "crapsules," can improve behavior and language production in those with autism. FMTs have been life-saving for those with *Clostridium difficile (C. diff)*,[61] a stubborn bacterium that has become antibiotic-resistant. Doctors process the donor stool (usually from a relative) by extracting and cleaning the bacteria, then pack the bacteria into triple-coated gel capsules that won't dissolve until they reach the intestines. No stool is left—just bugs. Stool can also be freeze-dried, which doesn't kill the bacteria, so this "medicine" can be stored and shipped to a needy patient.

Remember Dr. James Adams? A member of the Arkansas think tank in 2014, he has been studying a more complex *Microbiotic Transfer Therapy* (MTT) for patients with autism. People are clamoring to get into his study.

MTT is an intensive procedure involving a two-week antibiotic treatment, a bowel cleanse, and then an extended fecal microbiota transplant (FMT) using a high initial dose followed by lower daily maintenance doses for seven to eight weeks. This protocol produced improvements in gut and behavioral symptoms that persisted for at least eight weeks after treatment ended, suggesting a long-term impact.[62] Using this method, Adams has become a hero in the autism community.

Adams believes that p-cresol sulfate, the nasty metabolite he discovered in 90% of his autistic subjects, also caused autism symptoms in his laboratory mice. When treated with FMT, their autism symptoms vanished![63] FMT is only routinely used in hospital settings for *C. diff* infections or under a research study. So, if you'd like to try this with your child, you would need to consult a knowledgeable practitioner.

Helminths

Parasitic worms called helminths inhabit the intestines of mammals. These critters used to live in almost everyone. Though they have been largely eliminated in industrialized countries, they still inhabit the guts of most residing in countries without clean water and with contaminated food. Because those populations experience far less inflammatory illness, including allergies and autoimmune disease, some believe these helminths actually serve a positive purpose. Some doctors now implant them to support the ecosystem of the human body.

While this crazy-sounding intervention doesn't work for everyone, helminthic therapy, during which benign species of these worms are purposefully introduced into the intestines, has shown some potential in autism[64] and significant improvement in neuropsychiatric issues, including attention, receptive language, learning and focus, mood, and OCD behaviors.[65] A company in the UK markets these helminths. See BiomeRestoration.com

Obviously, this therapy is extremely controversial. For those who want to know more, read join the Helminthic Therapy Support Facebook group, and proceed with caution!

Take-Home Points

After making basic lifestyle changes, addressing digestive dysfunction is next. Circular thinking might conclude that individuals have gastroenterological problems because they have a neuropsychiatric condition, when just the opposite is true: They have behavioral and other symptoms because their digestion and other bodily systems are dysfunctional. Misunderstanding the relationship between digestion and behavior can take patients down diagnostic blind alleys and hinder proper treatment.

Improving digestion starts with taking away the bad stuff and putting back the good stuff, which often includes supplements like high-quality digestive enzymes, prebiotics, and probiotics. Most importantly, it requires the sometimes-overwhelming step of dietary modification. The next chapter will help you use food as medicine. That's the approach taken by Hippocrates over 2000 years ago, and it is still the best choice.

CHAPTER 8

DIETARY DOS AND DON'TS

Nothing is more fundamental to behavior than having a happy tummy from eating nutritious food.

This chapter begins with what comprises a healthy diet and then moves to an overview of the dietary modification process. The next sections look at what might be bothering someone's body and brain and steps to take for elimination of those factors. Chapters 9–11 cover supplementing missing nutrients, such as sources of calcium in dairy-free diets.

Then I outline the components of an eating plan that contains *nothing* that could possibly disturb even the most sensitive gut, and a sensible plan that eliminates gluten-, glyphosate- and dairy-laden foods. Next are descriptions of the many special diets that can help those with neurodevelopmental disorders digest more efficiently.

The diets are listed in order from the least restrictive to the most limiting. Finally, we look at picky eating, one of the banes of staying well-nourished. Obviously, our ultimate goal is optimal nutrition for optimal functioning.

How can you tell if someone is well-nourished? Nutritionist Kelly Dorfman makes these comparisons:[1]

Well-nourished bodies	Poorly nourished bodies
Have good coloring	Have a pasty complexion/yellow or grey pallor
Are well most of the time	Have three or more illnesses a year
Are interested in a variety of foods	Have restricted eating habits
Have fairly consistent responses to therapy	Are erratic and unpredictable
Have clear eyes	Have dull eyes
Have good breath	Have sour or bad breath

Bottom line: anyone who *looks* unhealthy probably *is* unhealthy.

Remember the Autism Research Institute (ARI) tracking of parent reports of various interventions for over 40 years, mentioned in Chapter 1? Over 50%

or more than 27,000 parents reported that removing targeted foods elicited significantly more positive behavioral changes than medications, and without any worsening of symptoms.[2]

Cleaning Up the Existing Diet

The first step is deciding what to include and what to eliminate. In her brilliant book *Cure Your Child with Food*, "nutrition detective" Kelly Dorfman states that behavioral problems related to food follow what she calls the Binary Law of Nutrition: either something is in the diet that is bothering the body, or something is missing that the body needs.[3]

Given that many diets are already so self-limited, the fear is that removing yet more foods will result in starvation. No worries. I promise I've never seen it happen!

Put in the Good Stuff

In order to improve the diet, first evaluate your definition of food. What is "food"? Do goldfish crackers, Twinkies, and organic tofu hot dogs qualify as food?

Michael Pollan, food guru and author of dozens of books, synthesizes what a good diet entails into seven simple words: "Eat food . . . not too much . . . mostly plants." He adds, "If your grandmother didn't eat it, you and your kids probably shouldn't eat it either."[4] Unfortunately, it's not as simple as it sounds, especially when attempting to feed a picky eater.

Eat real food. Strive for menus based on a balanced variety of unrefined, unprocessed food. Most unprocessed foods are likely to be safe and are much less likely to cause problems than processed ones with many ingredients, including chemical additives.

Buy and eat mostly organic, non-GMO foods in season. Minimize eatables that come in jars, boxes, bags, or cans. Read labels. Real food requires no labels.

Vary high-quality proteins, carbohydrates, and fats. Eat plants, which are complex carbohydrates. Eliminate trans fats and include good fat sources like cold-water fish, nuts, seeds, oils, and vegetables that supply essential omega-3 fatty acids. Rotate different grains, meats, fruits, and vegetables.

Expensive, you say? Not if you eat *real* food: protein from beans and animal sources, vegetables, and grains. Check out Aldi's, Costco, and Target, which all sell reasonably priced organic produce. If you must stray, stick with the Environmental Working Group's "Clean 15": asparagus, avocados, cabbage, carrots, onions, honeydew, kiwis, mangos, mushrooms, papaya, and watermelon.[5] (Be careful, however; most corn and papaya are now genetically modified.) Hard to do? Maybe a little at first.

Take out the Bad Stuff

Begin by replacing what you typically eat with a better-quality version of the same product. I know . . . the family is label sensitive. Try putting the new product in the old familiar box or bag. Some will be fooled; others won't.

Avoid the "Dirty Dozen," listed in chapter 4.

Say "No" to sugar. Avoid replacing a nutritionally bereft junk food diet with "organic," sugar-laden junk food. Today, almost every processed product from peanut butter to ketchup, soups, and even French fries are sweetened. These sweeteners can be disguised by more than 100 different names. Look for products with natural sweeteners like monkfruit, xylitol, and stevia.

Fruit juice, often considered "healthy," is one of the most common and over-looked sources of sugar. While juice is sweet naturally, too much natural sugar isn't good either. Dilute all fruit juices and limit to one cup a day.

No colors, "flavors," or preservatives. Since the 1960s, artificial flavors, preservatives, excitotoxins (like MSG), and sugar have crept into foods to enhance their taste, smell, and/or shelf life. Additives are poisonous and addictive!

Food is colored to be more appealing to the eye. Colors hide mostly in junk foods: potato chips, soft drinks, baked goods, cereals, ice cream, candy, and even shampoo. Studies have shown that artificial colors all pose health risks, ranging from mild hyperactivity and allergies to ADHD and cancer.[6]

Most processed foods contain unnamed "natural and artificial flavors." Did you know that *vanilla* is natural and comes from a class of orchids, but *vanillin* is artificial, and derived from phenol? "Vanilla flavoring" can include either one or both!

Ditch the GMOs. Processed foods (those made in a plant, not grown on a plant) and genetically modified (GM) foods (those whose genes have been artificially manipulated to withstand bugs or herbicides) have crept into our refrigerator and onto our tables. Processed and GM food is not real food.

Remember why genetically modified foods were developed from Chapter 2? Many were to enable spraying crops with large amounts of the herbicide glyphosate. Take the time to go back and reread this important message. Glyphosate wreaks havoc with the gut. Read about the relationship between glyphosate in GMO foods and gut health in *What's Making Our Children Sick?*[7]

Recall the story of Katie Reid, also in Chapter 2? Her daughter showed behavioral changes from eating foods containing excitotoxins like MSG and aspartame. Processed foods containing these additives can contribute to behavioral or learning problems.

Remove Targeted Offending Foods

One of the pioneers in recognizing some foods as causing problematic behavior is the late Doris Rapp, MD, author of the now classic book *Is This Your Child?*[8] For at least 50 years Rapp was an active advocate for dietary modification as a first-line therapy for improving learning and behavioral outcomes. She developed clear guidelines for implementing weeklong single food elimination diets for the most common offenders: dairy products and casein-containing foods, gluten- and wheat-containing foods, eggs, and sugar. Not convinced? Watch some of Rapp's videos on YouTube, showing angelic children turn into devils after ingesting a food to which they react.

Doing a single-food elimination diet before jumping into a more restrictive plan (which may be unnecessary) may be wise, because the impact of food on behavior can be profound. Foods that bother digestion are often the very same ones people crave, namely wheat and dairy. Remarkable positive changes in behaviors, such as decreased anxiety, increased language, and improved relatedness are often seen after removing offensive foods.

Researchers are beginning to understand that individuals differ markedly in their reactions to foods and that other factors, like genomics and timing of environmental exposures, must be taken into consideration. Applying Total Load, infrequent small amounts of a certain food may not be a problem, but if the food is eaten with other reactive foods, it can cause symptoms. In order to benefit fully from eliminating potentially reactive foods, eat them by themselves.

Food Allergy Reactions

Before adopting an elimination diet, many people choose to consult an allergist to determine whether dietary restrictions are absolutely necessary. Surprise! Extensive scratch testing often reveals no food allergies. Some choose to eliminate gluten and casein anyway and find they get improved attention, sleep, and/or behavior.

How is this possible if the allergy test was negative? The answer lies in understanding the different types of allergic reactions described in Chapter 5. Please reread that section carefully; it's important.

Comprehensive Special Diets

An allergen-free nutritious diet is foundational to health for everyone. For those with mental health issues, it can be life-changing—and hard. Daily cooking is mandatory because eating out or relying on pre-prepared foods at home make complying with any diet extremely challenging. When picky eaters move to a special diet, they are more likely to choose processed comfort foods over kale and broccoli, at least at first.

Want guidance? Contact Julie Matthews, a seasoned internationally respected Certified Nutrition Consultant. Her website NourishingHope.com offers considerable information on every diet included in this section as well as books, videos, workshops, and individual guidance for families and practitioners.

All members of the immediate family, especially in cases of split households, including grandparents, stepparents, aunts, uncles, cousins, and anyone else who spends any time with a child on a special diet, must be "on board" for it to work. If just one person feels sorry for the relative who cannot eat ice cream and cake at a birthday party, that individual can undermine the program.

All members of the school family, including teachers and their assistants, therapists, and shadows must also comply, including using acceptable substitutes for positive food reinforcers. Out with the gummy worms and Cheerios, and in with dairy-free organic chocolate chips. Fortunately, once everyone sees the results, compliance usually gets easier.

The Autism Community in Action (TACA) has put together a beautiful explanation about the importance of special diets. Founded as Talk About Curing Autism in 1999, by dynamo autism mom Lisa Ackerman, this organization remains a trusted resource for all things autism. According to their website, TACANow.org, special diets can

- Improve overall health,
- Expand food choices in those with limited diets,
- Alleviate sleep issues,
- Decrease temper tantrums and other challenging behaviors, including self-injury, and
- Increase safety awareness.[9]

Here are almost a dozen special diets to consider. They are listed in approximate order from least to most restrictive. This chart can guide you on which to consider for various conditions.

CONDITIONS

	ADHD	ASD	OCD	ANXIETY	BIPOLAR	SLD	SPD	NVLD
ALLERGY ELIMINATION	X	X	X	X	X	X	X	X
GF-CF	X	X	X	X	X	X	X	X
FEINGOLD	X	X	X	X	X	X	X	X
YEAST-FREE	X	X	X	X	X	X	X	X
SCD	X	X	X	X	X	X	X	X
GAPS	X	X	X	X	X	X	X	X
KETOGENIC	X	X	X	X	X	X	X	X
BODY ECOLOGY	X	X	X	X	X	X	X	X
LOW OXALATE	X	X	X	X	X	X	X	X
LOW GLUTAMATE	X	X	X	X	X	X	X	X
LOW FODMAPS	X	X	X	X	X	X	X	X

Adapted from *The Rainbow Blueprint: A Clinical Guide to Integrative Medicine and Nutrition for Mental Well-Being*, p.120–121, Leslie Korn

Allergy Elimination Diet

A simple food elimination diet is the best way of "taking out the bad stuff." Refer back to how Dr. Doris Rapp recommended this simple measure with great success, before high-tech laboratory testing was available. For some lucky people, eliminating a targeted food or two does the job, but for most, implementing a more comprehensive, stringent plan is necessary, especially if allergy testing shows intolerances to more than one or two foods.

A subset of those for whom elimination diets are indicated is folks with a history of ear infections. A now-classic study showed the elimination of chronic ear infections and fluid in 86% of children when milk, wheat, corn, eggs, soy, and chocolate were removed.[10]

Gluten-Free, Casein-Free Diet

The basic special diet many start with is the same one Karen Seroussi discovered: the *gluten-free, casein-free* (GF/CF) diet. This diet decreases gut inflammation and the emotional and behavioral effects of opiates as well as leaky gut syndrome.

Literally hundreds of studies since Seroussi's discovery find that removal of gluten and casein from the diet can make a marked and often immediate difference not just for autism,[11] but for many mental health conditions, including anxiety and depression, as well.[12] One classic study shows that 91% of people with ASD showed improved behavior on a strict gluten/casein-free diet that also eliminated soy.[13] Clinical evidence suggests that a cross sensitivity to soy may exist in up to 50% of those who react to casein in dairy products because the chemical structure of the soy protein is similar to casein. It is thus inadvisable to substitute soy milk for cow's milk during elimination trials.

Removal of all soy products is highly recommended, so it becomes a GF/CF/SF diet. For an extensive treatise on the controversies concerning the use of any soy products in human diets, read *The Whole Soy Story* by Kayla Daniels.[14]

Why do so many individuals respond so very favorably to the removal of dairy from their diets? It may be because of a condition called cerebral folate deficiency (CFD), a byproduct of impaired methylation due to MTHFR variants. Dairy foods block folate receptors. Removing it allows increased transport of active folates into the central nervous system, and removing gluten-containing grains reduces exposure to synthetic folic acid.[15] Read more about CFD in the next chapter.

Most clinicians recommend a trial period of three months to chart behaviors and symptoms. This amount of time allows symptoms to improve gradually as the opioid peptides leave the body.

The best candidates for the GF/CF diet are those who

- already eat a limited diet of mostly gluten- and casein-containing foods;
- have a history of ear infections or colic;
- get sick easily or have chronically loose stools, and
- are difficult to engage.

Anecdotally, clinicians I know report that approximately one-third of patients in their practices improve dramatically when gluten and casein are eliminated. Another third show improvement in secondary symptoms such as poor sleep. An unanticipated result is that for some, seizure frequency diminishes with a GF/CF diet.

While full compliance is optimal, 100% may not be necessary for improvement. Because gluten and casein masquerade under many names, avoiding every single food containing gluten or casein could be difficult at first. In most cases avoiding the major offenders will be adequate as a start. Stricter adherence comes with practice and after seeing small gains. Make sure that menus include naturally gluten- and dairy-free products, such as vegetables and fruits, and avoid the trap of switching from empty-calorie wheat and dairy products to empty-calorie GF/CF foods.

Healthcare providers now recognize that removing foods must be accompanied by nutritional supplementation to address the inevitable vitamin and mineral deficiencies. Those implementing a gluten-free, casein-free diet always recommend supplementing it with potentially deficient nutrients, especially calcium and vitamin D.[16]

Help going GF/CF is available from many online sources that detail gluten- and dairy-free cooking, along with thousands of recipes. Cookbooks abound. The 25th anniversary edition of Lisa Lewis's *Special Diets for Special Kids* has recipes that meet kid standards—such as GF/CF macaroni and cheese, pizza dough, and chicken nugget coating—as well as adult-friendly healthy casseroles and imaginative breakfasts.[17] Pete Evans's *Healthy Food for Healthy Kids: 120 Simple, Nourishing, Gluten- and Dairy-Free Recipes Your Whole Family Will Love,*[18] published by Children's Health Defense, is the new kid on the block.

Want a vacation from cooking? Most restaurants now offer gluten-free menus or at least a few GF choices. More and more companies sell products and mixes that make GF/CF cooking a breeze. Consider carefully chosen pre-prepared gluten-free foods, always watching for too much sugar and too little nutrition.

The demand for gluten-free is growing at a rapid pace. The global gluten-free products market was valued at $6.45 billion in 2022 and is expected to grow at a rate of 9.8% from 2023 to 2030 according to a 2022 report by Grand View Research, Inc.[19]

Feingold Diet

An early supporter of the food and behavior movement was Ben F. Feingold, MD, Chief of Allergy at the Kaiser-Permanente Medical Center in San Francisco in the mid-1960s. After removing some foods and additives in his adult patients, he began implementing the protocol he dubbed the *K-P diet* for kids. Guess what? Their hyperactivity and impulsivity lessened, while attention and focus heightened. This relationship has been proven scientifically over and over again.[20]

Feingold's K-P diet has morphed into the *Feingold diet*, or simply the *Feingold program*. It eliminates artificial colors and flavors, synthetic sweeteners, dyes, and three preservatives (BHA, BHT, and TBHQ). Though most dyes and preservatives come from petroleum, "crude oil" is obviously not listed among a processed food's ingredients.

Another potentially problematic food ingredient is *salicylates*, which occur naturally in some fruits and vegetables and can cause irritability and hyperactivity. Salicylates are in apples, grapes, berries, and oranges. Apple and grape juice are both high in sugar as well, giving susceptible patients a double dose of potential poisons. Tomatoes, peppers, and cucumbers are also high in salicylates. And *acetyl-salicylic acid* is the main ingredient in aspirin, making it illegal on the Feingold diet.

The liver breaks down salicylates using the same detoxification pathways that target drugs. Research supports implementation of the Feingold diet for those with impaired detoxification capacities.[21]

Feingold's legacy is the Feingold Association of the United States (FAUS). For well under $100 per year, members receive tremendous support, including menu planning, guides to buying supplements and eating out, and an extremely informative newsletter. FAUS meticulously researches thousands of products and reports when ingredients change.

Jane Hersey, a veteran Feingold parent and volunteer, has written a wonderful book, *Why Can't My Child Behave?*,[22] about the advantages of the Feingold program. The Feingold diet, along with sugar reduction, is one of the first steps for treating ADHD without drugs.[23] Learn more at Feingold.org.

Yeast-Free Diet

Healthcare practitioners often recommend also going "yeast-free" when on the GF/CF and Feingold diets. The name is actually a misnomer, as the idea is to not just eliminate yeasts but to eliminate foods that *feed* yeast as well. This significantly more restrictive diet rules out many otherwise acceptable additive-, allergy-, gluten-, and dairy-free baked goods and products. Many guidebooks are available for support.

A yeast-free diet eliminates almost all baked goods—such as breads, crackers, pastries, pretzels, cakes, cookies, and rolls—because most contain copious amounts of simple carbs. Sugar and high glycemic carbohydrates like potatoes and rice, which turn into sugar, feed yeast. Yeasts ferment sugars into alcohol, resulting in drunk, hungover, unfocused, or hyperactive behavior.[24] Stevia and xylitol are two natural sugars that do not increase yeast.

This diet also eliminates fermented foods such as pickles, foods that are naturally moldy like cantaloupe and peanuts, and natural fungi like mushrooms. Almost 60% of parents responding to the aforementioned ARI survey reported that their kids' behavior improved on a yeast-free diet.

Recall the findings of Dr. William Shaw, who discovered those brothers with extremely high levels of toxic yeast byproducts in their urine.[25] Restricting yeast minimizes these secondary irritants. Because of Shaw's groundbreaking work, the yeast-free diet has become a staple of biomedical intervention.

Another important component of the yeast-free diet is supplementation with probiotics. If you missed the discussion on this digestive therapy in the last chapter, go back and learn how these good bacteria are used to replenish gut microflora by crowding out the yeast. In many cases, identifying and removing irritants, along with sugar, and adding probiotics is sufficient to rebalance the gut. However, in severe cases, substances to kill candida and other yeasts are also required.

Natural yeast killers, such as oregano oil, garlic, olive leaf extract, caprylic acid, uva ursi, and various herb combinations, might work. Stubborn cases of candida require those available only by prescription, such as nystatin, fluconazole (Diflucan), and ketoconazole (Nizoral). Antifungal therapy is least effective with individuals who have normal or marginally elevated yeast.

Even though antifungals remove the yeast from the blood, they may transfer it to the tissues, causing more intensified yeast overgrowth which can demand further intervention sometime in the future. *Hyperbaric oxygen treatment* (HBOT) with pressurized oxygen in hard chambers can be an effective way of treating systemic yeast in the tissues. See more about HBOT in Chapter 14.

Specific Carbohydrate Diet (SCD)

The SCD takes GF/CF and yeast-free a step further. The goal is to stop the vicious cycle of malabsorption and microbe overgrowth by eliminating both the microbes and those foods upon which the microorganisms feed.

Like so many other treatments and diets, the Specific Carbohydrate Diet was developed by a mother, the late biochemist and cell biologist Elaine Gottschall, to heal her child, in this case, of ulcerative colitis. Gottschall's pioneering work is detailed in *Breaking the Vicious Cycle: Intestinal Health Through Diet.*[26]

The discovery of the SCD's potential for helping those with autism came many years later when yet another mother, Judy Chinitz, searched for a way to help her son, Alex, almost 20 years ago. Judy shares her remarkable story of the diet that "saved our lives" in the book she collaborated on with her son's physician, Sidney Baker, MD: *We Band of Mothers: Autism, My Son, and the Specific Carbohydrate Diet.*[27]

The SCD starts with an introductory plan for about a week or two, consisting of proteins and a limited selection of specific carbohydrates. Complex carbohydrates, cereal grains, processed meats, soy products, cow's milk, sugars, and canned fruits and vegetables are prohibited because they sit in the intestines and provide a breeding ground for yeast and bacteria.

SCD includes a specially fermented homemade goat milk yogurt as a source of good bacteria; culturing changes the structure of the casein and renders it harmless. This yogurt is well tolerated if you start out with a tiny amount and gradually increase it.

An incredible resource is *The SCD for Autism and ADHD: A Reference and Dairy-free Cookbook for the Specific Carbohydrate Diet.*[28] Other informative sources to understand thoroughly the progression of allowed foods and to implement the SCD with many delicious recipes include BreakingTheViciousCycle.info, PecanBread.com, and SCDRecipe.com.

Gut and Psychology Syndrome (GAPS) Diet

The GAPS diet was derived from the SCD by physician Natasha Campbell-McBride, the parent of a child with learning disabilities. Campbell-McBride believes in an absolute link between learning disabilities, mental health, physical health, our food and drink, and the condition of our digestive system.

The GAPS diet is designed to reduce inflammation, support the gut lining, and restore microbial diversity by focusing on removing foods that are potentially damaging to gut flora and replacing them with nutrient-dense foods, thus allowing the gut to heal. This diet is appropriate for anyone experiencing chronic digestive symptoms.

GAPS proceeds in six stages. A quick start introduces *bone broths,* partnered with *probiotic foods*, all homemade. Stages 2–6 gradually add eggs, additional fermented foods, homemade yogurt and kefir, and ghee, a special healthy fat made from clarifying butter. Then comes avocado, almond flour pancakes, roasted meats, fruits and vegetables, and their juices. Finally, some baked goods are included.

The basics of this diet premiered in the original 2010 book, *Gut and Psychology Syndrome Diet,*[29] which focused mostly on mental health. It has recently been updated and expanded. Translated into two dozen languages, *Gut and Physiology Syndrome: Natural Treatment for Allergies, Autoimmune Illness, Arthritis, Gut Problems, Fatigue, Hormonal Problems, Neurological Disease and More*[30] now includes a chapter on epilepsy and an expanded section on fats and covers the gamut of conditions in this book.[31]

Dr. Campbell-McBride is *the* world's leading expert on healing the gut. She has trained practitioners all over the world. Her knowledge is boundless, and her energy matches.

Bottom line for me: while the GAPS diet is extremely sensible, it is also very restrictive and time consuming. My experience is that this very healing diet feels foreign to some Americans. Campbell-McBride uses a good number of British products, language, and measurements, which do not translate well, despite her efforts to speak a universal language. Campbell-McBride has a huge following and works closely with the Weston A. Price Foundation (WestonAPrice.org) to teach people how to make these foundational, healing foods.

To learn more, check out the official GAPS website at GAPS.me. Excellent information is also available at GAPSdiet.com.

Body Ecology Diet

An alternative to GAPS is the *Body Ecology Diet* (BED), developed by nutrition consultant Donna Gates more than 25 years ago. It too is appropriate for anyone with chronic digestive issues and is compatible with both the GF/CF and SCD programs. Like GAPS, the BED is not condition specific because it fights yeast and heals the gut by removing the most problematic edibles and nourishes it with nutritious foods to allow organ systems to rebuild.

BED pushes organic dark, leafy green and root vegetables, animal proteins, four gluten-free grains (quinoa, millet, amaranth, and buckwheat), unrefined oils and fats, and selected fruits. It also includes many signature fermented foods, made from young, white coconuts. Donna adds raw butter, that is very high fat, and whey protein, neither of which contains significant casein. Like GAPS, BED borrows another food staple from the Weston Price Foundation: bone broths. These all aid in digestion, detoxification, and healing. (Be aware that bone broth and fermented foods are very high in histamine, which can be an issue for many.)

The BED follows seven unique principles that allows for individuality as well as for always-changing nutritional needs. Adjustments are recommended for different seasons and blood types[32] and to balance expansive (sweet) with contracting (salty) foods and acid with alkaline. BED recognizes that keeping blood in a slightly alkaline condition is vital to restoring health. Too much acid encourages yeast, viruses, parasites, and other unhealthy critters to thrive. A good reference book for this principle is *Alkalize or Die*, by Theodore Boroody.[33]

Another principle is 80/20. Every meal should contain 80% alkaline-forming foods and 20% acid-forming foods. 80% of food should be land or ocean vegetables. The remaining 20% is animal protein (fish or poultry), an acceptable grain, and starchy vegetables. Eat until your stomach is 80% full, leaving 20% available for digestion. Eating to 100% full slows digestion and furthers yeast overgrowth, leaving little room for the digestive juices to do their job.

A unique BED principle is about combining foods. Because different nutrients require different enzymes to be properly digested, BED recommends always eating

- protein with non-starchy or ocean vegetables;
- allowable grains and starchy vegetables with non-starchy or sea vegetables; and
- fruits alone, on an empty stomach.

BED focuses on continual cleansing to remove the toxins of modern-day living, and is unique in recommending regular colon hydrotherapy, restful sleep, and other lifestyle tweaks.

The BED can be a very different way of eating for many people, especially those who subsist on a standard American diet replete with wheat and sugar. However, not only can it help with depression, anxiety, and other emotion-based conditions, the whole family is guaranteed to become healthier. See BodyEcology.com.

Ketogenic Diet

First introduced in 1921, the *ketogenic diet* has been used widely to control seizures since the 1970s; it is now the new darling for all things behavioral and mental health. Sometimes shortened to simply *keto*, this diet is rich in high-quality fats, moderate in protein, and very low in carbs.

Most healthcare professionals understand that fat does not make you fat, and that ingesting fat is essential to a healthy nervous system and brain, which are 60–70 percent fat. Nerves are covered with a sheath called *myelin* that is made from fat—and not just any fat will do. Myelination must come from ingested omega-3 fats, inherent in fish, nuts, seeds, and a few fruits and vegetables. Grandma was right to push cod liver oil! When nerves demyelinate, they resemble frayed electrical wires, and they spark, causing seizures. That's why building a richer myelin sheath stops epilepsy.

The ketogenic diet restricts most carbohydrates, including grains, fruits, sugar, honey, starches, and beans. It uses copious amounts of coconut products (including oil, crème, and sugar) as well as other healthy oils for everything from salad dressings and cooking to making "fat bombs," a delicious dessert treat, containing only coconut oil, sweetener, and flavoring, like strawberries, chocolate, or peppermint—all organic, of course! Keto fat bomb recipes abound online.

Like the SCD, keto severely restricts carbs to "break the vicious cycle" of producing sugars that feed yeasts and other nasty bacteria. By reducing carbs even farther than SCD, the body is induced into a state known as *ketosis*, during which the liver breaks down fats into ketones. Because the body has few carbs to burn for energy, its main fuel becomes these ketones.

I first encountered keto in my very first job out of graduate school in the late 1960s working for a renegade neurologist in Boston who prescribed it successfully as a first-line defense against seizures. At that time, our only sources of fat were butter and whipped cream! We watched in amazement as kids with uncontrollable epilepsy stopped seizing when we piled mounds of these dairy products on their food!

Keto showed up in my world again around 2013, when innovative autism docs Martha Herbert, MD, and Julie Ann Buckley, MD, customized a gluten-free casein-free ketogenic diet for a 12-year-old girl with autism who developed seizures during puberty. The girl started out severely autistic but lost her diagnosis and had a documented rise in tested IQ of 70 points over 14 months on keto. Their primary source of fat was *medium-chain triglycerides* (MCT) from coconuts.[34]

In the past five years, integrative psychiatrists and neurologists have been touting keto as beneficial for patients with everything from depression and anxiety to mood and bipolar disorders, schizophrenia, and even cancer, for whom available pharmaceuticals were ineffective or inappropriate. Because it has such wide-ranging effects, they call it keto *therapy*, not diet. I will follow their lead from here forward. This revolution has spawned the new discipline of *metabolic psychiatry*, introduced in the first chapter.

Metabolic psychiatry is at the intersection of nutrition, metabolism, biology, and mental health. In 2015 doctors at Stanford University established a metabolic psychiatry clinic after a four-month clinical trial of ketogenic therapy for serious mental illness showed remission in 80% of those who adhered to it.

Stanford doctors found that over 40% of individuals with severe mental illnesses have metabolic energy deficits in the brain, even before a diagnosis of mental illness is made. The website MetabolicPsychiatry.com has links to more than 50 articles demonstrating the benefits of ketogenic therapy in ADHD, addiction, anxiety, bipolar disorder, depression, eating disorders, obesity, schizophrenia, and more.

Harvard psychiatrist, Christopher Palmer, MD, lucked into keto in 2016 when he was helping a longtime patient with *schizoaffective disorder* (a cross between schizophrenia and bipolar disorder) lose weight. Not only did the man drop the 150 pounds he had gained on psychiatric medications, but long-standing hallucinations, delusions, and mental anguish disappeared! Palmer is now sold on this therapy as essential for his patients. Read about this newfound approach to psychiatric conditions in his 2022 book, *Brain Energy: A Revolutionary Breakthrough in Understanding Mental Health—and Improving Treatment for Anxiety, Depression, OCD, PTSD, and More.*[35]

In 2017, Matt Baszucki, age 21, son of Silicon Valley philanthropists, became manic. In fewer than five years, he was treated by 41 clinicians and prescribed 29 meds before his psychiatrist pronounced his illness "treatment resistant." That's when his desperate parents discovered Dr. Palmer and ketogenic therapy. Within four months, Matt's mood stabilized. He later graduated from college and now works full-time in tech, producing electronic music.

In 2022, the Baszuckis held the first ever metabolic psychiatry conference. Soon thereafter, they founded Metabolic Mind to fund neuropsychiatric research. This is an amazing resource for anyone interested in metabolic health. They have a YouTube channel, newsletter, dozens of testimonials, and a multitude of supportive research at MetabolicMind.org.

If you have read these early chapters in sequence, you will recall that our initial treatment target is the body's agitated and dysregulated nervous system. Chapter 6 focused on calming it down and moving from fight-and-flight to feelings of safety, so the gut could heal. High-fat ketogenic therapy is the perfect partner for accomplishing this goal. Starting it early on and maintaining it as new therapies are added is showing efficacy for many conditions.

Keto is now the best-studied dietary intervention for its effects on the brain. Another organization that promotes its benefits is The Charlie Foundation. It was founded in 1993 by the Abrahams family after their 11-month-old son became seizure-free within a month on keto when nothing else worked. After five years on keto, he has remained well. The Charlie Foundation has continued advancing awareness of the advantages of ketogenic therapeutics in over 200 hospitals worldwide. CharlieFoundation.org.

According to Palmer, keto not only stops seizures, balances the microbiome, and lessens psychotic behavior, but it also decreases inflammation, affects neurotransmitters, and improves mitochondrial function. More on those later. Kudos to those neurologists and psychiatrists who have traded their pharmaceutical prescriptions for fat bombs!

Some Other Specialized Diets

Low FODMAPs Diet

This acronym stands for *fermentable* (rapidly broken down in the bowel) *oligosaccharides* (fructans), *disaccharides* (lactose), *monosaccharides* (fructose) *and polyols* (sorbitol, mannitol, maltitol, xylitol, polydextrose, and isomalt). These strange words all represent sugars that some bodies absorb poorly. A result of poor absorption is that they tarry too long in the intestines, where they become food for our gut bacteria and yeasts instead of feeding us. In an imbalanced gut, eating foods containing these sugars can contribute to further imbalances, overgrowth, inflammation, and leaky gut.

Developed in Australia and originally prescribed for irritable bowel syndrome (IBS), this gluten-free (but not dairy- or soy-free) diet is showing promise in reducing anxiety and for those with depression, schizophrenia, and autism.[36]

Many forbidden foods are generally considered healthy, like high-fructose fruits (peaches, pears, and watermelon) and fructons (onions, garlic, and asparagus), as well as cashews, pistachios, lentils, and chamomile tea. Fats, oils, and animal-based protein, including eggs, chicken, and fish are all allowed.

Sticking to a low FODMAPs diet requires becoming familiar with some novel substitutes: arrowroot for thickening, soba and buckwheat noodles for pasta dishes, high-fat cheeses such as Brie and Camembert, and lactose-free ice cream, yogurt, and kefir. The good news is that many fruits, such as bananas, berries, melons, citrus fruits, and grapes are allowed. Permitted vegetables include broccoli, green beans, lettuces, chard, spinach, squash, and the green part of scallions.

Working with a nutritionist on this one is almost mandatory. Lots of generalized support is available from books and online at TheLowFODMAPDiet.com and ALittleBitYummy.com.

Low Oxalate Diet

Oxalates and *oxalic acid* come from three sources: foods; fungi such as aspergillus, penicillium, and possibly candida; and normal human metabolism. Oxalic acid is the most acidic organic acid in the body; it is so acidic that car mechanics use it commercially to remove rust from car radiators! Some foods especially high in oxalates are spinach, beets, chocolate, peanuts, wheat bran, tea, cashews, pecans, almonds, and berries. Neither meat nor fish contains significant oxalates.

Yeast overgrowth, commonly associated with antibiotic usage, sometimes leads to increased oxalate production. High levels of oxalates slow mercury elimination. Too many oxalates can also cause inflammation and pain.[37]

Researcher Susan Owens proposed that the use of a diet low in oxalates could reduce symptoms in children with autism. In 2005, she founded the Autism Oxalate Project under the auspices of the Autism Research Institute.

Owens's research shows that those with autism have five to six times the normal amount of oxalates in their urine. None of the children on the autism spectrum had elevations of the other organic acids associated with genetic diseases of oxalate metabolism, indicating that oxalates are high due to external sources.

Go slow on this one; a dump of oxalates from the tissues into the bloodstream can make you feel worse. Also, a product from ION Gut Health (formerly Restore) can help keep dietary oxalates out of the bloodstream in the first place. See ZachBushMD.com.

Reported benefits of the low oxalate diet are more focus and calm; better gross and fine motor skills, sleep, receptive and expressive language, and sociability; and reduced self-abusive behavior and bed wetting. Nutritionist Julie Matthews offers help on managing this quirky diet at NourishingHope.com.[38] Another informative website is OxalateFacts.net. Finally, a number of private Facebook groups also offer support. "Trying Low Oxalates (TLO)" has more than 70,000 members.

Reduced Excitatory Inflammatory Diet

This low-glutamate diet, called the *REID Program*, was invented out of necessity by Katie Reid to treat her daughter's autism symptoms. On the REID Program, high-fiber vegetables fall at the bottom of the food pyramid, stressing their high nutrient value first and foremost. The Reid Program recommends 75% raw and cooked vegetables, 15% protein and fats together, and 10% fiber-resistant starches and fruits. What is eliminated are any processed foods containing excitotoxins or free glutamate. Refer to Chapter 2 to learn more about these additives. Katie's Approved List for Eating (KALE) is available on her website UnblindMyMind. org. Her 2023 book *Fat, Stressed, and Sick: MSG, Processed Food, and America's Health Crisis*[39] details how glutamates are implicated in many inflammatory conditions: obesity, diabetes, autism, addiction, depression, and even cancer.

Picky Eating

You may be reading this chapter with a huge amount of skepticism. Sure! You want me to get my family to eat what?

You are not alone! Anecdotal evidence continuously confirms that most sick people are picky eaters with extremely limited diets. In fact, the DSM-5 added a new diagnosis for these folks: *avoidant restrictive food intake disorder* (ARFID). This condition was recognized about 10 years ago and describes an *eating* disorder which can occur throughout the lifespan, whereas it was formerly diagnosed only in infants and children as a *feeding* disorder.[40]

Amazingly, they are all eating the same things: pizza, mac and cheese, and bagels. That's right: wheat- and dairy-containing processed foods in every combination. Not a vegetable (unless you count ketchup), fruit, or fish in sight!

Don't even consider changing brands—the smallest change in packaging or the slightest shift in color can cause a picky eater to abandon a previously well-loved food choice. Forget sneaking in some crushed up or liquid vitamins; they know instantly!

Although caregivers and therapists usually interpret picky eating as a behavioral issue and treat it with behavior modification, frequently other causes are at play. Picky eating, like poor eye contact, is a symptom that something is not right. The adults in a child's life must dig deep to determine what factors might be involved, and by being detectives they can determine which solution is right.

Nutrition guru Kelly Dorfman, cited at the beginning of the chapter is an expert on picky eating. She recommends a team approach because picky eating has many possible causes. Some hospitals have feeding clinics featuring multi-disciplinary teams composed of oral-motor specialists like occupational thera-pists, speech-language pathologists, and nutritionists. Picky eaters often have the following:

- *Weak digestive function.* (See Chapter 7.)
- *Nutritional deficiencies.* Refer to the chart at the opening of this chapter to check on this possibility. Improper eating creates nutritional imbalances and malnutrition, which can contribute to disinterest in food, leading to further nutritional imbalances. Unfortunately, no one can be forced to eat the proper diet to correct malnutrition, so supplements are often neces-sary. Nutritional deficiencies often begin as early as in the preemie nursery where tubes taped to the faces of tiny babies produce sensitivities of the lips, tongue, cheeks, and face.
- *Sensory misreading in the mouth.* When nutritional issues combine with sensory problems, toddlers reject foods because of their flavors, odors, color or textures, and picky eating results. Babies can emerge from the neonatal intensive care unit (NICU) with tactile defensiveness, low tone, or hyposensitivity in the very areas responsible for their survival. Nursing from a breast or a bottle may be difficult. As feeding becomes a challenge, infants and mothers become anxious. Dorfman believes that those with significant sensory issues, refuse food because this is one way they can con-trol their surroundings in order to lessen their anxiety. The train has just left the station for a possible lifetime of eating disorders.
- *Hypersensitivity in the mouth leads to oral-motor issues.* A combination of sensory and oral-motor issues, such as limited lip and tongue mobility, difficulties swallowing, chewing, and teeth grinding are common. Read more about these in later chapters. Without early expert oral-motor ther-apy, many infants are destined to become picky eaters.
- *Drug side effects,* like reduced appetite and nutritional deficiencies. And the cycle continues.

Despite their strong need for sameness in foods, getting picky eaters to broaden their choices *is* possible. Caregivers must combine a stalwart approach with the utmost patience and perseverance.

What's for Breakfast?

One of the most frequently asked questions from anyone new to special diets is, "What's left to eat for breakfast when you have eliminated juice, cereal, toast, pancakes, bagels, and pastries?" You'd be surprised!

What is a good breakfast, and why is it important? It's all about glucose levels.[41] Fasting for 12 or more hours between dinner and awakening in the morning triggers *hypoglycemia* (low blood sugar) and *acidosis*. Remember the BED principle that an acidic gut encourages yeast overgrowth, viruses, parasites, and other unhealthy cells to thrive.

Research has long shown the importance of a good breakfast. One British study found that kids who consumed a low-glycemic breakfast performed significantly better on tests of memory and attention and showed fewer signs of frustration than those who had consumed a high-glycemic breakfast.[42]

Here are some guidelines:

- *Think protein.* At least 10 grams. Protein in the morning is the best gift to give your family; any animal or vegetable protein will do. Serve dinner for breakfast! Last night's leftovers, a turkey burger, lamb chop, salmon patty, scrambled eggs, lentil soup, or a bean burrito.
- *Think dense nutrition.* Donna Gates suggests a breakfast soup made with bone broth, fennel, broccoli, garlic, and parsley. Delicious!
- *Think quick and easy.* Look in the freezer section for good quality gluten- and dairy-free waffles and pancakes. Up the protein with a nut butter spread. Try GF/CF cereals made with amaranth or quinoa and added flaxseeds. Make high-protein muffins; if necessary, use premade mixes from excellent suppliers.
- *Think portable.* Try some high-protein GF/CF breakfast bars for a fast breakfast on days when time is a problem. A huge variety is available from good health food stores; sample several until you find some you like.
- *Think Essential Fatty Acids (EFAs).* Add flax, hemp, flavored cod liver, pumpkin seed, or other high-quality oils to whatever you make. They enhance texture and are close to tasteless.
- *Think drink.* Use nut and grain milks with protein and fruit/vegetable powders to blend a quick, nutritious, colorful, drinkable breakfast smoothie.

Rotate four or five of the above breakfasts, never having the same breakfast two days in a row. Soon, breakfast will be your favorite meal of the day!

Take-Home Points

Improve digestion with dietary modification, one of the most powerful therapies you can implement for anyone with any type of disorder. At the risk of using a cliché, how well a car runs is limited by the quality of the gasoline in its tank. The human body, like a car, *must* have excellent fuel to run on. Do your family members have high-test fuel in their tanks?

Take a big step by choosing and committing to one or more of the above diets, and stick to it. Move to a more stringent one if changes do not show up in a couple of months. Get support from family members and the many resources available if you are having trouble.

Remember, everyone is unique, and one size does not fit all. The ultimate goal is to remove load factors and free up energy for talking, relating, and learning. Bon appétit!

CHAPTER 9

THE IMMUNE SYSTEM: WHAT GOES WRONG

Healing the chaos in the gut for those with mental health disorders is only the beginning. Next is figuring out the complex relationship among environmental toxins and the gastrointestinal and immune systems, and their effects on function and behavior.

Of all the possible stressors associated with Total Load in neurodevelopmental disorders, dysfunction of the immune system is the one that has received the most attention. With increasing frequency, medical professionals in many disciplines are recognizing that most chronic diseases, including allergy, auto- and neuroimmune conditions, and even cancer start as immune disorders. Researchers in the autism world proposed that idea about 15 years ago,[1] and more recently, psychiatrists and psychologists have begun applying this theory to those with many of the other disorders in this book.

Because it is *so* crucial that we understand what goes wrong with the immune system and possible ways to strengthen it, this and the following two chapters are arguably the three most important chapters in the book. (I think I said that about the chapter on digestion, as well!)

Chapter 9 begins with a basic lesson on the immune system and the stressors that cause it to become inflamed and imbalanced. Next, it looks at common infections and the comorbidity of mental illness and autoimmune disorders. Then, mitochondrial disorders, including cerebral folate deficiency and a blood disorder called pyroluria, are described. This chapter ends with a discussion of the role of antibiotics in weakening immunity. The complex connection between vaccination and immune disorders is covered in the next chapter.

Immunology 101

Humans and infectious microbes have coexisted for as long as humans have walked the earth. The human immune system has developed complex and very efficient ways of protecting itself from and dealing with infections caused by viruses, bacteria, and other organisms. The ideal immune system recognizes all foreign organisms efficiently, rapidly destroys invaders, prevents further infection, and never causes damage to itself.

The body's immune system is divided into two complementary and inter-dependent functions: the *cellular* or "innate" component and the *humoral* or "learned" part. In healthy people, cellular immunity is the primary immune defense system, while humoral immunity plays a secondary role. Receptors for cellular immunity are located in the mucous membranes of the gastrointestinal and respiratory tracts and in their respective lymph nodes, while humoral immunity originates in the bone marrow.

Both sides of the immune system have important roles in keeping our bodies safe. A well-functioning immune system requires a delicate balance between cellular and humoral immune responses. The natural sequential process of a cellular response to infection followed by a humoral response matures both parts of the immune system.[2]

Foreign substances, like bacteria, viruses, or toxins—known as *antigens*—enter the body naturally through the skin and mucous membranes of the gut and lungs. In the presence of an infectious microorganism, the body's first line of defense is an acute inflammatory response from the cellular immune system.

Inflammation and Immunological Stress

Chapter 6 introduced inflammation and its consequences. An acute inflammatory response is the immune system's attempt to deal with an antigen. Fever, redness, swelling, and/or mucus are all the symptoms that are hallmarks of inflammation. When working properly, this inflammatory reaction stimulates the cellular branch of the immune system to clear all signs of infection—as well as the inflammation. All symptoms, including the inflammation, should be temporary, and once the infection resolves, should disappear.

Inflammation can show up in every bodily system from the gut to the brain. The English language has a suffix for these inflammatory conditions: *-itis*. Inflammation of the joints is *arthritis; gastritis*, the stomach; and *meningitis*, the spinal cord and the membranes surrounding the brain.

Encephalitis

Inflammation of the brain is called *neuroinflammation* or *encephalitis*. As the brain begins to swell, symptoms such as headaches, sound and light sensitivity, and even seizures appear, disrupting mood, sleep, focus, and memory and negatively affecting how one thinks, feels, and behaves.

Inflammation and Antibodies

While acute inflammation can feel terrible, it is a good thing. It is the beginning of the body's innate defense program against threats. This program serves three purposes:

- it contains and eliminates the threat;
- initiates repair and resolution; and
- restores function once the threat has cleared.

One way inflammation works to eliminate threats is by signaling the production of Y-shaped proteins that bind to bacteria, viruses, toxins, and parasites in order to neutralize them. These Y-shaped proteins are the immunoglobulins we call antibodies. Recall the discussion in Chapter 5 about the IgE, IgG, and IgA antibody classes. Each of these is a measurable immune response to a perceived threat (though in the case of allergic IgE responses, the threat is rarely real).

IgA antibodies dominate at mucosal surfaces, like the nose, which are the usual entry points for "invaders." They are critical for neutralization without excessive inflammation. An additional type of response is immunoglobulin M (IgM). IgM antibodies are more immediate and more generalized than the more-specific IgG antibodies that follow, establishing resistance to future infection with the same pathogen.[3] (Note that specific IgG antibodies are the desired result of vaccination but are only a small part of the body's natural immune response to infection.) IgM antibodies, found in blood and lymph fluid, make up only about 5–10% of the antibodies in a body with a healthy immune system. A higher number of specific IgM antibodies usually indicates either an acute infection or an ongoing one.

If a threat doesn't resolve—either because it is ongoing, it has overwhelmed the body's defenses, or body tissues are too similar to the initial threat and are attacked as well—inflammation persists and becomes chronic. The body never gets to the "restores function" phase—and that's a bad thing. The adaptive, humoral part of the immune system dominates, continuously trying to resolve an unresolvable situation.

A higher number of specific IgM antibodies indicates either an acute infection or over-activity, both of which can produce acute inflammation. If inflammation persists and becomes chronic, the humoral part dominates, continuously trying to resolve a sometimes-nonexistent invader. This prolonged and inappropriate response causes yet more inflammation. IgM is the first antibody to respond to an infection, followed by *immunoglobulin G* (IgG), which confers longer-term immunity.

Cytokines. The immune system is made up of a huge network of various types of cells that communicate with each other through special signaling proteins called *cytokines.* Dr. Neil Nathan likens cytokines to bullets fired by white blood cells against pathogens.

Sometimes the immune system gets carried away and the body produces too many inflammatory cytokines and not enough cytokines that control inflammation. This unbalanced barrage of bullets results in what is known as a *cytokine storm.*[4] The inflammatory cytokines start "storming" out of control, without enough action from anti-inflammatory cytokines.

During the COVID pandemic, a cytokine storm was the most feared and serious complication of the virus. Too many pro-inflammatory TH1 and not enough TH2 anti-inflammatory cytokines went into the circulation.[5] TH1 cells monitor cellular immunity, and TH2 cells monitor humoral immunity. Once one subset of cytokines dominates, the body's response does not shift easily to the other. This is important.

Macrophages. Also of great importance to this discussion are white blood cells called *macrophages*, derived from the Greek words meaning *big* and *eat*. Think "Pac-Man." Macrophages gather debris of all kinds, from damaged or dead cells to foreign invaders like aluminum, bacteria and viruses.[6]

In healthy individuals macrophages are constantly at work. Their sole purpose is to clean up and remove foreign, potentially harmful substances of all kinds, thus regulating the body's immune response. Let's call them the "garbage men."

GcMAF. When foreign invaders of any kind cause inflammation, our bodies produce a substance called *granulocyte macrophage activating factor* (GcMAF). Molecules of GcMAF, along with vitamin D, activate macrophages. The sequence of events is as follows: toxin > inflammation > vitamin D activation > production of GcMAF > macrophage activation > neutralization of offending toxin > inflammation controlled.[7]

The late autism doctor Jeff Bradstreet and his colleagues were working on the role of GcMAF in their patients at the time of his death in 2015. They discovered that an enzyme called *nagalase* was elevated in their patients with autism. When treated with GcMAF, a majority showed improvement and nagalase levels dropped.[8]

Immunological Stress

An imbalance between the two parts of the immune system stresses it out, and the immune systems of the people with the disorders covered in this book are stressed! An immune response is appropriate against invaders like bacteria, yeast, viruses, parasites, and toxins, but the immune systems of those with mental illness respond to almost anything, including, but not limited to, molds, pollen, chemicals, metals, common foods, food additives, and incompletely digested particles of food.[9] In fact, such immune overreaction is being recognized as a hallmark of ADHD, autism, and other disorders.

Remember our discussion of genetics and how certain polymorphisms (SNPs) in some genes can interfere with function? Some SNPs turn on inflammation. A SNP in the gene for tumor necrosis factor (TNF), a cytokine that determines whether something coming into the body is harmful or not, can cause it to misidentify neutral or beneficial intruders as threatening, increasing inflammation.[10]

When chronic inflammation persists, the immune system is imbalanced with humoral dominance. Symptoms of humoral dominance include, but are not limited to: seasonal allergies; reactions to common foods;[11] frequent infections; rashes; hypersensitivity to dyes, chemicals, perfumes, and/or medications; cold hands and feet; and most likely at least one bad vaccine reaction. Even depression can signal that the immune system is stressed and out of balance.[12]

The Immune System in Autism and ADHD

Almost 20 years ago, Paul Ashwood, PhD, an immunologist at the MIND Institute in California reviewed many studies and concluded that autism most probably has an immunological basis.[13] He declared the immune system "a new frontier in autism research."

Going as far back as the mid-1980s and '90s, Ashwood discovered that most children with autism have significant immune system abnormalities, with inflammation the primary marker.[14] The same is true with bipolar disorder. A more recent article reported a Chinese study which also found comorbidity in a meta-analysis of ten studies of patients with bipolar disorder and autoimmune illnesses.[15]

According to longtime autism researcher, Sudhir Gupta, MD, PhD, a professor of neurology, pathology, microbiology, and molecular genetics at the University of California, Irvine, individuals with autism have more TH2 cells than TH1 cells when compared to neurotypical children. Gupta believes that the fewer number of TH1 cells may explain why these children are more susceptible to viral and fungal infections.[16] The dominance of humoral (TH2) immunity over cellular (TH1) immunity, along with persistent inflammation, is emerging as one of the hallmarks of ASD.

Repeated vaccination may also be instrumental in causing chronic and uncontrolled inflammation in many individuals with and without these disorders.

Remember, a natural stimulation of the immune system can make it stronger and better able to maintain good health. When the immune system is stimulated *unnaturally* through vaccination, it can get "stuck" on inflammation that leads to chronic illness. Read ahead for more about this process.

In a 2005 Johns Hopkins study in researchers reported examining tissue from three different regions of the brain in 11 people with autism, ages 5 to 44, who had died of accidents or injuries. They also compared the number of inflammatory proteins in the cerebrospinal fluid and brains of normal controls with those of six living patients with autism, ages 5 to 12. Those with autism had abnormally elevated cytokines, evidence of an ongoing inflammatory process.[17]

Defense versus Tolerance. Most people think about our immune system as the defender against things that can kill you. While it does act like the military and send out its soldiers to fight invaders, it also actively develops tolerance of harmless substances.

Tolerance is an important (and active) function of the immune system, even though it is meant to *prevent* our immune systems from reacting to things that are harmless. It's when tolerance breaks down that the immune system reacts to harmless things like pollen. People whose immune systems have developed a high tolerance for pollen don't "turn on" in the springtime. When your body has low or limited tolerance, it develops allergies.

Allergies and autism go together, as Kenneth Bock, MD, states in his 2008 book *Healing the New Childhood Epidemics: Autism, ADHD, Asthma, and Allergies.*[18] The exact number of children who have both autism or ADHD *and* allergies is unknown, but anyone who works with this population knows it is high. The commonality is immune system dysregulation.

Inflammation and the role of immune dysfunction are topics of major interest in autism research today.[19] The comprehensive 2010 textbook *Autism, Oxidative Stress, Inflammation and Immune Abnormalities*, with over 400 pages of contributions from experts worldwide, covers this subject in great depth.[20]

Determining Immunological Total Load Factors

In order to determine how dysfunctional the immune system is, an extensive medical and developmental history is necessary. This is the only way to figure out which load factors are at play. As noted earlier, this must be an exhaustive look at prenatal, natal, and post-natal social and environmental factors.

Some environmental assaults may be obvious, such as from an old house with peeling lead paint; others may be subtler, such as toxic lawn treatments tracked into the home. One family I know became sick when the fertilizer they had been using on their farm was unknowingly cut with recycled radioactive material. Extraordinary efforts were necessary to trace the cause of their children's illness and developmental problems to this source.

Other exposures to consider are medications, vaccines and their ingredients, pesticides, chemicals, tobacco, toxic building materials, pet products, travel, changes in living environment, and impure water. A complete history should also include a comprehensive food diary as well as how much sleep, screen time, and exercise the patient is getting. When was the last time a doctor asked you about those?

Remember that exposures from air, food, water, dust, and all other sources are cumulative and incrementally add to an individual's Total Load. Although history-taking alone is sometimes adequate to determine an appropriate treatment plan, previously described laboratory testing is frequently necessary to evaluate individuals' response to toxic agents, their gut flora, and their nutritional status.

Many individuals with neurodevelopmental disorders can trace gut, neurological, and behavioral symptoms to exposure to specific infectious agents. Sometimes their immunological reactions, such as pollen allergies, come and go with the seasons. More often their inflamed immune systems are attempting to respond to specific intruders.

When the body fails to defend successfully against microbes and other intruders, infections occur. Our exhaustive questioning should tell us whether we are talking about infection caused by an outside source like a virus, bacteria, mold, or other toxin, or from the body attacking itself, especially the brain (called *autoimmune encephalitis*, or *AE*). (See below.) Let's review those bugs that cause infections.

Infections from Viruses, Bacteria, and Mold

Viruses

Viruses are strands of DNA or RNA that are encased in a protective protein shield and looking for a host. They are packets of proteins that cannot function outside the body or reproduce on their own, so they get cozy in the cells of their hosts where they can sometimes live for years. Unlike bacteria (see below), viruses do not respond to antibiotics. Most colds and the flu are caused by viruses. Viruses have the ability to mutate and shift within the body, wreaking havoc with function.[21]

Viruses in autism. The role of viruses in autism is well-documented.[22] Borna disease virus,[23] congenital rubella,[24] Epstein Barr,[25] herpes simplex,[26] and measles virus[27] are all considered to be possible contributors to the behaviors we call "autism."

The understanding of the role of retroviruses in autism comes from the work of veteran scientist, Judy Mikovits, PhD, known for her pioneering research on HIV/AIDS during a 20-year career at the National Cancer Institute. There, in 1999, she discovered how human retroviral infections modulated DNA and developed the concept of inflammatory cytokines and chemokine signatures of infection and disease.

In 2006, Dr. Mikovits became interested in patients with autism and other little-understood neuroimmune diagnoses, such as chronic fatigue. She was one of the first to demonstrate the relationship between environmentally acquired immune dysfunction, chronic inflammation, and autism.

After discovering Ashwood's and others' research supporting the premise that autism is a neuroimmune disorder,[28] Mikovits kept looking for viruses and other pathogens. Sure enough, she found them. Dr. Mikovits's story about her discovery and the rabbit hole it led her down is documented in the book *Plague: One Scientist's Intrepid Search for the Truth about Human Retroviruses and Chronic Fatigue Syndrome (ME/CFS), Autism, and Other Diseases.*[29]

Klinghardt believes that having an outdoor, vaccinated cat or dog that licks you is sufficient to infect you with retroviruses. He is also adamant that environmental factors, such as cell phone radiation, glyphosate in food, mercury from fish, and dental amalgams, all of which drive inflammation, are in part responsible for the explosion of retroviruses in our bodies today.[30]

Many vaccines for animals are manufactured by using cell lines from other animals, which are known to produce retroviruses. However, the significant risks of infection from one species to another through vaccination have been ignored.

COVID-19. In December 2019 everything changed when a pneumonia-like virus later named *Coronavirus Disease 2019*, or *COVID-19*, emerged from Wuhan, China. By spring 2020 all 50 states in the U.S. had reported at least one case, and by mid-2020 cases were being reported worldwide.[31]

Physically, this virus caused flu-like symptoms. Emotionally, it elicited frozen fear, locking the nervous systems of millions, if not billions, into fight-or-flight, as schools, businesses, and the world shut down. Fears were rampant among adults and children alike. Even those without formal mental health diagnoses experienced unprecedented fear and anxiety—not just about COVID, but about almost anything. An epidemic of mental health disorders had arrived and is still with us. Dr. Darin Ingels reports that he saw many kids with autism who regressed after COVID, even when they did not have classic COVID-19 symptoms.

By August 2020, medical professionals recognized that COVID-19 was, in part, driven by autonomic dysregulation, specifically sympathetic overdrive.[32]

Many of those who survived could not shake fatigue, brain fog, and other symptoms. Others acquired allergies to substances that had never bothered them before; still others experienced exacerbated symptoms of previously diagnosed conditions, such as autoimmune, degenerative, and psychiatric disorders. Now referred to as *long-haul COVID*, this cluster of symptoms followed not only the disease itself, but also vaccination against the disease, or both.

Infections from Bacteria

Bacteria are one-celled organisms that abound in and on our bodies' skin, guts, and environment. Bacteria can be "bad," triggering uncomfortable immune reactions such as rashes, diarrhea, and more, or "good," such as those used to ferment foods. Everyone is familiar with the damage that bacteria can do to the body by causing a variety of infections, some of which can be life-threatening. Washing hands and assuring that water is clean eliminates most pathogenic bacteria, and our contemporary society pushes all types of commercial anti-bacterial products to kill the remaining bad (as well as some of the good) critters we might have missed.

Ear infections. Frequent ear infections are found in the health history of many of our depressed and anxious patients, a sign of an unbalanced immune system. Sometimes, but not usually, ear infections can be a result of a bacteria, and that is why I include them here.

A classic 1991 study by Talal Nsouli, MD, an allergist at Georgetown University Hospital, found that 78% of early childhood ear infections are related to food allergies. By limiting the diet of his patients, Dr. Nsouli avoided the unnecessary use of antibiotics or insertion of ear tubes,[33] which current research now shows is rarely beneficial.[34]

In a 1994 survey of parents by my nonprofit Developmental Delay Resources (DDR), 75% of children with diagnoses of at least one developmental disorder had had five or more ear infections.[35] Several studies confirm this anecdotal evidence.[36] One researcher even targeted early ear infections as a major contributor to autism.[37]

Maryland nutritionist and DDR cofounder Kelly Dorfman coined the term *post-traumatic ear infection syndrome* to describe a condition in children who have had continuous ear infections followed by gradual behavioral deterioration, including hyperactivity, distractibility, and lessened relatedness.[38] She believes that many treatments with antibiotics, both during infections and prophylactically, may further exacerbate their immune system dysregulation by increasing vulnerability to damage from toxic metals. See below.

Clostridia. One family of nasty bacteria that can be both "good" and "bad" is *clostridium*. Some of over 100 strains of clostridia are normal inhabitants of the intestinal tract; others are potentially fatal. *Clostridium botulinum* causes botulism, *Clostridium tetani* causes tetanus, and *Clostridium difficile*, or *C. diff*, causes watery, bloody diarrhea.

Clostridia bacteria give off toxic metabolic byproducts, including *proprionic acid* (PPA), which has been found to be common and troubling in autism. PPA is a short-chain fatty acid found in high quantities in the guts of those with autism; it is also contained in many food preservatives. Calcium propionate, an FDA-approved food preservative widely used in wheat, bread, cheese, juices, and dried fruits, contains it.[39]

Dr. Derrick MacFabe and his Canadian colleagues became particularly interested in PPA in autism and decided to investigate. They injected normal adolescent rats with PPA to see what happened. Guess what? The rats demonstrated "autistic" behaviors, such as impaired socialization and cognition, as well as increased repetitive behaviors, like spinning.[40]

Dr. William Shaw, founder of the Great Plains Laboratory (now Mosaic Labs) studied Clostridia in autism. He believes that it is an often-missed culprit not only in autism spectrum disorders, but in many diseases as well, including Parkinson's, multiple sclerosis, chronic fatigue, bipolar, and others. By looking at some additional byproducts of *C. diff* and their effects on neurotransmitters, especially dopamine, he has developed a unique method of diagnosing and treating resistant clostridia infections.[41]

Strep. "Do you feel like your child has been stolen from you overnight?" asks naturopathic physician Jill Crista, a mother who has been there. If tics, abnormal movements, a decline in fine motor skills, picky eating, or OCD show up after a strep infection, and a subsequent infection causes those symptoms to worsen again, *pediatric autoimmune neuropsychiatric disorder associated with streptococcal infections* (PANDAS) should be suspected. In PANDAS antibodies to strep bacteria cross-react with proteins in the basal ganglia of the brain and cause inflammation that leads to these symptoms.[42]

An abrupt onset or worsening of OCD symptoms can be triggered not just by a streptococcal infection, but also by other infectious bacteria, viruses, or parasites carried by spiders, mosquitos, and other bugs. This variation is called *pediatric acute-onset neuropsychiatric syndrome* (PANS).

Although previously dismissed with "it's all in your head," PANDAS and PANS are getting more and more attention. Chapter 12 can help you tame this monster.

Lyme. The bacteria carrying Lyme disease are very prevalent, while the disease is very under-diagnosed and often overlooked. Lyme disease is caused by several species of bacteria in the *Borrelia complex,* including *Borrelia burgdorferi* (*Bb*), *Borrelia afzelii, Borrelia miyamotoi,* and *Borrelia garinii.*[43] Like syphilis, *Bb* is a spirochete that has several forms, at least one of which is not susceptible to antibiotics, so it's difficult to diagnose and control, especially in young children according to Klinghardt.[44]

This insidious disease can be passed between sexual partners and from mother to unborn child. Even without evidence of tick bite, followed by the telltale red bull's-eye rash, Lyme should be ruled out in absolutely everyone with psychiatric or brain-based symptoms.[45]

Diagnosing and treating Lyme is extremely complex and controversial. Many top doctors have designed very complex diagnostic and treatment procedures and protocols. Klinghardt, along with Dr. Marco Ruggiero, developed an elegant procedure published in the *American Journal of Immunology.*[46] Named the RK Protocol, it combines two safe and noninvasive diagnostic techniques: Klinghardt's unique muscle testing, Autonomic Response Testing (ART), and Ruggiero's unique use of ultrasound.

Using the RK Protocol, Klinghardt discovered that seven out of eight patients test positive for Bb, even though this did not always show up on labs. Therapeutically, instead of guessing which treatments would benefit a patient, the RK Protocol allows prioritization of medications, supplements and other therapeutic interventions in Lyme, as well as in other chronic conditions including autism.

Lyme is really complicated. In order to diagnose, treat, manage, and heal from Lyme, one must read material from many experts, listen to webinars about the newest techniques, and work with a "Lyme literate" professional. Chapter 11 will help you treat the havoc wreaked by Lyme.

Infections from Mold

Mycotoxins are poisonous substances produced by specific types of mold related to water damage. You read about them along with electromagnetic radiation in Chapter 2 as an environmental stressor.

Mycotoxins can suppress the immune system by inhibiting the production of antibodies. Furthermore, they produce chronic inflammation by destroying the integrity of the body's mucosal barriers, including in the gut, causing dysbiosis and leaky gut. They also increase susceptibility to and severity of other pathogens, including bacteria, viruses, and parasites.[47] Add electromagnetic radiation from

5G cell towers, which potentiates the growth and virulence of mold, and you are in *big* trouble, says functional medicine doctor, Jill Carnahan, MD.[48]

No wonder that finding the bacteria that cause Lyme disease and identifying mold as a trigger in illness revolutionized the way Dr. Nathan treats his patients, as described in *The Sensitive Patient's Healing Guide*. He believes that underlying mycotoxins are one of the most missed causes of all types of illness.

Nathan's colleague Jill Crista, a naturopathic physician, has become a trusted voice not only about PANS/PANDAS, but also about mold. Her comprehensive book, *Break the Mold: 5 Ways to Conquer Mold and Take Back Your Health*[49] includes a questionnaire to see if you're an unsuspecting target of mold, which foods to avoid, and more. Learn more at DrChrista.com.

Autoimmunity

When the body loses tolerance for its own tissues, rather than something from the environment, it can turn on itself. That condition is called *autoimmunity*. Diseases like *ulcerative colitis*, *rheumatoid arthritis*, and *Hashimoto's thyroiditis* are autoimmune conditions that are frequently found in those with the behaviorally diagnosed conditions discussed in this book. Integrative physician Ty Vincent, MD, describes this phenomenon beautifully in a chapter in Neil Nathan's *Sensitive Patient* book mentioned above.

Remember that inflammation is a useful and appropriate immune response, but it should be temporary. If it doesn't resolve and becomes chronic, autoimmunity results. In autoimmunity, the body stays inflamed because the immune system is dysregulated.

Inflammation, immune dysregulation, and autoimmune diseases are rampant in family members of those with neurodevelopmental disorders. Their parents frequently have allergies and autoimmune diseases such as Parkinson's, lupus, multiple sclerosis,[50] fibromyalgia, and chronic fatigue or were vaccinated prior to or during their pregnancy. A 2009 Danish study of over 3,000 children found correlations between family history of type 1 diabetes and/or maternal history of rheumatoid arthritis and celiac, and autism diagnoses.[51]

Their own developmental histories almost always include several of the following signs of immune system distress:

- digestive problems, including colic, reflux, vomiting, constipation, and diarrhea;
- skin problems, including eczema and pallor;
- dark circles under the eyes (allergic "shiners");

- red ears or "apple" cheeks;
- recurrent ear, sinus, or strep infections;
- history of vaccine reaction(s) or vaccines given simultaneously with antibiotics and/or acetaminophen (Tylenol™);
- chronic unexplained fevers;
- respiratory problems, including asthma and bronchitis;
- repeated use of antibiotics;
- febrile seizures associated with vaccination reaction and Tylenol™ ingestion;
- regression in function between 15 and 30 months;
- seizures in puberty;
- hyperactivity; and
- sleep disturbances.

Bottom line: many of today's generation also have comorbid autoimmune disorders in which their overactive humoral immune systems are continuously fighting their chronically inflamed bodies.

Researchers are also finding a link between ADHD and autoimmune disorders. One comprehensive article exploring this connection looks at exactly the components of Total Load Theory: genetic susceptibility, environmental triggers, and inflammation.[52]

While I did not research each and every diagnosis touched upon in this book, I suspect frequent comorbidity. What is interesting is the potential reversal of a cause-and-effect relationship. Remember that gastroenterologists blamed constipation and other digestive issues on autism when rather the opposite appears to be true. Likewise, I believe that autoimmunity precedes brain-based disorders, not vice versa.

Testing for Autoantibodies Due to Strep, Lyme, COVID, and Other Infections

Chapter 5 details testing for many biological conditions, including some trusted sources for Lyme and PANDAS/PANS. Because these conditions are so complicated, in 2013 two experts in infection-induced autoimmune disorders, Madeleine Cunningham, PhD, and Craig Shimasaki, PhD, founded Moleculera Labs to test a host of autoimmune diseases more thoroughly. They developed the *Cunningham Panel*™, a complex diagnostic instrument designed specifically to determine possible underlying causes of neuropsychiatric disorders.

This group of complex blood tests has been expanded to include COVID and is now known as *The Autoimmune Brain Panel*™. The panel measures the levels of specific autoantibodies that target the brain and their ability to trigger neuropsychiatric symptoms. Each of these targets has been selected because of their association with various neurologic and psychiatric symptoms, including tics, OCD-like behaviors, ADHD, anxiety, and depression.

An elevated level indicates that symptoms may be due to an infection-driven, autoimmune process. Test results can assist in directing appropriate treatment. To learn more about using this test go to moleculeralabs.com.

Mitochondrial Disorder

Mitochondria are the power centers inside cells that are essential to converting food to energy. Well-functioning mitochondria create readily available and sustained energy, resulting in toned and well-formed muscles. When these muscles are exercised, they get stronger steadily and predictably.

In *mitochondrial disorders*, also known simply as *mito*, the ability of the mitochondria to generate energy is damaged. Environmental toxins such as heavy metals, pesticides, and aggressive antibiotic use can all injure the mitochondria, resulting in loss of muscle tone, stamina, and other functions.

Cerebral Folate Deficiency and Pyroluria

Remember the MTHFR polymorphisms from Chapter 1? These interfere with detoxification, resulting in toxins piling up and interfering with immune function.

Some children with autism develop *folate receptor autoantibodies* (FRAs), throwing them into a state of autoimmunity called *cerebral folate deficiency* (CFD). FRAs develop because of an abnormality in the synthesis of hemoglobin; a byproduct is created called kryptopyrrole, a neurotoxin that circulates in the bodies of many patients with psychiatric symptoms. FRAs disrupt folate synthesis by binding to and depleting vitamin B6 and zinc, thus preventing their use in neurological function in the brain and body.[53]

Enter pediatric neurologist Richard Frye, MD, at the Arkansas Children's Hospital Research Institute, who has studied children with mitochondrial issues and autism for years. You met him in Chapter 4 because he headed a group of eminent researchers studying the microbiome in 2014. As Frye read about more and more cases of cerebral folate deficiency in the literature, it became clear to him that symptoms of CFD were consistent with those of autism.

So, he and a colleague tested children with autism to see if they also had FRAs in their blood. Bingo! FRAs were present in 75% of the 93 children studied. Not only did these kids have a very high prevalence of FRAs in their blood, but some also had low levels of folate in their central nervous system, just like the kids with CFD.

A urinary kryptopyrrole test that requires a doctor's prescription is now available from Mosaic Labs. It can determine whether pyrrole disorder is present in patients with ASD, ADHD, bipolar disorder, schizophrenia, and Down syndrome.

Supervised treatment usually includes supplementing with zinc, vitamin B6, and other nutrients that may be depleted. What if this were the magic bullet for one of these disorders?

Leucovorin

Because vitamin B9 (folic acid) does not cross the blood-brain barrier, folinic acid (called Leucovorin commercially), a vitamin with a long history of safety and the ability to get to the brain in high doses, can be used to help overcome the effect of folate-receptor antibodies. Treatment with folinic acid addresses not just behaviors, but rather a key underlying biological abnormality that might be at the root cause of autism symptoms for some people. If treatment is successful, autism symptoms should improve.

Frye purposefully recruited children with autism who had language impairments because he believed that folinic acid would have the greatest impact on verbal communication. At the onset, he measured the severity of their symptoms, including verbal communication skills, in great detail using well validated standardized tests. After 12 weeks of treatment with either folinic acid or a placebo, he repeated those same evaluations in exactly the same manner to determine differences.

The subjects were treated with up to 50 mg per day of an activated form of folinic acid over four months. Findings: one third of those who received folinic acid went from "moderately severe" to "much improved" in verbal communication, receptive and expressive language, attention, and stereotypical behaviors. They also demonstrated significant improvement in daily living skills and lessened irritability and hyperactivity compared to those in the placebo group.[54]

Relating FRAs to autism and mental health disorders is a hugely significant medical breakthrough because Frye discovered that early detection of folate receptor autoantibodies is a biomarker for early onset and regressive autism.[55] CFD is thought to be a form of mitochondrial disorder, for which a dairy-free diet also works extremely well because eliminating cow's milk can regulate FRAs. Refer back to the previous chapter on how to implement this.

The Role of Antibiotics

Because antibiotics kill dangerous pathogens that used to kill people, most people view antibiotics as lifesaving medicine. Yet, *antibiotic* translated literally means *against life,* and they kill microscopic critters just as poisons kill larger undesirable critters like roaches and termites our homes. Antibiotics work most of the time, at least in the short term; they are a double-edged sword, however. In addition to killing life-threatening bugs, they indiscriminately kill friendly bugs as well, and we've already established the importance of gut biome diversity to overall health.

The choice of antibiotic delivery system is an important factor. External antibiotic creams, used to combat skin infections, kill on the body's surface and do not typically disturb the gut. However, taken orally, antibiotics kill not only bad bugs that cause infections, but also knock out virtually all intestinal bacteria, including beneficial varieties. Many people experience digestive problems after taking an oral antibiotic for this reason.

As we learned in the previous chapters, the gut is host to hundreds of types of flora that live together cooperatively in the healthy person. Antibiotics usually disrupt this symbiotic environment. The good bacteria exist to control the growth of yeasts and other fungi in the digestive tract. In their absence, yeasts colonize and their usually small colonies can then proliferate.[56] Intestinal overgrowth of yeast and bad bacteria is a well-documented outcome of taking broad-spectrum antibiotics.[57] Many parents report that their children regressed into autism following antibiotic treatment.

As bacteria have become resistant to first-generation antibiotics such as penicillin, doctors have moved to stronger ones like Ciprofloxacin. Today's antibiotics are "atom bombs" compared to the "water pistols" of the previous generation. No one knows their long-term effects to the digestive, immune, and other systems.

According to New York microbiologist Martin Blaser, MD, the average child in the developed world has taken 10 to 20 rounds of antibiotics by age 18![58] Blaser and many other physicians are very concerned about the role of antibiotic overuse in the dysregulation of the immune system.[59]

If a baby develops any infection, allowing its immune system to fight the invader may actually strengthen immunity in the long term, because when the same invader reappears, the body recognizes it and is usually able to fight it off. If, instead, an antibiotic fights the infection, some believe that the antibiotic will suppress the immune system, causing it to fight less vigorously the next time around.[60]

By the fourth or fifth infection, the immune system might not even recognize an invader as a threat because it now depends on the antibiotic to do its job.

Boyd Haley, PhD, a world authority on mercury toxicity, found that it takes less mercury to do harm in the presence of the antibiotics ampicillin and tetracycline.[61] Allergies, ear infections, mercury, and antibiotics are a potent cocktail for triggering immune system reactions and accompanying auditory-processing difficulties, learning disabilities, and attention deficits.

Take-Home Points

This chapter adds immune system dysfunction to the Total Load causing behavioral symptoms related to previously described nervous system and gastrointestinal problems. Viruses, bacteria (including Lyme and strep), mycotoxins, and other infectious critters contribute to the inflammatory process and, if unresolved, further compromise immunity. Inflammation and lessened immunity are huge load factors, caused and exacerbated by environmental toxins and medical treatments such as antibiotics and, for some, vaccination.

Inflammation is a natural part of the healing process. Acute inflammation is good, but if it becomes chronic rather than resolving, it can lead to brain dysfunction of varying degrees of severity. Brain inflammation, called encephalitis, can turn on the body, resulting in autoimmune disease and distress. We have now entered the world of "mental illness."

How vaccines and their additives add to the Total Load is covered in the next chapter. Keep reading.

CHAPTER 10

WHAT DO VACCINES HAVE TO DO WITH IT?

"Do vaccines cause autism?" Whenever I lecture or meet someone new, as soon as I disclose that I have written three books on autism, inevitably, that is the question I am asked. I have to quickly decide on my response depending upon where I am, who queried me, their assumed bias, and the suspected reaction to my response.

"Yes, they can," is usually where I start, since vaccines are a huge trigger in Total Load. For many, a wellness visit to the pediatrician and the vaccines received that day are the proverbial "straw that breaks the camel's back." You now understand it all depends on a child's Total Load.

Almost no subject is as polarizing as vaccination. It is dividing spouses, family members, and patients from their trusted doctors. The question of whether, when, how, and what to vaccinate against consumes prospective, young, and seasoned parents, as well as adults. New vaccines for "protection" during pregnancy, travel, aging, and everything else are being pushed on us in the media, by our doctors, siblings, and friends who work hard to convince us that all "vaccines are safe and effective," when they are neither completely safe nor completely effective.

Vaccine History

In over 50 years working with families of children with disabilities, vaccination has become an increasingly hot topic. We barely heard mention of it in the sixties and seventies. By the 1980s we began raising our eyebrows as we heard occasional reports of vaccine reactions, and the movie *Rain Man* introduced us to autism. The 1985 book *DPT: A Shot in the Dark* by Harris Coulter (1932–2009), an internationally recognized homeopath, and Barbara Loe Fisher, cofounder of the National Vaccine Information Center (NVIC), raised the first concerns I had ever heard about the risk of neurological damage from vaccines. Barbara continues to speak publicly about how her son Chris, now an adult, sustained medically obvious symptoms of a vaccine reaction leading to a later significant learning disability and attention deficits.

In the late '80s, kids whose development had regressed following vaccination showed up in my counseling practice. Home videos of them happily waving at the camera on their first birthday were in marked contrast to those at age two, showing glassy affectless eyes, allergic shiners, and red cheeks. Today I know hundreds—yes, hundreds—of children, many now adults, who were "those kids."

In the early 1990s, before you could get instant information with the click of a mouse, you had to go to a library and do research without Google because there were few books on vaccine science. My blossoming nonprofit, Developmental Delay Registry (DDR, later renamed Developmental Delay Resources), naively surveyed parents as to what they thought was the cause of their children's behavior, learning, and physical problems. Was it genetics? Toxins? Antibiotics? Ear infections? Allergies? Birth trauma? No. Maybe it was vaccines? Many parents thought so.

In 1998, the autism world was highly agitated. British gastroenterologist Andrew Wakefield (along with 12 colleagues) discovered a novel form of inflammatory bowel disease associated with autism. The majority of parents in his original case series reported that their children's symptoms began shortly after they received the MMR vaccine. Wakefield's team later discovered measles virus in the intestinal tissues of 75 of 91 of these patients and only 5 of 70 controls.[1] Wakefield's group did research on health histories of adults with inflammatory bowel disease (ulcerative colitis and Crohn's disease) and found strong associations with early measles and mumps infections that were close in time.

Wakefield was the keynote speaker at several autism conferences and given a standing ovation as a hero. Parents of these kids were redeemed. They were not crazy. Something had happened to their smiling, happy kids, and it had something to do with their shots.

The medical world went on the offense. The American Academy of Pediatrics, strongly subsidized by pharmaceutical companies,[2] told their members that vaccines were "safe and effective." Doctors told parents that vaccines were "safe and effective," and if they didn't comply with the AAP vaccine schedule, they were "fired." Andrew Wakefield was mercilessly attacked (to the point where "Wakefielded" became a verb to describe pharma attacks on doctors who threatened their bottom line).

Good doctors who were cautious about injecting their tiny patients with seven pathogens in the same day were also persecuted by their local medical boards, who revoked their licenses.

In 2012 and then again in 2015, I held conferences in Pittsburgh I called "Vaccination Conversations." My friend Barbara Loe Fisher, mentioned above, was my keynote speaker. I invited local pediatricians to come hear her speak, but none came.

Fast forward to March 2020, when the unthinkable happened. The world shut down to protect everyone from this virus from China that could kill millions. No school, no celebrations, no going to work, nope—it was too dangerous! A vaccine against the virus was our only hope.

With Project Warp Speed, scientists produced several vaccines in record time. Within a year, many hoped we could all stick out our arms and feel safe again. Others knew that wasn't going to happen unless this miraculous vaccine was *very* effective. Unfortunately, it was not. Still, schools reopened, those working remotely went to the office a few days a week, and we could celebrate Grandma's 90th birthday in person.

At the same time, the media started censoring anyone who dared to question the safety and effectiveness of the mRNA vaccines. Their new mechanism for preventing COVID-19 was touted as safe and effective, although we have learned it is neither. Their effectiveness is exceptionally short-lived, and the serious adverse events are through the roof. Many were forced to make the extremely difficult decision to refuse vaccination, go with one using a traditional method of immunizing, or accept the new hyped mRNA. Those who knew vaccine adverse events are much more common than medical agencies admit said, "No, thank you."

Anyone who did not vaccinate became a pariah. A longtime friend told me she "abhorred" me (and has not spoken to me since) for my "selfish" decision to not vaccinate that put her at risk. You couldn't go to church or synagogue without a vaccine card. Some went underground to get that treasured document with made-up dates and lot numbers.

"Antivax" became the new slur used to demonize and dehumanize a large segment of the population. Anyone who questioned whether this new vaccine was safe or effective was "antivax." Any mother who petitioned her doctor to spread out the vaccine schedule was "antivax" and fired from the practice. Media pundits maligned anyone who was "antivax."

Bobby Kennedy, founder of Children's Health Defense, announced he was running for President. He was denied security even though that is the norm for presidential candidates, and Kennedy's status as a threat to pharma's bottom line made it more likely he would be at risk, not less. Kennedy became the poster boy for vaccine choice. He is not "antivax." What he so strongly believes in is vaccine choice. If you and your family want to vaccinate, that should be your right, but vaccines—like all drugs—do not affect all people the same way, and people should have the right to decide what is right for their families.

As the world returned to life, it was no more "normal" than those kids who had regressed following their MMR booster. Only one side of the story was told, the pharma-sponsored side. It was no longer possible to discuss vaccination in a civil fashion.

No one wants either the return of killer and crippling diseases, nor does anyone want to watch a loved one react to vaccination. One side uses emotional arguments by reminding us of disturbing times with photos of children in iron lungs. I get it.

The other side pulls up graphs showing that mortality from many contagious diseases had virtually disappeared even before vaccines arrived to save the day.[3] Did vaccines *really* save us from these dreaded diseases? Or have we been overreacting to the threat of infectious disease for decades? And at what price? Recall from Chapter 1 how many of us have immune-related conditions. Have we traded acute contagious infections for chronic illness?

Fear is the primary motivation for both vaccinating and for not vaccinating. While fear rising from a mother's intuition can save her child's life, fear imposed by the media is simply wrong. In Chapter 6 you learned that any type of fear is a sympathetic response and a poor motivator that stresses out the immune system.

Who to believe, what to do? Everyone is different, and everyone must decide what is best for themselves and their families. Education is the only answer. Let's separate vaccine science from myth so you can make rational decisions that are not based on fear.

Increasing Numbers

For about 150 years after the smallpox vaccine was introduced by Edward Jenner in 1796, additional shots were developed and recommended mostly for children and travelers. The history is too long and complicated to go into here. But most did not prove safe and/or effective, so were not generally adopted. By 1962, the total number of recommended vaccines had increased to only five and included polio, diphtheria (D), pertussis (P), and tetanus (T). Then things accelerated markedly.

In 1964, the CDC developed the Advisory Committee on Vaccines after an ad-hoc advisory committee recommended measles vaccines. Until then it was the AAP making broad recommendations. Within the next 20 years, the CDC and the American Academy of Pediatrics (AAP) recommended adding childhood vaccines for measles, mumps, and rubella, bringing the Total Load to 22 doses of seven vaccines (DPT, polio, MMR) by age six. The second dose of MMR wasn't recommended until 1989. The first vaccinations were delivered at two months, depending upon the health or birth history of the baby.

Between 1980 and 2006 the number of diseases targeted on the childhood vaccine schedule doubled, from 7 to 14. Those additional vaccines added 26–28 doses to a schedule with 22 doses already. Vaccines for chicken pox, Hib, rotavirus, pneumococcal, influenza, and hepatitis A and B were added in the 1990s.

In 2016, the CDC and AAP recommended that children receive 69–79 doses of 16 vaccines by age 18, with the first dose given at birth and additional doses given to the mother during pregnancy. Almost half of those doses occur before the age of 15 months. Additions included the HPV against the human papillomavirus for preteens at age 11–12, the meningococcal vaccine for preteens and a booster at 17–18, and annual flu vaccines.

Changes for the 2025 schedule include a reduction from three doses of Gardasil to two (if given "on time") and the addition of a single dose of an RSV monoclonal antibody product. COVID vaccines that were never approved for children under 12 were briefly on the schedule but are no longer generally recommended. By age twelve, fully vaccinated children would receive more than five dozen doses of 17 vaccines.[4] Sometimes as many as nine different injections, covering 13 different antigens are given at a time.

Table 10–1 shows the escalation in the number of vaccine doses during the past 45 years.

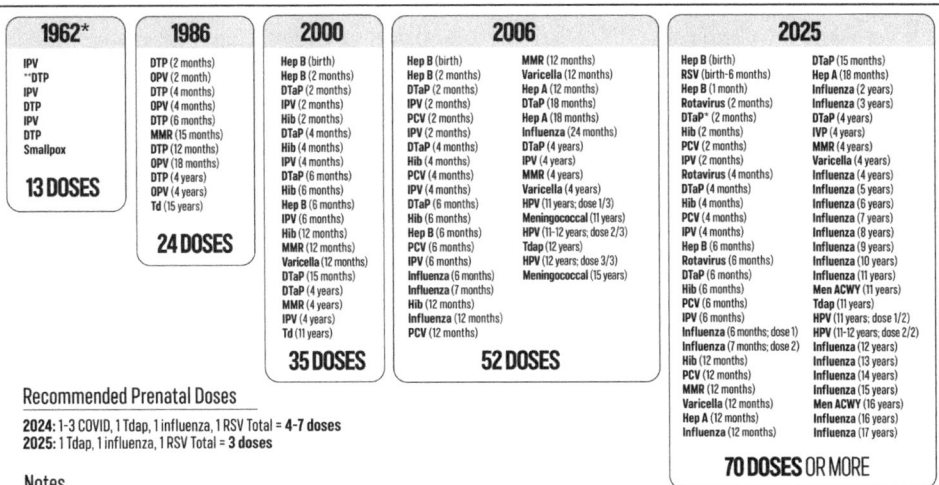

1962*	1986	2000	2006		2025	
IPV	DTP (2 months)	Hep B (birth)	Hep B (birth)	MMR (12 months)	Hep B (birth)	DTaP (15 months)
**DTP	OPV (2 month)	Hep B (2 months)	Hep B (2 months)	Varicella (12 months)	RSV (birth-6 months)	Hep A (18 months)
IPV	DTP (4 months)	DTaP (2 months)	DTaP (2 months)	Hep A (12 months)	Hep B (1 month)	Influenza (2 years)
DTP	OPV (4 months)	IPV (2 months)	IPV (2 months)	DTaP (18 months)	Rotavirus (2 months)	Influenza (3 years)
IPV	DTP (6 months)	Hib (2 months)	PCV (2 months)	Hep A (18 months)	DTaP* (2 months)	DTaP (4 years)
DTP	MMR (15 months)	DTaP (4 months)	IPV (2 months)	Influenza (24 months)	Hib (2 months)	IVP (4 years)
Smallpox	DTP (12 months)	Hib (4 months)	DTaP (4 months)	DTaP (4 years)	PCV (2 months)	MMR (4 years)
	OPV (18 months)	Hib (4 months)	Hib (4 months)	IPV (4 years)	IPV (2 months)	Varicella (4 years)
13 DOSES	DTP (4 years)	PCV (4 months)	PCV (4 months)	MMR (4 years)	Rotavirus (4 months)	Influenza (4 years)
	OPV (4 years)	Hib (6 months)	IPV (4 months)	Varicella (4 years)	DTaP (4 months)	Influenza (5 years)
	Td (15 years)	Hep B (6 months)	DTaP (6 months)	HPV (11 years; dose 1/3)	Hib (4 months)	Influenza (6 years)
		IPV (6 months)	Hib (6 months)	HPV (11-12 years; dose 2/3)	PCV (4 months)	Influenza (7 years)
	24 DOSES	Hib (12 months)	Hib (6 months)	HPV (12 years; dose 3/3)	IPV (4 months)	Influenza (8 years)
		MMR (12 months)	PCV (6 months)	Meningococcal (15 years)	Hep B (6 months)	Influenza (9 years)
		Varicella (12 months)	IPV (6 months)		Rotavirus (6 months)	Influenza (10 years)
		DTaP (4 years)	Influenza (6 months)		DTaP (6 months)	Influenza (11 years)
		MMR (4 years)	Influenza (7 months)		Hib (6 months)	Men ACWY (11 years)
		IPV (4 years)	Hib (12 months)		PCV (6 months)	Tdap (11 years)
		Td (15 years)	Influenza (12 months)		IPV (6 months)	HPV (11 years; dose 1/2)
			PCV (12 months)		Influenza (6 months; dose 1)	HPV (11-12 years; dose 2/2)
		35 DOSES	**52 DOSES**		Influenza (7 months; dose 2)	Influenza (12 years)
					Hib (12 months)	Influenza (13 years)
					PCV (12 months)	Influenza (14 years)
					MMR (12 months)	Influenza (15 years)
					Varicella (12 months)	Men ACWY (16 years)
					Hep A (12 months)	Influenza (16 years)
					Influenza (12 months)	Influenza (17 years)
					70 DOSES OR MORE	

Recommended Prenatal Doses

2024: 1-3 COVID, 1 Tdap, 1 influenza, 1 RSV Total = **4-7 doses**
2025: 1 Tdap, 1 influenza, 1 RSV Total = **3 doses**

Notes

Only vaccines generally recommended for children are listed; e.g., hepatitis A was originally only recommended for "high-risk" areas and does not appear until universally recommended.

**Though the CDC did not yet recommend vaccines in 1962, the AAP recommended diphtheria, pertussis, tetanus, polio, and smallpox vaccination. "Fully vaccinated" children received the "primary series" of 3 doses DTP, 3 polio, and 1 smallpox (as listed), but most did not do so until entering school, if at all.

** Multivalent vaccines count as 1 dose per disease; thus, DTP, DTaP, Tdap, & MMR count as 3 doses, and Td counts as 2.

***The RSV shot given to infants is technically not a vaccine as it provides passive immunity, but as it is covered by the Vaccines for Children program and is listed on the CDC's immunization schedule it is included here. It is not recommended for infants whose mothers received the RSV vaccine during pregnancy.

Most people are unaware of the number of times we bombard the immune system in the name of prevention that this table depicts so graphically. My friends, who received only a very limited number of vaccines as children, are the grandparents of today's sick kids. "We were vaccinated, and we are fine," they state boldly, refuting my distrust of vaccination, even though a majority of them either has a chronic inflammatory condition or has had cancer. I am one of those grandparents, and I was one of Jonas Salk's guinea pigs in kindergarten in my public school in Pittsburgh.

Myth 1: Immunity through vaccination is superior to natural immunity.

Vaccination fools the body into believing it has encountered a real infectious microorganism, although it does not exactly mimic the body's natural infection process. Instead of allowing pathogens to enter the body randomly through the gut or lungs, antigens and their additives are usually injected. This process stimulates humoral immunity first, bypassing cellular immunity, which should be the body's primary defense against infection.

After vaccination, just like after infection, the body produces both IgM and IgG antibodies. Recall that IgM antibodies should constitute only about 10 percent of the antibodies in the blood. IgM antibodies, the body's first responders to invaders, appear, and levels typically increase before tapering off. IgG antibodies take longer to appear and stay in the bloodstream indefinitely, providing protection against reinfection. IgG should be the most common immunoglobulin, constituting around 80% of the total antibody concentration.[5] But that is not the case in those with autism, where IgG antibodies can be high or low.[6]

Newer vaccines contain only parts of infectious organisms. This makes them less likely to cause an infection, but it also means that the antibodies produced are very specific and not protective when the antigen evolves to escape them. We saw this with the spike-protein focused COVID vaccines. Infection-induced immunity is generally much broader and more likely to be protective against future strains.

Another thing: determining the number of doses necessary to stimulate sufficient antibodies to be deemed "immune" is similar to figuring out a table for life insurance; it is based on averages. My immune system might *seroconvert* with the first dose and produce sufficient antibodies for years, while yours never does at all. It all depends on our individual combination of genetic predisposition plus load factors. The number of doses in a primary series is based on how many doses it takes to get almost all people to seroconvert, and the number of boosters is based on the *average* of the amount of time it takes before antibody levels drop. So, I might get shots I don't need, thus hyperstimulating my immune system as well as raising my levels of pathogens and additives unnecessarily, while you may require ever more shots and never even reach the desired level.

Vaccines also hijack the amygdala, the body's processor of emotion that plays a key role in our responses to threats, including fear and anxiety. Veteran scientist Laura Hewitson showed this 15 years ago, before the pediatric vaccine schedule had accelerated to today's level.[7] The baby monkeys in her experiment showed fearful behaviors when following the pediatric schedule of the time. Imagine what today's far more intensive schedule is doing to today's human infants. And we learned how important feeling safe and reducing the cell danger response are in Chapter 6.

Now, the last and maybe the most important thing: suffering through a bout of measles, chicken pox, rubella, and other viral illnesses, as I did as a child, almost always ensures permanent immunity. Encountering the pathogen again later in life (when it is usually more dangerous) does not result in reinfection. Vaccine-induced immunity, however, is not permanent.

TRUTH: Natural immunity acquired through disease is more complete and longer lasting than that obtained through vaccination.

Myth 2: The rise in autoimmune diseases such as diabetes is a mystery and unrelated to vaccines.

The late anthroposophic physician Philip Incao, MD (1941–2022), maintained that overuse of vaccines to suppress all acute, externalizing inflammation early in life can set up the immune system to respond to future stresses and infections by developing chronic, internalizing disease later in life.[8] I was fortunate to know this remarkable doctor who, according to the eulogy at his funeral, was almost a saint.[9]

The elimination of most natural experience with infectious "minor" childhood diseases, such as measles and chicken pox, through mass use of vaccines may be resulting in some unintended consequences. By disrupting the balance of cellular and humoral immunity, mass vaccination may be causing the cellular immune system to atrophy from disuse, instead of becoming stronger.

Elevating the humoral system to dominance over the cellular system through vaccination reverses the natural immunologic scheme our bodies have followed for thousands of years. Many believe this is an important overlooked factor in understanding immune issues in those with chronic illness. A key clue is that many of the bacteria, viruses, retroviruses,[10] and metals found in the bodies of our patients can only have originated from vaccinations.

Two reviews of the impact of vaccines on the developing immune system are available: *Vaccine Illusion* by Tetyana Obukhanych, PhD, an impeccably credentialed immunologist,[11] and *A Commentary on Current Childhood Vaccine Programs* by veteran physician, Harold Buttram, MD.[12]

For a very scientific deep dive into this subject read *Vaccines and Autoimmunity*, edited by Yehuda Shoenfeld and Lucija Tomjlenovic.[13] It explains what Shoenfeld calls *ASIA* (for *autoimmune/inflammatory syndrome induced by adjuvants*).

TRUTH: Vaccination is implicated in the epidemic of autoimmune disease in both children and adults.

Myth 3: Vaccines have saved the world from dread disease.

In 2009 I viewed a PowerPoint presentation at a conference showing graphs of the rise and fall of mortality rates for "dread" diseases, including pertussis and measles, against which we now have vaccines. Each graph depicted the incidence of deaths from the 1800s until the present. Virtually every one showed that mortality had declined more than 90 percent (and was still declining) by the time the vaccine was developed and marketed. This was also true for diseases for which we have never vaccinated, like tuberculosis and scarlet fever.

Today's scientists demonstrate how history "proves" that vaccines save lives by cutting off the tail at the end of the graph, starting just before the vaccine entered the picture, and blowing up that small part of the graph to make their point. If the 100+ years preceding the vaccine aren't shown, no one sees this classic way of lying with statistics I learned about 60 years ago in my undergraduate statistics classes.

TRUTH: No, vaccines did *not* save the world. In almost all cases, mortality from the disease in question had almost petered out *before* widespread use of a vaccine.

Myth 4: Vaccines are rigorously safety tested.

This is the refrain of Big Pharma. Is it true?

Many seasoned doctors and their patients are rethinking the safety and effectiveness of vaccines. One of the first was Larry Palevsky, MD, a longtime pediatrician in New York who has been educating his patients and the public about vaccine risk since the mid-1990s.

Another was Richard Moskowitz, MD, a Harvard-educated family physician with decades of experience. In his 2017 book, *Vaccines, A Reappraisal*, he provides scientific evidence for his clinical impression that vaccination comes with substantial risk. He argues that without a public health emergency, a heavy-handed mandatory vaccination schedule is unnecessary and "one of the most reckless and wildly expensive medical experiments ever undertaken!"[14]

Yet another is Paul Thomas, MD, the author of the recently published *Vax Facts*.[15] He opened his own medical practice because he believed firmly in informed consent. Parents were free to vaccinate or not. He then gave up his medical license when the Oregon Medical Board went after him for publishing his data that indicated that, not only was his recommended schedule "as safe" as the CDC's, not vaccinating at all was actually much safer! Read more about his vaxed versus unvaxed study later in this chapter. Much of the material in this chapter comes from his book.

In 2022, the publication of the English translation of an anonymously written Israeli book titled *Turtles All the Way Down: Vaccine Science and Myth* shook both the pro-vaccine and vaccine-hesitant world. Reviewers in Israel were aggressively attacked for finding this vaccine-critical book to be "well-written, serious, scientific, and important." An incensed Israeli ethicist and philosopher offered a substantial cash prize to anyone who could refute the arguments in the book's first chapter. To this day, no one has been able to demonstrate that any of the childhood vaccines approved in Israel or the United States had been tested in a true placebo-controlled clinical trial.

The English translation of *Turtles All the Way Down* was edited by attorney and Children's Health Defense President, Mary Holland, and Zoey O'Toole, co-founder of Thinking Mom's Revolution (TMR). This "must-read" tome is the source of some of the information that follows. Note that every reference in the original book comes from a mainstream source, such as a leading government agency or scientific journal.

The comical title refers to an approximately 150-year-old story depicting a conversation between a famous scientist and a simple-minded, elderly woman who believed the Earth rested on the back of a giant turtle, which in turn rested on an endless column of turtles. In *Turtles* it refers to how claims of vaccine safety should rest on pre-approval clinical trials that compare new vaccines to placebos. But, instead, the trials compare new vaccines to older vaccines or other biologically active substances, which were themselves never tested against placebos. If the number of adverse events of the new vaccine is even remotely comparable to a "control" that is considered "safe," then the new one is also. Yep, it's "turtles all the way down," leaving nothing but vapor supporting the claim that vaccines are safe. Everyone's jaw should drop when reading this 500+ page book with more than 1,200 references.

Not-previously-skeptical renowned medical doctors like Peter McCullough, Pierre Kory, Robert Malone, and Paul Marik "saw the light" during the COVID pandemic when they strayed from standard-of-care practice, were maligned, and punished for speaking out publicly. They reluctantly jumped the fence to support "medical freedom." Then, after reading *Turtles* (2022–23), they became outspoken critics of vaccine science in general.

In addition to the many profitable vaccine safety lies told to unsuspecting consumers, the world learned from *Turtles* that using "herd immunity" to justify vaccine mandates is more than a little misleading.

TRUTH: Medical science has never proved that childhood vaccines are safe.

Myth 5: Vaccine benefits outweigh their risks.

Ethically, we have a right to choose what goes into our bodies and those of our children. Before vaccination, doctors are required by law to inform all parents of the benefits of each vaccine *and* to discuss the risks, signs, and symptoms of a vaccine reaction. This knowledge would allow adults to monitor themselves and their children closely after vaccination and recognize when a reaction has occurred. In reality, doctors rarely provide a true picture of the risks.

Reactions vary from obvious to subtle. Symptoms within 24 hours, such as redness or a rash at the site of the injection, high-pitched screaming, inconsolable crying, high fever, paralysis, seizures, or loss of consciousness, can be sure signs of a cause-and-effect relationship.

Delayed symptoms, such as behavioral changes, sleep disturbances, flapping, loss of relatedness or eye contact, developmental lags in talking, or even death that occurs after 24 hours, lack a sure cause-and-effect relationship. It can even take several years after vaccination for an autoimmune disease to become clinically apparent. Given the wide range of time between the trigger and onset of symptoms, how in heaven's name can you pin *any* of these on a doctor visit that included several injections on a certain day? Difficult, right?

Vaccine reactions in infants can be misinterpreted as *sudden infant death syndrome* (SIDS) or *shaken baby syndrome*. Read Neil Miller's review on this subject.[16] Such was the famous case of Alan Yurko. His infant son received a hepatitis B shot within hours of birth and a round of vaccines 13 days before he died. The hep B vaccine is known to cause brain hemorrhaging and death in some patients.[17]

He was given heparin at the hospital, which can also induce abnormal bleeding. Yurko was accused of child abuse and jailed for eight years until finally acquitted.[18]

Often what is called "autism regression" is a vaccine reaction, which can be accompanied by other symptoms of immune dysfunction, such as new food and environmental allergies, asthma, digestive problems, skin disorders, and chronic respiratory and ear infections. Together, these vaccine-induced brain and dysregulated immune system symptoms may persist. Eventually, a team of specialists will make a diagnosis of a learning disability, attention deficit (as with Barbara Loe Fisher's son), or autism (like so many I know).

Those who suspect a vaccine reaction can report it formally to the Vaccine Adverse Event Reporting System (VAERS). If they can prove cause, adults and their children who have suffered injury or death following vaccination can receive damages from the federal Vaccine Injury Compensation Program (VICP) created under the historic National Childhood Vaccine Injury Act of 1986 (PL99–660). Even though few vaccine injury victims are compensated, as of January 2025 this program has awarded over **five billion dollars to more than 9,000 people** who have died or been injured after vaccine reactions.[19]

Unfortunately, this law also removed liability for harm from the pharmaceutical companies who manufacture the vaccines, so no one can sue them. A later Supreme Court decision, *Bruesewitz v. Wyeth*, ruled that because vaccines were "unavoidably unsafe," manufacturers couldn't be sued for design defects. Here is what they said:

> No vaccine manufacturer shall be liable in a civil action for damages arising from a vaccine-related injury or death associated with the administration of a vaccine after October 1, 1988, if the injury or death resulted from side effects that were unavoidable even though the vaccine was properly prepared and was accompanied by proper directions and warnings.[20]

The case of Hannah Poling, daughter of an MD/PhD neurologist/biophysicist father and a lawyer/nurse mother, is an important example. In 2000, when she was 19 months old, Hannah received five "catch-up" shots, with vaccines for nine different diseases, in a single day because she had missed some recommended doses due to a series of chronic ear infections.

At the time, Hannah was interactive, playful, and communicative. Two days later, she was lethargic, irritable, and febrile. Ten days after vaccination, she developed a rash; months later she was exhibiting delays in communication and relatedness and behavior consistent with an autism spectrum disorder.

Hannah's parents and doctors believed her vaccines had triggered her autism. They petitioned the VICP for compensation, and the Department of HHS conceded the case.[21] When the news was leaked to the press, officials pretended Hannah's encephalopathy was caused by a "rare" "underlying" mitochondrial disorder and refused to acknowledge her autism or that it was caused by her vaccines. (*Encephalopathy* is medicalese for *brain inflammation*.) Yet, neither Hannah's medical records nor her parents' observations indicated that prior to her catch-up vaccinations she had any symptoms of mitochondrial disorder.

Anyone interested in hearing from experts on the possible role of vaccination in autism and other disorders should watch the nine-part documentary series *Vaccines Revealed*, available online at VaccinesRevealed.com. Twenty-four of the world's leading doctors, scientists, researchers, lawyers, and parents, including Robert F. Kennedy Jr., Stephanie Seneff, Barbara Loe Fisher, Suzanne Humphries, Terry Wahls, Mary Holland, and others, discuss the science and their experiences with the above topics. Although, admittedly, some claims are exaggerated, I think you will be blown away by what they share.

Dr. Thomas urges everyone to weigh the risks of *both* vaccinating and not vaccinating, especially for those diseases like polio that have essentially disappeared in developing countries. The Vaccine Information Sheets (VIS) that a doctor is supposed to share with you prior to vaccinating minimize risk and emphasize benefit. That is why educating yourself is so important.

TRUTH: Vaccines have myriad risks that outweigh the benefits for many, especially for those with already over-burdened Total Loads.

Myth 6: Vaccine reactions are rare and completely unpredictable.

One in a million! Yep, that's what doctors tell their patients. Even if that were true, if that one were you or yours, you'd want to do everything you could to avoid that possibility. But as we have seen that doesn't even begin to cover the range of negative health effects vaccines can cause.

For over 200 years the medical literature has shown that neurological and immune system dysfunction can be caused not just by infectious diseases, but also by vaccines created with those same viruses and bacteria. Depending on the number and type of specific load factors, your risk of an adverse reaction to a vaccine is individual and different from mine. You and I each have a unique load of stressors, as did those "kids" I have known, some of whom are now adults requiring 24-hour care.

Although integrative practitioners are getting better at identifying which individuals are at a higher risk than others for suffering a vaccine reaction, you need to calculate your *personal* risk versus benefit ratio for not being vaccinated as well as from taking one. Too many shots, especially so close together, and especially without a strong ability to resolve inflammation, may not be a good idea for your family.

Do you even know how to recognize a vaccine reaction? Both immediate and subtle-but-delayed reactions are common and often unrecognized. In fact, a study that was meant to improve adverse event reporting concluded that 2.6% of all vaccinations resulted in reportable events within 30 days. That's 26,000 times more common than "one in a million." Thus, doctors frequently inform patients of benefits, but rarely of risks.[22]

TRUTH: All vaccines carry some risk of injury or death.

Myth 7: The rise in anaphylactic food allergies is a mystery.

In her groundbreaking 2011 book, *The Peanut Allergy Epidemic: What's Causing It and How to Stop It,*[23] a Canadian mother, Heather Fraser, whose child had an anaphylactic reaction to peanut butter at 13 months of age, convincingly demonstrates the relationship between vaccination and peanut allergies. This information, although based on sound science, has been rejected by mainstream medicine. The following material comes from her extensive research.

Prior to 1941, there were no reports of peanut allergies in adults or children. A survey of people self-reporting peanut allergies found 0.3% of those born in 1944–47 and 0.6% for those born between 1959–67. Articles published in the late 1950s and early 1960s show a growing awareness of peanut allergy. No formal study of peanut allergy in children was done until 1973.

In 2021, researchers estimated incidence of peanut allergy was 5.5% of children born in 2019. Doctors watched the mysterious rise in peanut allergies, but few asked why.[24]

French Nobel laureate and immunologist Charles Richet discovered that injection of minuscule amounts of undigested proteins resulted in *sensitization* to the proteins. A second injection after several weeks (the time required for an animal to develop IgE antibodies) resulted in *anaphylaxis*, a life-threatening allergic reaction. Once sensitization occurred, any leakage of matching proteins from the digestive tract could also result in deadly reactions. Vaccine antigens are often grown in media with food proteins like egg whites. Those proteins are not completely screened out in the manufacturing process. Thus, the huge increase in vaccination, combined with the breakdown of the gut lining in recent decades, inevitably resulted in a huge spike in food allergies.[25]

Interestingly, in 1964 Merck & Co. announced the development of a highly effective vaccine adjuvant that combined up to 65% peanut oil with aluminum stearate. Scientists chose plentiful, inexpensive peanut oil as a carrier in the vaccines to replace cottonseed oil because, from the 1930s through 1950, injection of cottonseed oil in penicillin caused allergies to both products. The inventors of Adjuvant 65–4 knew allergic sensitization to peanut was a distinct possibility and considered allergy an inevitable outcome of vaccination. Their job was simply to balance potency and safety. Potency won out. Adjuvant 65–4 was never used in an approved vaccine, but patented vaccines can contain undisclosed ingredients considered "trade secrets." Peanut oil is *generally recognized as safe* and could potentially be among those undisclosed ingredients. Another well-known possibility is *molecular mimicry*, that is injection of proteins that resembles peanut proteins, causing cross-reactions.

The possibility that hundreds of thousands of children were sensitized to common foods through ingredients in one or more routine pediatric vaccinations is just too much to conceive. But it is too obvious to deny. The real clue is the sudden rise in peanut allergy following the escalation of the pediatric vaccine schedule.

TRUTH: Today's epidemic of anaphylactic allergies was caused by vaccines.

Myth 8: Vaccinated children are healthier than the unvaccinated.

Dr. Paul Thomas developed a "Vaccine-Friendly Plan" (VFP) that he felt would benefit his patients by delaying, spreading out, and eliminating unnecessary vaccine doses. He later published this plan, along with Jennifer Margulis, an award-winning science journalist unafraid of controversial subjects.[26]

This welcome book for concerned consumers brought Thomas to the attention of the Oregon Medical Board, who began hounding him for not complying with the carved-in-stone CDC schedule. They demanded he prove that his schedule was safe.

Thomas recognized that his pediatric practice had yielded over ten years of data on over 3,000 patients, 2,763 vaccinated and 561 unvaccinated. He contacted science researcher James Lyons-Weiler, PhD, mentioned above, to analyze this wealth of information. It was the perfect opportunity to carry out a longitudinal study of health in his vaccinated (modified schedule) and unvaccinated patients. The bottom line: the highly vaccinated are sicker than the unvaccinated. While a vaccine may reduce the chance of getting a specific disease, it makes you more vulnerable to other infections.

Drs. Thomas and Lyons-Weiler's analysis of the data was published in a peer-reviewed journal.[27] It is the largest real-world study of the vaccinated/unvaccinated carried out to date. The journal that published the study retracted it against normal procedure, but not before the Oregon Medical Board reacted by suspending Dr. Thomas's medical license. Rather than risk the hundreds of thousands of dollars it would cost to defend himself, he retired and relinquished his medical license in December 2022. Read about this witch hunt in Jeremy Hammond's book, *The War on Informed Consent*.[28]

Lyons-Weiler, however, reanalyzed the data with Russell Blaylock in another study that confirmed the original conclusions and addressed the previous journal's assumption of bias. The results may surprise you.

- Office visits for asthma, allergic rhinitis, behavioral issues, ADHD, eczema, and anemia were four to eight times as common in those who had followed his vaccine-friendly plan as in the unvaccinated;
- Even vaccinating slowly wasn't as safe as not vaccinating at all;
- Children whose parents stopped vaccinating at any time had half the health challenges of those who continued vaccinating.[29]

How many practicing pediatricians do you think know about these findings?

TRUTH: Unvaccinated children are healthier than vaccinated children.

Myth 9: What's in vaccines cannot hurt you.

Few know that, in addition to the pathogens, vaccines contain a multitude of potentially toxic additives, and multidose vaccine vials (which includes COVID shots) contain preservatives (to keep the vaccine sterile).

Excipients is the scientific term for these agents, which can be dangerous individually, and the toxicity of their interaction is virtually unexplored. Vaccine excipients include mercury (as the preservative thimerosal), aluminum, formaldehyde, phenoxyethanol, gluteraldehyde, sodium chloride, monosodium glutamate (MSG), yeast, sodium borate, propylene glycol, human and bovine albumin, gelatin, and others.

Let's start with the preservative ***thimerosal***, which is 49.6% *ethylmercury* by weight. It has been used in vaccines and other pharmaceuticals like eye drops for more than 50 years. Adding thimerosal allows medical personnel to use one vial of vaccine for more than one person, thus reducing the cost. The more expensive single-dose vials do not require a preservative to remain sterile.

Multidose vial vaccines have many advantages for doctors, medical facilities, and the environment. They are cheaper, more convenient, and make less waste for sure. However, they present considerable possible danger to the patient if CDC precautionary measures[30] are not followed to the letter. Absolutely no one is assured of getting the "right" amount of each ingredient. A researcher who tested this hypothesis found that later doses were likely to have greater amounts of mercury.[31] Pretty interesting, huh?

In 1930, Eli Lilly conducted a single study on 22 patients seriously ill with meningitis to test whether thimerosal was safe for injection into humans. All subjects died, but their deaths were attributed solely to meningitis and not to any effects of thimerosal. Eli Lilly then pronounced thimerosal safe for human use.[32]

After a series of government hearings beginning in 1998, the FDA banned thimerosal from most products, including children's vaccines, because of potential toxicity and many adverse reaction reports.[33] Still, thimerosal was allowed to remain in multidose vials of certain vaccines, including those against the flu. Did you know that those thimerosal-laced vaccines are sometimes given to pregnant women and infants over six months of age?

Although pharmaceutical companies want you to believe that thimerosal has been *totally* removed from *every vaccine*, that is not true. This topic is too big to cover in this book; if you want to know more about it, watch the film *Trace Amounts: Autism, Mercury, and the Hidden Truth* and read RFK Jr's *Thimerosal: Let the Science Speak: The Evidence Supporting the Immediate Removal of Mercury—a Known Neurotoxin—from Vaccines*.[34] Fortunately, at long last the Advisory Committee on Immunization Practices has *just* voted to recommend only single-dose, thimerosal-free flu vaccines.

Adding mercury is only one way vaccines contribute to the Total Load. One study showed that, although mercury was removed from many vaccines, other culprits, especially aluminum, may link vaccines to autism.[35]

Adjuvants: Other vaccine additives called *adjuvants* (from the Latin *adjuvare*, to enhance) stimulate the immune system, upping the body's production of antibodies to a pathogen. Adjuvants reduce production costs because the vaccine maker needs less of the expensive antigen.

The late immunologist Charles Janeway (1943–2003) called vaccine adjuvants "the immunologist's dirty little secret."[36] On their own, adjuvants have the potential to cause not only an increased immune response, but also inflammation. When added to a vaccine, they increase the risk of allergies and other adverse effects. Neurosurgeon Russell Blaylock, MD, argues that repeatedly stimulating the immune system with adjuvanted vaccines can result in intense inflammatory reactions.[37]

Aluminum, long known as a neurotoxin,[38] has dramatic negative effects on the immune system and is the most common ingredient in vaccine adjuvants.[39] Injecting it into the bloodstream is different from ingesting it in food. Yet, that fact has been ignored.

Aluminum appears in vaccines as aluminum gels or aluminum salts such as aluminum hydroxide, aluminum phosphate, and aluminum potassium sulfate, which are in vaccines against hepatitis A and B, diphtheria-tetanus-pertussis (DTaP, Tdap), *Haemophilus influenzae* type b (Hib), human papillomavirus (HPV), and pneumococcus infection. The worst one is, *amorphous aluminum hydroxyphosphate sulfate* (AAHS), which is in the HPV vaccine, Gardasil, as well as Recombivax and Vaxelis.

Symptoms of aluminum poisoning include personality changes, progressive speech disorder, stuttering, apraxia, tremors, myoclonic jerks, seizures, inflammation, abnormal EEG, psychosis, and dementia.[40] No, aluminum used as a vaccine adjuvant isn't safe![41]

The results of a Canadian study confirm that aluminum adjuvants are also suspect in autism. Children from countries with the highest ASD prevalence had the highest exposure to aluminum from vaccines. The increase in exposure to aluminum adjuvants significantly correlates with the increase in ASD prevalence in the United States observed over the last two decades, and a significant correlation exists between the amounts of aluminum administered to preschool children and the current prevalence of ASD in seven Western countries.[42]

Mainstream media is still preaching that the benefits of using aluminum as an adjuvant outweigh potential dangers. The very same week I am writing this chapter, two newspapers published articles declaring aluminum in vaccines safe: *The New York Times,* "Yes, Some Vaccines Contain Aluminum. That's a Good Thing,"[43] and the *Daily Tribune,* "Robert F. Kennedy Jr. and the Debate over Aluminum in Vaccines."[44] In an immediate response, highly respected science researcher, James Lyons-Weiler, PhD, set the record straight point by erroneous point, declaring the articles "dangerously misleading" for ignoring decades of research showing aluminum's toxic potential.[45]

J.B. Handley, father of Jamison, born in 2002, and cofounder of the nonprofit, Generation Rescue, believes aluminum from vaccines is strongly implicated in his son's autism. He cites numerous papers from scientists around the world (interestingly, none from American scientists) that illuminate every aspect of a mechanism by which injected aluminum adjuvants can and do cause autism.

One of these studies[46] comes from Christopher Exley, PhD, a British scientist who has been evaluating the effects of aluminum on health for over 35 years. He and his colleagues at Keele University in England found exceptionally high levels of aluminum in brain tissue from people who died with autism. This finding implies that aluminum had somehow crossed the blood-brain barrier.

The method of transport and location of the aluminum in the brain was of particular interest to Exley. Both blood and lymph cells were "carrying a cargo of aluminum" into the brain, where it was deposited in the very locations known to be inflamed in people with autism. His conclusion was that the source of this aluminum was "probably vaccines containing aluminum adjuvants and that the presence of aluminum in inflammatory cells in the meninges, vasculature, grey and white matter is a standout observation and could implicate aluminum in the etiology of ASD."

Two years later Exley examined the amount of aluminum that would be absorbed from a single dose of the vaccine Infanrix Hexa, given to infants in the UK at eight weeks of age, and found it to be equivalent to 25 times the accumulation of aluminum that would occur from daily doses of even the most contaminated infant formulas. And this is only one of *three* doses recommended in the pediatric schedule. When this vaccine is given along with two other aluminum-containing vaccines on the same day, the combined total load of aluminum far exceeds safe levels, causing acute exposure.[47]

Amazingly, these remarkable pieces of research, which took two years of dedicated work to complete, were totally ignored by the American press and only circulated through social media and blogs such as Handley's. Professor Exley was as unprepared for the "din of silence," as he called it, as he was for criticisms of the journal for publishing the research, of the brain bank for providing him with the tissues he studied, and of the research methods used in obtaining the data.[48] A year after that study was published, his funding was terminated,[49] and then he was forced out altogether in 2021.

Want to dig deeper into aluminum? Read Exley's 2020 book *Imagine You Are an Aluminum Atom: Discussions with Mr. Aluminum.*[50]

The Handleys, father and son, tell their remarkable story in the 2021 book, *Underestimated: An Autism Miracle,*[51] which recounts how Jamie emerged from his prison of silence using a method called Spelling to Communicate (S2C). Learn more about this method and Jamie's breakthrough in Chapter 20.

Aluminum compounds used to be the only adjuvant used in vaccines in the United States, but in the past few years FDA has been approving vaccines with several new (untested) adjuvants. What will they add next?

TRUTH: Vaccines contain ingredients that are potentially toxic and tilt the immune system in favor of chronic brain inflammation, which can result in autism and other neurological diagnoses.

Myth 10: Vaccines do not cause autism.

While the government continues to want us to believe there is no relationship between vaccination and autism,[52] parents and some doctors have been convinced of a link for decades. The late research journalist David Kirby (1960–2023), author of *Evidence of Harm* (2005),[53] and activist Louise Kuo Habakus and CHD President Mary Holland, editors of *Vaccine Epidemic* (2011),[54] convincingly reviewed evidence for a cover-up by government health officials and vaccine manufacturers of data linking autism with vaccines. Since then, many more scientific studies have confirmed the vaccines/autism connection.

One, a 2013 study, demonstrated more than ten years ago that children who received mercury-containing vaccines were significantly more likely to have autism.[55] A current study of almost 50,000 nine-year-olds who receive Medicaid in Florida found that the more vaccinations a child received, the more likely they were to be diagnosed with a neurodevelopmental disorder. This study included not only autism, but ADHD and learning disabilities as well.[56]

Proposed biological mechanisms for thimerosal-induced autism have focused on mercury-containing vaccines and genetic factors interacting to create

- oxidative stress leading to impaired methylation and lowered levels of cysteine and glutathione,[57]
- DNA damage and death of brain cells,[58]
- brain inflammation,[59]
- mitochondrial damage,[60] and
- autoimmunity.[61]

A mechanism involving aluminum adjuvants has been fully demonstrated. Nanoparticles of aluminum are engulfed by macrophages, which can cross the blood-brain barrier to the microglia. If the macrophages die while there, they deposit their aluminum and provoke long-term inflammation. The aluminum causes certain cytokine levels to rise, particularly IL-6, which interrupts brain development in laboratory animals and leads to autistic behaviors. And we know that aluminum causes autoimmunity (ASIA), including *autoimmune encephalitis*, a condition that often presents as autism.

Empirical evidence for a vaccine-autism connection has continued to grow even as doctors and scientists inside industry, government, and medical organizations persist in denying a vaccine-autism relationship. And autism is not the only outcome. Autoimmune disease, mood disorders, attention deficits, learning disabilities, and mental health disorders can all often be traced back to too many vaccines, too close together in susceptible individuals. Officials' continued denial of the truth perpetuates a myth supported by "turtles all the way down."

"Impossible," said their doctors. "Improbable," declared the grandparents. "Inevitable," concluded those who watched their sickly kids regress.

TRUTH: Vaccines can and do cause autism in those whose Total Load exceeds their limit.

Take-Home Points

The science of vaccine safety and effectiveness is scant and flawed. Vaccines have not saved the world from dread disease. What is in vaccines can hurt you. Vaccine reactions are common and can be lethal. Natural immunity is more complete and longer lasting than vaccinated immunity. We have exchanged temporary contagious disease for chronic autoimmune disease. Unvaccinated children are healthier than vaccinated ones. Vaccines can and do cause autism, ADHD, and other neurodevelopmental disorders. The debate is over. *Period*!

If you are a prospective parent or a grandparent with adult children considering having children, read more about vaccines in Chapter 23. You'll be glad you did.

CHAPTER 11

A THREE-PRONGED APPROACH TO HEALING THE PHYSICAL BODY

We are finally ready to address the Total Load triad of medical problems that is sickening our loved ones with neurodevelopmental disorders. After calming down the agitated nervous system, a synergistic program that simultaneously balances the immune system, eliminates infections, and fights inflammation is the next step. Remember that a majority of the immune system lies in the inflamed gut, so this is a monumental step.

Personalizing each unique patient's needs necessitates taking an exhaustive history, running traditional and nontraditional testing with measures described in previous chapters, and a thorough physical exam, not often available in these days of telemedicine. The more information we have, the more precise the treatments.

Before reading further, return to Chapter 2 to review the body's two major biological load factors: oxidative and immunological stress. Reducing dysfunction in these basic arenas is a priority because they make the body vulnerable to all categories of pathogens and systemic inflammation. This chapter focuses on lessening immunological stress, while toning down oxidative stress is part of the chapter on detoxification.

We begin with immune boosters that should be a part of everyone's lives: whole foods, good quality supplements, and some specialized products. This chapter then moves to addressing mold, the silent and oft-missed culprit that Dr. Nathan has identified as an underground immune suppressor hiding in his sickest patients. Whether mold could be an underlying root cause for you or your loved ones will emerge during comprehensive history-taking and testing. Simply said, no one can get well from any condition if living or working in a moldy building, period.

Next comes mold, Lyme, and PANS with their combination of mycotoxins, bacteria, and autoantibodies that cause brain encephal*itis*, the most important "itis" disorder that, by definition, implies inflammation and is responsible for many of those DSM diagnoses. The underlying havoc caused by mold, Lyme, and PANS is often missed, misunderstood, and mistreated and cannot be overemphasized.

Following a prescribed biomedical plan that targets lurking possible root causes can frequently diminish or eliminate symptoms, allowing suffering loved ones to return to an almost normal life. Let's get started.

Strengthening Immunity

About 400 BC, Hippocrates proclaimed, "Let food be thy medicine and medicine be thy food." Over 2,000 years later, that advice is still sound.

Diet: The best way to boost anyone's immune system is to choose nutrient-dense, anti-inflammatory real foods like fresh fruits, vegetables, seeds, nuts, beans, gluten-free grains, spices like turmeric and ginger, good quality fats, and plant and a few animal proteins. Remember we always aim to "take out the bad stuff and put in the good stuff."

A few specialty foods you may not have heard of are earning high marks as immune boosters:

Broccoli Sprouts are a favorite "go to" solution for just about everything according to Dr. Dietrich Klinghardt. They contain sulforaphane, a powerful antioxidant that enhances the body's ability to detoxify. Grow your own, cutting them when they are about two inches tall, or buy them at the supermarket. Alternatively, drink a heaping tablespoon of powder in water twice a day. Research from Harvard Medical School shows this simple and inexpensive treatment increased social interaction and decreased several behaviors characteristic of autism, such as repetitive movement and speech, in teenage boys.[1]

An additional plant that Klinghardt likes is Mediterranean *Cistus incanus*, a powerful antiviral found in Mediterranean countries. Research shows its effectiveness against not only mold, but also the toughest of the tough: Lyme, Ebola, and HIV.[2]

Klinghardt recommends making cistus tea from the leaves. Sweeten with whole-leaf stevia tincture (not commercial clear stevia extract). Leaves can be reused twice more. Order the highest quality tea leaves and stevia from KiScience.com.

Camel Milk: one of the most interesting is highly nutritious camel milk. No joke! Nomads in the desert have recognized the power of camel milk for centuries.[3] With five times the vitamin C, ten times the iron, and more omega-3 fats and calcium than cow's milk, it may just be the perfect food since the type of casein it contains is nonreactive for even the most sensitive kids.

Camel milk targets all three steps: it can not only strengthen the immune system, but also eliminate food allergies and gut problems, fight viruses and bacteria, *and* lessen inflammation![4] Its pathogen-fighting qualities are truly amazing because of the small size of the immunoglobulins that can penetrate deep into the tissues to fight infection.[5]

After a 2005 research study showed that camel milk was like intravenous immunoglobulin therapy (IVIG; see below) for children with autism,[6] mothers jumped all over it. The latest study shows that one of the reasons camel milk is so beneficial is that it decreases oxidative stress by increasing the antioxidant glutathione.[7] (More about that in the next two chapters!)

Learn more about the benefits of camel milk in *Camel Crazy: A Quest for Miracles in the Mysterious World of Camels*[8] from my friend Christina Adams. Want to try some camel milk without traveling to a desert? Order online from DesertFarms.com and ColoradoCamelMilk.com. Camel milk is available fresh, frozen, or powdered. If you are serious about using this product, join the Healing with Camel Milk closed group on Facebook with over 18,000 members.

Supplements: If a child will not eat nutritious, immune-boosting foods, second choice is using immune-boosting supplements. All supplements are not created equal. Buy the best you can afford: pharmaceutical grade with no fillers, colors, or anything except what is therapeutically necessary. The now-classic 1989 book *Superimmunity for Kids,*[9] by Leo Galland, MD, is an excellent guide and a keeper for a lifetime, as chapters move sequentially from prenatal to adult immunity with both food and supplement options.

Here are some special immune boosters that can benefit anyone—with or without poor immune function:

Vitamin D is a fat-soluble vitamin that is produced naturally as D3 or *cholecalciferol,* when sunlight hits the skin. The body makes what it needs, accumulating and storing significant reserves in the tissues, liver, spleen, bones, and brain; it can then be available during darker months. Vitamin D acts more like a neuro-steroid hormone than a vitamin, directly affecting brain development and regulation of behavior. It is crucial in immune system function.[10]

Research suggests that vitamin D offers neuroprotection, antiepileptic effects, and immunomodulation of several brain neurotransmitter systems and hormones. It enhances the body's ability to fight inflammation and destroy dangerous microbes.[11]

This "sunshine vitamin" made headlines during the COVID-19 pandemic as a factor in survival, severity, and treatment. Bottom line: those with vitamin D deficiency were more likely to be badly affected by the virus. The lower the level, the more severe the disease and the higher the possibility of dying. Treatment with supplementation raised survival and recovery rates.[12] Yes, vitamin D levels are important.

Evidence supports a vitamin D deficiency theory of autism.[13] In 2012, researchers at the Gillberg Neuropsychiatry Centre in Sweden found the vitamin

D–autism connection so strong that they called for "urgent research."[14] In a
Dutch study, being born to women who had low blood levels of vitamin D while
pregnant more than doubled children's risk of autism.[15] Maybe the high autism
incidence among Somalis living in Minnesota is due to the fact that their bodies
have adapted to the strong sun in their native land making them unable to manufacture enough vitamin D in their new home.[16]

Integrative psychiatrist James Greenblatt, MD, has been studying root causes
of his patients' mental health disorders for over four decades. In a recent article he
examines the role of vitamin D deficiency in depression, ADHD, anxiety, suicide
rates, PTSD, and more.[17] His impeccable research shows an inverse relationship
between vitamin D levels and symptoms in all of these plus other mental health
disorders. Good news: supplementation can decrease and even eliminate symptoms. Convinced yet?

Blood levels of vitamin D should be 50–100 ng/ml. Ask your doctor to test
your level twice a year, or buy an inexpensive home test kit online. Stick a finger
to get a few drops of blood and send the kit back. The results are mailed to you
without involving a physician.

Work with a healthcare practitioner to find a supplementation program that is
right. Some bodies absorb drops more easily than tablets or caps. Others require
high dosages for a short time to reach acceptable levels. Monitoring is essential to
avoid toxicity. Do you know your vitamin D level?

The Mitochondria Cocktail: refer back to Chapter 9 on the immune system
to refresh your memory on *mitochondria*, those Energizer batteries in each cell.
Recall, too, that it was a mitochondrial disorder that convinced the "Vaccine
Court" to award damages to the family of Hannah Poling, and according to the
scientists who study them, they are much more common than the Department of
HHS implied at the time.

For over a dozen years, mitochondrial disorders have been implicated in neuropsychiatric disorders, including bipolar, depression, schizophrenia, and anxiety.[18]
Energy impairment is an underlying mechanism driving abnormal behaviors.[19]

These powerhouses get damaged in a number of ways, including the overuse of
antibiotics and attack by destructive molecules called free radicals, which can hurt
their fragile cell membranes and disrupt energy production. Our mitochondria
need a varied diet of vitamins and other nutrients for efficiency and protection
from further damage. Nutritional guru Kelly Dorfman offers a basic cocktail of
nutrients critical for mitochondrial function and repair:

B vitamins: thiamine (vitamin B1) and riboflavin (vitamin B2) are both
required cofactors for different parts of energy making. A typical mitochondria

formula may contain 50–100mg each of vitamins B1, B2, and/or B3 (as niacina-mide). Balancing B's isn't easy, so they should be added one at a time, with close medical supervision if agitation occurs.

Co-enzyme Q10 (CoQ10) is a fat-soluble vitamin that is found in every cell of the body. Available as ubiquinol (or the less absorbable ubiquinone), its unique enzymatic function promotes the energy metabolism needed to increase muscle tone and protect mitochondria from free radical damage.

Vitamin E (mixed tocopherols) is another important anti-free-radical agent for protecting and healing cell membranes. Vitamin E is well-tolerated and has no known toxicity. Natural vitamin E is usually derived from wheat or soy, so in rare cases it may cause allergic problems.

Acetyl-L-Carnitine (ALC) metabolizes the fat from food and turns it into energy. L-carnitine alone can help maintain the integrity of the membranes of the mitochondria and facilitate the transport and utilization of fats so they can be used to make energy. But L-carnitine can sometimes cause irritability or stomach distress. It may also increase the risk of seizures in children with a history of epilepsy. Adding the acetyl group to it makes it more bioavailable to the brain, thus supporting cognitive function and energy, according to autism doctor Darin Ingels ND.

Transfer factors (TFs) and Colostrum are fantastic naturally occurring immune boosters that contain an individual's immune history. They aren't vitamins, minerals, or herbs. Rather, they are messenger molecules made by white blood cells that travel to the maturing mammary glands and release the donor's immune history into maturing mammary tissues.

Colostrum, the liquid breast-fed babies receive postpartum for the three or four days before a woman's milk supply comes in, is a transfer factor. As the newborn nurses, he receives his mother's immune history via her immune factors, hence the name. If one thinks of vitamins and minerals as the bricks and mortar that build immune system cells, TFs are the blueprints that determine where the bricks will be placed.

An amazing combination of substances that strengthen and support the immune system, colostrum potentiates the development and repair of cells and tissues, ensuring the effective and efficient metabolism of nutrients. It includes cytokines that inhibit inflammation and growth factors that promote maturation of the immune system.[20]

Bovine colostrum has been used successfully to boost immunity in many immune-deficient illnesses.[21] Transfer factors are now available in capsules, chewables, and powders. Some manufacturers offer chewable colostrum in flavors like chocolate, orange, cherry, and pineapple.

Immunologist Hugh Fudenberg, MD (1928–2014), is the researcher whose name is usually associated with TFs. In a small study of children with autism, all but one responded positively to TF.[22]

Kenneth A. Bock, MD, uses transfer factors as part of his treatment protocols. He believes that TF is an effective immune modulator, especially for those with chronic viral infections and certain autoimmune disorders.

Colostrum as an immune system modulator is beginning to be investigated in mental health disorders. A recent Iranian study showed benefit especially in patients with anxiety, depression and sleep disorders.[23]

Transfer factors should not be combined with digestive enzymes. For more information, read *A Guide to Transfer Factors and Immune System Health.*[24]

Remediating Mold Toxicity

After addressing nervous system dysfunction, mold sickness must be ruled out. Many molds with funny names like *aspergillus, cladosporium,* and *stachybotrys* are toxic to the immune system. Their spores can colonize in the sinuses and gut, insidiously producing toxins that raise the Total Load for months or even years without clear symptoms.

Inspect for mold. Return to Chapter 4 for details on how to take this important step both informally and professionally. Look everywhere and anywhere mold can lurk, in warm and wet places. At home, in and under the fridge, carpet, or wallpaper, in the dishwasher, near a leaky toilet, in the basement (especially if "finished"), at school, at work, in the car, in air ducts. Because our eyes may not detect tiny spores, formal testing is necessary. You can use home testing kits (envirobiomics.com), or hire someone to evaluate suspect places.

Measure mold mycotoxins. The most accurate way of making a mold toxicity illness diagnosis is a simple urine test. Savvy doctors send samples to MosaicDX. com, RealTimeLab.com, and MyMycoLab.com. These tests vary in the information they provide about the dozens of dangerous mold byproducts that can hurt you, even at low levels. Urine mycotoxin testing measures only excretion. MyMycoLab, which also measures IgG and IgE antibodies against mycotoxins, better reflects exposure and mast cell activation. You need a medical referral to order one of these tests.

Evaluate symptom etiology. Mold toxicity causes dozens of symptoms, including, but not limited to, overwhelming fatigue, brain fog, pain, anxiety, depression, OCD, gastrointestinal dysfunction, and chronic sensitivity to sound, light, food, and EMFs. Dr. Jill Crista has developed a questionnaire to assist you and your doctor to determine whether mold toxicity is a possibility. Access it at DrCrista.com.

Eliminate and Treat Mycotoxins

Foods: some foods including peanuts, mushrooms (a fungus itself), cantaloupes, and cheese are naturally moldy and add to the Total Load. Other fruits, vegetables (and juices made from them), and herbs and spices are antioxidants that act like Pac-Man, gobbling up the free radicals that cause oxidative stress. Tomatoes, broccoli, artichokes, garlic, red beans, pecans, walnuts, green tea, blueberries, pomegranates, acai berries, cinnamon, cardamom, turmeric, cumin, mustard seeds, oregano, and parsley all fall into this category.

Binders: depending upon which molds are present, Drs. Nathan and Klinghardt believe that binders are absolutely necessary to attach themselves to the mycotoxins, so they can be urinated or pooped out.

Nathan provides a chart on page 122 of his *Sensitive Patient* book on herbs and other supplements to consider. These include quercetin, curcumin, resveratrol, and more. He also recommends binders like activated charcoal, bentonite clay, chlorella, and *saccharomyces boulardii,* a beneficial fungus that is frequently used to balance out "bad" fungi.

Klinghardt also likes *zeolite* and **Matrix Minerals**, a peet extract made of composted organic material taken from deep under the earth's surface in ancient freshwater lake beds. Matrix Minerals contains fulvic and humic acids. Fulvic acid is death to many viruses and detoxes glyphosate.

Dilute in water and spray throughout the entire house twice a week to infuse the body and brain with many beneficial nutrients, such as trace minerals and electrolytes that optimize cell absorption and normalize the pH. Matrix Minerals can also be taken internally. Start with two drops away from food and increase to a full pipette. Available from biopureus.com.

Propolis: one of my favorite all-natural weapons against mold is *propolis*. A waxy balsamic resinous substance that forms inside beehives, its potent antimicrobial properties were discovered several centuries BC.

The word *propolis* is made up of two Greek words: *pro,* meaning *in front or in defense of,* and *polis,* which means *town.* Propolis is from the town of the bees, or a beehive. Propolis kills molds and has fantastic long-term anti-inflammatory effects.[25]

Propolis can be taken orally or applied externally. Or heat it in a vaporizer to release it into the air at home or in the car. Find it easily online or even in high-end pharmacies. I found a spray and "shot" bottle by *Beekeeper's Naturals* at Trader Joe's!

Treating Lyme, PANDAS/PANS, and Other Infections

Lyme Disease

Along with mold, undiagnosed Lyme disease underlies many brain and body ills. As you read in Chapter 9, Lyme disease comes from bacteria in the Borrelia Complex and their partners in crime the bacterium *Bartonella* and the malaria-like parasite *Babesia*.

Symptoms of mold toxicity and Lyme are similar and include everything from unrelenting fatigue to cramping and brain fog. Once again, exhaustive history-taking accompanied by specialty lab testing is absolutely necessary. Check out the testing chapter for recommendations on laboratory Lyme testing.

Much controversy exists about not only what Lyme is, but how to treat it. Each Lyme-literate doctor has their own tweaked method. Klinghardt has developed a very complex Lyme cocktail that emphasizes herbs rather than the traditionally prescribed antibiotics. See the endnotes for a link.[26]

Neil Nathan, MD, covers Lyme extensively in two of his books: *Toxic: Heal Your Body from Mold Toxicity, Lyme Disease, Multiple Chemical Sensitivities, and Chronic Environmental Illness*[27] and *The Sensitive Patient's Healing Guide*. Nathan refers to his colleague Richard Horowitz, MD, a pioneer in Lyme treatment, for details on treatment. They are cofounders of the International Lyme and Associated Diseases Society (ILADS) and have treated more than 15,000 cases of tick-borne diseases. Horowitz's 2013 classic *Why Can't I Get Better? Solving the Mystery of Lyme and Chronic Disease*[28] and his newer *How Can I Get Better? Treating Resistant Lyme & Chronic Disease*[29] are excellent resources.

Another of my favorite doctors is Darin Ingels, ND, the author of *The Lyme Solution: A 5-Part Plan to Fight the Inflammatory Auto-Immune Response and Beat Lyme Disease*.[30] Ingels is a seasoned naturopath in California who has walked in your shoes. He contracted Lyme when living in Connecticut and has treated chronic Lyme not only in thousands of patients, but also in himself.

Like Klinghardt, Ingels recognizes that the conventional way to treat Lyme with antibiotics does not get to the root of the problem. Why? Because Lyme is a complex medical problem that causes immune system dysfunction, inflammation, circulation problems, and hormone imbalances. His Total Load approach using diet, nutrients, herbs, homeopathy, and immunotherapy is totally compatible with this book.

The Lyme Solution offers a five-step immune-boosting plan: strengthening the immune system, eating an immune-boosting diet, targeting infections using natural substances, tweaking lifestyle and environment, and reducing stress levels. How we all wish that it was not so difficult!

A couple other excellent Lyme resources:

- Daniel Kinderlehrer's Recovery from *Lyme Disease: The Integrative Medicine Guide to Diagnosing and Treating Tick-Borne Illness*.[31] Kinderlehrer is a nationally recognized physician with expertise in the fields of nutrition, allergy, environmental medicine, and Lyme disease and cofounder of The New England Center for Holistic Medicine in Massachusetts.
- The classic book by the late Steven Buhner (1952–2022), *Healing Lyme: Natural Healing of Lyme Borreliosis and the Coinfections Chlamydia and Spotted Fever Rickettsiosis*.[32]

Strep Infections and PANDAS/PANS

If after remediating mold, someone presents with abrupt OCD and other unusual behaviors, think PANS or PANDAS, oft-missed autoimmune reactions to infections that can cause brain inflammation, psychiatric symptoms, and more. Sudden onset obsessive-compulsive behaviors, tics, and handwriting and sleep issues can be triggered by strep and/or other infectious bacteria, viruses, or parasites carried by ticks, spiders, and mosquitos.

While the usual signs of strep (such as a sore throat) may or may not be present, unusual symptoms can also appear almost everywhere: sudden stuttering, a red rectal rash or ring, an eating disorder, pupil dilation (indicating a sympathetic nervous system stress response), air hunger, night sweats, a bumpy red "strawberry tongue," or even hallucinations and "Alice in Wonderland" syndrome, where everything looks distorted.

These can all result as these critters trigger antibodies that attack the basal ganglia, the part of the brain responsible for controlling movement, emotional regulation, and learning and behavior, causing overwhelming brain inflammation.[33] Whoa—pretty important, wouldn't you say?

PANS is *so* important, that a 2017 special issue of the *Journal of Child and Adolescent Psychopharmacology* was devoted to it.[34] Despite this recognition, its identification, definition, and treatment remain controversial. Many veteran docs have stepped into the fray, including Drs. Klinghardt, Jill Crista, Kenneth Bock, and Nancy O'Hara. What follows is a combination of their recommendations.

Begin with testing: a complete physical exam followed by lab work. Refer back to Chapter 9 and Moleculara Labs' Autoimmune Brain Panel™; it has emerged as the gold standard in testing. Their combination of tests looks for underlying biological markers of psychiatric symptoms: hormones, enzymes, carbohydrates, and others common in autoimmune disorders.

Knowing what nutrients are missing and what toxins or enzymes are present that should not be there allows treatments to target root causes with precision. An alternative to the Moleculara panel is Klinghardt's unique Autonomic Response Testing, described in Chapter 5, now being used around the world by his mentees.

Treatments focus on the three-pronged approach of this chapter: immune support, identifying and stopping the infections, and taming inflammation. In addition, they target problematic chronic symptoms (e.g., insomnia) with natural products (melatonin and valerian), as appropriate, until things calm down. Each doctor has their favorite products and protocols. Choose your practitioner with care; look for longtime experience, good diagnostic skills, and an open mind. Find one near you at PandasNetwork.org.

Helpful reading on this subject includes Kenneth Bock's *Brain Inflamed*,[35] Dr. Jill Crista's *A Light in the Dark*,[36] Dr. Nancy O'Hara's *Demystifying PANS/PANDAS*,[37] *Brain under Attack*,[38] and a very scientific chapter in a textbook on strep.[39] Healing is gradual, flares are common, and patience is required.

Targeting Other Infections

If antibiotics work against the body's ability to heal itself, what works in concert with the body? From ancient times, healers have prescribed herbs and other natural substances to eliminate all types of pathological organisms living within the human body.

Here are some recommendations to fight against other specific bugs, mostly from Klinghardt.

- **Borna virus.** Klinghardt teaches that this nasty bug attacks the limbic system and is thus responsible for psychiatric symptoms, especially manic depression. High levels of Borna viruses are apparent in mania. He uses herbal antivirals, like olive leaf and mushroom extracts for at least four months.
- **Giardia, amoebas, and protozoa.** Klinghardt treats these and other bugs with rizoles, a new class of bioactive compounds created by ozonating plant oils, including clove, wormwood, marjoram, cumin, and others, for several weeks. Ozone, a form of oxygen therapy, kills anaerobes—bugs that live without oxygen.
- **Herpes viruses, including Epstein-Barr, varicella (chicken pox), and cytomegalovirus.** Treatment depends upon which type is present. While

some doctors choose pharmaceutical medications such as *valacyclovir hydrochloride* (Valtrex), Klinghardt prefers chlorella, freeze-dried garlic, and ozonated oils. Others like using turmeric and milk thistle as antivirals.

- *Measles virus.* Klinghardt treats measles with vitamin A palmitate in a prescribed regimen of 400,000 units in three divided dosages for two consecutive days only. Dosage is repeated every six weeks several times, then once annually, just to make sure. Klinghardt says that using this protocol immediately following a vaccine reaction to the MMR can possibly prevent future problems.

- *Parasites like roundworms, threadworms, tapeworms, and liver flukes* are all very hearty bugs that should never be underestimated. Rizoles are good for treating all worms, cysts, and larvae. Klinghardt also uses Alinia (which crosses the blood-brain barrier), albendazole, and Biltrocide in high doses to attack these critters.

Immunotherapy

For the past 50 or so years, doctors have investigated ways to reverse or lessen the food and environmental reactions that add to Total Load. Several novel techniques under the umbrella of *immunotherapy* are making a huge difference in reducing the immune sensitivities and reactions leading to neurodevelopmental diagnoses.

In Chapter 9 on the immune system, the difference between an immune system that *defended against* and one that *tolerated* harmless substances was explained. Low tolerance to these substances causes sensitivities and allergies.

Intravenous immunoglobulin (IVIG): Dr. Sudhir Gupta, whom you met in Chapter 9, is one of the pioneers of immunotherapy. He discovered that individuals with autism have more TH2 cells than TH1 cells when compared to neurotypical children. In the early 1990s he used intravenous immunoglobulin for the first time to "retrain" abnormal antibodies in a patient with autism. The results of this method of rendering *passive immunity* were overwhelmingly positive, and it gave families hope when Cindy Goldenberg's son Garrett fully recovered from regressive autism.

Since then, doctors have developed some new types of immunotherapies that target allergic reactions. What if injection (or ingestion) of foods, inhalants, and chemicals could help the immune system adapt to substances instead of reacting to them by developing antibodies against them?

Sublingual immunotherapy (SLIT) is a modified traditional allergy-shot approach that starts with a moderate dilution of the allergen and then escalates to force the immune system to adapt. It is particularly useful for anyone terrified of needles. Daily sublingual (under the tongue) drops desensitize allergic reactions to foods and environmental triggers.[40]

Enzyme Potentiated Desensitization (EPD) was brought to the United States in the 1990s by William Shrader, who renamed the technique *low-dose allergy* (LDA) *therapy*. Every seven weeks, doctors injected a mixture of minute amounts of a specific allergen with an enzyme called *beta glucaronidase*, a natural biological response modifier that upregulates the immune system. This process gradually shifts the immune system from an allergic state (TH2) to a less allergic state (TH1), or from reactive to a state of tolerance.[41]

Low Dose Immunotherapy (LDI) is today's choice for teaching the immune system to be less reactive. Developed by Ty Vincent, MD, an integrative physician in Hawaii, LDI goes beyond LDA and EPD by mixing foods, antigens, and chemicals in minute doses. After a comprehensive interview to identify symptoms and specific allergens, treatment goes through three phases, ending in maintenance.

Dr. Vincent doesn't claim to "cure" allergies, just keep them at bay. Consistent with Total Load Theory, if a patient endures additional stress of any kind, the immune system can again overreact and require a "booster."[42] See below for a couple of other allergy elimination techniques.

Nambudripad's Allergy Elimination Techniques (NAET®) is a complex, sequential intervention that combines applied kinesiology, chiropractic, acupuncture, and nutrition, the various professional degrees of its developer, Devi S. Nambudripad, PhD, MD, DC, LAc. Using muscle testing, the therapist locates blockages related to a specific allergy on one of the acupuncture meridians. Allergens are "cleared" in a specific sequence, one at a time, and only one item per day. The substance must then be completely avoided for 25 hours following treatment, at which point it may be permanently neutralized. If an individual has exceptionally high sensitivity, the process is repeated.

When an NAET practitioner eliminates blockages in the presence of the allergens, the brain learns to respond differently. A yearlong controlled research study on 60 children with autism who were treated with NAET for 50 common allergens showed improvement in 30 of 35 symptoms, including communication skills. Some were able to function in regular classes for the first time following treatment.[43]

Since 1983, Dr. Nambudripad has trained thousands of medical practitioners in NAET procedures all over the world. Read *Say Good-Bye to Allergy-Related Autism*[44] or go to NAETAutismTreatmentCenter.com to learn more.

BioSET™, like NAET, is an acupressure technique that utilizes energetics and enzymes to clear or reprogram an individual's immune response. BioSET (for

Bio-energetic Sensitivity and Enzyme Therapy) balances meridians blocked by allergens. This natural, noninvasive desensitization was developed over 25 years ago by Ellen Cutler, DC, who also trains practitioners worldwide. Read *The Food Allergy Cure: A New Solution to Food Cravings, Obesity, Depression, Headaches, Arthritis, and Fatigue*[45] or go to DrEllenCutler.com.

Taming Inflammation

Everyone knows the symptoms of inflammation: swelling, pain, redness, warmth, and discoloration. For our population, most of these symptoms occur internally, especially in the gut and brain, and are invisible. A little inflammation is important to healing, but too much for too long can cause chronic pain, long-term physical damage, and emotional symptoms, including depression.

Bugs and the infections they cause mobilize the body's immune system and trigger an inflammatory reaction. Thus begins the cycle of an overresponsive immune system and increased inflammation as the response persists; in other words, instead of triggering resolution as it should, inflammation triggers more inflammation.

The third part of boosting and balancing the immune system is to put out the uncontrolled fire of inflammation in various parts of the body. To tackle this dilemma doctors are investigating possible treatment alternatives that bypass the digestive system because so many medicines further compromise the already-disturbed gut.

Low Dose Naltrexone (LDN), an opiate blocker, is one of the most-studied anti-inflammatories.[46] LDN was developed in the 1960s and approved by the FDA in the 1980s for treating opioid addiction. Since then, lower doses have proved effective for various autoimmune, neurological, and mental health conditions.[47]

LDN appears to correct the imbalance between TH1 and TH2 safely, with few side effects and no known toxicity. Benefits are improved sleep, decreased pain, better digestion, and general resolution or improvement of behavioral symptoms due to brain inflammation.[48] It is administered orally or as a transdermal cream (TD-LDN), which is easy to apply nightly before bed. Studies have shown that LDN is safe for use in children.[49]

Curcumin, the active ingredient in turmeric, an Indian spice, is known to have strong anti-inflammatory effects.[50] It is showing good success in reducing both gut and neurological inflammation. Curcumin is one of the most well-researched natural anti-inflammatories and is available as oral capsules or liquid. It should be taken with food to improve absorption.

Seizures

A seizure is the result of too much electrical activity in the nervous system. Picture an electric wire that is frayed and sparks. That's a seizure. A seizure can be a small blip on the screen, like an eye blink, or an earthquake that causes the entire body to shake.

Neuroinflammation due to allergy, excitotoxicity, environmental, or other stress can trigger a seizure. Yep, that's Total Load in action. Reducing the overall load by treating the brain inflammation can reduce seizure activity.[51]

Seizures are common in autism, especially in puberty when hormones combine with metals and bugs. That can be a lethal combination. Yes, death from a seizure can occur. A grand mal seizure is what killed Nick D'Amora at only 23. Known in the autism world as "The Changer," Nick was an indefatigable advocate for nonspeakers like himself.

I knew Nick from childhood. His mother Barbara turned his special life into a mission through the Grace Foundation in Staten Island, NY. Since his death in 2023, she continues to advocate for nonspeakers at CrimsonRise, where "autistic people have all the support they need in every sphere of life."

An estimated 11–39% of children with autism have seizures,[52] with numbers increasing as the more impaired children get older.[53] In one study of 86 patients with childhood-onset seizure disorder, 42% also had ASD.[54]

Psychedelics

Since ancient times, healers like shamans and medicine men have used roots, bark, seeds, and leaves of plants and fungi to heal sickness as well as alter consciousness. Some readers may have enjoyed their mind-altering properties in the sixties and recall tales of Harvard psychologist Timothy Leary's "trips" made famous by the Beatles. I do.

Well now, over 50 years later, we are beginning to understand the biochemical mechanisms underlying marijuana, LSD, "magic mushrooms", and other psychedelic drugs for medicinal purposes. The Johns Hopkins Center for Psychedelic and Consciousness Research is the world's largest center exploring innovative treatments using these fascinating natural resources. Let's enter the world of psychedelics. Caution: strict supervision is essential when using these substances!

Cannabis is a family of plants that includes hemp and marijuana. The legalization of medical marijuana in most states by 2016 peaked interest in hemp (as oil) and cannabis for reducing symptoms of autism, anxiety, depression, bipolar disorder, and schizophrenia.

No, this is not your grandparents' pot. *Cannabidiol* (CBD), a fat soluble cannabinoid found within the oil of marijuana and hemp plants, is the drug of choice because it does not result in users feeling "stoned." For many, it is not only calming, but it also reduces seizure activity.

Hemp oil contains high amounts of cannabidiol (CBD), but *little tetra hydro cannabinol* (THC), the substance associated with marijuana's high. The oil itself is the delivery mechanism that results in the alteration of the endocannabinoid system, a previously little-known biochemical cell communication network in the body that plays a crucial role in immune system function and mood regulation.[55]

One study on autistic children showed highly activated cannabinoid receptors, thus triggering an immune response.[56] This opened the door for using cannabis as a treatment option.

Tampa pediatrician, David Berger, MD, FAAP, founder of Wholistic Pediatrics & Family Care and its affiliated medical cannabis clinic, Wholistic ReLeaf, adds small amounts of THC to the CBD to treat severe aggression, self-injury, and intractable seizures in his patients with autism. Even though ASD is not explicitly listed as a qualifying condition under the state of Florida's medical marijuana law, comorbid conditions such as OCD or insomnia often qualify these patients for this treatment.[57] Learn more by watching a video.[58]

Erica Daniels, an autism mom who saw the benefit of cannabis for her teenager, founded HopeGrowsForAutism.org. Many companies offer CBD products, so buyer beware: pharmaceutical grade is a must!

Although many people with ADHD, anxiety, and mood disorders are attracted to using cannabis, for some it can exacerbate the very symptoms they are trying to treat, like paranoia, panic, and fluctuating mood.[59] It may not be the best choice for them!

Medicinal mushrooms have been of interest to those pursuing ways to alter their moods since the beginning of time. Psilocybin, a naturally occurring chemical found in *magic mushrooms*, or *shrooms*, penetrates the central nervous system to reduce inflammation and anxiety, as well as the symptoms of depression.[60]

Other "fabulous fungi" that do this and more include the following:

- *Reishi*, the "mushroom of immortality," is stress-reducing and calming.
- *Lion's Mane* boosts cognitive function and mental clarity.
- *Cordyceps* has energy-boosting, stress-reducing properties.
- *Chaga* is rich in antioxidants and an immune booster.[61]

Foodie Michael Pollan explores both the physical and spiritual effects mushrooms in *How to Change Your Mind* (2019).[62] This addition to his ever-expanding offerings joins some of my favorite books on the benefits of plants, cooking, and more.

Using mushrooms is easy: add them to salads, soups, and just about any recipe, fresh or dried. Take them as supplements in capsules, powders, and extracts. Put them in coffee, tea, or a smoothie. Check out options at specialty food stores, vitamin suppliers, and online.

Ketamine, like CBD, has moved from illegal recreational use to acceptance by medicine. Originally—and still—used as an anesthetic, increased research into ketamine has shown possible benefits for relieving depression, as well as comorbid OCD, anxiety, and panic, by targeting inflammation.

The link between inflammation and depression is well accepted, and anything that can harness a cytokine storm and put out the fire of inflammation will also decrease depression. Unlike most prescribed antidepressants, ketamine is fast-acting, with notable results in about an hour, as compared to weeks.

Dr. Nathan's *Sensitive Patient's Healing Guide* includes an entire chapter on the benefits and risks of ketamine, especially for Lyme and other tick-borne illnesses, where both physical and mental anguish is so profound. Inpatient facilities and psychiatrists are using it as a part of comprehensive treatment plan that also includes dietary and other lifestyle modifications.

Stem Cell Therapy

Another emerging intervention that is both controversial and expensive is *stem cell therapy*. Scientists started studying stem cells in the 1950s but have been investigating their use in mental health for only about 10 years.

Stem cells send out chemical signals that target old, damaged, or injured, inflamed cells and restore them, thus optimizing the body's capacity to heal. Different types of stem cells are found almost everywhere in the body, including in blood vessels, skin, teeth, the brain, heart, and most abundantly in umbilical cord blood, bone marrow, and fat tissue. *Mesenchymal stem cells* (MSCs) get the most attention because of their ability to replicate and differentiate into various types of cells for repair and maintenance and their anti-inflammatory and immune-boosting potential.

Emerging research is promising on the use of umbilical cord-derived MSCs in autism. One 2013 study showed significant improvements in visual, emotional, nonverbal, and intellectual arenas.[63] In another using cells from bone marrow, substantial improvements in behavior, social relationships, speech, language, and communication were seen in 96 percent.[64]

"Warrior mom" Tracy Slepcevic credits MSCs for resolving her son Noah's motor skills deficits that persisted despite occupational therapy. She details her story in the book *Warrior Mom: A Mother's Journey in Healing Her Son with Autism.*[65] Her nonprofit AutismHealth.com educates caregivers and families.

Several years ago, clinical trials at the Duke University Center for Autism and Brain Development using umbilical cord stem cells caught the attention of Eric Weiss, MD, a plastic surgeon and father of a son with autism. This trial on 25 subjects with autism showed promising results, so, after pursuing many other therapies, he used it with his son. He and his wife tell their success story with stem cells in *Educating Marston: A Mother and Son's Journey through Autism.*[66] Dr. Weiss now offers regenerative stem cell therapy for patients with autism in his Florida practice.

Not surprisingly, many parents of children diagnosed with autism are looking at stem cell therapy as a "magic bullet." They are flocking to clinics in Panama, the Dominican Republic, and elsewhere.

I have concerns about jumping on the stem cell bandwagon without doing the necessary foundational work on the nervous system, digestion, diet, and other biomedical and structural problems. *Please* do your homework first! If you are *really* interested in this option, join one of the closed Facebook groups where stem cell users share their experiences. All support groups offer both helpful and erroneous information. This is one area where extreme caution should be exercised because the amount of money required is enormous and the outcome is unknown.

The Nemechek Protocol™

Osteopath Patrick Nemechek developed a comprehensive rehabilitation program focused on repairing, reversing, restoring, and maintaining the central nervous system. It is included here, rather than in Chapter 6 because of its systems approach.

According to Nemechek, during the first 18 years of life, the body naturally and gradually reduces the approximately 100 billion neurons that children are born with by half in a process called *synaptic pruning*. If pruning is disrupted by any stressor, excessive brain inflammation can occur. Then toxic encephalopathy from the overproduction of proprionic acid (PPA), a byproduct of clostridium in the gut, aggravates the situation, resulting in behavioral symptoms.

Phase One targets the gut by reducing PPA through the reversal of bacterial overgrowth. It utilizes inulin powder, an over-the counter prebiotic fiber derived from natural plant sources—such as agave, chicory root, or Jerusalem artichokes—dosed at 1/8 to 1/2 teaspoon twice a day. Inulin acts like a fertilizer for friendly bacteria.

Propionic acid has a sedating effect. Once propionic acid levels decline, patients become more present and seizure activity often reduces as well.

Phase Two targets inflammation using the healing omega-3 fats DHA and EPA from fish and ALA from plants. Not only do these fats suppress inflammation, but Nemechek says they also promote stem cell growth. At the same time that omega-3 fats are increased, the protocol advises limiting harmful omega-6 fats from margarine, soybean, grapeseed, safflower, and sunflower seed oils. This reduction is accompanied by substituting extra virgin olive oil, an omega-9 fat, for all cooking.

The Nemechek protocol could also include vagus nerve stimulation as described in Chapter 6. Read *The Nemechek Protocol for Autism and Developmental Disorders*[67] or go to NemechekConsultativeMedicine.com.

Take-Home Points

Treatment of immune system dysregulation involves a three-step plan of boosting the immune system, fighting the infections, and taming the inflammation naturally. In the past decade protocols that combine these essential components using special foods, herbal supplements, and oils as natural anti-inflammatories that bypass the disturbed gut are showing promise. Allergy elimination and desensitization is a lengthy and often worthwhile process that can help reduce the Total Load for those with severe reactions.

Anyone using the methods in this chapter must become knowledgeable about available noninvasive options and work with experienced functional doctors and healthcare practitioners who understand, respect, and support the body's ability to heal itself. A fantastic book, written by a physician and mother of four is *The Holistic Rx: Your Guide to Healing Chronic Inflammation and Disease.*[68] The book is arranged by specialty, such as gastroenterology, pediatrics, and dermatology. In each section, the author offers suggestions for herbs, supplements, homeopathic remedies, and acupressure points to heal many conditions. We are on the road to healing!

CHAPTER 12

THE ENDOCRINE SYSTEM AND HORMONE DISRUPTION

The gut is healing; we are boosting immunity and decreasing inflammation. The Total Load is lessening.

The next step is dealing with the endocrine system and the inevitable disruption caused by metals, mold, and other stressors. Understanding the extremely complex role of the glands, hormonal secretions, neurotransmitters, and their amazing interactions is crucial in almost every diagnosis in the DSM, including bipolar disorder,[1] autism, schizophrenia, and more. In fact, this complicated interaction is *so* important that both Drs. Christopher Palmer and Neil Nathan dedicate entire chapters to this system in their books.

Chapter 2 enumerates the frightening load of chemicals that can set off a cascade of unnatural reactions that interfere with the synthesis, transport, binding, and action of hormones. Interference with any aspect of hormonal function during any stage of life, especially the first two years, impairs brain development. Take heart! Keep reading about ongoing research that is making it possible to figure out what to do to intervene.

Chapter 5 includes a section on endocrine, hormone, and neurotransmitter testing. A complete battery of the right tests is essential to understanding the incredibly complex interactions inherent in this system.

This chapter begins with a short lesson on the endocrine system and its key glands: the pituitary, hypothalamus, thyroid, and adrenals. Glands are the organs in the body that secrete hormones.

Then we look at how they and neurotransmitters determine how our bodies respond to stress, by either calming us down or putting us into physical and emotional tizzies. Balancing all this is tricky, so measuring levels is essential to restoring healthy function and homeostasis. Let's get started.

Endocrine System 101

The endocrine system is a convoluted network of glands that regulate many of the body's organs and functions. These glands regulate growth, development, mood, metabolism, immunity, and reproduction by releasing minute amounts of molecules called **hormones**. Our hormones flow into the bloodstream and fluid surrounding the cells and mix with chemicals from the environment and nutrients from foods. This amazing soup tells our organ systems what to do.

Human hormones are too numerous to list. Some, like *estrogen*, *testosterone*, and *insulin*, are well-known; others, such as *melatonin* and *oxytocin*, are becoming more familiar as the public learns from scientists about their roles in sleep and feelings of well-being. Appropriate hormone levels are critical for controlling and regulating all of the body's functions throughout the lifespan.

Neurotransmitters are what scientists call those chemical messengers that do their jobs at nerve endings and synapses. Thus, they are very short-lived. The names of some neurotransmitters—like *serotonin*, *dopamine*, *epinephrine* (adrenaline), and *norepinephrine* (noradrenaline)—are also entering the lay vocabulary with increasing frequency as medical professionals check their levels in patients presenting with mood swings, learning problems, disrupted sleep, and behavioral symptoms. Unremitting biological and environmental stress from toxic heavy metals, toxic chemicals, viruses, and other immune aggravators can produce a chronic state of neurotransmitter imbalance.

Note that hormones and neurotransmitters are distinguished by their functions, not their structures. In fact, some molecules (e.g., epinephrine and norepinephrine) count as both depending upon whether they end up circulating in the bloodstream or crossing synapses between nerve cells.

Let's look at some of the endocrine glands and the molecules they secrete. The relationship among organs, hormones, and neurotransmitters is **so** complex that extreme oversimplification is necessary in this short introduction.

You met the **hypothalamus**, the **pituitary,** and their companions, the **adrenals,** in Chapter 6. They are the triad making up the HPA axis, which receives a signal from the limbic system when the body perceives it is under threat. The hypothalamus and pituitary sit side-by-side in the middle of the brain, where their jobs include controlling body temperature, hunger, thirst, appetite, circadian rhythms (day/night cycling), and emotions.

The hypothalamus and pituitary have many important functions, including stimulating the **thyroid gland** and producing the hormones *oxytocin* and *vasopressin*, while regulating several others.

Oxytocin (OXT) is a hormone that signals, modulates, synchronizes, and sensitizes the body's cells to talk to and work with each other. OXT bridges communication between the physical body, its energetic fields, and emotions.[2] It plays a major role in childbirth, when the hypothalamus and pituitary work together to produce and release large amounts during labor to facilitate birth, and afterwards in letdown, when milk is released into the breasts. Scientists are now investigating oxytocin's role in many other behaviors and emotions.

Vasopressin (VP) is a hormone that enhances muscle tone, peer bonding and memory. It is also known as the *antidiuretic hormone*, or **ADH,** because it increases water permeability and absorption in the collecting ducts of the kidneys. Like OXT, most of ADH is stored in the pituitary, where its release plays an important role in social behavior and bonding.

Neuroscientist C. Sue Carter, PhD, from The Brain-Body Center at the University of Illinois at Chicago has studied the implications of depressed oxytocin and vasopressin (VP) in autism spectrum disorders. VP levels are dependent on testosterone and are thus of particular importance to male behavior. Carter believes that disruptions in the VP system, which result from epigenetic modifications during early development, could contribute to the vulnerability of males for ASD.[3]

The *thyroid* is a butterfly-shaped gland that lives in the front of the neck, attached to the lower part of the larynx (voice box) and to the upper part of the windpipe. Its name comes from the Greek word for *shield* because of its protective role in health.

The thyroid secretes hormones that act on *every* cell in *every* organ of the body, especially in the brain. It plays a profound role in energy metabolism, growth, development, temperature regulation, and more. It receives chemical messages from the pituitary and hypothalamus that signal it to produce its own hormones that regulate metabolism by stimulating the mitochondria, which have thyroid hormone receptors. Getting a signal from the thyroid, they go to work producing more ATP to make more energy available.[4]

Remember those mitochondria and their importance in regulating immune, inflammatory, and endocrine responses to stress? If they are not fed properly with nutrients and hormones, every behavioral response will be affected.[5]

When the thyroid is under- or overactive, dysfunction occurs. Low levels of thyroid hormones, called *hypothyroidism,* and high levels, called *hyperthyroidism,* account for many bodily symptoms—like fatigue, depression, and brain fog—that accompany neurodevelopmental conditions. In fact, hypothyroidism is often due to an autoimmune disorder and is also linked to bipolar disorder, schizophrenia, and dementia.[6]

Testing thyroid function is tricky because it is very misunderstood and incomplete in many patients. Many doctors test only for *thyroid stimulating hormone* (TSH), the pituitary hormone that stimulates the thyroid to produce its own hormones. But that is only part of the picture. More savvy doctors add on the thyroid hormones, *T3* and *T4*. Now we are getting somewhere. But most functional medicine doctors know that the **reverse T3** level is also important.

Research strongly confirms that a pregnant mother's thyroid function is a factor in both the physical and emotional health of her unborn baby. Children born to mothers with hypothyroidism were found to be at risk for many conditions, including adverse birthing events such as preeclampsia and diabetes,[7] as well as later DSM diagnoses, including autism, ADHD, and intellectual impairments.[8]

In today's toxic world, thyroid cancers have more than doubled since the early 1970s, making it the fastest-growing cancer in the United States[9] expected to be the fourth most common cancer by 2030.[10]

The thyroid must have sufficient iodine to do its job properly; too little iodine disrupts thyroid function. Any iodine will do—if the thyroid is unable to get adequate supplies from food or water, it takes iodine from the air, even if it is contaminated with radiation from nearby power plants. That's why families living adjacent to nuclear reactors are given clean iodine supplements to supercharge their thyroids so that this gland does not rely on iodine from the reactors.

Learn more about this important gland from the master of thyroid biochemistry, David Brownstein, MD, in one of his 15 books, including his most recent *A Holistic Approach to Viruses*.[11] A classic on hypothyroidism is *Why Do I Still Have Thyroid Symptoms When My Lab Tests Are Normal?*[12]

The **adrenals** are a pair of glands located on top of the kidneys. Their name denotes their location: *ad* is Latin for *near*, and *renes* for *kidneys*. They produce the stress hormone **cortisol** as well as a class of hormones called **catecholamines,** which act as chemical messengers in response to emotional or physical stress. Catecholamines prepare the body to take action and include **epinephrine** and **norepinephrine**.

Cortisol is critical in helping the body and brain deal with stress; it is classified as a *glucocorticoid* because its primary job is to raise blood sugar (glucose) and to suppress immune activity. Proper regulation of glucocorticoids is essential for normal brain development, especially vigilance and cognition. Cortisol also prevents the release of substances that cause inflammation.

Cortisol is made by the body during sleep. Levels vary during a 24-hour cycle—peaking in the early morning, falling during the day, and bottoming out three to five hours after falling asleep, or between midnight and four a.m.[13]

Elevated cortisol levels have been associated with various mental disorders, including anxiety, depression, and post-traumatic stress disorder.[14] Both abnormally high and low cortisol levels are seen in psychiatric disorders. British researchers found that adolescent males with Asperger's do not experience expected morning cortisol peaks.[15] More recent research suggests that cortisol levels were inversely related to function level for individuals on the spectrum: those with the lowest level skills had the highest cortisol levels.[16]

Because of its variability, cortisol testing requires the same 24-hour cycle, with saliva testing four to six times throughout the day to catch its ebb and flow. If the cycle is off, reaching peak levels in the middle of the night, for instance, sleep could be disturbed or daytime energy affected.

Dopamine, epinephrine, **and** ***norepinephrine*** are adrenal catecholamines. Their functions and interactions with each other as well as with other neurotransmitters are both complex and critical to many functions, including motor control, arousal, and motivation.[17]

Very simply, dopamine is one of the "feel good" hormones, considered the "pathway to pleasure."[18] Too much dopamine can lead to addiction, while low levels can trigger depression. Epinephrine is the fight-or-flight hormone. It is useful in a real emergency but can be inappropriately triggered leaving a person feeling stressed when there is no real danger, thus disrupting sleep and triggering anxiety when awake.[19] It can also increase heart rate and blood pressure, break down fat, and increase blood sugar levels to provide more energy to the body.

Another hormone produced by the adrenal glands is ***dehydroepiandrosterone,*** **(DHEA)**, which is also made in the brain and skin. DHEA is a steroid hormone that boosts immunity. DHEA can easily become depleted. The body uses DHEA to make testosterone, progesterone, and estrogen; when it is deficient, so are they. That is a common occurrence in many patients who have been ill for a long time, so DHEA testing is very important. Supplementing with easily obtained capsules can make a huge difference because that simple intervention can raise estrogen levels and thus reduce cortisol.

DHEA prevents the destruction of ***tryptophan***, an amino acid that the body uses to synthesize proteins. Bottom line: low DHEA levels are a sign of chronic stress. Low DHEA, low immunity. Low DHEA, low tryptophan. Low DHEA, low estrogen.

Want to raise your DHEA levels naturally? Try some quinoa, a gluten-free grain that can also raise melatonin production (see below) and thus improve sleep quality.

Serotonin is a neurotransmitter that affects basic psychological functions, such as mood and anxiety, in interaction with many other neurotransmitters, including tryptophan and dopamine. It is not produced by an endocrine gland, but rather by the brain and in the gut. Thus, dysbiosis and gut inflammation can disrupt its production through a TH2 cell called interleukin 13, a mediator of allergic inflammation that is implicated in bowel disease.[20]

Other Important Hormones

Estrogens are hormones, secreted mostly by the ovaries, that stimulate and control female characteristics. Doctors measure levels of various estrogens, especially at menopause, and look at their relationships to each other as well as to **progesterone**, estrogen's female companion. This is important information since deficiencies can cause crazy-making mood changes and other symptoms.

Some foods, including soy, are *phytoestrogens*, which means they mimic the effect of estrogen and can lead to more female features and traits.[21] Dr. Chris Palmer relates that estrogen therapies are sometimes used to treat mood disorders.

Testosterone, mainly from the testes in men and the adrenal glands in women, is an androgen—a hormone that stimulates and controls the development and maintenance of male characteristics. Mercury is androgenic and raises the body's testosterone levels, resulting in more masculine features and traits, such as facial hair and aggression.

British researcher Simon Baron-Cohen has studied the role of testosterone in autism for more than ten years. He and others have concluded that high testosterone levels are a load factor in both males and females with autism. Females with autism are frequently found to have abnormally high levels of this male hormone.[22]

In a 2010 study, Baron-Cohen followed 235 mother-child pairs over eight years, periodically completing questionnaires designed to measure autistic traits. None of the children in the study had yet received an autism diagnosis, but Baron-Cohen found that those who had been exposed to higher testosterone levels in the womb—measured via amniocentesis during pregnancy—had a greater chance of displaying autistic traits.[23]

A follow-up study, published in 2015, provided support for this epigenetic risk for autism and/or other sex-biased neurodevelopmental conditions occurring during periods of fetal brain development. Those later diagnosed with autism were exposed to higher amounts of testosterone and other steroidal hormones.[24]

Melatonin is a powerful antioxidant that is vital to health. It is a hormone that is secreted by the tiny pineal gland in the middle of the brain and is best known for governing our circadian rhythms or sleep-wake cycles by inducing drowsiness, lowering body temperature, and calming the nervous system. When our retinas sense low levels of light, they send messages to the hypothalamus, which tells the pineal gland to send some out to almost every cell in the body.[25]

Low light does not occur just at night, but also during those gray days of winter. Many people with melatonin deficiencies suffer from a condition called *seasonal affective disorder*. The pineal gland needs darkness to secrete melatonin. Klinghardt believes that many individuals with sleep disorders have melatonin deficiency because the pineal gland is not getting the darkness it requires.

So, once again, with a melatonin deficiency we have the perfect formula for depression: insufficient melatonin => poor sleep => disturbed circadian rhythms => depression.[26] The benefit of melatonin supplementation is well documented. See below.

Endocrine and Hormone Disruption in Neurodevelopmental Disorders

Endocrine function can be influenced by both genetics and the environment. Refer back to earlier chapters to understand how SNPs in certain genes negatively affect the body; the endocrine glands are no exception. In addition, their ability to produce hormones and neurotransmitters can be epigenetically altered.

At the end of 2013, a multinational study in the *Proceedings of the National Academy of Sciences* confirmed that a polymorphism of OXTR, the gene that encodes the receptor for oxytocin, affects not only social interactions such as mother-infant bonding, but also the ability to recognize faces.[27]

Dr. Emily Gutierrez implicates three genes that affect neurotransmitters. First is the MAOA gene, which helps the body break down serotonin. A variance in the *MAOA* gene can cause behavioral symptoms, including defiant and oppositional behavior.

Second is the *GAD1* gene, which converts the amino acid glutamine into the neurotransmitter GABA, crucial in calming and reducing stress. As already noted, glutamate is excitatory and, if not inhibited, the body can become overstimulated, difficult to calm down, and anxious. In fact, the symptoms of this missing process are very similar to those of ADHD. Dr. Gutierrez recommends CBD oil, described in the previous chapter, as a helpful remedy for behaviors caused by this genetic condition.

Third is the *COMT* gene, which acts to degrade catecholamines, preventing the body from getting flooded with emotions. Those with variances in the COMT gene need longer periods to cool off from meltdowns and other emotional outbursts.

A study headed by George Washington University researcher Valerie Hu, PhD, revealed that the excess testosterone in individuals with autism suppresses a gene called *RORA*, which lowers aromatase, an enzyme that helps convert testosterone to estrogen. Excess levels of testosterone both stifle the activity of the gene and increase the likelihood of autism.[28] These findings were duplicated by Chinese investigators.[29]

The good news is that all of these genetic variants are treatable. Supplementation can help through the magic of nutrigenomics.

Toxins, chemicals, metals, bacteria, mold, Lyme, and most recently corona-viruses also disrupt the endocrine glands and their ability to produce hormones and neurotransmitters.[30] Surreptitiously and nefariously, they set off a cascade of unnatural reactions interfering with the synthesis, transport, binding, action, and elimination of hormones.[31]

When stressors alter any of the endocrine glands' ability to produce hormones, any functions can be "off." In a recent lecture Dr. Klinghardt cited research show-ing that the COVID spike protein does damage to the pituitary and pineal glands and possibly others as well. Resulting hormone production is thus affected, as are functions like metabolism, learning and memory, which have become com-promised in many people.[32] He reports that a majority of his patients are now overweight, without eating more or exercising less. Furthermore, those with latent illnesses, including Lyme and mold, are seeing recurrences.[33]

According to mold guru, Jill Crista, ND, some mold toxins are estrogenic, and not the good kind. Under the influence of a water-damaged building, mold forms mycotoxins that promote a cancer-causing estrogen overload.[34]

In addition, excessive exposure to electromagnetic fields (EMFs), especially from cell phones and Wi-Fi, encourages mold growth, causing it to increase its production of mycotoxins. EMFs also disturb the pineal gland's production of melatonin, making falling and staying asleep more difficult.[35]

Endocrine disruptors are ubiquitous, creeping up not only in food and water, but also in an extremely diverse collection of consumer products, including cos-metics, shampoos, detergents, sunblocks, perfumes, and pharmaceuticals. Did you know that the magic that makes some medications, including prenatal vitamins, "time-released" is an endocrine disruptor? These chemicals enter the body through inhalation, absorption, ingestion, and placental transfer if taken during pregnancy.[36]

Much current research is focusing on how PBDEs, which you read about in Chapter 2, and other chemicals, such as perchlorates in drinking water, bisphe-nol A (BPA—a byproduct of making plastic), phthalates, solvents, and plasticiz-ers, disrupt the entire endocrine system.[37] The Endocrine Disruption Exchange Inc. provides a searchable database with more information about the activity of approximately 870 endocrine disruptors that assists professionals in identifying possible culprits in various types of dysfunction.[38]

Endocrine disruption in autism is well-documented. More than ten years ago, doctors found that serotonin, dopamine, and norepinephrine were directly or indirectly linked with different aspects of social development associated with autism.[39]

Mothers of children later diagnosed with ASD had decreased melatonin production, altering their circadian rhythms. Serotonin—melatonin's precursor—diminishes, contributing to this dysregulation. Hypothalamic-pituitary dysfunction is next, leading eventually to low production of thyroid hormones.[40]

Weak thyroid function is very common among mothers of those later diagnosed with autism and other disorders.[41] Recall the discussion of cerebral folate deficiency (CFD), where folate receptor autoantibodies (FRAAs) prevent folate from crossing the blood-brain barrier. Results of one study suggest that fetal and neonatal exposure to maternal FRAAs affect the development of the thyroid of their offspring, since thyroid dysfunction is also common in their children.[42] Researchers continue to investigate whether thyroid dysfunction in ASD may be related to blocking FRAAs, and generally contributory to overall endocrine pathology in autism spectrum disorder.

Puberty can be a stage of significant endocrine dysfunction. High testosterone is a little-known cause of precocious puberty.[43] The late Mark Geier, MD (1948–2025), a geneticist, and his son David found that approximately 80% of boys and girls with autism experienced precocious puberty. Curiously, affected girls seem to have even higher testosterone levels than the affected boys, leading the Geiers to conclude that high testosterone was necessary to overcome the naturally protective effects of estrogen.[44] Another study revealed both testosterone-related medical conditions and delayed puberty in women with high-functioning autism and their mothers.[45]

The combination of testosterone and mercury can be especially problematic. The volatile combination of testosterone and mercury might explain why so many children with autism develop seizures in adolescence. One paper reports that while about 13 percent of young children with autism have epilepsy, that number doubles to 26 percent during puberty.[46]

The Geiers believed that mercury also *raises* testosterone levels and that the high levels of testosterone block the body's ability to make glutathione, which is necessary for detoxification (see the next chapter). Furthermore, mercury binds to glutathione, thus inactivating whatever stores the body may already have.

The Geiers prescribed the drug Lupron "off-label" to lower testosterone levels in autism patients like Wesley, the son of the Reverend Lisa Sykes. Lupron is approved by the FDA for treating precocious puberty. Tests showed that Wesley's already-high testosterone levels were pushed from "the stratosphere into orbit" by additional testosterone production in puberty.[47] To many parents, including Wesley's mother, it made sense that adolescent seizures and aggressive behavior were related to the high testosterone. They were ecstatic when their son's testosterone levels dropped and his seizures and aggressive behavior lessened.[48]

However, authorities were not amused by the Geiers' off-label use of Lupron, and the Geiers were accused of "chemically castrating" their young patients. In 2011, authorities suspended Mark Geier's license to practice medicine in several states.[49] Dr. Geier successfully fought the charges against him, stating that his practice should be judged on the very positive results parents report. Without Lupron, many of the adolescents and young adults who showed extreme aggression would probably have required psychotropic drug intervention and even institutionalization.

On December 7, 2017, a Montgomery County, Maryland judge awarded $2.5 million in damages to the Geiers, ordering 14 medical board appointees, the board's lead attorney, and the lead investigator on the Geier case to pay half of the damages out of their own pockets.

Paralleling thyroid imbalance is extreme adrenal dysregulation. Cortisol production becomes deficient, and adrenocorticotropin hormone (ACTH) production rises significantly due to imbalanced melatonin and ACTH.[50] Two common symptoms of ASD, hypersensitivity and insomnia, have a strong connection to this imbalance in the adrenal system.

Sometimes, the adrenal glands are so busy in patients with neurodevelopmental disorders that they too get stressed out in a condition called "adrenal fatigue." Too much stress and insufficient hormones equals burnout. Time to add an endocrinologist to the team.

Balancing and Repairing Hormone Production

Medications often prescribed for depression and anxiety, like Abilify, Luvox, Paxil, Prozac, Tofranil, and Zoloft, are *serotonin reuptake inhibitors*, which means they allow the body to recycle (rather than use up) its precious supplies of serotonin to fight depression. Stimulants like Adderall, Cylert, and Ritalin increase levels of the fight-or-flight neurotransmitters norepinephrine and dopamine.

Instead of relying on drugs, which can have numerous deleterious side effects, many healthcare practitioners are prescribing hormones and their precursors. Hormone therapy for depression, anxiety, and mood disorders is becoming more and more common.[51]

Melatonin

High on the list of prescribed hormones is melatonin because it addresses the devastating sleep problems rampant in this population. When one member of a family isn't sleeping, the whole family's sleep can be disrupted. Melatonin is very safe, even for children.[52] In addition to capsules, it is available as a liquid and as a cream, so it is easy to use. Note that melatonin is responsible for falling asleep and is not as useful in people who have difficulty staying asleep and wake frequently in the night.

In a study of over one hundred children with autism, melatonin was beneficial in improving sleep in over half. Dosages range from one to six milligrams, starting low and working up if needed, given a half-hour before bedtime.[53]

At a 2017 Sleep Congress in France, members evaluated the efficacy of melatonin for mental illness. Their conclusion was that melatonin could be used as an adjuvant treatment when insomnia symptoms were present in mood disorders (bipolar, major depressive, and seasonal affective), in ADHD, anxiety, and schizophrenia.[54]

Klinghardt believes that melatonin is a highly misunderstood and underrated antioxidant. He is extremely enthusiastic about melatonin supplementation in high doses not only for sleep, but for anyone with neurological symptoms and even for anti-aging.

Furthermore, he teaches that the best pathway to the brain is through the skin. He recommends transdermal delivery of melatonin, as well as other products. He suggests that melatonin cream be applied between dinner and bedtime. Learn more about his uses of melatonin from his videos on YouTube.

Another of Klinghardt's favorites is a nicotine patch. Yes—the same one used for smoking cessation, but at its lowest potency. Applying it out-of-reach between the shoulder blades prevents kids from pulling it off and adults from fiddling with it. This off-label use of nicotine was discovered during the pandemic because, believe it or not, smokers were less affected by the virus than nonsmokers according to a study done by the tobacco industry.[55] Those who lost their smell and taste found that these senses were renewed with nicotine.[56]

Yet another population has responded positively to this approach: kids with attention deficits. A pilot study of ten ten-year-olds showed a great reduction in hyperactivity and other symptoms, albeit with some unpleasant side effects, including nausea, stomachache, itching under the patch, and dizziness.[57] Nicotine is highly addictive and should only be used under the supervision of a physician.

SAMe, Tryptophan, and 5HTP

Dr. Amy Yasko, an autism specialist in Maine, has delved deeply into neurotransmitter anomalies, especially in autism. Her testing often indicates excessive norepinephrine, for which she suggests *S-adenosyl methionine* (*SAMe*, pronounced "sammy"). *SAMe* assists in changing neurotransmitters into other substances. It supports the MAOA gene variant discussed above, especially when used with lithium orotate, and helps convert serotonin into melatonin and norepinephrine into epinephrine (adrenaline). Access her supplements at HolisticHeal.com.

Another option is supplementing *tryptophan* and *5-hydroxytryptophan* (**5HTP**), which combine and convert to serotonin. Dr. Jared Skowron combines melatonin with zinc and the above serotonin precursors to rebalance circadian cortisol rhythms. He also recommends the herb holy basil (ocimum sanctum) to even out cortisol spikes. His supplements, primarily for autism and ADHD are available at SpectrumAwakening.com.

DIM

DIM, short for *diindolylmethane* is a compound created when you digest cruciferous vegetables, such as broccoli or brussels sprouts. Research suggests that DIM can lower estrogen levels in the body. That's why some practitioners are recommending it, especially for young males who have been exposed to estrogen-enhancing mold.[58]

Oxytocin Treatment

Researchers have been looking at supplementing oxytocin, sometimes nicknamed the "anti-stress hormone," to address impairments in reciprocal social interaction for at least 20 years. A 2014 systematic review by Italian doctors of seven randomized controlled studies of over 100 mostly male subjects from January 1990 to September 2013 reported potentially promising findings in measures of emotion, facial recognition, and eye gaze.[59]

Another review of the literature, published by Australian scientists in early 2016, found that while oxytocin may optimize social interaction by enhancing motivation and learning, current evidence of therapeutic benefit from extended oxytocin treatment still remains very limited.[60] Both studies agreed that further investigation is needed.

Psychiatrist Martha Welch, MD, founding director of the Nurture Science Program at Columbia University Medical Center, has been studying the role of oxytocin in the biological phenomenon she has termed *emotional co-regulation* for almost her whole 50-year-long career. More than 20 years ago, she demonstrated that dysregulation of oxytocin combined with oxidative stress in the gut disrupts the whole gut-brain network. Supplementing oxytocin decreases social anxiety and increases social behaviors, including trust, empathy, eye contact, generosity, facial recognition, and bonding.[61]

A deficiency of oxytocin could explain many of the symptoms and behaviors of those with relationship issues. However, the big, important question is: why would these individuals have too little?

The oxytocin-Pitocin connection. The answer appears to stem from something that happened at birth: a gene called OXTR may have been turned off.[62] By what? Pitocin, the synthetic oxytocin often administered to induce and accelerate labor.

Veteran autism researcher Eric Hollander, MD, of New York's Mount Sinai School of Medicine discovered that while only about 20% of all births are assisted by Pitocin, 60% of the patients with autism in his clinic had been exposed to Pitocin in the womb. That is quite a difference and is statistically significant no matter how you figure it.

Hollander has tracked 58,000 children whose mothers' procedures were monitored during pregnancy. The oxytocin/Pitocin connection could be a perfect example of epigenetics in action; in genetically vulnerable infants, administering Pitocin at a critical time when the brain is still developing could downregulate the oxytocin system, leading to developmental problems. To test this hypothesis, Hollander and his colleagues used oxytocin and found that it consistently reduced repetitive behaviors in his patients with autism.[63]

University of Utah scientists do not agree. Their 2015 review of a study sample composed of 2,219 children with ASD and 166,361 children without ASD, belonging to the 1998, 2000, 2002, 2004, and 2006 birth cohorts at local hospitals, found no relationship between induction and/or augmentation with Pitocin during childbirth and increased odds of an autism diagnosis in childhood.[64] Maybe lack of genetic diversity in Utah could account for that finding.

Another theory is that Pitocin interferes with the natural rhythms of birth and can cause enormous pressure on the baby's brain and cranial nerves during the birthing process, compromising the developing nervous system.[65] If that is true, any baby enduring a Pitocin-assisted delivery would benefit from the therapies for structural impediments and birth trauma described in Chapter 6.

One more possible problem with Pitocin: a 2009 study shows that administering Pitocin during labor decreases glutathione (GSH) levels in newborns.[66] Read more about the essential role of GSH in detoxification in the next chapter.

Treatment with oxytocin: Recall that oxytocin appears to stimulate receptors in the regions of the brain that involve social memory and social affiliation. Taking it can improve those functions.

Oxytocin can be delivered in several ways: in sublingual drops, a nasal spray, or in Klinghardt's preferred method, a transdermal cream. The spray is absorbed into the bloodstream more rapidly, and the cream gets past the blood-brain barrier. Klinghardt combines it with a probiotic, *lactobacillus reuteri.*

Work with intranasal oxytocin by the Hollander team and others in Australia showed additional positive effects, including heightened ability to recognize emotions and show empathy.[67]

Dr. Angela Sirigu, an Italian neuroscientist, and her team at the Centre de Neuroscience Cognitive in Lyon, France, further confirmed that inhalation of oxytocin holds significant therapeutic potential for autistic and other individuals with impaired social skills to interact more effectively.[68] And, finally, a 2017 study demonstrated that intranasal oxytocin treatment improves social abilities in children, and that those with the lowest oxytocin levels at the beginning of the trial experienced the greatest improvements.[69]

Take-Home Points

The more researchers explore dysfunction in and interaction among the nervous, immune, digestive, and endocrine systems in those with autism, the more they are convinced that psychiatric disorders, especially autism, are metabolically based. Identifying which hormones are disrupted and balancing them requires careful testing, history-taking, and observation. As endocrine dysregulation lessens, so do gut problems, inflammation, and day-to-day issues such as sleep, eating, and behavioral concerns.

The endocrine issues in neurodevelopmental disorders are far-reaching and include not only estrogen and testosterone, but also lesser-known hormones such as oxytocin and cortisol, as well as neurotransmitters like melatonin, serotonin, dopamine, epinephrine, and norepinephrine. As doctors begin to understand interactions, new protocols are continuously emerging for treatment.

The interaction between the endocrine glands and the numerous hormones they secrete, as well as interaction among the body's organ systems, the genetically engineered foods we eat, and the chemicals in the environment, is extremely complex. That's why I call it a Total Load. Many believe that what is happening in the brains of those of us with anxiety, depression, ADHD, and autism is the outcome of some very confused hormones.[70]

The next chapter focuses on detoxification. Understanding why so many children are so sick is not just about knowing what they are exposed to, but also why their bodies are unable to rid themselves of the toxins and deleterious bugs. On to methylation and sulfation: getting the bad stuff out!

CHAPTER 13

DETOXIFICATION: GETTING THE BAD STUFF OUT

Detoxification is a household term today. At last check, Google yielded 15,400,000 hits and Amazon carried over 1,500 titles including this word. You can detoxify "naturally," "herbally," "quickly," "slowly," "sensibly," "safely," or "organically." It can be a "relief," a "miracle," or even a "transformation."

Why include this subject in a book about chronic illnesses? Because if the rate at which toxins accumulate is more rapid than the body can rid itself of them, you get sick, physically, emotionally, and spiritually.

Limiting exposures is only half the battle; having efficient detoxification pathways is the other half. No matter how diligently you have avoided toxins by eating a clean diet, filtering air and water, getting sufficient sleep, and avoiding traumatic situations, it is certain that microorganisms, metals, and trauma have crept into your cells. You must "clean house" frequently and do a thorough deep cleaning once in a while. Living in a toxic world requires continuous detoxification.

Detoxification and Neurodevelopmental Disorders

Many of us have lived a "perfect storm" of overexposure to chemicals, metals, food additives, radiation, and traumatic experiences—including abuse, neglect, abandonment, and poverty—coupled with the poor ability to detoxify. In such cases, the detoxification process can be life changing. Knowing where to start and how to get the bad stuff out is a complicated project that is both an art and a science.

Once again careful history-taking is crucial in assessing the origin and extent of toxic exposures. Intake evaluations must include questions about timing, duration, levels, routes, agents, and interactions because pinpointing a relationship between exposures and outcomes is extremely complex. In general, the most sensitive time for toxic assaults is during critical periods of rapid brain development, such as the pre-, peri-, and post-natal periods, when the blood-brain barrier is not fully developed.

Like vaccination, the topic of detoxification as related to neurodevelopmental disorders is controversial. The idea that anyone with one of the diagnoses in this book is "toxic" has been rejected by some mainstream doctors, who usually attribute positive behavioral and emotional changes to ancillary therapies or "coincidence." Thank goodness that is changing as a new generation of metabolic psychiatrists and functional doctors recognize the benefits of some of the detoxification strategies in this chapter.

For those who are "believers," focusing on the process of clearing out the bad stuff must wait until the nervous system and gut are fairly stable. Attempting to rid the body of the toxins that wreaked havoc there without a healthy gut only wreaks more havoc. Anyone who is vaccine-injured and/or has a limited diet of toast, French fries, and noodles is at an increased risk for gut problems during detoxification. That is why the subject of detoxification does not appear until halfway through this book.

For successful detox, a person must be eating sufficient protein and a fairly wide variety of foods including a minimum of three to four servings per day of fruits and vegetables, not including juice. Toxins are not excreted well when the diet is low in protein.

This chapter begins with a simplified explanation of the detoxification process, including a review of terminology. It then moves into essential vitamins and minerals that come from foods and supplements, including the essential role of the masterful cleaner-upper, glutathione.

Next come some genetic polymorphisms that interfere with the detoxification process and a review of foods and other substances that enhance detoxification. Precautionary detoxification guidelines follow. The chapter ends with ways to excrete toxins through the skin using external methods such as Epsom salt baths and a sauna. Energetic methods of detoxification, including homeopathy, homotoxicology, and family constellation therapy are in the next chapter.

Detoxification 101

Detoxification is the human body's way of neutralizing, processing, and eliminating poisonous substances. Efficient detoxification is crucial to good health for everyone; it is a natural process occurring regularly in the body.

As the body metabolizes different kinds of toxins in various ways, it has several major routes of detoxification: the digestive system through bowel movements and urine, the skin through sweating, and the hair and nails as they grow out. Several organs are involved, but the liver, an extremely complex organ involved in over 300 different functions, does most of the work.[1]

The basics of detoxification are the same as the lifestyle basics to staying healthy. Eat well, drink plenty of water, sweat often, move frequently, and sleep soundly. If these foundations are not in place, toxins accumulate.

Remember way back in Chapter 2 that stress occurring at a cellular level is called *oxidative stress,* one of the body's most significant load factors. This impediment is the outcome of a chemical process called *oxidation* (think rust), which overwhelms and weakens the cells' ability to fight off invaders known as *oxidants* and produces an excess of unstable molecules known as *free radicals.*[2]

The antidotes to oxidative stress are *antioxidants,* which gobble up free radicals like Pac-Man, allowing bodily processes to proceed smoothly. Antioxidants include vitamins and minerals from foods and glutathione—the body's most powerful antioxidant—and its precursors, which the body can use to produce or recycle this precious commodity.

Phases of Detoxification

Detoxification takes place in two phases. In their now-classic book *7-Day Detox Miracle,*[3] naturopaths Bennett and Barrie compare detoxification to a two-step wash cycle. In **Phase One** nutrients in the body convert toxins into less-harmful constituents and prepare them for elimination. First, cells must be cajoled into breaking down stored toxins into intermediate forms. Antioxidants act like scavengers by gobbling up oxidants produced during this process. A few toxins are ready for removal at this stage, but others need a second wash cycle. If the process stops after phase one, sickness and even death can occur because bad toxins can float around, impeding function in all systems of the body.

Phase Two is the actual elimination phase, during which the body attempts to excrete toxins safely, usually in partnership with an "escort" nutrient. Six steps—methylation, sulfation, acetylation, glutathione conjugation, amino acid conjugation, and glucuronidation—make up phase two detoxification. Five of the six interdependent pathways rely on efficient sulfur chemistry, which glyphosate guru Stephanie Seneff teaches is disrupted by that evil herbicide.[4] Since this process is so very complicated, it will not be completely covered here. Here are the highlights of two.

Methylation and Sulfation

A body's strong ability to *methylate* is the key to detoxification. During methylation, the sulfur amino acid *methionine* (considered "the queen of amino acids" by Dr. Sidney Baker) gives away a methyl group and becomes "horrible *homocysteine.*" It is then converted to *cysteine,* in a process dependent upon an ample supply of *methyl B12.* Cysteine is one of the raw ingredients for making the body's most powerful antioxidant, glutathione. Read more about this powerhouse of detoxification below.

Proper *sulfation* requires the pathway to the production of sulfates to be functioning optimally. This pathway takes the amino acid cysteine through numerous steps to make *sulfite*, which is then processed into *sulfate* through an enzyme called *sulfite oxidase*. Many nutrient interactions must take place for this pathway to function properly. Efficient processing of sulfur is essential for sulfation.

Sulfur molecules are "sticky," allowing cells to adhere to one another.[5] In other words, cells containing sulfur can grab onto heavy metals and other toxins and escort them out of the body. When sulfur metabolism or sulfation is lacking, cells "leak." Leakiness, or excessive permeability creates different problems in different parts of the body. Leakiness in the gut lining, or "leaky gut" is described in Chapter 7. Leaky skin creates eczema; leakiness in the joints leads to arthritis; leakiness in mucus membranes creates otitis, sinusitis, and rhinitis; and a leaky blood-brain barrier can lead to neurological issues.

Glutathione: The Body's Antidote to Oxidative Stress

Glutathione (GSH) is the body's most powerful antioxidant. It is created by every cell of the body from three amino acids: cysteine, glycine, and glutamic acid. The body makes about half the GSH it needs. *N-acetyl-cysteine* (NAC), methyl B12, folinic acid, *trimethylglycine* (TMG), and vitamin C are all important nutrients for increasing and maintaining healthy glutathione levels in the body by giving it the raw materials to build from.

Almost 20 years ago, S. Jill James, PhD, and her colleagues at the University of Arkansas carried out some game-changing research that increased our understanding of the impaired detoxification pathways in autism. James found that many of those with ASD have deficiencies in both cysteine and methyl B12, thus impeding phase two detoxification. They believed that the stores of GSH in the children with autism were depleted by their overexposure to mercury, specifically from thimerosal-containing vaccines.[6]

More recently researchers have looked at impaired GSH levels in schizophrenia. They conclude that low glutathione is clearly a biomarker for this devastating disorder.[7]

A body's glutathione level determines how many toxins it absorbs. Glutathione and its cofactors bind with hundreds of environmental toxins, including the metals cadmium, lead, and mercury, dragging them out of the blood to the liver, gall bladder, and gut, through which the body excretes them.[8]

Adequate levels of GSH are also necessary for many aspects of immune function. Each molecule of a toxin uses up a molecule of glutathione. Low levels impair immunity, which leads to infection. When levels of toxins are high, the body's natural store of GSH becomes depleted; as GSH levels fall, the body accepts more heavy metals.[9]

As described earlier, the body's poor response to infection causes inflammation and oxidative stress, which, in turn, lowers GSH. A vicious cycle perpetuates when there is inadequate GSH to offset oxidative stress, further reducing immunity and allowing opportunistic infections like yeasts and parasites to proliferate. GSH supplementation, discussed further on, is an essential part of any detoxification program because as GSH levels rise, the body is better able to excrete poisons.[10]

Detoxification and Microorganisms

The presence of the microorganisms enumerated in previous chapters, such as fungi, molds, bacteria, and viruses, complicates the detoxification picture. As a patient detoxifies, microorganisms can be liberated and sometimes proliferate. Refer back to Chapter 11 on remediating mold and Lyme. Remember, this step is absolutely essential before addressing other toxins.

Today, Dietrich Klinghardt, MD, PhD, whose wisdom has guided me for many years, is one of the few doctors in the United States who understands the complex relationship among inflammation, toxins, and microorganisms and why flare-ups of previously hidden infections occur regularly during detoxification. His most recent lectures on "The New Era of Post-Covid Medicine" emphasize the role of the spike protein in reawakening dormant Lyme, chronic fatigue, and even cancer. He believes that this demon has caused premature aging and even autism-like symptoms in young adults. Furthermore, for some, lingering fears of contamination, anxiety, and even panic from the pandemic remain locked in their cells. Learn more about these phenomena on his website, KlinghardtInstitute.com.

Detoxification and Genetics

Knowing one's genetics is vital to understanding why some aspects of detoxification aren't working as well as they should, thus allowing toxins to build up. Refer back to Chapter 5 to review genetic and genomic testing options. Which tests you run, and the results you receive will determine how deeply you can dive into the genetics of detoxification.

One gene is responsible for production of an enzyme called *methylenetetrahydrofolate reductase* (MTHFR) that helps the body detoxify. A mutation in this gene may prevent the body from converting folate into its active form *L-5-methyltetrahydrofolate*. If your eyes are not rolling back yet and you want to learn about MTHFR, go to MTHFR.net, the website of naturopathic physician Dr. Ben Lynch.

MTHFR is becoming well-known, especially in the autism community, because many families have discovered they have mutations that prevent them from methylating efficiently.[11] If you are clever and can visually add a few letters to MTHFR, you can guess the obscenity that they call this gene.

While severity varies dramatically, up to 60% of the population may have at least one copy of a problematic MTHFR mutation. Mental health professionals are recognizing that MTHFR mutations can also be associated with depression, bipolar disorder, schizophrenia, and ADHD, as well as miscarriage and migraines.[12] Be aware that just because someone has a mutation in this gene does not necessarily mean the gene does not work. It suggests it has a disposition to not working properly. Fortunately, you can measure homocysteine in the blood; elevated levels would mean the gene was being expressed.

Another mutation commonly seen in neurodevelopmental disorders is one for *sulfite oxidase,* or *SUOX.* While the MTHFR mutation inhibits methylation, the SUOX gene mutation impedes sulfation. When this marker is present, an individual cannot properly process sulfur-containing foods and turn the sulfites into less-toxic sulfates.[13]

Related to the SUOX mutation is another "coincidence": the timing during the 1980s of a recommendation by the medical community to replace aspirin with acetaminophen for alleviating pain, especially for children with viral infections. Acetaminophen is the main ingredient in Tylenol™ and some other popular painkillers. Tylenol depletes glutathione,[14] so taking it is counter-indicated for everyone!

According to William Shaw, founder of Great Plains Laboratory, acetaminophen toxicity can overload the defective sulfation pathway catalyzed by *phenol sulfer-transferase* (PST), leading to overproduction of the toxic metabolite *N-acetyl-p-benzoquinone imine* (NAPQI). Increased levels of NAPQI reduce an individual's ability to detoxify toxic chemicals, thus increasing oxidative stress and cascading into protein, lipid, and nucleic acid damage from free radicals.[15]

In her previously mentioned 2018 book,[16] Dr. Emily Guiterrez enumerates several additional genetic polymorphisms that can interfere with the body's ability to produce certain enzymes essential to taking out the garbage.

The *glutathione s-transferase pi 1 gene (GSTP1)* is a big player. A polymorphism in GSTP1 makes an individual more susceptible to mercury and other xenobiotic toxins because it slows down the body's ability to produce glutathione.

Superoxide dismutase 2 (SOD2) aids in cleaning up the mitochondria. A polymorphism in this gene causes a buildup of oxygen free radicals, and benefits from supplementation of antioxidants (see below).

Cystathionine gamma-lyase (CTH) is a gene that helps produce cysteine, one of the precursors to glutathione. Supplementing with n-acetylcysteine can be extremely beneficial, especially for those with PANDAS.

N-acetyltransferase 2 (NAT2) helps the body metabolize perfumes, chemicals in cleaning products, and even scented candles. Yep, those with this genetic mutation are our "chemically sensitive" patients. As their toxic burdens rise, they become increasingly more sensitive. Despite these mutations, we can do a great deal to help those who have them poop, pee, and sweat out their toxic loads.

Foods and Supplements for Detoxification

As in healing the gut, the first-choice weapons for detoxification are foods. Some fruits and vegetables are natural detoxifiers because they are highly antioxidant and support the body's detoxification pathways. Other options are spices and condiments, which are natural antioxidants, and add flavor and color.

Dr. Sharon-Hausman Cohen, cofounder of IntellXXDNA.com, is at the forefront of using nutrigenomics for detoxification.[17] Refer back to Chapter 5 to refresh your memory on her and her company.

Some highly detoxifying edibles are blueberries, acai berries, and pomegranates; dark leafy green vegetables such as kale, collards, dandelion greens, Swiss chard, and broccoli sprouts; walnuts, hazelnuts, and almonds; green tea and red wine; cloves, cinnamon, black pepper, and turmeric; and everyone's favorite, dark chocolate. Some of these are contraindicated on the special diets in Chapter 8, so cross-check carefully.

For those who are progressing well with adjunct therapy programs and whose laboratory tests show high metal levels, ingesting detoxifying foods and supplements may be all that is needed to make slow, steady progress. Few have negative reactions to this method, although changes may not be noticeable on a day-to-day basis. Over time, the nervous system becomes better regulated, the body and brain more responsive to therapy, and language more complex. Even gentle nutritional detoxification using foods and supplements requires strict supervision and should take place only under the guidance of a healthcare practitioner.

Supporting Detox with Vitamins and Minerals

Heavy metal excretion depletes minerals and antioxidants; thus, safe detoxification requires extra vitamins, minerals, and other nutrients, even in the presence of a nutritious diet, to assist in the mobilization of metals. Natural over-the-counter products, made up of either large quantities of food substances or vitamins and minerals derived from antioxidant foods are readily available as an additional boost.

Nutritionist Kelly Dorfman believes that adding extra nutrients during detoxification, even with a balanced diet, is essential. The poorer the diet, the more nutrients are needed, even with mild detoxification techniques. She strongly advises adding nutrients one at a time, three days apart, and carefully monitoring reactions, the most common of which is irritability.

A good nutrient program alone naturally encourages detoxification and gentle heavy metal displacement. Antioxidant supplementation alone often jump-starts the body's natural methylation and sulfation mechanisms. During any detox process, Klinghardt protects the brain and gut with electrolyte-enhanced drinking water.

Getting supplements into picky people can be challenging and may require some experimentation with delivery methods, such as liquids and gummies instead of pills. Oils mix well into puddings and pear- or applesauce. Sometimes just holding the nose and biting the bullet, with a great reward for compliance, is the easiest method! [18]

Chinese researchers have shown the potential of antioxidant supplementation for protection from and treatment of anyone with neuropsychiatric disorders, including schizophrenia, major depressive disorder, and anxiety disorders.[19] These antioxidants and minerals are the basics:

Vitamin C is a water-soluble antioxidant that immensely benefits the brain and body in a multitude of ways. Diarrhea occurs before the body can absorb too much, so there is little concern about toxicity. To bypass the gut, it is sometimes delivered in higher doses by IV.

Vitamin E is a fat-soluble antioxidant. Since cell membranes are basically a layer of fat, vitamin E is important for protecting membrane integrity. It also helps the mitochondria clean up debris for more efficient energy production.

Calcium enhances detoxification as it neutralizes excess aluminum. Individuals with calcium deficiency, perhaps due to a casein-free diet, may be irritable, hyperactive, sleep-disturbed, or inattentive, and have stomach and muscle cramps and/ or tingling in arms and legs. Though dark green vegetables and almonds are calcium-rich, in practice children rarely eat enough of these foods to cover their calcium needs without supplementation. Calcium must be properly balanced with magnesium, which improves absorption, and can cause loose stools. Dorfman recommends 800–1,000 mg per day for children ages one to ten.

Magnesium (Mg), the eighth most plentiful element on the planet and the fourth most abundant mineral in the body, is essential for over 300 biochemical bodily reactions, including digestion and kidney function.[20] Two major functions of Mg are stabilizing cell membranes and maintaining proper electrical balance.

Signs of magnesium deficiency include constipation, hypersensitivity to loud and high-pitched sounds, irritability, muscle cramps and twitches, cold hands and feet, insomnia, and carbohydrate cravings. Mg deficiency is most likely in those who eat many processed foods, overcook foods, and drink soft water. Deficiencies can develop when Mg elimination is increased by taking medications like anti-psychotics, because these products—as well as alcohol, caffeine, and sugar—leach Mg. Foods high in Mg are avocados, beans, molasses, almonds, Brazil nuts, cashews, pumpkin and sunflower seeds, whole grains, fish, kiwis, and leafy greens, especially spinach.

Molybdenum (Mo) is an essential element in human nutrition whose role was little understood until recently. Molybdenum is vital for the critical conversion of sulfites to sulfates. Sulfite/sulfate imbalance wreaks havoc not only with meth-ylation, but also hormonal balance and digestion. Low Mo, poor sulfation. If a person has the SUOX or CBS mutation, or is diagnosed with SIBO, then Mo is an especially important supplement.[21]

About ten years ago I started taking daily molybdenum after testing showed a SUOX SNP. No more "ulcerative colitis," crazy fluctuating mood swings, and best of all, I lost weight because as my body let go of toxins it no longer needed excess fat to protect my brain.

How does one get molybedenum? Some foods contain it. Your best bets are barley, yams, beef kidneys and liver, buckwheat, eggs, leafy greens, spinach, sun-flower seeds, oats, and potatoes. If those don't do the job, supplementation is necessary. Work closely with a licensed healthcare professional on this one!

Copper (Cu) and Zinc (Zn) must be balanced carefully. With too much copper and too little zinc, psychiatric symptoms can occur. Molybdenum is also important in maintaining the proper Cu/Zn ratio. Research on Cu/Zn imbal-ances is the work of William Walsh, PhD, an innovative biochemist who reports that 85% of his sample of 603 patients with autism showed a highly statistically significant Cu overload and Zn depletion when compared to healthy controls matched for age and gender. Imbalanced Cu/Zn impairs the hippocampus and amygdala, which monitor social-emotional function. Patients thus have a bio-chemical tendency for emotional meltdowns and attentional deficits.[22]

A *comprehensive mineral supplement* containing 30 mg of **zinc**, 100–200 mcg of the strong antioxidant **selenium**, and 2–5 mg of **manganese** is essential. Adequate zinc protects against adverse effects from heavy metals, including lead, cadmium, and copper. To minimize the number of pills try a mineral combination from a trusted source, such as Kirkman, New Beginnings Nutritionals, or Researched Elements.

Essential fatty acids (EFAs) are a crucial part of the structure of the nervous system, which is 60–70% fat. EFAs must be ingested because the body cannot produce them. Most people with nervous system imbalances are deficient in omega-3 fatty acids, with elevations of arachidonic acid (an omega-6 fat) and trans-fatty acids.[23] Grandma was correct: take cod liver oil! It's no longer stinky nor does it cause annoying burps because it comes in flavored burpless capsules, liquids, and gummies.

Symptoms of EFA-deficiency are hair loss, dry or peeling skin, eczema, fatigue, aggression, dry brittle hair, eating disorders, excessive or diminished thirst, gallstones, growth impairment, immune deficiency, hyperactivity, and impaired wound healing.

Essential fats are fragile, and when overprocessed or exposed to air, they can become rancid. Old fish oil is worse than no fish oil, as the body must deal with the results of the fish's oxidative stress! Refrigeration can slow deterioration. Quality in fish oils is of utmost importance; make sure your product comes from mercury-free fish and is soy oil free!

Whole books have been devoted to the importance of EFAs in brain and mental health. The reason that metabolic psychiatrists have so strongly embraced the ketogenic diet is because of its abundance of good-quality healing fats.

Dimethylglycine (DMG) and Trimethylglycine (TMG) are essential additions to the mental health arsenal because they both play critical roles in methylation pathways.[24] Before his death in 2006, Bernie Rimland, founder of the Autism Research Institute recommended the use of DMG (vitamin B15) for over 25 years and found it to be nontoxic and potentially helpful for improving language and socialization skills.[25]

Folinic acid, vitamins B6 and B12, zinc, and magnesium are often added to antidote to the irritability sometimes caused by DMG or TMG. Their synergistic action can be especially helpful in those with MTHFR mutations who also have a SNP in the *folate receptor 1 gene* (FOLR1) according to Emily Gutierrez, mentioned above.

Vitamin B12, also called **cobalamin**, is one of the important B vitamins for detoxification. B12 deficiencies are a well-known factor in many conditions, including autism[26] and neuropsychiatric conditions.[27] Dr. Guitierrez cites two genes, MTRR and GIF, with mutations that can negatively affect the body's ability to absorb B12.

Cobalamin comes in several forms; methylcobalamin, the active form of B12, is the preferred type for detoxification over the less-expensive, more-available cyanocobalamin.

James Neubrander, MD, in Edison, New Jersey, pioneered injections of pure concentrated methylcobalamin for those with autism. Injections, unlike oral B12, bypass the impaired gut and directly feed the nervous system.[28]

Klinghardt's preferred method is to deliver methyl B12 as a nasal spray with or without folinic acid. Klinghardt believes that combining folinic acid with methylcobalamin and sometimes TMG can be a winning formula.

This treatment is especially suited to those with a history of vaccine reactions and chronically loose stools. When combined with folinic acid, B12 facilitates the complicated methylation processes important for creating optimum metabolic balance, which eventually allows the body to detoxify itself.[29]

Alpha lipoic acid (ALA) is an antioxidant that neutralizes free radicals and has the unique characteristic of being able to function in both water and fat.[30] It is thus able to replenish antioxidants such as vitamin C and glutathione after they have been depleted. ALA also aids in the formation of glutathione.[31] It is plentiful in spinach, broccoli, peas, brewer's yeast, brussels sprouts, rice bran, and organic meats.

Glutathione Supplementation

Given the findings of Dr. James above, it is not surprising that supplementing glutathione has gained favor in autism.[32] In the original study supplementing GSH along with methyl B12 showed both improved speech and levels of cognition.[33]

Glutathione can be delivered in numerous ways: orally, by IV, as a liquid, nebulized, as a nasal spray, or in a cream. The cream surrounds GSH with fat droplets called *liposomes*, allowing maximum absorption and protection from oxidation.

The liposomal glutathione cream Essential Glutathione™ is available without prescription from Wellness Pharmacy, WellnessHealth.com. They suggest beginning with only one-quarter teaspoon twice daily per 30 pounds of body weight, always starting with an even smaller dose in sensitive children. Intravenous (IV) GSH is also available from them, although using needles with children who have sensory problems can be challenging.

Alex Zaharakis and Christian Bogner of Autism Is Biomedical (AiB) believe that cajoling the body into **making** glutathione by giving cells the raw ingredients is superior to **taking** glutathione. Their product Glutathione Genesis provides what is needed to assist natural processes in creating new glutathione.

Another one of their concoctions is Thioguard Lotion, which recycles gluta-thione, allowing continued methylation and detoxification and prevents toxicity and/or deficiency. Both are available on the AiB website. Many practitioners combine GSH with other antioxidants and essential nutrients to assist in breaking down and releasing stored toxins.

Specialty Detoxifiers with Good Safety Records

Here are some more of Klinghardt's natural favorites:

Cilantro (coriandrum sativum), also known as Chinese parsley, mobilizes mercury, cadmium, lead, and aluminum from the bones, central nervous system, and the cells.[34] Cilantro is *so* effective that it may flood the connective tissue with metals, causing "re-toxification." Use with caution, or only under the supervision of a doctor who understands its power. Cilantro is available in a number of forms. Chop it fresh and put in salads, soup, or tea, make up a tincture, or use it in a transdermal cream.

Sodium alginate, extracted from algae, is an excellent detox agent for cadmium, barium, lead, strontium, and mercury, as well as for detoxing environmental fumes.[35]

Chlorella vulgaris is a staple in the Klinghardt arsenal for anyone interested in maintaining good health, especially when combined with cilantro.[36] Chlorella, as can be surmised by its name, contains vast amounts of chlorophyll, nature's own detox agent. Without chlorella, neurotoxins can be reabsorbed on the way down the small intestine by the abundant nerve endings of the enteric nervous system located in the gut. It is available as capsules, a powder, and chewable tablets.

Chlorella growth factor (CGF) is a heated extract from chlorella that concentrates specific peptides, proteins, and other ingredients for detoxifying every existing toxic metal in a profound way. Some chlorella products are purer than others; try the BioPure brand, which is treated with ultrasound, is not oxidized, and is guaranteed to be free of metals and other toxins.

Garlic has been known to have healing powers for over 200 years. It contains numerous sulfur components which oxidize mercury, cadmium, and lead and make them water soluble. Garlic's alliin is enzymatically transformed into allicin, nature's most potent antimicrobial agent.[37] Two forms, allium sativum and bear's garlic (allium ursinum) protect the white and red blood cells from oxidative damage caused by metals in the bloodstream on their way out.

The half-life of allicin from crushed garlic is fewer than 14 days, so most commercial garlic products have little allicin-releasing potential. Freeze-dried garlic still contains allicin and is available in capsules.

Metal-toxic patients almost always suffer from secondary infections, which are often at least partly responsible for symptoms. Garlic also contains the most important protective mineral against mercury toxicity: bioactive selenium. Klinghardt believes that selenium from garlic is preferable to most selenium products that are poorly absorbed and don't reach the proper sites.

Dandelion tea: did you know that those leaves and roots from the bright yellow flowers popping up in your lawn each spring can make a delicious detoxifying tea? Klinghardt teaches that this simple beverage can even block the spike protein from the COVID virus.[38] Not convinced? Watch this interesting YouTube video: https://youtu.be/M4rxtjAF3KA. But you don't need to wait until April to harvest them; many organic dandelion tea products are available commercially.

Chelation

The word *chelation* is derived from the Greek word for *claw*, which describes the chemical binding of metal ions that occurs as a detoxifying agent attaches itself to toxic metals or other poisons, allowing the body to expel them. The heaviest and generally most toxic metals such as lead, mercury, nickel, and cadmium are the first to bond to chelating agents. The resulting chelated minerals are stabilized and don't cause damage.

Every mineral taken into the body, whether essential or toxic, binds to a chelating agent to stabilize it. The detoxifying substances mentioned above, while natural, are considered mild to moderate chelators because they bind to and escort toxins out of the body. The chelating agents in this section are much more powerful.

Chelation is a more aggressive last resort for many people whose testing shows the presence of toxic metals, and who are not making progress either with over-the-counter metal-removal agents or with therapies. Doctors have used chelation for many years to remove lead. Available only by prescription, medical chelation is *very, very, very* controversial, requires special training, and can be costly.

All chelation protocols must be supported by a diet rich in antioxidants and a strong nutritional program designed to alleviate oxidative stress and to replenish important minerals that chelators remove along with the toxic metals. Balancing the bowels is a must before starting chelation because heavy metals are partially excreted through the stool. Use dietary modification and nutritional supplementation to get the bowels functioning properly, heal the gut, and strengthen immunity.

Doctors are continually looking for safe and effective chelation protocols. Here are some chelators they are using:

- *Dimercaptosuccinic acid (DMSA)* has a long safety record and is an effective chelator for many metals.[39] It is delivered orally, through rectal suppositories, or in a transdermal cream, and it can be combined safely with cilantro and/or chlorella. Common negative effects include gastrointestinal disruption and the leaching of essential minerals, especially zinc.
- *Thiamine tetrahydrofurfuryl disulfide (TTFD)* cream is a synthetic version of vitamin B1 derived from garlic. The late Dr. Derrick Lonsdale (1924–2024), an expert in thiamine deficiency who lived two weeks beyond his 100th birthday, believed that TTFD has three sulfur-related mechanisms that benefit children with autism spectrum disorders.[40]
- *Ethylenediamine tetra-acetic acid (EDTA)* is a synthetic amino acid with a strong binding affinity to calcium and lead, and an effective chelator for aluminum and nickel. Traditionally, it has been used orally for lead poisoning; however, it can remove many metallic ions.[41] Doctors are using EDTA by IV; this procedure requires *very* close monitoring while being done.
- *Dimercaptopropane sulfonate (DMPS)* is a sulphuric compound that binds to heavy metals, especially mercury. DMPS is used off-label on children, as it is not FDA-approved. Some experience severe detox reactions and zinc loss. DMPS can be delivered transdermally (TD-DMPS), as a highly stable, oxidation-resistant lotion that is rubbed into the skin, and as suppositories. To avoid mineral depletion, many physicians start an aggressive two-week mineral repletion program prior to initiating treatment with TD-DMPS.

Recently, chelation has lost favor due to chronic pesky gut symptoms and reports of temporary regression following treatment. Those practicing chelation often prefer transdermal or intravenous delivery systems because they bypass the gut.

Detoxification Guidelines and Precautions

Invest in quality supplements. Supplement quality is equally as important as the quality of our food, air and water. Buy products only from natural health food stores, healthcare providers, and direct from manufacturers who take pride in their products and tell you what's *not* in them. The good ones say GMP certified, which is the nutritional supplement version of being "pharmaceutical grade." This ensures the company has done the proper quality control to make sure that ingredients and amounts on the label are what is actually in the bottle. Look at the origin of the supplement and watch for fillers. Many instances of "the supplement did not work" are due to poor-quality products off the shelves of grocery, health food, and big-box stores. Supplements are one instance where you get what you pay for.

Use the proper form of a vitamin or mineral to ensure it is absorbable. Ask your healthcare practitioner to recommend the form and delivery system.

Go slow and low. Err on the side of caution by starting with a small amount and moving gradually to a target dosage. Make sure that the initial dosage is not too powerful. Drop back or stop the supplement if negative symptoms such as irritability, diarrhea, or constipation occur. Decrease dosages until symptoms disappear.

Watch for possible detox reactions. Patients are unique in their responses to different detox methods, so this is one area where vigilance is essential. Some people go through the process without incident. For others, detox reactions occur if agents remove poisons too quickly. Reactions may be as benign as muscle aches that indicate the redistribution of toxins into the connective tissue and an insufficient nutritional support program. Other responses can be serious, such as rage, extreme hyperactivity, depression, headaches, seizures, and increased pain, indicating the redistribution of metals into the central nervous system, again, the result of a support program that is too weak. Eye and ear problems that occur during detox indicate redistribution of toxins into these organs.

During a detox program, a patient may also become temporarily "allergic" to a substance that carries out the toxins. Every time a patient receives a detoxifying substance, toxins emerge from their hiding places into the more superficial tissues of the body, where the immune system detects them. The immune system sometimes believes that the detoxifying substance itself is the enemy and reacts to it. This reaction typically resolves spontaneously within six weeks after detoxing.

Look for reappearance of signs of candida. Using probiotics and digestive enzymes can be helpful in avoiding this side effect.

Use only one detoxifier at a time. Each detoxification agent has a primary place of action, which determines when, how much, and for how long it is used. In choosing individual detox agents, Klinghardt considers the part(s) of the body where metals are stored, relying upon autonomic response testing (ART). For example, he finds that chlorella is ideal for removing virtually all toxic metals from the gut, but has too little effect on mercury stored in the brain. Intravenous glutathione may reach the intracellular environment, even in the brain, but is fairly ineffective in removing mercury from the gut.

Taking multiple toxic compounds simultaneously increases the opportunity for a synergistic effect from which a patient may have difficulty recovering. Using multiple detox agents sequentially (rather than simultaneously) can address various forms of mercury in the body because mercury is bound in the tissues, cells, and organs in virtually hundreds of ways, and no single agent can clear all toxins.

Additional Novel Methods of Detoxification

Epsom Salt Baths

A soothing tried-and-true support for detoxification is *magnesium sulfate*, also known as Epsom salts. Their efficacy is linked to the phenyl-sulfotransferase (PST) enzyme, which causes a change inside the cells by adding the molecules *adenosine* and *phosphate* to sulfate before any sulfotransferase enzyme can use it. The molecular additions turn sulfate into its "activated" form. Pour one cup of Epsom salts into a tub of warm water and soak for approximately 30 minutes per day. Epsom salts can also be used to supplement both magnesium and sulfur as they raise their levels in the blood. Too much magnesium in the tub can cause loose stools as it is easily absorbed through the skin and magnesium is a natural muscle relaxant, so cut back on the amount of Epsom salts if this occurs.

Sweating in a Sauna

Sweating is one of the oldest, safest, most natural, and least expensive ways to detoxify. Saunas are finding their way into the homes of many health-conscious people so they can sweat daily all year round.

A far infrared sauna duplicates the same frequencies as normal body heat. Far infrared heat rays penetrate the body to a depth of 1.5–2 inches. The body's tissues selectively absorb these rays as water in the cells reacts in a process called *resonant absorption*, which causes toxins to be released into the bloodstream. These toxins are then excreted in sweat, feces, and urine. Because no chemicals are added and the heat forces the use of sweat as the major mode of toxin elimination, this method puts less stress on the kidneys and liver.[42]

Saunas are available for two, three, or four persons; the best are made of poplar, which—unlike cedar, redwood, spruce, or pine—does not outgas any chemicals. Some are equipped with stereo. Find one that is totally natural, with no off-gassing or toxic glues. Two great sources are HighTechHealth.com and HeavenlyHeatSaunas.com.

Ionic Foot Bath

What if you could remove toxic metals, pesticides, and other poisons from your body by soaking your feet? According to A Major Difference, their patented IonCleanse® does just that, painlessly, in a noninvasive fashion.

Subjects place their feet into a plastic tub, to which a little salt and an electric array have been added. When the array is plugged in and turned on, it delivers positively and negatively charged ions into the water. No current is felt by the patient. Since opposites attract, these ions neutralize oppositely charged toxins and pull them out of the body through the skin. The end result is a tub full of nasty-colored water, and a calmer, more relaxed person.

Members of the Thinking Moms Revolution (TMR), an international non-profit, decided to test this equipment on 24 children with autism, age two to 19 in two studies. Subjects' behaviors were measured using the Autism Treatment Evaluation Checklist (ATEC), before and after treatments. Sessions ranged from 15–30 minutes, depending on the children's age. 100 percent of the subjects showed gains in socialization, language, cognition, and other areas. Overall, average reduction in ATEC scores was 35% in the first study and 55% in the second study, which included more teenagers, over a 120-day period. The greatest average reduction in ATEC scores was in the oldest kids. The IonCleanse foot bath has a unique patent on their technology, so other foot baths may not produce the same results.

Take-Home Points

Toxins make people sick. When a body's Total Load of toxins exceeds its ability to detoxify, the poisons eventually affect the nervous system, including the brain. For individuals with neurodevelopmental disorders, a comprehensive detoxification program is essential. Using a variety of techniques and tools, practitioners can gradually chip away at the stored aggravators and diminish the load in patients of all ages.

Each healer has his or her favorite methods with which to approach this intervention. Mild, natural herbs and foods used daily can help a body detox slowly. Chelation is a more controversial way of removing toxins from the body and is being practiced with less frequency than in the past.

Anyone undergoing detoxification should be monitored by a professional; a home-based program is not a substitute for appropriate individual medical care. Detoxification is a continuous, lifelong process. It can be an elegant, direct route through the maze of treatment options or a frustrating sequence of blind alleys and dead ends. Supervision by an experienced and qualified practitioner is absolutely essential for success.

CHAPTER 14

BIOREGULATORY MEDICINE

A group of treatments that achieve an optimal healthy functioning of the whole body falls under the broad topic of *bioregulatory medicine*, or simply *biomed*. While most allopathic physicians specialize in a single system like cardiology, and their medicines suppress, block, and inhibit symptoms like fever and mucus, most bioregulatory practitioners are multisystem docs who honor the body's ability to self-heal by tapping into its "vital force," known as *chi* or *qi*. I thank Mary Coyle for introducing me to this innovative holistic approach.

As this book has emphasized again and again, all bodily systems are related, and treatment of one system affects others. We've already looked at the nervous, digestive, immune, and endocrine systems, and you now know they do not work in isolation. I introduce the oft-ignored and underappreciated lymph system in this chapter. Remember the HPA axis that combines the efforts of the hypothalamus, pituitary, and adrenals to make balance? That is only one of the many multiorgan feedback loops at work keeping us alive and well.

Bioregulatory medicine practitioners determine the best nutritional, dietary, immune regulation, inflammation reduction, and detoxification strategies for each unique individual and combine them with some out-of-the-box, mostly invisible energetic methods from German and Chinese medicine. The targeted result is a state of balance called *homeostasis*.

I am excited to share these remarkable less well-known interventions with you. Although you may never have heard of some of them, many have long-standing histories of safety and efficacy despite being censored and maligned by allopathic medicine for five decades or more. Unfortunately, most are unknown to mainstream doctors, so don't bother seeking their stamp of approval; as the old saying goes, "People are down on things they're not up on."

This chapter begins with a short course in *homeopathy* and its variations, *homotoxicology* and *CEASE therapy*. Almost extinct in the 1980s, homeopathy used to be extremely popular, according to Massachusetts homeopath Jerry Kantor, as he details in his fascinating book *Sane Asylums: The Success of Homeopathy before Psychiatry Lost Its Mind*.[1]

Believe it or not, in the United States, in the late 1800s and early 1900s there were more than 100 homeopathic hospitals, at least 1,000 homeopathic pharmacies, and 22 homeopathic medical schools. In fact, homeopathy was the treatment of choice for most medical conditions before pharmaceutical companies emerged so powerfully. It is still huge in Europe, especially in Germany, where many homeopathic medicines are made.

After homeopathy comes a short discussion of *neural therapy*, another treatment of German origin and a favorite of Klinghardt-trained healers. Detoxifying the intangible psychological, emotional, and past-generational load factors comes next, with a discussion of Family Constellations and the emerging field of *psycho-neuroimmunology*.

This might be a good time to open your mind and heart—any one of these programs could be a game changer for someone you know.

The Lymphatic System

The *lymphatic system*, intricately connected to both the immune and circulatory systems, is an oft-ignored major player in keeping us healthy. A convoluted network of organs, tissues, and vessels—including the tonsils, thymus gland, and spleen—it runs throughout the body and into the brain.

This superhighway carries a cleansing fluid called *lymph*—full of white blood cells and antibodies, as well as metabolic cellular waste, destroyed bacteria, and other garbage—to endpoints called *nodes*. Lymph nodes are small bean-shaped structures strewn throughout the body in clusters, with most strategically located in the armpits, neck, groin, and abdomen.[2] When about three quarts of lymphatic fluid a day pass through the nodes, this toxic soup is filtered and microorganisms are neutralized by *lymphocytes* before they can do any harm. When a foreign organism is detected, the lymph signals immune cells and triggers an inflammatory response.[3]

Keeping lymph moving is essential to staying healthy. Unlike the circulatory system, the lymphatic system does not have a pump. You are in charge of making it work properly, and the only way to get lymph moving is to nudge it along with skeletal movement like walking or other exercise. When we are too sedentary, lymph backs up, resulting in swelling called *lymphedema*. Virtually **everyone** with a chronic illness has problems with the lymphatic system.[4] Good news! Help is here. Many books are available online for self-care.

Lymphatic Drainage Massage

Feeling fatigued? Think lymphatic congestion. Assisting lymph's movement along the superhighway with massage is a wonderful intervention that can make just about anyone feel better.

Luscious creams containing detoxifying herbs can be applied anywhere on the body to increase lymph (and blood) flow. Look for oils and creams with mullein, red clover, and pokeweed, a potent go-to for moving stubborn lymph when other herbal remedies fall short. Reject products that are petroleum-based or have artificial scents or preservatives. Lymph drainage massage has shown efficacy in both autism[5] and psychiatric disorders;[6] it is in the toolbox of many bodywork healers listed in Chapter 6. Again, a multitude of sources can help you find a therapist who is talented in this area.

Transvascular Autonomic Vasculation Modulation (TVAM)

A related condition, where blood and lymph remain too long in the brain and have trouble getting back to the heart, is called *chronic cerebrospinal venous insufficiency (CCSVI)*. A delay in deoxygenated blood leaving the head can cause *hypoxia*, a lack of oxygen in the brain. Plasma and iron from blood deposited in the brain tissue can also be very damaging, allowing iron and other unwelcome cells to cross the crucial brain-blood barrier.[7]

Addressing this impeded blood and lymph flow in the neck with TVAM—allowing increased blood flow to the brain—originated with Italian vascular surgeon Paolo Zamboni, MD. He applied his professional knowledge and skills to his wife, who was diagnosed with MS. After the doctor dilated vascular restrictions that showed up on a special imaging device, she made dramatic improvement.

Since 2012, Klinghardt and others have been applying this concept to autism. TVAM involves stretching the jugular vein with small catheters; accompanied by a combination of microbe killers and detoxifying agents, it can open up and clean out the veins and arteries for two to three years. This is one of those "stay tuned" therapies.

Tonsils

The tonsils are a group of five lymphatic tissues found at the back of the throat, as a part of Waldeyer's ring. Klinghardt calls them the "toilet of the brain."

One of the first active lines of defense for our immune system, the tonsils' primary job is protecting the body and brain against pathogens and toxins that originate in the nose or mouth. When tonsils detect any unfamiliar organisms, they signal an immune response and cause inflammation.

Infections, mostly in the teeth, gums, and sinuses, all affect the tonsils, which, like all lymph nodes, expand or swell when exposed to germs. Once illness passes, healthy tonsils shrink, but in chronic illness, low-grade infections, or allergies, sometimes they do not.

Klinghardt believes that chronic infections often reside in decaying tonsils. The tonsils and the gut also have a relationship because the tonsils are one of the first parts of the lymphatic system to come in contact with food. They prime the immune system by telling it whether certain foods are good or bad for gut health.

Because swollen tonsils can interfere with eating, breathing, and sleeping, they are often considered more troublesome than useful. The standard treatment for chronically inflamed tonsils is simply snipping them out. But this approach does not address *why* the tonsils became swollen in the first place. While a tonsillectomy may resolve chronic tonsillitis, it can also set the body up for continuous trouble as infections can then move and become pockets of pus around teeth and in sinus cavities. Furthermore, a new study suggests that early tonsil removal is a risk factor for later stress-related disorders and psychiatric illness.[8]

According to Klinghardt, scars left from removing the tonsils can interfere with the pineal gland's ability to produce melatonin, thus affecting sleep. Yet, another cascade of events that starts by treating symptoms (inflamed tonsils) without addressing causes (low-grade infections) that results in new troubles (sleep disruption).[9]

Instead of removing the tonsils, Klinghardt has established a protocol to heal them. It starts with injecting them and gargling with a fermented product to kill the infections, then using homeopathic remedies to build the immune system, and finally restorative cryotherapy to burn them off. Like flower bulbs, they regenerate, good as new. For more information on this treatment, which is available only in Germany, go to KryoPraxis.de.

Homeopathy

Homeopathy is a 250-year-old approach to healing currently used by a small number of medical and osteopathic doctors, many naturopathic physicians, and a growing number of lay professionals without medical licenses. Dozens of double-blind, placebo-controlled clinical research studies prove its efficacy.[10]

Dr. Samuel Hahnemann, a German physician searching for a safe, effective alternative to conventional medicine, founded it. The word *homeopathy* comes from combining the Greek words *homoios*, meaning *similar,* with *pathos*, meaning *suffering* or *disease*.

Homeopathy looks at the individual as a whole, focusing on the totality of symptoms, not a diagnosis. Like other practitioners, the classical homeopath takes a comprehensive history of an individual's symptoms. Unlike allopathic doctors, a homeopath asks questions about not just physical symptoms, but also about mental and emotional states, behaviors, beliefs, and even preferences such as handedness.[11]

Classical Homeopathy

Many professionals practice *classical homeopathy,* which focuses on determining the single best homeopathic medicine, or *constitutional* remedy, for a patient. Most homeopathic medicines, called *remedies,* are energetic signatures of plant, mineral, or animal substances that occur in nature.

Classical remedies usually contain one ingredient (combination remedies are common in other modalities) that stimulate the body's chi, thus enhancing its ability to heal itself. Remedies come as pellets, tablets, creams, ointments, salves, or liquids in varying potencies. Most contain only energetic traces of the substance from which they are derived.

Remedies are prepared through a process of dilution, which results in a potentized medicine. Counterintuitively, the higher the number of dilutions, the greater the *potency* and the more powerful the remedy. Homeopathic remedies, similar to vaccinations, may produce mild symptoms of a specific disease in a healthy person.[12] But unlike vaccinations, long-term negative side effects are minimal.

Homeopaths are particularly interested in the connection between symptoms of illness and vaccination. Many of their clients have "never been well since" a round of vaccines. Some use homeopathic alternatives for those skittish about vaccinating as a way to immunize as well as mitigate possible adverse vaccine reactions.

Homeopathy as a treatment for mild to severe autism and ADHD is well documented. Read *A Drug-Free Approach to Asperger's and Autism: Homeopathic Medicine for Exceptional Kids,*[13] *Ritalin-Free Kids,*[14] *The Impossible Cure*[15] from autism mom Amy Lanksy, and Kantor's *Autism Reversal Toolbox: Strategies, Remedies and Resources.*[16] Some practitioners have found homeopathy helpful for psychiatric disorders that include delusions, anxiety, and depression.[17] The classic book *The Homeopathic Treatment of Depression, Anxiety, Bipolar and Other Mental and Emotional Problems*[18] has many suggestions.

Miasms

Hahnemann wrote about *miasms*, including the profound connection between emotions and their physical manifestations in illness. A miasm is a complex of inherited imbalances similar to genetic inheritances, but they are energetic rather than tangible. These constitutional characteristics manifest as physical, mental, and emotional tendencies that predispose an individual to certain diseases and are passed down through generations like tall stature.[19] Homeopathic treatment opens a path to restoring balance.

Jerry Kantor considers miasms to be key to understanding illness. In his book *The Emotional Roots of Chronic Illness: Homeopathy for Existential Stress,* Kantor combines his knowledge of Traditional Chinese Medicine with homeopathy and acupuncture to address miasms. He compares inborn foundational emotions like anxiety and anger to tools, each designed to solve a stress-related problem. When a healer considers the physical, dermatological, pain, and personality manifestations of each of five possible miasms, he can go deeper into understanding the root causes of an individual's illness. Then homeopathic remedies for each miasm and its associated symptoms can be prescribed.[20]

Kantor asks existential questions related to each of five miasms. He then relates each miasm to one of the Chinese Five Elements, and to a sense, specific emotions, and one or two of the body's organs. For instance, the Tubercular Miasm is about the question, "Am I alone in life or do I act in synchrony with nature and with others?" This miasm is related to fire, touch, joy, and to the heart and small intestine. In addition to homeopathic remedies, Kantor lists some specific health-promoting activities like drumming, yoga, massage, and even philanthropy to counteract each miasm.

The beauty of homeopathy is that, potentially, a single homeopathic medicine can address all concerns at the same time. Often, though, an unwinding of factors requires the use of sequential or even concurrent remedies. Healing has gotten more and more complicated because of what we've done to our environment. Some, but not all, homeopathic practitioners are open-minded about combining their treatments with other therapies, especially nutritional supplementation and dietary modification.

Homeopathic medicines require training to use correctly. Computerized software programs are now available to *repertorize* or derive the best remedies for a patient.

As you would expect, Klinghardt includes homeopathy in most of his protocols, since his background is German. His courses include learning about many of these medicines, which are inexpensive and have many years of support behind them. Since homeopathy is an energetic modality, there are fewer side effects—although aggravations, usually mild and self-limiting, may occur as the body mobilizes and removes the toxins through the body's elimination pathways such as the colon or kidneys.

Homotoxicology

Homotoxicology, a bridge between homeopathic and allopathic medicine, was developed by German physician Hans-Heinrich Reckeweg, MD (1905–1985), in the 1940s. According to Dr. Reckeweg, all illnesses are merely expressions of an organism's attempt to react against toxins and expel them. When the body does not possess sufficient cellular energy to remove invaders fully, it must also attempt to counteract the damage manifested by these toxins.

Dr. Reckeweg theorized that this sequential process follows specific, increasingly pathological phases that should be familiar by this point in the book. In the early phases, illness is generally reversible using natural solutions. As damage becomes deeper and more complex, the organs and tissues degenerate, and it becomes much more difficult for the body to heal.

- At first the body tries to *excrete* the toxins, first through the primary pathways of pee and poop. If they are blocked or overwhelmed, then with reactions such as skin eruptions, mucus, fever, cough, runny nose, etcetera. These symptoms combine to create a cascade of increasingly deleterious health effects, including *inflammation.*

- If the accumulation of toxins continues to build, the results are the *deposit* and *impregnation* of the toxins, with symptoms such as warts, hemorrhoids, or cysts.

- Finally, if chronic inflammation persists, the body enters the *exhaustion stage* from fighting both the toxins and their deleterious actions, eventually leading to organ damage from *degeneration* and diagnoses of specific conditions.

- The last stage occurs when toxins cross the "biological division" from the extracellular to the intracellular phase and penetrate and damage cell membranes and enzymatic function. In essence, too many toxins came in and didn't go out fast enough. Hovering in this exhaustion state, patients become "too sick to get sick."

Homeopaths sometimes use lab tests to verify the release of toxins. Once toxins are excreted and inflammation decreases, the body can then continue the healing process. And as the body's health is gradually restored, it then has the energy and capacity to maximize the effectiveness of more aggressive therapies. For this reason, as well as for many others, homotoxicology is considered a great "first step" in healing.

Homotoxicology utilizes cellular drainage, nutrition, and other time-tested techniques to stimulate the self-healing mechanism. Practitioners using homotoxicology treat individuals in highly compromised states very gently with low-potency homeopathic remedies designed to activate and support the body's defense mechanisms.

Homotoxicology uses special remedies called *nosodes* and *sarcodes*. Nosodes are derived from the pathogens which cause a disease, while sarcodes originate from healthy endocrine glands or normal secretions of living organs.

Some remedies target specific pathogens or xenobiotics—such as yeast, parasites, or heavy metals—to alert the body to their presence so it can then expel them. All their homeopathic remedies are designed to intervene on multiple levels, thus regulating, balancing, and strengthening patients' immune, endocrine, and other organ systems, increasing the body's ability to dump toxins.[21]

Because a seemingly infinite number of remedies in potency and combination are available, homotoxicologists often use an EAV machine, described in Chapter 5, to test the healing power of combinations of different medicines. Note: classical homeopaths who use single remedies only are generally not in favor of combination products.

Homotoxicology has become a popular method of detoxification because its low-potency remedies give someone with an already weakened immune system the ability to remove toxins at a slow and steady pace. It is one of the least invasive, least costly, safest, and most efficient methods we have today. Read more about the man who conceived this idea in *Bizarre Medical Ideas . . . and the Strange Men Who Invented them.*[22] Testimonials from parents about how homeopath Mary Coyle uses homotoxicology to help their children are at RealChildCenter.com.

CEASE Therapy

The late Dutch physician, Tinus Smits, MD (1946–2010), treated more than 300 cases of autism. His conclusion was that 70% of autism was due to vaccines, 25% to toxic substances, and 5% to physical disease. Today his work continues through a program called *CEASE*, for *Complete Elimination of the Autistic Spectrum Expression.*

Smits coined the term *post-vaccination syndrome* to describe the relationship between vaccination and the physical, cognitive, and behavioral problems in those on the autism spectrum. He hypothesized that "vaccination damage can take months to express itself, sometimes starting very insidiously, almost imperceptibly."[23] Smits believed that homeopathic nosodes were a perfect tool to treat post-vaccination damage.

CEASE is a type of sequential homeopathy that is loosely based on the work of Jean Elmiger, MD, as detailed in his book *Rediscovering Real Medicine: The New Horizons of Homeopathy*.[24] It addresses all possible causative factors—not just vaccines. Step by step, it detoxifies medications, environmental exposures, and other poisons with homeopathically prepared, diluted and potentized, versions, which clear energetic imprints out of a patient's field. CEASE also uses supplements, especially Vitamin C, fish oil, and zinc, in the detoxification and healing process.

Practitioners trained in Smits's approach offer services around the world. To learn more about this method, read *Autism Beyond Despair*[25] and *Inspiring Homeopathy*[26] and go to CEASE-Therapy.com.

Acupuncture

Acupuncture is an ancient Chinese approach to enhancing health by treating the person's vital energy or chi, which is thought to circulate throughout the body on invisible highways called *meridians*. Physical, chemical, emotional, and other stressors block the flow of chi. Acupuncture, and several specific techniques that fall under the acupuncture umbrella and incorporate its principles, can be beneficial in addressing symptoms associated with neurological and psychological conditions.

Traditional acupuncture is done by puncturing the skin with hair-thin needles on specific points along the meridians, each of which corresponds to an organ system. These points can also be stimulated with massage (acupressure), heat (moxibustion), and/or herbs. Waking up the flow of energy along a meridian affects the body physically and emotionally. A little nudge goes a long way toward improving specifics like digestive function, sleep, depression, anxiety, headaches, and pain.

Emotional Freedom Technique (**EFT**) is a simple, effective, and long-lasting acupressure technique that doesn't require a doctor or needles. Developed in the early 1990s by healer Gary Craig, EFT uses tapping on acupuncture points to restore energy balance disrupted by anxiety or trauma. Numerous studies have demonstrated its efficacy in reducing stress.[27]

EFT involves tapping on nine acupressure points while repeating a phrase about reducing an identified issue or fear. When Craig retired in 2010, he released his method of tapping to the public domain. Go to EmoFree.com to download a free EFT manual with charts showing tapping points. Check out the many videos online demonstrating this amazing intervention.

Neural Therapy, a German form of acupuncture practiced by Klinghardt-trained practitioners and others, is a powerful invention that involves the injection of the anesthetic procaine, diluted medications, and homeopathics into scars and specific acupuncture points. Scars from surgeries such as cutting the umbilical cord, circumcision, and tonsillectomy can create abnormal electrical signals that cause them to become magnets for heavy metals. If not treated, these traumatized body parts can develop into toxic storage sites, especially for aluminum.

Injecting the scars with healing substances helps mobilize the toxins so excretory organs can eliminate them. The injections are superficial, safe, easy to learn, and effective for alleviating a wide variety of medical problems associated with toxicity.[28] To learn more from New York Klinghardt-trained physician Miriam Rahav, go RahavWellness.com.

Laser Energetic Detoxification (LED) is another technique developed in Germany and perfected in the United States by world-renowned integrative physician W. Lee Cowden, MD. First, a practitioner uses ART or other testing techniques described in Chapter 5 to determine the unique components of an individual's protocol. Then they add homeopathic preparations of those components to a vial and send laser light through the vial into the body in a process called *photonphoresis*. Combinations of drainage remedies, heavy metals, antibiotics, and other drugs, and even glyphosate, can be included.[29]

Remember biophoton theory introduced in Chapter 5, which states that our cells communicate with each other via particles of ultra-low-level light called biophotons? When cells lose coherence, or their ability to communicate with each other, beaming them with low-level light carrying energetic healing substances consistent with their functionality can restore their coherence. Instead of putting the medicine into a pill, liquid, or cream, laser light is the carrier. Brilliant, huh?

Beaming the light through the vial onto targeted acupuncture points on the ears, soles of the feet, and palms of the hands allows the body to receive the energetic information from the homeopathic remedy immediately and can elicit rapid and deep effects.[30]

LED speeds up and amplifies detoxification because the power of the laser light affects the body's biophoton field, stimulating it to specifically dump a toxin out of the cells and extracellular spaces. Coupling LED with oral chelating agents such as chlorella binds the toxins and shuttles them out of the body at a much quicker rate than with chelation alone. The beauty of LED is that it is very safe, especially when used along with a lymphatic drainage program. Side effects are greatly minimized because detoxification support is built into the protocol. Want to know more? Go to holistichealingjs.com.

Origins and Detox of Psychological Issues

The association between toxicity and psychological issues is frequently overlooked. According to Klinghardt, without addressing deep emotional blockages, a patient cannot get fully well.

Often, as physical detoxification takes effect, repressed emotional material emerges, causing the patient to experience psychological symptoms such as anxiety and anger. Healthcare professionals can mistakenly interpret these emotions as side effects of detoxification, or "detox reactions."

Klinghardt believes the degree of physical toxicity correlates to the number of unresolved psychological issues. The axiom that for each unresolved psycho-emotional conflict or trauma there is an equivalent of stored toxins and an equal amount of pathogenic microorganisms is Klinghardt's *Triad of Detoxification*. That's why an important component of detoxification is ridding the body of toxic emotions to regain balance and make room for further detoxification.

When detoxification programs do not include the simultaneous elimination of harmful unresolved trauma and emotions, with the elimination of infectious agents, the body stops releasing further toxins. The discrepancy between the infections and/or unresolved psycho-emotional material and the already-released physical toxins is simply too large. The toxin container is less full than the containers of bugs and emotions. Effective detoxification cannot progress without appropriately addressing all these issues simultaneously.

The Origins and Resolution of Psychological Issues

Surprisingly, traumas and unhealthy or estranged relationships that can surreptitiously undermine health and social-emotional function do not have to take place during the lifetime of the patient. Unresolved psycho-emotional issues and traumatic family issues from previous generations can be passed down as epigenetic imprints in the same way eye color is passed down genetically.

Each unresolved case from the past causes the body to lose some of its ability to recognize and successfully excrete toxic substances. In fact, using ART, Klinghardt has shown that where specific infectious agents and metals reside in the body can be predicted with a high degree of accuracy by knowing what type of unresolved psycho-emotional conflict a client has and at what age the associated event occurred!

Proper energetic psychological intervention can lead to a release of deeply stored toxins. The following are some of the most effective and accessible interventions.

EMDR and APN

Eye Movement Desensitization and Reprocessing (EMDR) therapy is a widely used mental health intervention that taps into the brain's neuroplasticity. Known best for its use in post-traumatic stress disorder (PTSD), it has also been shown to be beneficial in anxiety, panic, and phobic disorders,[31] as well as depression,[32] bipolar,[33] and other neuropsychiatric disorders.

EMDR originated in the late 1980s, when Francine Shapiro, PhD (1948–2019), then a psychology student, tried to shake off an upsetting memory during a stroll in the park by darting her eyes back and forth; the painful memory quickly faded.[34] It taps into the limbic system as the patient thinks about emotionally painful experiences while simultaneously moving the eyes in a prescribed fashion. While the ability to verbalize feelings can be helpful, a big advantage of EMDR is that it does not *require* language; thus, it can be done with nonspeakers.

EMDR is not a do-it-yourself therapy and requires working with a trained professional. To find one near you, go to EMDR.com.

Applied Psycho-neurobiology (APN)

Klinghardt developed a technique called *applied psycho-neurobiology* (APN), for which he received the Physician of the Year award from the Global Foundation of Integrative Medicine in 2007. APN is a form of muscle-biofeedback-assisted counseling that combines Klinghardt's deep understanding of psychoneuroimmunology, neurobiology, psycho-kinesiology (biofeedback-guided counseling and healing), EMDR, color therapy, and energetic tapping techniques. It is based on the assumption that all of life's events are accurately recorded by the subconscious.

Testing is done using ART to see which of the above interventions is appropriate for the patient at that point in time. As an example, one could use one tint from a set of colored glasses while tapping on acupuncture points and repeating an affirming phrase or combine any of those treatments with laser therapy.

Once the "prescription" is determined, treatment takes place with the client lying comfortably on a massage table for 30 to 60 minutes. At the end of the session, the client usually feels a sense of relaxation and a feeling that something important was just resolved. As the weeks progress, subtle improvements will be seen in areas such as allergies, elimination of stored toxins, better relationships, and overall well-being. APN is one of the most revolutionary, yet noninvasive and elegant healing techniques available today. Many Klinghardt-trained practitioners in several disciplines use it worldwide. To read about APN in depth, go to the link in the endnotes.[35]

Family Constellations

Another Klinghardt favorite for healing multigenerational psychological wounds is Family Constellations, conceived by the late Dr. Bert Hellinger (1925–2019), a renowned German psychoanalyst. Growing out of the family systems movement of the 1950s, this approach begins with the idea that dysfunction and suffering often relate to painful events in the *family's* past, instead of simply originating in an individual's life history from birth to the present.

While parents' deepest wishes are that their children thrive and their relationships work, Hellinger believed that sometimes an individual's future may be out of his or her control. An entanglement with a deceased relative from a past generation may be at play, sealing an unhappy fate, with a child as the victim.

Nothing is more important than belonging. Sometimes suffering like those who came before helps keep us connected. Anyone can become entangled in the difficult fate of a past family member and unconsciously draw unhappiness, failure, addiction, or illness into his own life.

Klinghardt notes that the large energetic fields of those with autism make them particularly vulnerable to disturbances in the energy field of the extended family, including unhealed transgenerational issues. Stored toxins prevent the physical body from receiving and processing perceptions and communications. Together, these factors further perpetuate their illness.

Read more about the expanded energy fields of some nonspeaking individuals with severe autism and how this benefits them in communicating telepathically in the upcoming chapter on communication. This recent finding is changing our perceptions of the competency of those who cannot speak.

The impact of unresolved transgenerational trauma in the family system as a psychological stressor is grossly underestimated. Family Constellations can break an energetic bond of pain and suffering by revealing hidden family dynamics and pointing the way toward resolution. Often therapy and biomedical interventions do not progress satisfactorily until a healing constellation is done by an experienced facilitator.

Constellations usually take place in groups. An individual chooses representatives for members of his or her family from the circle of participants and positions them in a way that seems right. In a short time, the representatives begin to experience physical sensations, emotions, or urges that belong to the family members they represent. It is as though they have become antennae, receiving information from a family soul mysteriously present in the room. Facilitators refer to this as the *knowing field*.

Through observations, questions, trial statements, and movements, the facilitator and client come to see the issue in a new way and create a resolution that enables the client to break his or her connection with difficulties in the family's past. As the hidden dynamic becomes clear and movements toward peace and reconciliation arise, the genuine love and strength in a family begin to flow in a healthy way. Klinghardt recommends two or three sessions of family constellation work to free up the system to release more toxins, so other interventions suddenly become more effective.

An example. The family history of one boy with autism was full of turmoil and pain on both sides. The father was excessively involved with his own mother; the boy's mother, an immigrant, had been unable to establish roots in her husband's country, and a decision to abort a child from a former relationship had been made with little respect or gravity to bring the event to a healthy closure.

As the constellation unfolded, the body language of the boy's representative showed how he was tied to those issues. To lighten the burden on their son, the father needed to free himself from his attachment to his own mother and become more available to his current family. The mother needed to develop a sense of rootedness in and respect for her new country. Together they needed to honor the soul of the aborted child by creating a place for it in their hearts.

Klinghardt suggests that families seeking help for conditions covered in this book put prodigious effort into healing through a constellation. By acknowledging the invisible members—such as children who have died young, aborted babies, husbands excluded after a divorce, mothers who died in childbirth, and men who died in wars—healing today's family members can occur more easily. Healing involves relating and communicating to everybody who is alive, as well as holding a loving memory of those who have gone.

Hellinger Institutes and independent groups offer family constellations across the United States. Hellinger's work, available in many books published by Zeig, Tucker & Theisen, evolved over a lifetime of rich experiences, including his years as a priest and living with the Zulu tribe in Africa. To learn more about Family Constellations, read Ulsmer's *Healing the Power of the Past*[36] and Mark Wolynn's *It Didn't Start with You: How Inherited Family Trauma Shapes Who We Are and How to End the Cycle.*[37] Visit Hellinger.com.

Oxygen Therapies

Hyperbaric oxygen therapy (HBOT), a powerful treatment previously proven efficacious for those who have had near-drowning accidents, stroke, and other brain injuries, is an increasingly popular treatment for neurological inflammation. Patients of all ages are entirely enclosed in a chamber in which they breathe oxygen at a pressure of 1.3–1.5 atmospheres.

Both the enclosed chamber and the pressure are necessary for HBOT to work.

Treatments last one to two hours. Oftentimes, 40–60 sessions are necessary to see significant benefit. HBOT delivers oxygen deep into the tissues of the body, where it attacks and kills yeasts and anaerobic (oxygen-hating) bacteria with a short-lived die-off that is not hard on the liver. The body must be supported with antioxidants and antifungals during this therapy because HBOT can flare yeast.

Pressurized oxygen has tremendous healing capabilities. HBOT increases blood flow to the brain, and the pressure seems to help decrease some of the inflammatory problems. Some physicians are very enthusiastic about HBOT as an adjunct therapy for individuals on the spectrum. Research shows benefit not only in reducing inflammation, but in increasing social skills and positive behaviors.[38] It also has potential for reducing anxiety and depression.[39]

HBOT can be dangerous if not monitored and used properly. Several instances of deaths have occurred. Make sure that precautions are taken at the center you use. Plastic blow-up chambers for the home are not recommended.

Ozone (O3) is a gas consisting of three molecules of oxygen, instead of the usual two, which combine with hydrogen to make water. Discovered in the mid-1800s, its safety and medicinal benefits in activating the immune system and inactivating anaerobic bacteria, viruses, fungi, yeasts, and other bugs is well documented.[40]

Ozone therapy can be administered in multiple ways: IV, rectally, nasally, vaginally, and into the ear. Almost any orifice will do! It has been used primarily in holistic dentistry, when after repairing a cavity or extracting a tooth, the dentist will fill the space with ozone to prevent infection. Check out books and videos online for more information.

Pulsed Electromagnetic Frequency Therapy (PEMF)

PEMF is an FDA-approved potent intervention for reducing anxiety, OCD symptoms, depression, and associated pain by stimulating blood flow and tissue repair. Open-minded practitioners are recommending it for use in Lyme, long-haul COVID, and other conditions with neuroinflammation.[41] Dr. Darin Ingels discusses this intervention in depth in *The Lyme Solution*.

Unhealthy high-level electromagnetic frequencies (EMFs) from Wi-Fi, computers, and cell phone towers disrupt our cells' ability to communicate. Alternatively, PEMFs are very low-level frequency pulses that resonate with our organs and cells, repairing and restoring their functioning. Ingels recommends two 75-minute sessions a week for six weeks to treat Lyme. He reports dramatic improvement after only a single session.

Devices such as mats, helmets, and patches delivering PEMF are available for both professional and home use, and length of treatment varies. If this intervention interests you, research further and try out a few delivery methods to see what might work for you.

Flower Essences and Essential Oils

Another intervention that can affect all aspects of functioning, from breathing and digestion to emotional and immunological healing, and profoundly change behavior are flower essences and essential oils. While smell is a primitive sense upon which humans rarely depend for information, some of our sensitive patients have strongly attuned noses and often *do* respond quickly to products that tap into olfaction.

For thousands of years people have looked to plants for healing. Outside of the United States, where we depend more upon pharmaceuticals than on natural products, these plant products are considered serious options in medicine. Ironically, most drugs are the result of a discovery of healing from a plant.

Extracting or steam-distilling the plant essences results in highly concentrated substances that can be ingested, inhaled, diffused, or absorbed by the skin. Studies show that, when taken in through the olfactory system, the strong effects of these products travel quickly through the body and can cross the blood-brain barrier, thus affecting us physically, emotionally, and even spiritually.

Single oils that may be useful include:

- *Bergamot essential oil* to improve motor coordination;
- *Crocus flower essence* for diarrhea, excess gas, and bloating associated with yeast overgrowth;
- *Calendula flower essence* for decreasing sensitivity to touch (by things or people);
- *Geranium essential oil* for anxiety and increased comfort with eye contact;
- *Lavender essential oil* to calm agitation, rage, or meltdowns; and
- *Neroli essential oil* for bed wetting.

Oils are *highly* concentrated. *One* drop is usually adequate to rub on your hands and smell. Depending upon the circumstances, offering a massage with an essential oil like lavender mixed with coconut oil at bedtime, sniffing the oil during a temper tantrum, or diffusing the oil to keep germs away can be all that is necessary. Rubbing a drop onto the wrist or on the bottoms of the feet during the day can also have a calming effect.

All oils are not equal, just as supplements vary in quality. To learn more go to LearningAboutEOs.com or refer to a guidebook on the subject. Flower essences and essential oils are available in health food stores and online at YoungLiving.com and MyDoTERRA.com. Speak to a knowledgeable aromatherapist before proceeding.

Take-Home Points

This chapter includes techniques from bioregulatory medicine that are powerful and readily available to almost any healthcare practitioner. They use a variety of modalities to increase the body's vital force, known as chi, and all address more than one bodily system. While virtually unknown to many allopathic doctors, they are becoming more mainstream as curious patients pursue them. They include homeopathy and homotoxicology, acupuncture and its related fields, oxygen therapies, flower essences, and other safe, effective methods that take training, expertise, time, and patience. As additions to the tool chest of a variety of practitioners, these very well could be the "icing on the cake" that finally helps you or your child to feel and function better. Maybe one is your missing piece!

CHAPTER 15

DENTAL DECISIONS

Everyone would agree that going to the dentist for the first time is a rite of passage of in childhood. Sit in the big chair, open wide, and let someone poke and prod in your mouth. For most people this is no big deal; for sensitive patients, it can be one of the most dreaded experiences of their lives.

Like everything else, dentistry has become a specialty, separate from the rest of mainstream medicine. Yet, common sense tells us that oral health must be intimately connected to general health. To understand what is happening in the brain and the rest of the body, you *must* know what is happening inside the mouth. Exploring every aspect of that cavity is essential to understanding an individual's Total Load. Teeth, gums, tongue, jaw, and bite must all be examined for structure and function.

Reuniting the mouth with the rest of the body is fundamental to bioregulatory medicine. About 100 years ago, Columbia University biochemistry professor William Gies, considered the founder of dental education, stated that dentistry should be more than "tooth technology."[1] Dentistry, and especially orthodontics, offers many more options than in the past.

Let's start with a tour of the mouth.

The Mouth

We put everything into that large orifice in the middle of our face: early on, a breast or bottle, our thumb and fingers as babies, pens and cigarettes later in life, and food throughout our lives. The mouth is a complicated muscle, driven by oral-facial reflexes, jaw, lips, tongue, and teeth. It allows us to chew, talk, kiss, taste, and digest. Using our mouth is the most common way to eat, stay hydrated, and communicate and an alternative way to breathe.

Structure

Teeth. A child's mouth holds 20 teeth, and an adult with a full set of choppers has 32 teeth of varying sizes and shapes, suited for different purposes. Teeth are hard, durable, living objects, coated in white enamel that hides their complex insides. In a healthy mouth, tiny tubules containing nutrients run through the enamel, feeding and nourishing each tooth. In the center are nerve endings, blood and tissue called *pulp*, and blood-filled arteries that, along with flexible ligaments, connect them to the rest of the body.

Gums surround the teeth, providing a covering for bone, in which each tooth is suspended. They should be pink, tight against the teeth, and not be inflamed or bleed. Receding gums can cause pain and sensitivity and allow bacteria to enter.

The *Tongue* in the middle of the mouth is important for many functions, moving up, down, and side-to-side to push our food around and to make distinct speech sounds. Tiny bumps on the tongue are sensory organs called *taste buds* that detect sour, sweet, bitter, and salty.

Traditional Chinese Medicine considers the tongue to be a holographic map of the body's organs, with the very tip being the heart, the lungs right behind, and the liver and gall bladder along the sides. Tongue diagnosis is an ancient art that evaluates color, lines, coating, and more. Have you ever looked at your tongue?

Everything in the mouth is connected: the teeth to the gums, the tongue to the bottom of the mouth, the lips to the cheeks that surround them. Even the connections have names. For instance, the tongue is connected to the bottom of the mouth by a band of tissue called the *frenulum*.

The Oral Microbiome. Recall the discussion about the microbiome in Chapter 7 on digestion. Well, the mouth has its own set of organisms inhabiting it, making up a separate oral biome. Generally, when the balance between the "good guys" and "bad guys" in the mouth is appropriate and the diet is low in sugar (food for bad bugs), no cavities, infections, or inflammation appear, and teeth stay healthy.

What Can Go Wrong in the Mouth?

Lots of things. Even though teeth might look solid and unchanging, they are alive and, like other parts of the body, are constantly adjusting to the changing environment they live in, the mouth. When the mouth is not supportive of health, teeth can decay and get holes in them called *cavities* or *caries* from destructive bacteria forming colonies called *plaque*. The bacteria in plaque form an acid which can eat away at the enamel and gums, leading to tooth decay. If the bacteria venture into the gums, bleeding occurs, leading to a reversible inflammatory condition call *gingivitis*. If deterioration goes unabated, possible damage to the bone and root can occur.

An out-of-balance oral microbiome can cause all kinds of problems. Fascinating new research shows that testing the oral microbiome may be a way to diagnose just about any disorder. Take autism, for instance. A study comparing 32 children with ASD, seven to 14 years of age, to 27 healthy controls demonstrated that the salivary and dental microbiota of those with ASD were highly distinct from those of healthy individuals. Those with ASD lacked diversity, contained an overabundance of pathogens such as *haemophilus* and *streptococcus* and too few bacteria of the genera *prevotella* and *alloprevotella,* known to have beneficial effects.[2]

If the connections of tongue, lips, or cheeks are too short, thick, or tight, conditions known as tongue-, lip-, or cheek-ties occur. Tongue-ties, known scientifically as *ankyloglossia,* can restrict tongue movement. Similar conditions can occur when the upper lip is tethered to the gum or cheek, called a *buccal-* or *lip-tie.*

These aberrant anchors can interfere with a baby latching onto bottle or breast, impede breathing, and later affect speech articulation. Want to learn more about tongue-ties? Read Dentist Richard Baxtor's *Tongue-Tied: How a Tiny String Under the Tongue Impacts Nursing, Speech, Feeding, and More.*[3]

Here is the interesting part: remember that MTHFR gene which can cause detoxification difficulties? Well, those same MTHFR polymorphisms are also associated with midline structural problems like tongue-ties. So, if your baby has a tongue-tie, genetic testing for MTHFR is essential, because, as you have seen, other abnormalities like cerebral folate deficiency (CFD) are associated with MTHFR SNPs.

Who Can Fix It?

Once cavities, gum disease, or a tongue-tie show up, a trip to a dental specialist is necessary. At this point, you are faced with some important choices.

What follows is a comparison guide between how conventional dentists might approach a patient's mouth, looking for symptoms of disease, and how biological dentists who understand whole-body health might work. Remember, the body is not a car! We cannot simply replace damaged and dysfunctional parts, and hope that the rest of the car runs smoothly. It is a complex, holistic system, with connections and feedback loops among the gastrointestinal, immune, endocrine, and nervous systems that are only recently beginning to be understood.

Conventional Versus Biological Dentistry

Today's traditional dentists know about hygiene: keeping teeth, gums, and the mouth free of bacteria, cavities, and disease. Their training is primarily focused on the health of the teeth and gums and dental maintenance through frequent checkups, which might include bite-wing X-rays and oral cleanings, with little regard to how dental procedures can affect overall bodily function.

New mothers visiting a pediatric dentist are lectured about the advantages of breastfeeding over the bottle and what to do about discomfort from teething. A pediatric dentist might check older children for cavities, teach how to floss, ask about how much candy a child eats, and maybe caution against thumb-sucking, which can affect tooth alignment or cause dental and speech problems.

Just as pediatric MDs are guided by the American Academy of Pediatrics (AAP), conventional dentists abide by the Code of Ethics of the American Dental Association (ADA). Just as conventional doctors abide by the AAP tenet that vaccines are safe and effective, conventional dentists abide by the ADA tenets that fluoride, amalgams containing mercury, and root canals are safe. Yet, safety concerns surrounding "silver" fillings and root canals are second only to vaccination among the issues debated in the holistic health community.

Biological Dentistry

Specially trained holistic dentists, also called "biological" or "mercury-free" dentists, recognize and understand that the mouth is the gateway to the gut, the first stop in digestion. They view the mouth as an integral part of the body. They recognize the symbiotic relationship between oral health and physical health, especially via the immune and lymph systems. They are familiar with the research conducted by renowned German physician Reinhard Voll, MD, a mentor to Dr. Klinghardt, that found almost 80 percent of all illness is related to problems in the mouth.[4]

Yes, a new breed of dental practitioner is here. They challenge the ADA's position that mercury amalgams, fluoride, and root canals are safe. They have seen too many sick patients who got well when these poisons were removed, procedures reversed, and alternative biocompatible procedures applied. According to "whole body" dentist Mark Breiner, DDS, heresy in dentistry raised its head way back in 1833 when the safety of amalgams was first questioned after they were introduced in the United States.[5]

A typical holistic dentist begins his exam with a comprehensive health and eating history, followed by a thorough examination of the mouth. Biological dentists look at teeth, gums, and tonsils (if still in place), those pesky lymph tissues that can harbor infection (see Chapter 14). They look at the size and shape of the jaw, the arch of the palate. Is it wide enough to hold all the baby teeth? When they come in, will they be crowded? When they fall out, will the jaw be big enough to hold all 32 adult teeth? In addition to conventional bite-wing X-rays, biological dentists often take a panoramic X-ray, which gives a 360-degree image of teeth, jawbone, and sinuses.

Biological dentists check for amalgams, root canals, chronic bone infections known as *cavitations* (see below), gum disease, jaw alignment, and bite. They may also use a variety of other tests of the body's immune response and procedures to look for possible infections and pain.

If biological dentists find disease, they investigate possible underlying causes like the MTHFR gene's relationship to tongue-ties. They know that what is happening with the mouth, teeth, and gums reflects what is going on in the rest of the body and that dental treatments should have as little impact on the immune system as possible. They thus promote treatments that are safe and effective, with minimal or no side effects, and dental materials that are nontoxic and biocompatible.

Biological dentists do not use fluoride. Rather, they usually stress the importance of a healthy lifestyle, including diet, nutritional supplements, and the use of nontoxic products at home. They practice bioregulatory medicine and work with functional medicine doctors, naturopaths, chiropractors, and other health care practitioners mentioned in earlier chapters. They use homeopathy, energy techniques, and special types of testing for both diagnosis and treatment.

The principles of biological dentistry are as follows:

- Eat nutrient-dense whole foods, properly grown and prepared;
- Avoid root canals and amalgam fillings. If you already have any, have a biological dentist evaluate them and remove or replace them safely;
- Any necessary orthodontics should include measures to widen the palate;
- Extract teeth only as a last resort, and then in such a way as to avoid leaving the jawbone with cavitations, which can be focal points of infection.

Most biological dentists treat patients with special needs, many of whom have had various traditional dental procedures that may be contributing to their Total Loads and are suffering from chronic health consequences because of them. When these patients show up in the dental office of a biological dentist, few have virgin mouths. Many have had fluoride treatments, and most have amalgam fillings and/ or root canals. Practically all have dysfunctional bites, tight jaws, high-arched palates, cut or undiagnosed tongue-ties, and other structural issues.

Several "heretic" dentists have published scientific research applicable to those with neurodevelopmental disorders. Here are a few.

Weston Price

Weston A. Price, DDS, (1870–1948) has been called the "Isaac Newton of Nutrition." In the 1930s, he researched possible causes of the dental decay and physical degeneration he saw in his Cleveland practice by traveling the world and studying the "primitives" in 14 traditional cultures. Price found that those with the best dental health also had excellent overall physical health, as well as sound nutrition, a healthy diet, and a stress-free lifestyle. They ate a combination of meats and animal organs, fish, shellfish, eggs, and raw milk, cheese, and butter, providing their bodies with plentiful calcium and other minerals and considerable fat-soluble vitamins A and D, found only in animal fats. Almost none had cavities; all had normal facial and dental bone development with room for all 32 teeth and did not require orthodonture. Most were "happy and contented" with a good sense of humor and superior IQ.

These findings led Dr. Price to the belief that dental caries and deformed dental arches, resulting in crowded, crooked teeth and an unattractive appearance, were signs of overall physical degeneration.[6] Read about his discoveries and conclusions in his classic volume *Nutrition and Physical Degeneration*, now in its eighth edition.[7]

In addition to his work on nutrition, Dr. Price conducted extensive research into the destructive effects of root canals. His out-of-print two-volume set on this subject is now available online.[8] Today, Price's conclusions, ignored by the orthodox dental establishment for over 50 years, are part of the teachings of biological dentists.

Price's legacy is the nonprofit foundation bearing his name. The Weston A. Price Foundation puts on an amazing annual conference, publishes a themed quarterly journal, has local chapters, and supports farmers offering healthy products all over the country. Find them at WestonAPrice.org.

Hal Huggins

Hal Alan Huggins, DDS, MS (1937–2014), was the "Elder Statesman" of holistic dentistry and a pioneer in identifying and treating medical problems caused by toxic dental materials. His master's degree was in immunology, allowing him to evaluate his patients' calcium, cholesterol, liver function tests, and white blood cells, using his knowledge of chemistry. He consistently found that a patient's blood cell counts changed after a dentist placed fillings in his mouth.

Dr. Huggins practiced and performed research for over 40 years. He presented over 2,500 lectures in the United States and 16 foreign countries. He authored many books and wrote over 50 articles on the dangers of mercury-containing amalgams, root canals, fluoride, crowns, and implants. His books include *It's All in Your Head: The Link between Mercury Amalgams and Illness*[9] and *Uninformed Consent: The Hidden Dangers in Dental Care.*[10]

Huggins's protocol taught thousands of biological dentists how to help their patients avoid and recover from ailments caused by harmful dental procedures. Before his death he founded DNA Connexions, a state-of-the-art laboratory specializing in the detection of microbial DNA that resides in root canal teeth, cavitations, implants, and other oral environments.

Dr. Huggins passed the torch and his businesses to Dr. Blanche Grube. Before Huggins's death, they designed the Huggins-Grube Protocol, an integrated system to enhance immune recovery. It starts with the removal of all toxic materials from the mouth and then fully restores the mouth as holistically as possible, using biocompatible materials. Dr. Grube founded the Huggins-Grube Institute, where she is continuing his legacy by traveling the world to educate dentists, medical doctors, and consumers on biological dentistry. HugginsAppliedHealing.com has a worldwide directory.

Several of Price's and Huggins's disciples have followed in their footsteps, teaching others how to use healthier dental practices through writings, lectures and online resources, including the following:

- **Andrew Hall Cutler,** PhD (1956–2017), a chemical engineer and mercury-removal expert, authored *Amalgam Illness: Diagnosis and Treatment. What You Can Do to Get Better. How Your Doctor Can Help* in 1999.[11] He co-wrote *The Mercury Detoxification Manual: A Guide to Mercury Chelation* with Rebecca Rust Lee.[12] Almost 10 years after his death, his detoxification protocol is still very popular with many people experiencing mercury-related illness. If you want more information, join his Facebook group, "Andy Cutler Chelation (ACC): Safe Mercury and Heavy Metal Detox," with almost 90,000 members. Rebecca Rust Lee, owner of MaybeItsMercury, teaches how to detox using the ACC protocol.
- **Robert Kulacz, DDS,** wrote *The Toxic Tooth: How a Root Canal Could Be Making You Sick*[13] with **Thomas E. Levy, MD, JD,** a retired cardiologist and attorney, author of 10 other books, including *Hidden Epidemic: Silent Oral Infections Cause Most Heart Attacks and Breast Cancers.*[14]

Resources on Biological Dentistry

Here are some helpful organizations that support and promote biological dentistry.

- **DAMS** is a 501(c)(3) tax-exempt nonprofit organization founded in 1990 for the purpose of educating the public about mercury amalgam hazards and other ways that dentistry can affect health. DAMS stands for **D**ental **A**malgam **M**ercury **S**olutions. Louise Herbeck, its founder, recovered from multiple sclerosis after her amalgam fillings were replaced with biocompatible dental materials. Visit Amalgam.org

- **International Academy of Biological Dentistry & Medicine** is a nonprofit network of dentists, physicians and allied health professionals committed to integrating body, mind, spirit, and mouth in caring for the whole person. See IABDM.org
- **International Academy of Oral Medicine and Toxicology (IAOMT)** is a global network of dentists, health professionals, and scientists who research the biocompatibility of dental products, including the risks of mercury fillings, fluoride, root canals, and jawbone osteonecrosis. Visit IAOMT.org
- **Mercury-Safe Dental Directory:** DentalWellness4U.com

Now let's look at some early dental stressors that may be impeding the overall health of an individual and should be considered as possible contributors to symptoms and behavior. If these occurred, bring a biological dentist onto the team to take a holistic approach and help reduce the Total Load.

Early Dental Stressors

Several red flags for dental issues can present themselves early in an infant's development. A well-trained structural professional, such as a chiropractor, who evaluates an infant at birth, is the professional of choice to note and remediate these with small adjustments. Dr. Klinghardt also recommends a bite assessment by a cranial osteopath or a very experienced orthodontist for anybody with a neurodevelopmental diagnosis.

Low Tone. A baby who has a weak suck and cannot latch onto the breast may be on the way to poor tooth development. A floppy or low-tone baby could have low tone throughout the body, not just in the limbs or trunk. Low tone in the face is common in premature infants who have endured days or weeks in the ICU with tubes taped to them. Their small size prohibits them from feeding traditionally, and they don't get the experience of working those facial muscles.

Sucking and chewing require the coordination of not only the mouth, but the cheeks, tongue, and lips, as well. These motor skills, developed by gaining sustenance through eating, are the foundation for talking.

The *Babkin Palmomental Reflex* is important in development. Have your ever watched a cat knead its paws? In humans, the hand and feet motion of this infant reflex is linked with movements of the mouth to improve the flow of milk. The Babkin is also related to bonding: feeling secure and nurtured. When this reflex does not integrate properly, aberrant food-related behaviors like the compulsion to chew on things can result. Some clues are nail-biting, sucking on sleeves, teeth grinding, or sticking out the tongue when concentrating on writing. If you know someone for whom this is a possibility, locate a resource to do reflex integration therapy as described in Chapter 6.

Nutrient-Deficient Diets. If mothers eat poorly during pregnancy or while breastfeeding, their babies may not get sufficient nutrients to build strong teeth. After teeth erupt and baby is eating independently, nutritious food is essential. Babies who go to bed with a bottle of formula, juice, or anything but water run the risk of getting cavities or, worse yet, having their teeth dissolved away by the pooling of liquid as they sleep.

Jaw and Facial Structure. Structural abnormalities in facial development involving the head, neck, and jaw are sometimes obvious, like a cleft lip and palate. Others, like a high-arched palate, are not so easy to recognize. These and other malformations can cause defective dentition, such as an overbite, as well as problems with the temporomandibular joint. Even slight misalignments can compromise blood and oxygen flow to the brain and/or impede blood flow and lymphatic drainage in affected areas.

Jaw and facial abnormalities at any age, including in infants, can also block breathing and disrupt sleep, which, in turn, impairs physical energy and cognition. Dentist Meghna Dassani has made sleep-disordered breathing her life's work. In *Airway Is Life: Waking Up to Your Family's Sleep Crisis*,[15] she offers guidelines to recognize when sleep goes wrong, why it does, and what to do about it.

With less energy, the body tries to compensate and may become distorted with poor posture. As the lower body tries hard to hold up the trunk, both the spine and nervous system can suffer. A sleepless brain does not work well, leading to a cascade of illnesses, including sleep apnea. Improving function in the jaw can result in generally better health overall because of an increase in oxygen, especially during sleep.

Toxic Dental Materials

Mercury

In Lewis Carroll's *Alice in Wonderland*, the distractible Mad Hatter runs around aimlessly, talking incoherently. His behavior was typical of victims of "hatter's disease" in the 18th and 19th centuries. The hatters' illness came from breathing mercury vapor while processing felt hats with mercuric chloride.[16]

Remember the 2001 paper "Autism: A Unique Form of Mercury Poisoning" mentioned in Chapter 2? Review that endless list of symptoms, system by system: neurological, cardiovascular, immunological, endocrine, psychological, and more. Many of today's sick people are contemporary Mad Hatters with a mouthful of mercury.

This dangerous, ubiquitous toxin occurs naturally in the air, the ground, and in our food and water. We accumulate it in our bodies by breathing, eating fish, and drinking water. So, our load is fairly high just by living on Earth, before we even enter the dentist's office and expose ourselves to amalgams. Because of its toxicity, any mercury purposefully placed in the mouth has the potential to cause damage to all bodily systems.

Mercury Amalgams

A dental amalgam is a composite of five metals—mercury, copper, zinc, tin, and silver—almost 50% of which is mercury.[17] Boyd Haley, PhD, is the world authority on the toxicity of mercury. Formerly the chairman of the Department of Chemistry at the University of Kentucky, and now the chief scientist and CEO of EmeraMed, located in Ireland, Haley has dedicated recent years to researching chelators that can cross the blood-brain barrier.

Haley believes that mercury vapor is continuously released from amalgams in measurable amounts from the moment the teeth are filled. In his "smoking tooth" video, available on YouTube, he demonstrates mercury vapor coming off a 25-year-old extracted tooth using dark-field microscopy.

Furthermore, whenever a person chews, drinks hot or icy liquids, or brushes their teeth, mercury escapes from amalgam fillings. Haley has shown through oral vapor testing an increase in vaporized mercury in your breath even after chewing gum. Mercury is also found in the saliva of people with dental amalgams. Old fillings analyzed after removal can contain over 25% less mercury than when they were installed.

Where does that mercury go? From the mouth it enters the lungs, linings of the digestive system, and subsequently the bloodstream. Autopsies show that the amount of mercury in a dead brain correlates with the number of mercury amalgams in the mouth.[18]

Mercury Amalgams and Developmental Disorders

The histories of many individuals with neurodevelopmental issues include mothers who either had dental amalgams prior to pregnancy or, worse, had dental work such as a filling or root canal done during pregnancy. In fact, even low-level prenatal exposure to mercury has been shown to have adverse effects on the developing fetus's brain and nervous system.[19] For this reason, dental work during pregnancy is strongly contraindicated.

If mercury from amalgams and previously injected vaccines was all that was required to cause behavioral symptoms, almost everyone would have one of our diagnoses. We now know that it is a serious load factor, and our response to it is related to our unique genetics, our detoxification abilities, and our Total Load. With better and better testing, we are getting a better understanding of who is at risk of mercury poisoning.

Mercury poisoning could very well be among the causes for any individual with a neurometabolic diagnosis. So, the precautionary principle must rule: when in doubt, leave it out! That leaves us with two big questions: first, if a tooth has a cavity, what can you do that will cause no harm? And second, if someone has a mouthful of amalgams, or even one or two, should they come out?

Alternatives to Mercury Amalgams

What are the choices? Fill the tooth with something else or pull it. Of course, the former is preferable, but fill it with what? The challenge is to find the least toxic alternative that has the strength to withstand the abuse of biting, chewing, saliva, and chemicals.

Testing Tolerance. Many biological dentists rely upon a dental materials reactivity test. From a blood sample, this laboratory test compares the strength of a person's immune reaction to over 150 dental materials, including metals, their composites, cements, porcelain, and ceramic products. All positive responses are measured against the ingredients in most name-brand options. Even this precautionary test is not foolproof because anything placed in the mouth permanently is toxic to some degree.[20]

Composites. Breiner prefers plastic/glass, tooth-colored composites that are biocompatible for most people. As you would expect, these are more expensive and rarely covered by insurance plans. He cautions that they also require a highly skilled dentist to get them just right. If not installed with utmost precision, they can leak or break, and the patient is back to square one, dealing with decay and disease. The dentist has a choice of how to make them, either directly in the patient's mouth, layer by layer, or indirectly, by taking an impression, and fabricating them in a lab. He prefers the indirect route, which has the advantages of fewer chemicals and better wear resistance.

Bridges and Partials. The best choice for some patients may be to pull, rather than fill, a sick tooth. This could be the case for the highly allergic or sensitive or when the tooth is beyond repair. A *bridge* requires one supporting tooth on each side, with a false tooth permanently bonded in the middle. A new type of bridge called First Fit™ is a unique patented technology that requires minimal preparations in the teeth that support the missing tooth. Made out of zirconia, it is very biocompatible.

A *partial* is a removable false tooth, anchored with clasps to adjoining teeth. Partials containing metal are generally undesirable. New plastic partials are less problematic, although all partials put considerable stress on the teeth to which they are attached. Clearly, a fixed bridge is more stable. On his website WholeBodyMed.com Breiner discusses all options.

Safely Removing Mercury Amalgams

Many people who know they have detox issues are choosing to have their mercury amalgams removed. *CAUTION*: removing amalgams without the following safety measures can recirculate mercury and retoxify the body, potentially making the patient significantly sicker than before. Almost every one of the experts on the dangers of mercury, including Dr. Klinghardt, personally experienced mercury toxicity after an unsafe amalgam removal that made them extremely sick. Their experiences led them to find ways to remediate the damage.

Prior to removal. Locate a qualified and experienced biological dentist. If you can't find one by word of mouth, go to HolisticDental.org. Make sure the dentist is very familiar with the following safety measures.

First, get an overall general health evaluation to assure that the body is well enough to withstand the stress of removal. Run hair and blood analysis and supplement with vitamins, minerals, and whatever else is necessary to facilitate excretion and protect against retoxification.

Make sure that the bowels are working well and elimination is regular. Breiner includes support for the kidneys and liver, our major detox organs, as well as glutathione, B vitamins, zinc, magnesium, and selenium. Klinghardt, and the dentists trained by him, use Chlorella Growth Factor (CGF) before, during, and after mercury removal. He believes that can make detoxification easier, shorter, and more effective. The research shows that children taking CGF develop no tooth decay and have near-perfect maxillary-facial development. They experience fewer illnesses, grow better, have higher IQs, and are socially more skilled.[21]

During the procedure. While all biological dentists are different, these are common practices:

- Use a rubber dam to isolate the tooth being worked on and a saliva ejector under the rubber dam to suction away mercury vapor or a suction device to contain the pulverized mercury and its harmful vapors.

- Frequently rinse the mouth with a slurry of chlorella to bind escaped mercury. Patient, dentist, and dental assistant should all wear a nose mask attached to a source of oxygen or fresh, clean air.
- Keep copious amounts of water running through the mouth.
- Place waterproof disposable draping over the patient's clothing to minimize mercury contamination.
- Everyone present should wear protective goggles.

Following the removal. Detoxify the system using more supplement protocols and various therapies like coffee enemas, colon hydrotherapy, and tons of chlorella to facilitate this process. Mercury remaining in the body after amalgam removal must follow a natural elimination pathway, via the bowel, the kidneys, or the skin. Elimination following mercury removal could take as long as a year or two. Detoxification must be tailored to a patient's individual tolerance for healing.

Benefits of Amalgam Removal

Unfortunately, not everyone experiences a miracle following amalgam removal; it's difficult to know why. The most common benefits patients report anecdotally are a gradual lifting of "brain fog" and an awakening or mental clarity, as well as a gradual lessening of fatigue and feeling more energetic. It all depends upon how much of the Total Load was mercury, and how well the body was able to excrete it.

Fluoride

The history of fluoride and how and why it got into our drinking water is also described in Chapter 2. Recall that fluoride is a neurotoxin that is even more poisonous than lead and, like lead, can cause damage at even very low concentrations. It harms the bones; immune, musculoskeletal, respiratory, circulatory, and digestive systems; and thyroid, liver, and brain function. That's why RFK Jr. has just ordered it out of the drinking water.

How Does Fluoride Harm the Body?

Too much fluoride at a young age, when the adult teeth are forming, causes discoloration and crumbling of teeth in a process called *dental fluorosis*, characterized by white lines or streaks on the teeth.[22] Fluoride also interferes with tooth formation and delays tooth eruption.[23]

Dr. Weston Price believed that better nutrition, not fluoride, is responsible for a decline in tooth decay. According to the watchdog group Fluoride Action Network, 97 percent of the population in Europe drinks ***non*-**fluoridated water, and their rates of tooth decay have declined at the same rates as those in the United States.[24] Want to learn more? Read Christopher Bryson's classic *The Fluoride Deception*.[25]

Dental Sealants

Dental sealants are plastic protective coatings applied temporarily and prophylactically to the premolars and molar teeth to prevent decay. Sealants act as physical barriers to the tooth surface for approximately five years. The sealant fills pits and fissures on the biting surfaces of these chewing teeth to prevent plaque and bacteria from collecting.

The American Dental Association recommends that kids receive dental sealants as soon as their adult teeth erupt. Anyone with poor oral hygiene, like a child with ADHD, is considered a particularly good candidate for sealants. Sounds good so far, doesn't it? Yep, until you learn what dental sealants are made of: plastic containing the poison bisphenol A (BPA), which you read about in Chapter 2. Recall that BPA is an endocrine disrupter that has been banned from baby bottles and other plastics. Why put it on a child's teeth, even temporarily?

Although sealants might prevent tooth decay, their failure rate is high. Why? For several reasons: they can crack, chip, or wear off and harbor bacteria beneath them.[26] Say "*no*" to sealants, just as you did to fluoride.

Root Canals and Cavitations

Anyone who has experienced the pain of an infected tooth knows that what you want most is quick relief. While conventional dentists traditionally recommend performing a root canal to save the tooth, biological dentists sometimes seek out other alternatives.

What Is a Root Canal?

A root canal is the removal of the tissue from within the root of a tooth. This tissue is comprised of nerves, blood, and lymph, as well as the critters causing the infection. After allegedly sterilizing the hollow space left by the tissue removal, the dentist then fills it with a gummy material called *gutta-percha*, which is supposed to cut it off from further infection. Sounds logical, doesn't it?

What Is Wrong with a Root Canal?

Unfortunately, this procedure ignores the actual physical properties of teeth. This fact was discovered by our hero, Weston Price, who studied root canals for about 25 years.[27] What he found was that even after the tooth has had a root canal, millions of tiny tubules in the enamel of the teeth are still working to transport nutrients, fluids, and, yes, bacteria from the saliva in and out of the enamel of that tooth. Because the canal has been sealed, the cleansing process is blocked, allowing toxins to build up.

If someone has a strong immune system and excellent abilities to detoxify, their body may be able to withstand the assault. In individuals with compromised immune systems, however, the body cannot fight the buildup of toxins, resulting in chronic infection and deterioration of health.

Dr. Price was way ahead of his time, and, like most geniuses, his work was ignored for 50 years, until one of his disciples, George Meinig, DDS, discovered and wrote about it. His book *The Root Canal Coverup*[28] details what happens when the gutta-percha hardens and shrinks, allowing significant numbers of bacteria to enter the area and flourish. A veritable petri dish of problems!

If all of this sounds like heresy, I suggest you do an informal survey. How many people do you know who have had serious disease diagnosed within a year of undergoing a root canal? I know at least a dozen with various forms of cancer, Parkinson's, and multiple sclerosis. We all know that association is not causation. But this might just be a perfect example of Total Load in action.

What Are the Alternatives to Root Canals?

The obvious alternative is to remove the infected tooth. Some dentists from both the conventional and alternative sides prefer this route. Unfortunately, even that choice can have hidden issues if not done with caution and care, because removing a tooth creates a hole in the bone and gums. The body's remarkable healing powers might fill in that hole just right . . . or it might not.

Cavitations

If the hole does not fill in fully, leaving any space, a new problem called a *cavitation* arises. When a dentist extracts a tooth, he has the choice of also removing the periodontal membrane or leaving it intact. If the membrane is not removed, an incomplete healing of the jawbone could take place. If the doctor took an X-ray of the site, it might show a shadow of the tooth that was pulled. Yes, a phantom tooth! This is almost always indicative of a cavitation, according to Mark Breiner.

The most common locations for cavitations are the four corners at the back of the upper and lower jaw, where wisdom teeth used to be. Dr. Klilnghardt believes that most people who have had them removed have cavitations in at least one of those spots.

Inside the cavitation is a whole world of bacteria, retroviruses, and other toxins, which can breed, placing stress on the immune system and overall health in general. Some of the unusual testing techniques mentioned in Chapter 5, including the EAV machine and autonomic response testing (ART), are the most reliable ways to diagnose a cavitation, because X-rays do not always show a shadow.

Scott Forsgren, a health coach, also known as the "Better Health Guy" has been studying underlying causes of Lyme disease, mold illness, and other possible root causes of chronic illness. He believes that cavitations are a missing link for many. In his articles, blogs, and podcasts based on over 25 years of personal experience fighting Lyme disease after being bitten by a tick, he has helped thousands of people recover.[29] BetterHealthGuy.com did an interview with Dr. Klinghardt on this subject.

Most cavitations require surgery to open the area and clean out the necrotic tissue. Many biological dentists combine this operation with natural products, such as clove oil, homeopathics, or low-level laser light to kill any viruses that were missed. Otherwise, the toxins and retroviruses can move into the lymph system, as well as travel to the brain via the vagus nerve, and cause neurological issues. Another option is to inject ozone into the area; however, many injections may be needed to kill all the problematic bugs.

Treating Dental Patients with Neurodevelopmental Disorders

Sensory Issues

Today's dental offices are replete with sensory overload. Machines make noises, cleaning materials have smells and tastes, and bright lights add to the chaos. For our sensitive patients, putting on a protective smock, hearing a drill, and tolerating a cleaning can be a sensory nightmare.

The next chapter details the myriad of sensory issues common in those with neurodevelopmental issues. You will learn that many have significant difficulty processing touch, sound, smell, taste, pressure, and vision. Their nervous systems can sometimes be under reactive, but most of the time, they are hyperreactive.

This aberrant sensory processing can result in avoidance and difficult behaviors in the dentist's office that are appropriate protective mechanisms to keep the nervous system feeling safe. Understanding sensory issues can almost always increase positive dental experiences. Fortunately, occupational therapists and others have developed tools that can be extremely beneficial in calming the nervous patient.

Sensory Tools

Some patients with sensory processing problems come equipped to a dental appointment with familiar objects, like a favorite stuffed animal, which work well to calm them. For those who don't, dentists can provide some inexpensive options.

- *Seating Options.* A wide range of adaptive seating and positioning options are available online from suppliers serving the special needs population.
- *Fidgets.* A fidget is anything that keeps the hands busy in order to distract the mind and the body from something else.
- *Weighted blanket or vest.* Deep pressure using a weighted blanket, a vest or lead apron can be very calming for an agitated patient.
- *Hats.* Something on the head provides pressure and is grounding. A baseball cap with a brim can also block out bright lights.
- *Headphones* can be a lifeline to both muffle disturbing sounds and to pipe in calming music.
- *Essential oils.* Olfactory reactions can be very strong. I have seen a little spray of lavender or a homeopathic product called Rescue Remedy calm the most upset children. Other calming scents are vanilla and rose. Always choose natural products.
- *Sunglasses.* Have several colors of tinted lenses available. Some prefer darkening, while others may choose blue, red, or even yellow.

Ways Providers Can Help Prevent Meltdowns

Understanding possible sensory processing issues is only half the solution. The other half is finding strategies that minimize potential freak-outs.

- Familiarize sensitive patients with dental procedures beforehand.
- Minimize wait time.
- Engage a parent's help in the dental process.
- Ask parents for suggestions on how to make the dental experience pleasant and positive.

Social Stories

Described at length in Chapter 19, Social Stories help some folks who do not do well with surprises and have difficulty with novel situations like going to the dentist feel prepared. A social story about going to the dentist uses pictures and words, step by step, to answer "who," "what," "when," "where," and "why" questions in a specifically defined style and format. Because accuracy is important for many people who have trouble with new situations, photos of the actual office and people can help.

Social stories displace fear and uncertainty by familiarizing patients with where they will go, why they will go there, who they will see, and what the people will do and why. Knowing about a trip to the dentist prior to experiencing it also promotes trust. The dental office should work with the family to explain procedures clearly so the parent or guardian can develop an accurate social story.

Anesthesia

Treating an uncooperative patient can be difficult and dangerous. So, sometimes using anesthesia is the only way to take care of essential dental needs. Putting any patient under anesthesia is always a risk; for those with special needs, it can be even more dangerous. Dentists' history-taking should include learning about risk factors that could result in an undesirable outcome.

Sym C. Rankin, RN, CRNA, has written and spoken extensively on this subject. A nurse anesthetist for over 25 years and the mother of a son with autism, she has provided the following information in a 2009 article in *The Autism File*, now available on the website of the Autism Research Institute (ARI).[30]

Rankin recommends asking the anesthesia provider to have a preoperative chat to discuss a patient's needs. Make sure the provider considers not just behavioral, but biological and metabolic issues as well, and evaluates how those problems should affect anesthetic choice.

Ask what methods of anesthesia delivery and which exact drugs are going to be used. All general anesthetics have a tendency to inhibit mitochondrial function. *Versed*® and *fentanyl* are considered relatively safe and short-lived. Others like *propofol* and *nitrous oxide* ("laughing gas") may present problems. Propofol contains both soybean oil and egg but is considered safe for those without allergies to those foods.[31]

Risk Factors for Anesthesia

The following put patients at greater risk for complications when anesthesia is considered:

- a history of seizures;
- preoperative respiratory problems, such as a diagnosis of asthma;
- possible mitochondrial disease;

- MTHFR gene polymorphisms;
- increased homocysteine levels;[32] and
- vitamin B-complex deficiency or B12 deficiency indicated by increased methylmalonic acid as the cause of increased homocysteine levels.[33]

Potential Complications

The following are undesirable outcomes that can occur:

- excessive time frame or difficulty in coming out of the anesthesia;
- developmental regressions, including loss of expressive and/or receptive language, gross motor skills, fine motor skills, or cognitive function;
- overall neurologic deterioration (in some children, skill loss may be permanent);
- excessive fatigue and reduced energy level—can be short term (a few days) or chronic (weeks or longer); and
- possible death.

Precautions to Take

If sedation for dental work is necessary, take these steps to reduce the possibility of harm. Discussing a patient's biomedical needs with all dental professionals is essential. Prior to any procedure, do the following:

- Ensure the patient is well hydrated prior to the procedure.
- Consider IV placement without sedation instead.
- Be the first patient of the day to minimize length of fasting.

Successful anesthesia is possible. Everyone wants the experience to go smoothly. This is another case of educating yourself so you can educate the professionals. Good luck!

Correcting Jaw and Bite Issues

Earlier in this chapter you read about jaw and bite issues as early stressors on health. A narrow palate; crowded, misaligned, ground-down teeth; and other malformations can compromise blood and oxygen flow to the brain and impede lymphatic drainage in the vagus nerve. Jaw and facial abnormalities can also block breathing and disrupt sleep, which, in turn, impede physical energy and cognition.

Why Jaw and Bite Issues Occur

In his anthropological studies, Weston Price found that native people in undeveloped countries had wide faces and palates, all their teeth, and few health issues. Today's children look different and are sick. Their faces are narrow and long, with underdeveloped lower jaws and weak chins. And one in six has a health issue. Why?

In his book, *Crooked: Man-Made Disease Explained,* Forrest Maready tells "The incredible story of metal, microbes, and medicine—hidden within our faces."[34] Maready blames the very same load factors presented in earlier chapters for today's facial asymmetries. Look at those you know with neurodevelopmental diagnoses. Are their smiles crooked? What about their eyes?

Why Jaw and Bite Issues Matter

For more than 40 years, the Pankey Institute for Dentistry has been educating dentists about the relationship between facial, bite, and jaw issues and pain, health problems, and disease. Miraculously, with the bite adjusted, digestive problems, cold fingers and toes, and endocrine issues can also vanish![35] Learn more about the Pankey Institute at Pankey.org. Total Load once again!

Correcting Jaw and Bite Issues

Simple option: Dietrich Klinghardt suggests trying a simple splint in the mouth to open up the airway. Did breathing improve? I have seen him demonstrate this in his workshops, using ART. By using a tongue depressor to widen the bite, the patient no longer experiences a blockage of oxygen. When a dentist builds up the height of the teeth artificially, it changes the bite and can improve both blood and oxygen flow to the brain.[36]

Myofunctional Therapy: sometimes all it takes is an exercise training program for the muscles around your face, mouth, and tongue. These exercises are designed not only to address jaw and bite issues, but also to improve issues with talking, eating, or breathing. Florida speech-language pathologist Dena Freedman-Muchnick treats individuals at all ages with a focus on tongue movements. Visit MyofunctionalSpot.com.

Traditional braces: the most common way to fix the bite is by traditional orthodonture. Many people reading this section have endured old-fashioned braces which were ugly metal and prevented kids from smiling in their school photos. Some, like me, wore braces more than once because the correction did not hold. We now know why: because you cannot force against the natural inclination of the body. Traditional braces are "old news" and thus are not included in this book.

Expansion appliances: we now know that the most effective way to fix the jaw is with the use of an expansion appliance, usually placed inside the upper palate. Expansion appliances are generally available only from functional or biological dentists or orthodontists, not conventional ones.

Upper jaw expansion is a form of orthodontic treatment that widens the circumference of the palate, thus creating more space for the teeth, avoiding overcrowding. The appliance also can correct the bite, which may be improper, with upper teeth biting inside, rather than outside the lower teeth.

An added benefit of expanding the upper palate is improved breathing. For many mouth breathers the problem is in the jaw or bite, making it difficult to breathe through the nose. Expanding the palate opens up the nasal passages along with the bite.

The gold standard treatment is the Advanced Lightwire Functional Appliance (ALF), invented by dentist Darick Nordstrom, because it is based on the body's natural rhythms of cranial osteopathy. If you missed reading about this intervention, return to Chapter 6 for a review. The ALF impacts a patient's cranial rhythm and movement through osteopathic alignment which integrates cranial mobility with the palate, teeth, and jaw alignment, thereby promoting better swallowing, breathing, speech, sleep, and nervous system response.

The ALF is a thin, invisible, removable appliance constructed from a very flexible, resilient, nonreactive metal alloy wire that is placed on the interior of the upper teeth from molar to molar, exerting a light force all around. The dentist bends loops into the arched wire where expansion or spacing is desired. Other options are possible to individualize the appliance to create specific molar movements depending upon the needs of the patient.

The ALF can be used at any age: preventively up to age seven, to remodel and redirect teeth from ages seven through 12, and to repair improper dentition and bite after that. Because complete individualization is possible, almost any bite and tooth problem can be fixed using the ALF. Unfortunately, this therapy is expensive. Most dentists offer financing.

Consumer beware: ALF copycats exist. To find a certified ALF provider, go to ALFEducationalInstitute.com.

Chinese Tooth Chart

For thousands of years the Chinese have used the teeth to understand illness and disease. According to Traditional Chinese Medicine, every organ, joint, gland, and body part lies on a specific acupuncture meridian, along which energy flows. Each tooth in the mouth is related to a specific body part and is connected through the meridians.

Search "Chinese Tooth Chart" or "Meridian Tooth Chart" online. Find an interactive chart where you can point to a tooth and discover which organ it affects. I think you will find it uncanny how close the relationships are between a diseased tooth and a problematic organ. For instance, I know someone who has long had difficulty with the two "large intestine" molars. First cavities, then mercury amalgams, then a crown, after the tooth broke off. Soon after experiencing the dental problems, she began to experience gastrointestinal problems. First the tooth goes bad, and then the organ suffers. Just another reason to be careful about how you treat your teeth! What organ are you affecting by installing that mercury amalgam or root canal?

Resources

Want to read more about dental issues? Here are two books to add to your library:

- *The Holistic Dental Matrix: How Your Teeth Can Control Your Health and Well-Being* by Nicholas J. Meyer, DDS.
- *Whole-Body Dentistry* by Mark A. Breiner, DDS.

Take-Home Points

Inclusion of a chapter on dentistry in a book about autism, ADHD, OCD, learning and behavioral problems, anxiety, and more may seem strange at first. However, given that the mouth is the beginning of the digestive system, and that a huge number of those with these diagnoses experience gastrointestinal problems, maybe we shouldn't be surprised. It begins to make even more sense when you look deeper and realize that early stressors involving the mouth, jaw, and bite can affect breathing, sleep, speech development, and cognition. Perhaps *this* chapter is the most important in the book.

The differentiation between traditional or conventional dentistry and the emerging field of biological or functional dentistry is similar to that between traditional physicians and functional medicine doctors made in Chapter 1. Many dental procedures we have accepted and assumed safe, may, in fact, be contributing to nervous system and brain disorders.

Special considerations are necessary for dentists who work with patients with special needs, who may need some easy modifications to deal with sensory overload. New and exciting methods of orthodonture that are in tune with the body's own biorhythms are making a huge difference by expanding the palate, thus allowing better oxygenation and blood flow to the brain.

Have you considered the mouth, teeth, bite, and jaw? Maybe this is *your* missing link.

CHAPTER 16

SENSORY PROCESSING AND
MOTOR ISSUES

According to Stephanie Marohn, author of *The Natural Medicine Guide to Autism*, "If the senses are musical instruments, in neurodevelopmental disorders they lack a conductor. They are not able to play together."[1] An estimated 80–90 percent of individuals with autism experience atypical and problematic sensory processing.[2] The American Occupational Therapy Association (AOTA) puts prevalence even higher at 93 to 96 percent.[3] Sensory processing issues run rampant in other neurological disorders as well. The next stage in defying any diagnosis is subtracting sensory and motor load factors from the Total Load. This is the last step before finally focusing directly on fluent, open communication and positive socialization.

Clinicians have long recognized sensory issues in autism; psychologist Bernard Rimland, PhD, and psychiatrist Edward Ornitz, MD, described these problems in the 1960s.[4] For many, the world is simply "too loud, too bright, too fast, and too tight." Inefficient sensory processing is second only to inefficient digestion in its contribution to the suffering and necessary lifestyle adaptations in our sensitive patients. See Chapter 3 for an introduction to the sensory connection in the Total Load and Temple Grandin's role in raising awareness of this important area.

This chapter is one of the densest in the book because sensory processing is *so* vital to all aspects of function. It starts with the processing of touch, pressure, movement, taste, smell, and sound. The sense of vision is so crucial that is has a chapter of its own, which follows. Next, the relationships among the senses and how they integrate is introduced and, finally, various methods for evaluating and treatment.

Sensory Processing 101

Our brains process millions of sensory experiences a minute using specialized cells in the skin, tendons, muscles, joints, inner ear, eyes, and throughout the body. As these messages travel along neural pathways, different parts of the brain compare, combine, and interpret the sensory experiences, storing them for future reference.[5]

Each of the sensory systems has two functions: *discrimination* and *protection*. The discriminative function is important for developing sustained attention, understanding and using language, and connecting socially, as well as for all aspects of vision. Discrimination also provides detailed information about texture, temperature, shape, and size. Accurate discrimination allows a person to use objects appropriately. The protective function keeps people safe. Efficient, consistent, and accurate interpretation of sensations is essential for survival.

How many senses are there? It depends on who you ask! Textbooks generally teach about *five* senses: taste, smell, touch, hearing, and vision. Rudolf Steiner, a brilliant educator influential in the Waldorf School movement, taught that there are twelve senses.[6] For the purposes of this book, I will discuss seven. In addition to the basic five, I include balance (also known as the *vestibular system*) and our "muscle sense," called *proprioception*.

Understanding how typically developing individuals use the seven senses to process sensory information is necessary to comprehend what can go awry.

Touch, the most basic of the senses, comes through the skin, our largest bodily organ. Deep touch is soothing; light touch can be alerting or even startling.

Proprioception allows the brain to interpret sensations from our muscles and joints. Have you ever picked up an empty cup that you thought was full and been surprised by how light it was? That was your proprioception at work. Our muscle sense tells us whether things are heavy, light, hard, or soft.

The **vestibular system** is one of the first systems to myelinate in utero.[7] Its receptors are physically located in the inner ear, where the labyrinths receive signals and detect the position and movement of the head relative to gravity, giving us our sense of balance.

This sense is particularly important for muscles in the neck to regulate eye position whenever the head moves.[8] The late Josephine Moore (1925–2016), an esteemed occupational therapist, described this interaction as the *vestibulo-oculo-cervical (VOC) triad*.[9]

Signs of vestibular problems are poor balance, low muscle tone, and slumped posture. An underdeveloped vestibular system can contribute to faulty digestion, delayed language, poor social interaction, and lack of eye contact. These issues might go back to confining a mother with a high-risk pregnancy to bed rest. A practice now being questioned,[10] long-term bed rest can result in a baby being born with an underdeveloped vestibular system because the mother's movements stimulate vestibular development in the fetus.

Disruption in any part of the VOC triad interferes with function in the others, resulting in compromised adaptive responses to environmental demands. One outcome is *gravitational insecurity*, which creates feelings of anxiety and fear in response to vestibular-proprioceptive stimulation such as walking on uneven surfaces. If the secondary emotional responses are treated without addressing the probable underlying sensory causes, further incidents are inevitable.

While all senses are important, the vestibular system is of particular interest because, physiologically, it is connected to the digestive tract, the language center of the brain, the limbic system, and to the eye muscles. A well-functioning vestibular system will thus contribute to healthy digestion, receptive and expressive communication, emotional bonding, and visual focus.

Consider what happens when an adult lifts a baby off the ground playfully, holds him overhead, and smiles and babbles at him. A neurotypical child returns the smile, makes eye contact, babbles back, and in some cases, throws up! What an amazing system it is! Some call it the body's internal GPS.[11]

Auditory processing refers to how the brain interprets sound; it is closely connected to language and communication. This chapter uses the term *sound processing*, which extends beyond the perception of sound for speech, language, and listening and includes sound sensitivities as well.

Sound processing difficulties manifest themselves as the impaired ability to perceive, process, or respond to sound appropriately. These can cause

- **speech and language issues** (difficulties with articulation; memory; auditory discrimination, sequencing, rhythm, and timing; learning a foreign language; stuttering; or dysfluency);
- **social/emotional problems** (excessive crying, inability to make and keep friends, or using inappropriate words); or
- **academic issues** (problems with decoding, reading comprehension, spelling, writing, or math).

As children get older, further issues become apparent: difficulties tuning out background noise and following conversations, lack of self-confidence, and/or poor organizational, public speaking, and singing skills.

Some appear deaf at times. Others may be under-responsive and experience certain environmental sounds, such as the humming of a fluorescent bulb, as intolerable or even painful. Birthday parties, shopping malls, restaurants, sporting events, and other gatherings can be particularly troublesome for them because of the unpredictability of sound.

Distorted processing of auditory signals impairs the brain's ability to focus on and give meaning to what is heard. Inconsistencies in hearing the various frequencies of sound, or lack of synchronicity of the two ears, can result in distracted, avoidant, hyperactive, inattentive, or even bizarre behavior.

Taste and *smell*, two primitive senses that are not high priority for learning and behaving for most humans, can be the source of great distress for some of our supersensitive folks. Many have highly attuned taste buds and noses upon which they depend for information because their auditory and visual systems are inefficient and unreliable. Their oversensitivities in these areas can cause all types of behavioral and emotional issues, especially related to food, resulting in picky eating (see Chapter 7).

And, of course, *vision*. Read more about how vision must become the dominant sense in Chapter 17.

The senses rarely function alone; most behaviors are a result of processing a combination of senses. The relationship between the vestibular and proprioceptive systems is particularly close. The brain combines incoming sensory information simultaneously, just as a conductor combines individual instruments of an orchestra to make one beautiful sound. Efficient sensory processing makes a symphony of focused attention, enjoyable learning, and appropriate behavior.

Sensory Integration and Modulation

Beginning in the 1960s occupational therapist A. Jean Ayres, PhD, was the pioneer in understanding how our senses work together. She coined the term *sensory integration* to describe "the organization of sensory input for use."[12]

Ayres Sensory Integration® (ASI) is now trademarked, and includes the original theory, assessment, patterns of dysfunction, and intervention concepts, principles, and techniques articulated by Dr. Ayres and applied by therapists trained in this approach worldwide. In 2016 first-generation students of Ayres cofounded a new organization called The Collaborative for Leadership in Ayres Sensory Integration (CLASI). CL-ASI.org.

Well-functioning individuals can monitor the degree and timing of their responses to sensory information. Ayres called this automatic ability *registration*, or *modulation*. The late psychiatrist Stanley Greenspan, MD, and psychologist Serena Wieder, PhD, found that 95% of 200 children with autism exhibited sensory modulation difficulties.[13] Poor modulation is also common in those with psychiatric diagnoses, including depression, bipolar disorder, and schizophrenia.[14]

Some of Ayres's disciples have expanded on her definition: "Sensory integration is a neurobiological process that forms the foundation for adaptive responses to challenges imposed by the environment and learning. Sensory integration theory considers the dynamic interactions between a person's abilities/disabilities and the environment."[15]

Sensory Processing in Neurodevelopmental Disorders

Sensory Integration Dysfunction, Sensory Processing Disorder, or Sensory Processing Differences (SPD)

If the integration process is disrupted, compensatory behaviors are inevitable. Ayres called this *sensory integration dysfunction*, abbreviated *SID*. Many of those with symptoms of sensory integration dysfunction have issues primarily in the areas of touch and movement that affect their modulation and registration. They thus have difficulty exerting consistent behavioral control over a sensory stimulus.[16]

Because of the possibility of confusing SID with the unrelated disorder sudden infant death syndrome (SIDS), in the 1980s practitioners began using the term *DSI* for *dysfunction in sensory integration*.

Then, a small group of occupational therapists, led by now-retired sensory integration researcher Lucy Jane Miller, PhD, OTR, suggested some new terminology: *sensory processing disorder*, or SPD, as an umbrella term encompassing several types of sensory issues. Finally, in recent years, the term SPD has been modified slightly to be less pathologizing, and it now stands for *sensory processing differences*. A long-term goal of this movement is to include SPD in the *Diagnostic and Statistical Manual of Mental Disorders* (*DSM*); so far this effort has failed.

Hypo- and hyperreactivity. Ayres speculated that those with sensory integration dysfunction fell into one of two categories: hyporeactivity and hyperreactivity. The former are under-stimulated and seek sensory experiences; the latter are overreactive to sensations and exhibit defensive behaviors.

Contemporary research supports Ayres's theory and suggests that inconsistent sensory responses are due to either overly high thresholds for sensory input (hypo-) or sensory defensiveness (hyper-).[17] She further proposed that the abilities of those in both categories to register and modulate sensory stimuli varied from time to time and in different environments.[18]

Table 16–1 shows possible offenders in the environment which could potentially trigger a defensive physical response.

Table 16–1 Frequent Environmental Offenders from which Hypersensitive Individuals Seek Protection (Copyright, The HANDLE Institute; reprinted with permission)

Their Own Clothing	*Clothing on Others*
• stiff tags • stiff fibers (e.g., jeans) • seams in socks • waistbands and belts • jewelry • hairbands • synthetic fibers	• synthetic fibers • intricate patterns • metallic look • reflecting accessories (e.g., sequins, watches) • noise makers (e.g., "bangle jangle" bracelets, watch alarms)
Odors	*Lighting*
• paints, varnishes, glues • cleaning products • room fresheners • cologne, perfume, aftershave • hair spray, gels, etcetera • clothes that have been dry-cleaned • fabric softeners applied in the dryer • orange peel, banana peel • synthetics (e.g., plastic food packaging) • fatty foods (e.g., broiled chops) • extremely sweet odors • extreme or unexpected body odors	• fluorescent lights • halogen lights • strobe lights • strobe effect of flickering sunlight (e.g., through leaves, blinds) • severe contrast (e.g., stage productions) • lighted mirrors • reflective materials • certain colors (esp. yellow-orange) • color contrasts (e.g., red:black) • white paper (esp. glossy magazines) • LCD (flat panel TV) signboards • automobile lights at night or in the rain • automobile lights in white-tiled tunnels
Sounds	*Body-in-Space Situations*
• unexpected loud sounds • high-pitched sounds • deeply resonating sounds • disharmonious sounds • background conversation	• light contact with seat, ground, other • slightly tipped/irregular surfaces • swivel chairs • open areas behind one's back • close quarters • remaining seated while others move past

Sometimes when multiple sensory stimuli compete for the body's attention, one sense may be hyperresponsive, while another may under-respond; there are an infinite number of sensory combinations. For instance, someone who is tactually defensive on a "bad" day and screams bloody murder at the hairdresser may tolerate a haircut if they are feeling more organized on another day.

Sensory processing disorders are rampant in those with ADHD, anxiety, and even panic disorders. In fact, these diagnoses are frequently given to individuals for whom an SPD diagnosis might be more appropriate. The fact that *so* many with other disorders have moderate-to-severe sensory processing disorders is probably one reason that SPD still has not become a stand-alone diagnosis in the DSM. It is simply considered a comorbid, or co-occurring, condition.

The etiology of these sensory issues is unclear and varies from person to person. If during infancy high levels of toxins were present, sensory dysfunction could very well be related to environmental factors.

Another potential pathway is the repeated inner ear infections that many experience, as well as the treatments used to ameliorate them. Immature senses compete for the body's energy during those crucial first two years of life. When the body is struggling with digestion, inflammation, and toxins, little energy is left for the demanding tasks of processing sight, sound, touch, and movement, let alone for language, feelings, social-emotional cues, academics, and organization.

Both hypo- and hyperreactions to the thousands of daily sensations from foods, clothing, and all aspects of the environment can be the result of sensory overload. Some are nauseated by the odors of a garbage truck; others insist upon wearing the same soft, loose sweatpants every day. Some crave deep touch and pressure; others cannot stand hugs. For young children these responses interfere with developmental milestones. If they persist into child- and adulthood, they can cause problems with eating, dressing, learning, relationships and just "being."

School can be a sensory nightmare for young students. Sitting at circle time, standing in line, and negotiating moving bodies in the hallways and on the playground can involve intermittent and unpredictable human contact. For the child who experiences tactile oversensitivity and has difficulty reading cues that tell him what to expect next, these situations are likely to create anxiety and discomfort.[19]

Table 16–2   Behaviors Resulting from Hyper- and Hyporeactivity to Each Sense

SENSE	HYPERREACTIVITY	HYPOREACTIVITY
HEARING	Covers ears; tunes out	Attracted to sounds
	Dislikes haircuts	Likes vibrations
	Crying or tantrums	Turns up volume
TOUCH	Avoids messy foods	Ignores food on face
	Ticklish; flinches	Self-injurious behavior
	Picky about clothing	Touches everything
	Won't walk on grass or sand	High pain tolerance
TASTE & SMELL	Gags at new foods	Prefers spicy foods
	Reacts to odors	Smells clothing or own body
VISION	Picky eater	Mouths objects
	Blinks excessively	Poor focus
	Covers eyes	Lacks awareness
	Picks up specks of dust	Fascinated by reflections
	Poor eye contact	Flicks fingers by eyes
VESTIBULAR/	Fearful of movement	Seeks deep pressure
PROPRIOCEPTIVE	Gets car-sick	Spins or rocks
	Fear of being upside-down	Wiggles and squirms

Many function at a primitive sensory level, relying on smell and touch; they use their mouths and hands, instead of ears and eyes, to gain information about the world. They might sniff, mouth, or touch inedible objects (or people) to get information about where they and objects are in space.

In the book *Asperger Syndrome and Sensory Issues*,[20] the authors include over 25 pages of behaviors—such as "has rituals," "won't eat certain foods," and "poor eye contact"—with possible sensory interpretations and suggested interventions. This section of the book is a great reference for understanding the role of touch, movement, vision, and the other senses in behavior.

Sensory Processing and Reflexes

See Chapter 6 for an overview of the importance of reflexes in feeling safe and in sustaining nervous system regulation. With over 100 inborn reflexes, dysfunction in the reflex system can affect almost any condition. However, dysfunctional reflexes associated with particular senses can result in sensory problems. Two examples:

The *Asymmetrical Tonic Neck Reflex (ATNR)* and *Auditory Processing*. The ATNR develops in utero and assists in the birth process. A non-vaginal birth bypasses the first bilateral integration activity of twisting and turning to help the infant emerge from the birth canal. This can affect future bilateral activities such as using both ears and eyes together and crossing the midline to read or write. When the ATNR is not integrated, the presence of new auditory information can cause confusion or shutting down of auditory processing.

The Tonic Labyrinthine Reflex (TLR) and *Vestibular Processing*. The TLR also develops in utero and enables the infant to respond to gravity. This "tonic" reflex influences muscle tone throughout the body. Remember that the digestive system is a giant muscle, moving food and its byproducts along for release; thus, "low tone" will have a negative effect on digestion.

The TLR is stimulated by the vestibular system and should integrate about four months postnatally. A lingering TLR interferes with all activities that depend upon balance, muscle tone, and visual-motor processing. An active TLR is implicated in reflux and any other gut issues. Later symptoms of an unintegrated TLR are fear of heights, stiff posture, toe walking, and motion sickness. Pretty amazing, wouldn't you say?

Jean Ayres had a real appreciation for the role of infant reflexes in sensory integration. Some of her early works mention how these built-in responses provide building blocks for the development of motor and later learning skills. Ayres's sensory integration theory stresses that the body's ability to process and interpret information coming in through touch, movement, balance, and body position lays the foundations for motor development.

Motor Development

If processing of any sense is inefficient, motor skills will be delayed. A now-classic movement analysis by Philip Teitelbaum and his colleagues found that children later diagnosed with autism showed persistent asymmetries in lying, righting, crawling, and sitting as early as the age of six months.[21]

Deficits in sensory integration contribute to problems in motor coordination, visual perception,[22] and balance.[23] One report found motor problems, including low muscle tone, motor planning problems, and other issues in over 80% of cases of autism.[24] Gross, fine, and visual-motor deficits are common.[25] Adults have significant difficulty with higher-level sensory processing tasks, such as organization. With extreme sensory overload, one or more of the sensory systems eventually shuts down.[26]

Motor planning progresses from the bottom up and the inside out. When motor development progresses smoothly, a child gains control over their body in the following order: lower body, upper body, shoulders and neck, upper arm, lower arm, hands, and fingers.

Picture a one-year-old child beginning to walk with a wide-based gait, arms raised for balance and stability. As lower limbs strengthen and the trunk becomes more stable, arms drop to the sides. At first a child is unable to both walk and hold something in their hand. Eventually, as the trunk stabilizes, walking becomes more automatic, and the hands and arms can perform actions without fear of falling. As the upper body strengthens, the arms and hands become more coordinated.

Watch a three-year-old paint at an easel. The legs are positioned in a wide stance, the brush is held in a fisted grasp, and the movement is with the whole arm. Usually, the eyes are not focused on the easel; if they are, they are looking at the finished product, not the active painting. Eventually, as the upper arm stabilizes, the grasp improves and the standing posture narrows and becomes more flexible.

As motor development progresses, a child can sit comfortably at a table to write or draw, with trunk held upright and not collapsed. Both sides of the body begin to work together as two eyes and two hands coordinate, one to execute and one to support the task at hand. The shoulders are still, an elbow may rest on the table, and the joints of the fingers work independently to engage the writing implement.

This predictable sequence of motor development allows a child to gradually gain control of the large muscles and use smaller and smaller muscles with good coordination. This ability develops with everyday sensory and motor experiences; without such experiences, many remain unable to differentiate sensory input and, even as adults, lack fine motor coordination.

Apraxia and Dyspraxia

Praxis

Praxis results from motor, sensory, and sensory-motor integration all working efficiently. Defining it as "the ability to do,"[27] Ayres wrote that praxis refers to the process underlying planning and execution of novel and complex motor patterns and sequences.[28]

Praxis involves ideation, timing, sequencing, initiating, and transitioning. Activities such as getting dressed, taking turns in a game, carrying on a conversation, and planning what to put in the backpack for a day at school or in a suitcase for a business trip are examples of praxis in action. Good organizational skills and appropriate social interaction are two valued results of praxis.

Dyspraxia. A dysfunction in praxis, called *dyspraxia,* can manifest itself in many ways: trouble taking turns, organizing one's body and possessions, or yes, in difficulty speaking. A therapist who does not recognize the role of motor planning in these end-product skills may be treating only symptoms.

Dyspraxia is becoming more widely recognized as one of the critical functional deficits in those with autism spectrum disorders. While gross-, fine- and oral-motor planning issues certainly present problems in isolation, their dramatic effect on functional skills cannot be underestimated. Dyspraxia probably accounts for many of the symptoms of autism; inappropriate social-emotional behavior (especially involving interpersonal relatedness), speech and language delays, and symbolic play are the areas most affected.[29] Turn-taking, the pragmatics of language, and reading social cues most certainly involve ideation, timing, and sequencing.

Apraxia, a.k.a. childhood apraxia of speech, or CAS. *Apraxia* literally means *without praxis*, meaning motor planning is completely absent, and there are several forms. The most important form for our purposes, though, is when the poor motor planning affects speech. This condition is called *childhood apraxia of speech* or *verbal apraxia* but is usually shortened to *apraxia*. There is usually some overlap between global dyspraxia and apraxia, as has recently become obvious in the autism community where many nonspeakers are apraxic. Not only do they lack speech, most also lack the ability to coordinate movements to write, move their eyes together on a target, skip, ride a bike, and other motor skills most adults do automatically.

A revolution is taking place as we realize that just because these previously "dumbed down" individuals don't talk that doesn't mean they have nothing to say. See Chapter 18 for how spelling with letterboards has blown the roof off our understanding of these silent Autistics.

Oral praxis. In order for speech to develop, the muscles of the lips, tongue, and cheeks must work together. That's what permits us to speak at all and to match our speech to our local accent. Facile tongue movement and placement are particularly important. Failure to recognize and cut ties could undermine speech emerging.

Oral praxis, like whole-body praxis, develops rapidly during the first year of life, so that "dada" and "mama" emerge at about the first birthday. As oral motor skills become more coordinated, a well-articulated vocabulary accumulates rapidly.

Sensory Diet

Many individuals whose sensory processing is compromised find it hard to achieve and maintain consistently appropriate arousal levels. Switching from sensory-seeking to sensory-avoidant behaviors can be problematic for authority figures even though it serves a purpose for the individual.

Just as a nutritionist recommends certain foods and supplements, an occupational therapist recommends a diet of readily available opportunities to be physical, and supplements them with a prescription of "just right" sensory activities. Essential ingredients of a rich sensory diet are physically heavy work, physical activity, muscle exertion, movement, and firm, comforting touch. Having enjoyed opportunities to rock, touch, move, and jump, a "satiated" child might be less compelled to act out inappropriately.

Including vestibular stimulation activities as part of the daily diet can be extremely powerful. Working against gravity on suspended equipment like a swing can stimulate the crucial areas of digestion, visual function (including eye contact), communication, and socialization.

Here are some ways to give kids of all ages and abilities more sensory experiences throughout the day:

- *Modify routine activities.* Add weight, move things out of reach, go the longer way over uneven terrain, put pressure on shoulders to activate neck receptors for proprioception while walking, give tighter hugs.
- *Modify objects to increase sensory demands.* Provide obstacles to climb over, crawl under, or walk around; make door catches tighter so one has to pull or push harder; use objects of differing sizes and weights to require

frequent readjustment of exertion; offer larger versions of tasks, such as writing on a chalkboard instead of paper.

- *Create new opportunities.* Take breaks for moving, pushing, or pulling; wrap ace bandages around limbs for extra touch pressure; do isometric exercises and chair push-ups; do Brain Gym® before challenging activities (see Chapter 20); breathe deeply.

- *Add sensation to a non-sensory activity.* Review academics while swinging or while balancing on a T-Stool, answer questions while pushing or pulling, sit on a vibrating pad or slightly inflated ball while reading or taking a test.

- *Create a sensory room as a refuge.* Several companies now offer free design services for schools and families to create rooms that can be customized to individuals' unique sensory needs. Equipment such as bubble tubes, fiber optics, sound boards, mirrors, and interactive microphones all enhance development. Find these at Experia-USA.com.

How Sensory Diet and Nutrition Interact

While a good sensory diet is critical to correcting sensory imbalances, its effectiveness is limited by the quality of an individual's neural connections. Typical children have strong, well-nourished nervous systems. For those with neurometabolic issues, the above sensory modifications alone may be insufficient to produce efficient sensory integration.

Thus, a rich sensory diet must be supplemented with those minerals, essential fats, B vitamins, and antioxidants discussed in earlier chapters. For instance, if someone has a magnesium deficiency, nerve signal transmission can be inefficient, resulting in poor sensory processing.[30]

Those with sensory issues have higher needs for both sensory input and nutritional intake than their typical peers. A limited sensory diet plus a self-limited food diet leaves the tank empty. Nutritionist Kelly Dorfman suggests removing the worst empty-calorie foods from the pantry and closing the nutrient gap with basic supplements. Changing the structure and function of the nervous system takes time, and direct results may be hard to measure or tie directly to nutrition.

The good news is that "nutritious" sensory stimulation can actually improve the body's ability to use vitamins and minerals! Deep pressure, using weighted blankets and vests, and rhythmic activities, such as dance, release oxytocin, the "feel good" hormone.[31]

Think "diet" in broad terms. Providing both a nutritious food diet and a healthy sensory diet strengthens and regulates the nervous system, which in turn supports efficient cognition and behavior.

Sensory Evaluations

Sensory evaluations can be administered by qualified occupational therapists, physical therapists, optometrists, and others working with motor and visual skills. However, they are often omitted from school testing that includes only end-product skills, such as handwriting and scissor skills.

A complete sensory battery should include both formal and informal assessments of tactile, proprioceptive, and vestibular processing; motor planning, kinesthesia, muscle strength and tone, and motor control; attention to task; and visual perception and visual-motor integration.[32] Therapists also depend upon their informal clinical observations, especially for those who cannot comply with the demands of formal testing.

Evaluating sensory processing is an ongoing process, as behaviors change day to day and in different settings with varying demands and people. See "inappropriate" behavior? Think sensory processing!

When evaluating any area, a comprehensive history is essential. At what age did the first over- or under-sensory responses occur? What triggered and alleviated them? Observations and history should be gathered from all environments: home, school, work, and any other setting where the individual spends time, such as at a relative's. Include other parents, teachers, friends, coworkers, etcetera. Sometimes informal assessments, questionnaires, and checklists yield more information than formal testing.

Forms, checklists, and questionnaires for taking sensory histories are readily available. They can identify environmental issues, such as noises from heating and air conditioning, difficulties with lighting, or excessive visual stimulation. Altering any of these factors can help.

Two popular instruments are the *Sensory Profile™ 2,* appropriate from birth through age 14, and the *Adolescent/Adult Sensory Profile™* for age 11 and up. Both are available from Pearson Assessments. Studies are ongoing using these measures for both differential diagnosis and remediation recommendations. PearsonAssessments.com has a bibliography of hundreds of articles on sensory processing.

An informal assessment tool completed by an observer is the *Sensory Processing Measure (SPM-2)™.*[33] Now in its second edition, this comprehensive instrument for ages and stages from infant through teen is also available in

Spanish. For each sense, the SPM offers descriptive clinical information on processing vulnerabilities, including under- and over-responsiveness, sensory-seeking behavior, and perceptual problems. Comparison of sensory functioning at home and at school is available.

Formal tests for those able to comply with the demands of standardized testing, coupled with the examiner's clinical observations, can elicit a great deal of information about responses to sensory stimulation, posture, balance, coordination, and movement activities. These reactions are compared to age norms that allow a therapist to make recommendations.

The longtime standardized instrument of choice for children ages four through eight is the *Sensory Integration and Praxis Test (SIPT)*.[34] Specially trained and certified occupational or physical therapists can conduct a formal evaluation using this test to establish a baseline of functioning in each of the sensory systems. The SIPT is an extremely comprehensive measure consisting of 17 computer-scored subtests.

In a study all subtests of the SIPT discriminated significantly between children who were developing typically and those with high-functioning autism. Those with autism demonstrated significant difficulties in praxis, bilateral integration, sequencing, and some aspects of vestibular function.[35]

For those who are outside the age norms of the SIPT or cannot comply with standardization procedures, OTs observe spontaneous play and use of equipment, coordination, tone, laterality, goal orientation, and initiation of actions. These clinical observations are sometimes equally or more valuable than the test results.

Interventions for Sensory Processing Issues

In the 1970s Carl Delacato, an educational psychologist, was one of the first to propose that sensory abnormalities were treatable.[36] Today, OTs are the most likely profession to treat them, although practitioners from a variety of disciplines, including physical therapy, audiology, and optometry, have entered this arena.

A pioneer in applying sensory integration theory to patients with autism and other developmental disabilities was the late Lorna Jean King, OTR/L, FAOTA (1923–2006), founder and director of the Children's Center for Neurodevelopmental Studies in Glendale, Arizona. Beginning with just one student in 1978, this model center has served thousands of children ages 3 through 22 using sensory-integration techniques. TheChildrensCenterAZ.org.

Sensory-integration-based occupational therapy consists of guided activities that challenge and enhance the body's ability to respond appropriately to one or more senses. SI treatment is not designed to treat a specific diagnosis, but rather to be a component of a comprehensive multidisciplinary program of services.

Certification is required for those performing evaluations but not for doing sensory-integration therapy. Therapists use controlled tactile, proprioceptive, and/or vestibular input to elicit the simplest adaptive responses. Some activities are aimed at improving the processing of an individual sense, while others facilitate components of praxis, such as initiation, sequencing, bilateral coordination, timing, and imitation.

Starting with purposeful, safe sensory activities, such as swinging, receiving deep pressure under pillows, and being touched with a soft cloth, therapy gradually moves into more stimulating activities such as merry-go-rounds, hair brushing, and new foods. Soon sensory-seeking behaviors like running in circles are replaced with socially engaging activities, such as swinging, that fulfill the same sensory need. As efficient, organized responses occur more frequently and become more consistent, they ultimately heighten an individual's ability to pay attention, relate, sit still, organize language, and focus.[37]

Therapy does not focus on specific skills because that would not resolve the underlying problems, and the body would not learn how to adapt to future similar activities, but rather just how to do that single task.

Table 16–3

Tools for Normalizing Sensory Processing
AUDITORY—Whistles, musical instruments, CDs, environmental sounds
TACTILE—Stretch fabrics, including Lycra & spandex; fingerpaints, Play-Doh, ball pits
TASTE & SMELL—Salty, sweet, sour, crunchy foods; aromatherapy oils
VISUAL—Slant boards
VESTIBULAR—Hammocks, swings, slings, scooters, teeter-totters, gliders, skates, bikes, rockers
PROPRIOCEPTIVE—Weighted vests, blankets, ankle and wrist weights, seat cushions, beanbag chairs

Research involving a small sample of children with autism diagnoses strongly shows the efficacy of sensory-integration-based occupational therapy. As a child's ability to modulate sensation and organize the body improved, behavioral regulation did as well.[38]

Sound Therapies

Utilizing sound to facilitate bodily change is not new. Almost every culture on earth has developed methods using sound and song to heal. Unfortunately, many of these tools are associated with superstition and mysticism. Today's sophisticated computerized technology can document what is happening in the body and brain, thus removing the mystery.

Modern sound-based therapies include interventions that address auditory, listening, and vestibular issues. Some call sound-based therapies *listening therapies* or *auditory retraining therapies*. However, they are much more than "listening" to music or "retraining" the ear. The term *sound-based therapies* embraces all components of the other terms and includes other therapies that utilize sound and vibration on the body through special equipment, modified music, and/or specific tones or beats.

Over 20 different sound therapies support the connection between the voice, the ear, and the brain first identified by Alfred Tomatis (see below) in the 1950s. In the past 30 years, an increasing number of professionals from a variety of disciplines, including audiology, occupational therapy, psychology, and speech-language pathology, are trained in them.

For more than 50 years, the world's foremost sound therapist, Dorinne Davis, MA, CCC-A, F-AAA, has used sound to rehabilitate auditory processing difficulties. Davis is credentialed in every sound-based therapy, including Tomatis®, Berard, Solisten™, and others.

Davis developed the ***Diagnostic Evaluation for Therapy Protocol (DETP®)*** which allows therapists to determine which sound-based therapies may be appropriate for a client. While all sound therapies have efficacy, Davis believes that long-standing changes are possible only when therapies are administered in the correct order. To learn more, read one of Davis's 12 books; her newest is *Say it with Sound*.[39] Now semi-retired, she works only by appointment from her home in Syracuse, NY.

Two of the most common methods of sound therapy are the Tomatis® Method, named after French otolaryngologist and the father of sound-based therapies, Alfred Tomatis, MD (1920–2001), and Auditory Integration Training (AIT), which was founded in 1992 by Guy Berard, MD (1916–2014), a French physician who studied with Tomatis.

The Tomatis legacy is in its third generation, with a longtime collaborator, Thierry Gaujarengues, managing the company Tomatis Développement, based in Luxembourg with offices on five continents. Technology continues to improve, allowing the Tomatis® Method to become more accessible, innovative, and effective.

While anecdotal reports of improvement in language, sleep, social skills and other areas are common, research is mixed on the efficacy of this therapy.

The Tomatis® Method is appropriate for anyone interested in improving balance, coordination, motor skills, sensory integration, academics, communication, attention, memory, or organization, or for those just desiring to enhance overall development and personal growth. It is a unique listening program that works through several devices with special technology that instantly processes the voice and music to produce faithfully the Tomatis® effect.

Therapy takes place over two 13-day sessions of two hours of therapy each day, with a break of four to six weeks in between. A couple of older studies with small samples show that severely autistic children show improved speech, language, and behavior with Tomatis intervention.[40]

TalksUp® is the latest innovative portable device equipped with the Tomatis® system. Settings can be individualized based on an assessment. This device replaces the older Tomatis® Audio Pro (TAP) and the Solisten® programs. You can find a certified Tomatis provider at Tomatis.com.

Canadian psychologist Dr. Paul Madaule, a disciple of Tomatis, has modified the Tomatis method. For him, the ear is the conductor of the entire sensory orchestra.

Madaule has written about his experiences and his life's work practicing and adapting the Tomatis method in his book, *When Listening Comes Alive: A Guide to Effective Learning and Communication.*[41] It is also available in Spanish as *Terapia de Escucha.*[42] Madaule's adaptation of the Tomatis Method is called Listening Fitness Training or LiFT®. Madaule owns and operates the Listening Centre in Toronto, Canada, which he founded in 1978. A chapter in *The Brain's Way of Healing,*[43] features his work with individuals with autism, ADHD, and sensory processing disorders at the Listening Centre. ListeningCentre.com.

Another important Tomatis disciple is Pierre Sollier, who studied with Tomatis and has translated his works. His book *Listening for Wellness: An Introduction to the Tomatis Method*[44] is a great way to learn about this intervention. ListenWell.com.

Finally, read Sharon Ruben's heartfelt story about her daughter who recovered from autism using the Tomatis Method: *Awakening Ashley: Mozart Knocks Autism on Its Ear.*[45]

Berard Auditory Integration Training (AIT) is appropriate for those with hypersensitivity to sound, lack of awareness of sound, and the inability to discriminate sound differences. Originally called *Auditory Training*, the word *integration* was added later, making it AIT. This intervention randomly introduces low- and high-pitched sounds to the auditory system; these sounds increase blood flow to the brain, adding to the overall positive effect. AIT retrains the acoustic reflex muscle in the middle ear, allowing better transmission of clearer sound to the cochlea and subsequently to the brain for comprehension. Therapy consists of twenty half-hour sessions, twice daily for ten days.

Berard postulated that the quality of the perception of sound that one hears directly influences the behavior of the individual. He formulated his own theory that "human behavior is greatly conditioned by the way one hears." Berard's theory evolved further as he observed how the audiograms of children with disabilities change along with their behavior.

Berard originally described his method as one of "hearing re-education" in *Hearing Equals Behavior*,[46] which was recently updated and expanded in collaboration with Sally Brockett, MS.[47] Brockett trained under Berard, and is currently director of the IDEA Training Center in Connecticut.

Annabel Stehli introduced AIT for those with autism in the early 1990s after her daughter Georgiana's remarkable improvement.[48] She later founded and still runs the Georgiana Institute, a nonprofit organization and resource for "everything AIT." In 2015, Georgie released her own memoir, *Overcoming Autism*.[49]

In 2004, Stephen M. Edelson, PhD, and the late Bernard Rimland, PhD, of the Autism Research Institute (ARI) reviewed 28 published studies investigating the effectiveness of AIT in autism spectrum disorders. Most research shows that AIT clearly benefits a significant proportion of those with ASD or ADHD. Positive changes include reduced hyperactivity, social withdrawal, restlessness, anxiety, and sound sensitivity; increased attention span and language output; and improved speech perception. One study also showed biochemical changes, specifically, an increase in norepinephrine and decrease in serotonin levels.[50] Abstracts of all studies are available online.[51] Visit BerardAITWebsite.com.

Unyte Health, distributor of the *Safe and Sound Protocol* described in Chapter 6, was launched in 2007 as ***Integrated Listening Systems (ILS)***. The original ILS, developed by a multidisciplinary team of OTs and speech pathologists to improve cognition and emotional health, continues to be one of their products.

ILS—based on Dynamic Listening Systems (DLS), a therapeutic music and movement program developed by psychiatrist Ron Minson, MD—is applicable for depression and to improve general brain function.

Various length programs are available, from a 10-hour calming package to an intensive 40+ hour optimal performance program to improve sensory, motor, and auditory processing, concentration, attention, and reading. Today the ILS community includes thousands of professionals worldwide. IntegratedListening.com.

Interactive Metronome® (IM) is not a true sound therapy but is included here because it combines sound with a motor response. It is a true sensory-integration therapy! Created by James Cassily, a sound engineer, for over 30 years IM has helped thousands of individuals improve cognition, attention, focus, memory, speech/language, executive functioning, and comprehension, as well as motor and sensory skills.

A subject wears a special headset that emits signals. The IM computerized program challenges the participant to precisely match the computer's rhythm by tapping hand or foot sensors. During a full training, a person could respond over 35,000 times.

IM takes fifteen days, listening one hour per day over three to five weeks. Some people need up to as many as ten additional sessions. Interactive Metronome is now in use in more than 5,000 clinics in North America. See InteractiveMetronome.com.

The Sensory Learning Program (SLP) is an innovative approach for both children and adults that unites three modalities (visual, auditory, and vestibular) into one intervention, allowing individuals to better integrate sensory messages. It was developed in 1997 by the late Mary Bolles (1943–2014), a mother seeking help for her nonspeaking son who exhibited behaviors consistent with autism. Mary discovered that combining three individual modalities into one multisensory experience provided the positive results she had been seeking. Since her passing, Mary's work has been carried on at the Sensory Learning Institute by her daughter, Julie Stoots.

SLP is designed for those with ADHD, learning disabilities, autism, brain injury, and behavioral and nonverbal learning issues. The visual component uses syntonics—colored lights that pulse on and off about six times per minute (read more about it in the next chapter). The vestibular stimulation is delivered on a motion table that rotates in various planes. The auditory part uses headphones that contain modulated music.

Over 30 Sensory Learning Program centers around the world provide clients with two 30-minute sessions each day for 12 consecutive days. Each session is an individual sensory experience that simultaneously and sequentially engages visual, auditory, and vestibular systems to work in an integrated way. After the sessions, an individual returns home with a portable light instrument to continue the program. SensoryLearning.com.

The Squeeze Machine was invented by Temple Grandin, PhD, probably the most well-known and successful adult with autism in the world. As a child she regularly visited a relative's farm, where she was attracted to the machine used to restrain cows while they were branded. She loved the deep pressure the machine provided and found that it calmed her and decreased her severe tactile defensiveness. This life-changing experience led to her career as the world's authority on livestock handling equipment.

The squeeze machine appears to lessen tension and anxiety in those with modulation issues by delivering deep touch pressure to the trunk. For improvement, six to twelve sessions are necessary.[52]

A version of Grandin's design, as well as "snug hug" blankets in child and adult sizes and several colors, is available from Therafin.com. The Squeeze Machine allows the individual to control the amount of and need for pressure. According to the website, "This ingenious system is used for deep touch stimulation and produces a calming effect."

Carol Kranowitz, MA, author of the bestseller *The Out-of-Sync Child*,[53] now in its third edition, has expanded her reach with *The Out-of-Sync Child Grows Up*[54] and her latest offering, *Good Times with Out-of-Sync Grandkids: Activities for Grown-ups and Children with Sensory Processing Differences*.[55] Kranowitz has also partnered with perceptual motor therapist Joye Newman, MA, to produce *Growing an In-Sync Child: Simple, Fun Activities to Help Every Child Develop, Learn, and Grow*[56] and accompanying In-Sync activity cards.[57] They understand that kids and adults with sensory issues can benefit greatly from climbing, jumping, moving, and balancing and know how to make it fun.

Movement Programs

The Anat Baniel Method

The Anat Baniel Method, known as *NeuroMovement*™, focuses on the remarkable abilities of an individual's brain to change and learn until death. When Baniel was young, she observed movement genius Moshé Feldenkrais teaching in her parents' living room in Israel. When she was an adult, they reconnected, and Feldenkrais became her mentor.

Shortly before her mentor's death, she began teaching Feldenkrais professional training programs in the United States and around the world. After three decades of professional work, she tapped into her background in clinical psychology and dance, and developed her own hybrid program that combined kinesthetic, body-based movement with psychological work.

Baniel's gentle approach focuses on stimulating the brain to create new connections. It uses movement with attention to communicate with the brain and help it build new pathways. "Movement is the language of the brain," states this remarkable woman. As a result, her students connect more fully with their body and feelings, thus awakening their ability to make sense out of the world.

Baniel describes Nine Essentials for helping clients to discover easier and more efficient ways to move, develop clearer and more creative thinking, and attain greater vitality and well-being. These are powerful tools that are foundational to overcoming physical and psychological limitations. Starting with moving with intention, they form the core of the Anat Baniel Method.

This program is very subtle and does not subscribe to the expression "no pain, no gain." Baniel emphasizes moving slowly and with enthusiasm to get the brain's attention and stimulate the formation of rich, new neural patterns. Slow, intentional movement lets us feel and experience life at a deeper, more profound level. Enthusiasm informs the brain as to what is important, amplifies and exaggerates it, and infuses it with powerful energy.

Flexible rather than specific goals reduce anxiety, increase creativity, and result in success, vitality, and joy. The action of generating awareness, which Baniel calls "awaring," is becoming knowledgeable about what the brain and body are doing, sensing, thinking, and experiencing. Awaring is the opposite of automaticity and compulsion. When you are aware, you are fully alive and present.

In her book, *Kids Beyond Limits*,[58] Baniel describes working with the Nine Essentials to help those with autism, attention deficits, cerebral palsy, and complex genetic syndromes. *Move Into Life: NeuroMovement for Lifelong Vitality*[59] focuses on movement for adults. Visit AnatBanielMethod.com.

The Brain Balance Program®

The Brain Balance Program® is an individualized, comprehensive remedial program that integrates physical and cognitive exercises with effective educational and behavioral methods, and easy-to-follow dietary changes. Chiropractic neurologist Robert Melillo, DC, believes that an imbalance and communication breakdown between the two hemispheres of the brain is the foundation of dysfunction in those with neurodevelopmental disorders.

The program's goal is to correct the brain's imbalance and thus promote optimum brain and body function. Created almost 30 years ago, Brain Balance improves each individual function and then progressively integrates them.

Both child and adult programs combine the best of many modalities: motor, sensory-motor, language, cognitive, reflex, and other therapies. It is an intensive thrice-a-week commitment for one-hour periods over three months that includes home exercises, biomedical testing, dietary changes, and nutritional supplementation.

In his many books Melillo describes his program as reconnecting "discon-nected" kids with autism, ADHD, dyslexia, and other neurological disorders.[60] About 70 franchised Brain Balance Achievement Centers are open in the United States, run mostly by chiropractors. See BrainBalanceCenters.com.

Autism Movement Therapy (AMT), including the Aut-erobics® DVD, is a movement and music integration program developed by Joanne Lara, MA, a California special educator and dancer. AMT combines motor patterning, move-ment, auditory processing, rhythm, and sequencing in a unique and fun way with the goal of improving behavioral, emotional, academic, social, speech, and language skills. Available in three levels designed for those with "emerging," "developing," and "proficient" skills, each group goes through warm-ups, stationary movements, locomotion, improvisation, and cool down. Read about it in *Autism Movement Therapy® Method: Waking up the Brain!*[61] Visit AutismMovementTherapy.com.

Fitness Programs

If you are clumsy, can't catch your breath, and don't know where your body is in space, how do you function? Mostly by hanging out on the couch, not moving much, and socializing online.

Fitness is important for health and life. Like learning to parent, it is missing from most education curricula. In the "olden days" when I went to school, we had gym class every day. No longer! Daily gym has been replaced by weekly physical education (PE), and general fitness has been replaced by sports. But what if you cannot play competitive sports?

Some schools have adaptive physical education for students with special needs, but again they meet once or twice a week. Everyone, no matter their age or ability, needs to move every single day! Ultimately, klutzy children and adults can actu-ally gain joy in movement and healthy motor activity that can last for a lifetime.

Here are two local model programs that combine professional evaluations, individual goal setting, and a customized programming of coordination, sensory, and motor activities to improve fitness, motivation, self-esteem, self-advocacy, socialization, and overall physical wellness.

Autism Fitness on Long Island, NY, founded 20 years ago by Eric Chessen, MS, YCS. To Chessen, fitness is at the foundation of all human performance. It includes the ability to navigate successfully through everyday motor challenges, and includes life skills such as getting dressed independently, taking out the gar-bage, and, yes, playing sports.

How does Chessen get couch potatoes off the couch—away from their beloved video games—and help them overcome their movement deficits? By establishing a relationship and setting up reasonable expectations. His number one goal is to alleviate frustration and anxiety.

Chessen has developed a set of activities, which he offers as a DVD, that incorporate five essential movement patterns: pushing, pulling, rotating, squatting, and locomotion. These might use sandbells, stability balls, hurdles, or other equipment. He also has a distance mentoring program for fitness professionals, as well as an *Autism Fitness* eBook, designed to provide parents, educators, fitness professionals, and therapists with a blueprint for creating a successful fitness program for individuals or groups. See AutismFitness.com.

Fitness for Health in Rockville, MD, founded by Marc Sickel, ATC, to promote fitness for those with autism and other special needs. Equipment includes a laser maze that teaches athletes where their bodies are in space and builds motor planning, coordination, and balance as they crouch under and step over laser beams. They receive instant auditory feedback if they err in their movements.

Participants can play virtual reality sports games with actual body movements, scale an 80-foot transverse climbing wall at different levels, harness up and climb a 10-foot-tall, glow-in-the-dark climbing wall, bounce on a 30-foot trampoline while playing mini-basketball, negotiate a ropes course, bat in a batting cage with an indoor pitching machine, play glow-in-the-dark soccer, test skills against a high-tech reaction and coordination game, and be introduced to a variety of games and equipment involving visual perception and motor planning. Visit FitnessForHealth.org

Animals as Sensory-Integration Therapy Tools

"People and animals are supposed to be together," states Temple Grandin in *Animals in Translation*.[62] Thus, partnering with animals in sensory therapies is a natural! While almost any animal—including bunnies and turtles—will do, the most common therapeutic animals are dolphins, dogs, and horses.

Dolphins are true sensory-processing machines! Laboratory studies confirm that they use sound, taste, touch, and vision to communicate and navigate their environment.[63] Several years ago, I swam with dolphins at the Curacao Dolphin Therapy & Research Center (CDTC), located in the southern Caribbean, about 30 miles off the coast of Venezuela. What an incredible sensory experience! CDTC provides personalized dolphin-assisted therapy and rehabilitation programs to over 400 patients with special needs each year. Two-week-long programs in English, Dutch, German, and Spanish include the whole family as an essential part of their therapy concept. See CDTC.info.

A dog can provide considerable sensory stimulation, most of which is comforting. Just the smell of a familiar dog can be calming. Enlisting the help of dogs for people with disabilities is not new; Seeing Eye dogs have helped those with limited sight since 1929. Well-trained service dogs have become trusted companions for vulnerable individuals, including those with seizures. Read more about both types of dogs in chapter 18, where they are included because of their strong impact on improved communication and social interaction.

Therapeutic horseback riding, or *hippotherapy*, is practiced by specially trained occupational, physical, and speech therapists. Therapists' goals do not focus on specific riding skills, but rather on establishing the neurological foundations for improved function and sensory processing. Participants groom, dress, ride, care for, and love their horses but do not control their movements.

The horse's walk provides proprioceptive, vestibular, tactile, visual, auditory, and olfactory stimuli that are beneficial to the rider through movement which is variable, yet rhythmic and repetitive.[64] Through adapting to the horse's movements, the rider develops better balance and coordination, as well as flexibility, posture, muscle strength, mobility, and overall physical function.[65] These skills often generalize to other areas, including increased independence, which fosters improved performance in a wide range of daily activities. To find a stable near you, go to PathIntl.org, the website of the Professional Association of Therapeutic Horsemanship International.

Aquatic Therapy

The pressure of water offers a safe, soothing, and supportive environment in which to move without the pull of gravity and gives therapeutic proprioceptive and tactile input that improves range of motion, body awareness, and balance.[66] Simply moving in a tub or a swimming pool is a possible avenue to strengthen muscles and relieve tension.

Watsu is a specific aquatic therapy created by Harold Dull after studying some Asian bodywork techniques. Derived by combining the words *water* and *shiatsu* (a Japanese massage technique), it is performed in a 95° therapy pool.

The watsu therapist continuously supports a patient who floats calmly on their back, face above the water. The warm water's therapeutic benefits and freedom of movement make it an ideal medium for passive stretching. The support of the water takes weight off the spine and allows the body to be moved in ways impossible on land.

Gentle, gradual twists and pulls relieve the pressure a rigid spine places on nerves and helps undo any dysfunction this pressure causes to organs serviced by those nerves. To learn more about the benefits of watsu, read Dull's book *Watsu: Freeing the Body in Water.*[67]

Music

According to neurologist Oliver Sacks, MD, pathways for music occupy more areas of the brain than those for language.[68] Reflex specialist Sally Goddard calls music "one of life's earliest teachers." Prenatally, the fetus reacts to music with movement; infants respond to music and can imitate simple rhythms like patty-cake before they develop speech. Music supplies the architecture for many aspects of learning.[69]

Music can be very organizing and assist in the remediation of any skills by providing calming or alerting background sounds. It can enhance cognitive function, language comprehension and usage, and perceptual, gross, and fine motor skills. I always have background music, mostly Mozart, when I write.

Music Therapy

Music therapy, quite different from sound therapy, "provides a variety of multisensory experiences in an intentional and developmentally appropriate manner."[70] Certified music therapists use a variety of instruments, including their voices, to deliver concrete auditory, visual, tactile, proprioceptive, and vestibular stimulation to both hemispheres of the brain.

Several types of music therapy are effective, especially with individuals on the autism spectrum.[71] Music can offer stress relief, promote relaxation, produce positive emotions, enhance mood, decrease anxiety and depression symptoms, and improve self-esteem. What else does all that? Visit MusicTherapy.org.

Movement to Music

OTs incorporate music for both calming and alerting; several who have composed songs to enhance sensory integration are

- **Genevieve Jereb, OTR,** an Australian whose rhythmic songs and activities support learning, transitions, sleep, attention, focus, and self-regulation;
- **Aubrey Carton Lande, OTR,** and **Lois Hickman, MS, OTR, FAOTA,** with musician Bob Wiz offer "SongGames for Sensory Processing" for ages 3–8;
- **Lucy Jane Miller PhD, OTR,** collaborated with singer/songwriter Coles Whalen to produce "Songs for Sensational Kids: The Wiggly Scarecrow" and "I'm OK, Olé!"

All are available online.

- **Eve Kodiak** is a classically trained musician and motor development and reflex specialist who has developed many original albums and accompanying materials to integrate reflexes and move. She also blogs on the importance of movement. Visit EveKodiak.com.

Rhythmic Movement Training (RMT)

A reflex-integration approach that has its roots in Sweden is called *Rhythmic Movement Training*, or simply RMT. It originated with Kerstin Linde, a self-taught Swedish bodywork therapist who developed a series of rhythmic movements inspired by movements infants make spontaneously before they learn to walk.

Harald Blomberg, MD, a Swedish psychiatrist, met Linde in 1985 and introduced her method in the psychiatric outpatient clinic where he worked. During the 1990s Dr. Blomberg began treating children diagnosed with attention and learning difficulties using rhythmic movements, adding material about integrating primitive reflexes from a variety of sources. His method, Rhythmic Movement Training, can be magical for enhancing focus and regulating emotions without medication.

In his travels Blomberg met Moira Dempsey, an Australian kinesiologist. Together, they wrote a book about combining RMT with reflex-integration therapy.[72] Blomberg and Dempsey experienced a rift in their relationship that clouds the rest of the story about what belongs to whom.

RMT is easy and effective and getting more attention in the United States lately due to Sonia Story, a reflex-integration specialist in Washington State Resources include BlombergRMT.com, RhythmicMovement.org, and MovePlayThrive.com.

Musical Giftedness

Many individuals with sensory differences are intrinsically attracted to music. Some are musically gifted, like David Helfgott, featured in the movie *Shine*. Matt Savage, called the "Mozart of Jazz" by the late Dave Brubeck, was diagnosed with autism at the age of three and in his teens became a professional jazz pianist with his own trio. Now 32, he has played all over the world alongside the best adult musicians of our time since he was a teenager. Read his story in *Off the Charts: The Hidden Lives and Lessons of American Child Prodigies*.[73] Kaylee Rogers won everyone's heart with her magnificent rendition of Leonard Cohen's *Hallelujah* on the television show *Little Big Shots* in March 2018.

Stephen Shore, PhD, a professor at Adelphi University, is a very accomplished adult on the spectrum. He teaches music to those on the spectrum and has written eloquently about the role music played in his childhood. In his autobiography, *Beyond the Wall*, he explains how, when he could not learn the Hebrew necessary for his Bar Mitzvah, he perfectly mimicked an audiotape of a cantor singing the prayers, which he accompanied with the necessary ceremonial, rhythmic rocking called *davening*.[74]

Band Together Pittsburgh uses music as an instrument for change. It was founded in 2016 when John Vento and bar owner "Moondog" Esser recognized musical talents in their sons and produced open mics at the bar on Sundays. Anyone, any age is invited to step up and sing. Dozens show up each week. Instruments and lessons are available for free. So are instructions on becoming a DJ. Enthusiastic parents and friends cheer even the most out-of-tune! It's a lovefest that has launched several talented teens and young adults who perform at local theaters and concerts. See BandTogetherPgh.org.

The latest music and autism story involves composer Justin Morell, who wrote a full-length symphony incorporating the vocalizations of his nonspeaking autistic son, Loren. The music was recorded and released as an album in 2021. Two years later, the Cincinnati Symphony Orchestra, multi-Grammy-winning trumpet soloist John Daversa, dancers, a wordless choir, projection mapping, and a captivating, creative set design brought this phenomenal production to life. The performance was captured in the heartfelt documentary *All Without Words*, which chronicles how the Morell family found new meaning in and acceptance of Loren's autism. Visit AWWLive.com.

Sensory Connections to Other Issues

Sleep disturbances,[75] picky eating, and toilet training, three of the most problematic issues, especially in autism spectrum disorders, all have sensory connections. Parents frequently call upon therapists to intervene in these troublesome areas. Read about the former two in previous chapters.

Toilet Training

Moving from diapers to the bathroom is one of the most elusive milestones to reach. Sensing that the bladder or intestine is full, and acting upon it, requires a high level of sensory integration. When calm, kids may be able to feel the sensations that they need to urinate or defecate, but if they experience sensory overload, they may not be able to feel that urge. Thus, sometimes these kids can use the toilet correctly; at other times they will not.

Dr. Temple Grandin believes that severe sensory processing problems play a large role in toileting issues. Some children like to flush the toilet repeatedly as pleasurable sensory stimulation. For others with severe hearing sensitivities, the sound of the toilet flushing may hurt their ears or simply terrify them.

Grandin states two major causes of toilet-training problems: children are either afraid of the toilet, or they do not know what they are supposed to do. Sometimes the highly sensitive child can learn to use a potty chair located a short distance from the frightening toilet. Most need to see someone else use the toilet in order to learn.

Several excellent references enumerate management strategies for toilet training. A chapter is devoted to this subject in *Autism: A Sensorimotor Approach to Management*.[76] *Toilet Training for Individuals with Autism and Related Disorders* is a complete book on the subject.[77] Both focus first on habit training and determining readiness, then on sensory strategies and generalizing to unfamiliar environments.

Take-Home Points

Like nutritional issues, sensory problems are often responsible for disruptive behaviors. Instead of trying to extinguish them, therapists take a sensory approach to evaluate these behaviors, always looking for possible underlying causes.

Once specialists determine which senses are hyperreactive, which are hyporesponsive, and how the integration of sensory processing is being affected, they can design an appropriate intervention program. When sensory issues are resolved, coordination, sleep, picky eating, and even toilet training can be less difficult. Targeting problem senses such as touch, sound, balance, and pressure can result in more organized and integrated sensory processing and more appropriate behavior.

CHAPTER 17

FOCUSING ON VISION: THE DOMINANT SENSE

Do you know someone who has "vision?" What do we mean when we say, "I see"? If someone is "myopic," are we referring to his glasses prescription, his thinking, or both? Vision words are a part of our everyday vocabulary!

For over 40 years vision and its relationship to behavior and learning have fascinated me. Why do so many individuals with autism have poor eye contact, flick their fingers in front of their eyes, or demonstrate poor coordination? No, it is not because they are autistic! They are autistic, in part, because their vision malfunctions, resulting in these behaviors.

My training as a mental health professional did not satisfy my thirst for information about vision, so I turned to other disciplines for answers. In the 1980s I was fortunate to meet several brilliant optometrists who opened my eyes (literally and figuratively!) to how the eyes, brain, and body work together. They became my mentors, my friends, and my family; without them, my career would have been much less exciting, and this chapter would not have been possible. My last book was dedicated to them.

This chapter begins with defining vision as different from eyesight and explains what happens in typical development. Vision, like dentistry, deserves its own chapter because—in addition to being a strong interest of mine—it plays a huge, generally ignored and misunderstood role in psychiatric and other diagnosed conditions. Simply said, visual processing problems add a crucial, mostly unrecognized stress factor that increases the Total Load for many individuals.

Finally, the chapter describes how vision therapy using motor activities, lenses, and prisms can often ameliorate underlying visual problems, just as detoxification removes bugs and metals as load factors. Remediating visual issues frees up energy for immune system function, communication, and social interaction. In many cases behavior becomes more flexible, language emerges, allergies disappear, and social-emotional skills become more typical. If you are going to defy diagnoses, you must have vision!

Eyesight Versus Vision

Vision is the learned, developmental process of focusing on and giving meaning to what is seen. The operative words in this definition are *learned* and *developmental*. Vision depends upon motor and sensory experiences. Learning to use one's visual system takes place in a predictable, sequential manner. More on that later.

Vision involves the eyes, brain, and body; it is much more than 20/20 eyesight, which describes only the clarity of what we see at a distance. Vision allows us to attend to, focus on, organize, understand, and interact with the world around us; it is both perceptual and conceptual.

Vision Is Learned

We learn to use our vision in the same way that we learn to walk, talk, and write: through sensory experiences that teach us how to select which targets to touch, pick up, identify, think about, and talk about.[1] According to pioneer optometrist G. N. Getman (1914–1990), children learn how to use their bodies by fully participating in and exploring their world for the first two or three years of their lives.

During this time hands, eyes, mouth, and ears—the entire body—are tools for learning. According to Getman, "the movement of self through space, the manual and visual exploration and inspection of the world and its contents, vocalization of names, labels, needs and desires are the full-time occupation of the small child." Little children spend every waking moment visually inspecting their three-dimensional surroundings.[2]

For Getman and his colleague, developmental psychologist and physician Arnold Gesell, MD, PhD (1880–1961), the visual experiences of our early years lay the foundation for our ability to be successful in society. Gesell is the founder of the world-renowned child development center at Yale University. Note that the director of the Gesell Institute of Child Development from 1978–1985 was Sidney MacDonald Baker, MD, one of the founders of the Defeat Autism Now! movement. Baker, a brilliant physician with a heart of gold, is still one of the most respected leaders in the field of autism treatment today.

Getman and Gesell could never have imagined that two-dimensional tablets, computers, and smartphones with their touchscreens and cursors would replace three-dimensional wooden blocks, pots and pans, and balls as the playthings of childhood. They would probably be horrified by, but not surprised about, the less-than-positive outcomes these modern-day technological wonders have on the visual skills of today's young children. For these developmentalists, visual experiences in three dimensions drive imagination, creativity, and even many types of intelligence.

Vision Develops

Vision develops every time the brain integrates sensory input from the tactile, proprioceptive, olfactory, gustatory, and balance receptors, as well as from the ears, eyes, body, and brain, eventually emerging hierarchically as the dominant sense in well-functioning people. Gesell, along with his colleagues at Yale, pediatrician Frances Ilg, MD (1903–1981), and psychologist Louise Bates Ames, PhD (1908–1996), understood how babies learn. Their impeccable research on the predictable sequential development of children from infancy is immortalized in *Infant and Child in the Culture of Today*, first published in 1943 and revised in 1971.[3]

Getman, the eye-care professional on the Gesell team, wrote about the importance of sensory experiences a generation ago in his classic, *How to Develop Your Child's Intelligence*.[4] This tiny book, still available, comes with an amazing chart documenting developmental patterns in many areas.

At each age level, from six months through almost 14 years of age, the chart shows expected motor, visual-motor, visual-tactual, visual-language, and cognitive guidance patterns, giving developmental guidelines to parents and educators. Getman maps rhythm, emotional, ethical, hygiene, eating, and play guidance patterns for older children. The book is well worth owning for this unique chart alone.

I would be remiss not to comment on the recent move from the American Academy of Pediatrics to normalize delays in the motor, language, and social skills on this chart. For instance, walking alone was moved from 12 months to 15 months and saying a first word was moved from 12 months to 15 months. What happened? Because our educational and healthcare systems simply cannot handle the number of kids with delays, it just moved the bar, so fewer kids are considered delayed and require early intervention. Longtime pediatrician Dr. Larry Palevsky deplores this trend and believes that we cannot ignore this problem. Listen to him rant online.[5]

All of these pioneers undoubtedly recognized that during a typical child's development, the motor system stabilizes from the inside out: from the body's core and the midline outwards, and from large muscles to smaller ones—head, neck, shoulders, arms, elbows, wrists, hands, fingers, joints. Vision develops from the bottom up: from whole body to lower, then upper body, trunk, neck, head, and eyes. Each body part, including the eyes, eventually functions independently of the core, head, and upper body. Enter Swiss developmental psychologist, Jean Piaget (1896–1980).

Piaget's Developmental Hierarchy

Piaget, who practiced for another 20 years after Gesell's death, devoted his career to investigating stages of child development. Most courses in education and psychology include his classic work. Piaget argued that motor and visual development take place in tandem, in a predictable hierarchy. He conceptualized a series of progressive stages from sensorimotor (age birth to two), to preoperational (age two to seven), to concrete (age seven to eleven), to abstract (age twelve and up), leading to higher and higher levels of cognition.[6]

The *motor stage* comes first, during which individuals move primarily without purpose as the receptors for touch, proprioception, balance, vision, and hearing are maturing. At this stage, the baby is simply a motor being: the motor system drives vision. As tone heightens, sensory integration takes place, and movement becomes more directed by the body and brain.

The *motor-visual stage* follows. For months, the typical infant's and toddler's movements take the eyes along for a ride as the brain learns to control the upper and lower body. The eyes process what they see wherever they are directed.

Next comes the ***visual-motor stage***. Progressing to this level is a major hurdle in development. Vision now directs movement. Instead of moving purposelessly through space, children at a visual-motor stage take inventory of the space around them, focus on a target, and move toward it. Movement and drawing become purposeful. Preschoolers see the block corner, recall the tower they built the day before, and race there to reexperience the pleasure it gave them. They take the crayon and draw a person, one body part at a time. The motor and emotional experiences merge into one.

During this very critical time frame, vision begins to dominate the movement system, to coordinate the proprioceptive, vestibular, and tactile systems. If there is faulty information processing in any of the primitive sensory systems—perhaps further complicated by retained primitive reflexes—visual dysfunction is inevitable. Because their visual systems are inefficient, these patients still need to touch and move to be secure in their environment.

The last step is the ***visual stage***. At this level, people no longer need the actual bodily movements for experience; rather, they can "move" in their mind's eye. A child at this stage tells his mother, prior to going to school, that when he arrives, he will go right to block corner to make a tower. His sensory, motor, visual, and emotional brains have memory of the experience and want to have it again. This step is the beginning of the ability to plan, organize, conceptualize, and ideate: all results of the senses and motor abilities integrating efficiently.[7]

For a child to be academically "ready," the above steps take place in sequence, resulting in the development of a dominant eye and hand (not always on the same side) and the ability to use the two sides of the body together. Success at end-product skills, such as holding down the paper with one hand and writing with the other, or using paper and scissors together, are examples of bilateral tasks that developmentally emerge as the motor and visual systems work together.

Applying Piaget's Developmental Hierarchy

From a Piagetian point of view, many people with developmental delays, regardless of chronological age, are stuck at the **motor-visual stage**. They "look" with their whole bodies. Ask them to follow a moving object and observe that they move not only their eyes, but their heads and trunks as well. They read with head and body movements from side to side.

Students at the motor-visual stage may wander aimlessly around their classrooms, finding themselves somehow in the block corner, where they may line up instead of build with blocks. When offered crayons or markers at this stage, they scribble, sometimes not even looking at their hands or what they are drawing. They may start out with an idea of drawing a person but end up with a dog as their motor skills override their visualization abilities.

Self-stimulatory behaviors at a motor-visual stage, may actually serve a purpose. Arm flapping, for instance, allows interaction with the world, albeit in a strange way that tells the brain where the body is in space.

To move to the next step, the **visual-motor stage**, the body must know where it is in space and move automatically. If someone doesn't know where they are in space, it is extremely difficult—maybe, even impossible—for them to localize objects, such as a backpack, homework, or letters on a page. Many kids and even adults with underlying visual dysfunction are "lost in space" and experience resultant organizational difficulties.

The **visual stage** is the long-term target goal for organized, efficient functioning. Only a very few, like Temple Grandin, have figured out how to "think in pictures."[8] Others have photographic memories and can draw whole cities or recall where a word is on a page. Usually, though, these talents are splinter skills that do not generalize to real life or allow them to organize themselves, conceptualize ideas, and think abstractly.

I realize that this is quite a lengthy description of visual development, but I believe it is ultra-important in understanding behavior. Keep reading to see how it can be applied over and over again to avoidant, fearful, obsessive, and compensatory behaviors in learning disabilities, attention deficits, and sensory-processing differences.

The Two Parts of the Visual System

As the body and eyes work together sending messages to the brain, magic happens. Neurological connections form, allowing the visual system to develop two parts:

- *Focal or "central" vision,* which permits an individual to determine "What is it?" Optometrists sometimes call this *parvo vision.* Focal vision is primarily a conscious function, allowing one to see clearly, to recognize objects, and eventually to read.
- *Ambient or peripheral vision* which answers "Where am I?" or "Where is it?" Optometrists sometimes call this *magno vision.* It is a subconscious function, and its role is to orient an individual and object in space.

When vision is functioning normally, magno and parvo vision integrate. Dysfunction results when a person uses one part more frequently than the other or has trouble engaging and disengaging between the two systems, as do many people with autism and emotional vulnerabilities.

Fundamental Visual Skills

During normal development, many important visual skills emerge that are necessary for all aspects of functioning. Table 17–1 lists essential visual skills.

Table 17–1 Essential Visual Skills
Accommodation—Ability to activate focus for near-visual space to attain clarity and to shift focus between near point and far point to maintain clarity
Acuity—Sharpness or clarity at both far and reading distances
Binocularity—Ability to move the two eyes together as a team—up, down, in, and out—to coordinate their messages and send one distinct image to the brain
Convergence—Ability to move both eyes together inward simultaneously toward the nose (cross the eyes)
Divergence—Ability to move eyes together outward away from the nose
Focusing—Ability to maintain clarity while changing visual distances
Eye Tracking and Fixation—Ability to look at and accurately follow an object
Binocular Vision or Fusion—Ability to use both eyes together efficiently

Eye Teaming—Ability to aim, move, and work the eyes as a coordinated team
Visual-Motor Integration—Ability to combine visual input with input from the other senses and respond motorically
Visual-Spatial Understanding—Ability to know where one is in relationship to objects and people

When the two eyes are working together properly, the brain tells the eyes what to look for, the eyes take in information, and the brain stores and uses it efficiently. Language emerges as we conceptualize what we see.[9]

Skeffington Circles

The father of modern developmental optometry, A. M. Skeffington, OD (1890–1976), meshed knowledge about vision and the other senses in the 1960s and '70s. According to developmental optometrist Leonard Press, OD, FAOO, FCOVD, what Skeffington conceived was the interplay between the ambient and focal processes, as well as between the eyes, brain, and body, to generate the basic "what," "where," and "how" of vision.[10] Figure 17–1 shows this interaction.

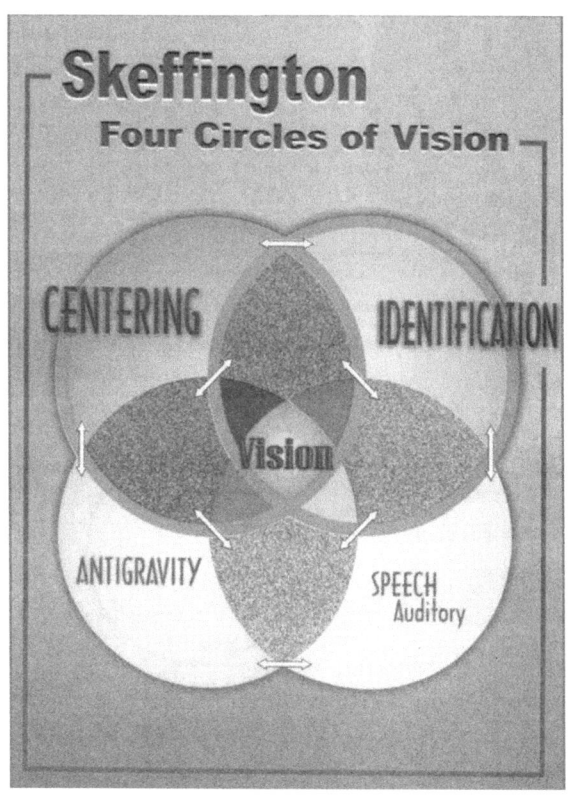

Skeffington's brilliant model demonstrates that what we call "vision" results from the overlap of the four circles, which represent the integration of the senses. In this diagram, *Centering* is ambient vision, *Identification* represents focal vision, *Anti-gravity* describes balance and spatial orientation (vestibular function), and *Speech/Auditory* is self-explanatory. An understanding of this model demonstrates how vision eventually emerges as the dominant sense.

Optometrist Rochelle Mozlin, OD, FCOVD, retired faculty from the S.U.N.Y. College of Optometry, reminds us that, because the circles overlap, one circle affects the others. Deficits in one area will bring down another; likewise, improvement in a domain will spill over into another.[11]

A. M. Skeffington was to developmental vision what A. Jean Ayres was to a sensory integration approach to occupational therapy. Even though they were contemporaries, no one knows if they ever met; if they did, they certainly would have had some very interesting conversations!

Reflexes and Vision

As has been described, inborn primitive reflexes are the foundation for all things motor. Since motor and visual development take place as a team, it should be no surprise that reflex integration is crucial to visual development. In Chapter 6, the tendon guard reflex was introduced. Integration of this key reflex is essential to feeling safe because it is important in focus. When not integrated, several of the skills in Table 17–1 are affected, including convergence, divergence, binocularity, and visual perception.

Today's safety measures such as back sleeping (even when remediated by "tummy time"), car seats, backpacks, strollers, Bumbo seats,[12] and baby exercisers can all impede the emergence of important reflexes. Svetlana Masgutova teaches an entire course on visual reflexes. This important topic cannot be ignored when discussing our dominant sense. For a more in-depth look at how poor reflex integration is implicated in neurodevelopmental disorders, refer to her seminal book, *PTSD Recovery: Gentle, Rapid, and Effective Treatment with Reflex Integration.*[13]

Vision and Neurodevelopmental Disorders

Visual Symptoms

Unintegrated reflexes are not the only thing that can interfere with vision development. Genetic and environmental factors which undermine health and interrupt the emergence of reflexes, motor development, and sensory processing, also impair vision. At some point, someone notices one or more of the following vision symptoms:

- Has difficulty making eye contact
- Tilts head when observing closely
- Squints or closes an eye
- Is fascinated by lights, spinning objects, shadows, or patterns
- Looks through hands
- Flaps hands, flicks objects in front of eyes
- Looks at objects sideways, very closely, or with quick glances
- Shows sensitivity to light (photophobia)
- Becomes confused at changes in flooring or on stairways
- Pushes or rubs eyes
- Widens eyes or squints when asked to look
- Bumps into objects or touches walls while moving through space

What Causes These Visual Symptoms?

Not surprisingly, there is a dearth of research on the fundamental visual skills of our target disorders. Most studies, including the more recent ones, focus on autism and have small sample sizes even though these individuals abound. Pioneer behavioral optometrist John Streff, OD (1927–2010), was one of the first to write about optometric care for children with autism in 1975 at the aforementioned Gesell Institute.[14] Because of little interest in autism for the next 20 years, no known articles on vision and autism appeared again until the early 1990s.

Studies on those with psychiatric conditions look not at comorbidity of visual skill deficits, but rather at disease, such as macular degeneration and glaucoma.[15]

While it is difficult to separate nature from nurture, nearsightedness (myopia)[16] and a turned eye (strabismus)[17] are believed to run in families. Some contemporary societal practices that restrict babies' and young children's movements for safety reasons may also be responsible in part. An increased early use of electronics such as computers, TV, and video games that force visual focus at inappropriate distances and for extended periods of time may also contribute in subtle ways, according to the American Optometric Association.[18]

The eyes are profoundly affected by vitamin deficiencies. In fact, the eyes are one of the most nutritionally demanding organs of the body and are often the first place disease, such as diabetes, appears.[19] In the late 1990s, Mary Megson, MD, discovered that a deficiency of vitamins A and D played a role in regressive autism.

Megson saw great progress when supplementing the diets of her patients with cod liver oil, an excellent source of the deficient vitamins.[20]

Heavy metals, many medications, and the foods we eat and drink—especially aspartame, alcohol, and excitotoxins such as MSG—quickly affect vision. Visual side effects include blurring, double vision, or even blindness. Did you know that dilated pupils are a sign of possible mercury poisoning?[21]

Oxygen deprivation from many sources can affect vision. Whether it is from anoxia at birth or shallow breathing at any time of the day or night, visual function will suffer.

The late optometrist Melvin Kaplan, OD (1930–2023), has written the ultimate go-to reference on vision issues in autism. *Seeing through New Eyes: Changing the Lives of Children with Autism, Asperger Syndrome and Other Developmental Disabilities through Vision Therapy*[22] is the source of much of the information in this chapter.

Kaplan believed that a high percentage of patients on the autism spectrum exhibit dysfunctional breathing patterns. Their inability to integrate movement and breathing stresses their muscles and nervous systems and reduces attention.

Return to Chapter 3 to understand the profound impact of inefficient breathing. Because constraints on breathing affect every aspect of performance, Kaplan believed that an effective vision remediation program must incorporate teaching appropriate breathing patterns. Muscle tension and a stressed nervous system impair vision, leading to mismatches in how the brain interprets where things are relative to where they appear to be.

Difficulty integrating central and peripheral vision. Another significant issue is coordinating central and peripheral vision. Some people, especially those with autism, tend to use one part of their vision, but not both simultaneously and efficiently. Some who are more *focal* may play with specks of dust, obsess over details, or demonstrate savant abilities. Others are *ambient*—fascinated by contrast, lights, shadows, and shiny objects. Without visual integration, the body is under stress.

If the eyes are not working together as a team or are processing information inefficiently, the brain must decide what to do with the partial information. Kaplan thought hyperactivity might just be a survival mechanism for those who are attempting to establish the limits of "self" by locking onto objects in space with such intensity that they lose awareness of self when standing.[23]

Likewise, poor integration of ambient and focal vision could cause other behavioral symptoms such as extreme behavioral inflexibility. These visual problems, along with poor binocularity, inefficient and slow focusing, poor integration of vision with other senses, and poor visualization, come crashing together to manifest themselves as poor attention, memory, spatial awareness, visual-motor integration, perception, or visualization! Inflexible behavior? Think inflexible vision!

Convergence insufficiency (CI) is the inability to coordinate the eyes for any length of time at a near distance, such as in reading.[24] The ability to converge is an essential skill that keeps the eyes binocular and allows the brain to receive and interpret one integrated image.

Convergence insufficiency is not uncommon; it occurs in about 5% of the population.[25] More than ten years ago, eye doctors at the University of Alabama in Birmingham found CI in 15 percent of their patients with schizophrenia diagnoses.[26] Because CI interferes with focusing, it can easily be missed and is commonly misdiagnosed as attention deficit disorder.[27]

Unfortunately, it is rarely picked up in routine vision screenings, so those affected rarely know they have it and it goes untreated. According to the American Academy of Ophthalmology, CI can also be acquired later in life as a result of trauma, injury (like a concussion), or a disease affecting the brain, especially auto-immune disease.[28]

In those with CI, eyes tend to drift outward; crossing the eyes is difficult, if not impossible, to hold for more than a few seconds. Symptoms of CI include headaches, blurred vision, sleepiness while reading, and difficulty concentrating. It is one of the most common diagnoses for those experiencing learning problems.

Strabismus. A commonality of many individuals with neurodevelopmental diagnoses is an eye turn, called a *strabismus*.[29] While only 4% of the general population has this problem,[30] I have heard estimates of up to 50 percent of those with autism or other disorders suffer this functional abnormality.

In 1992, a Chicago team of Janice Scharre, OD, and psychologist Margaret Creedon, PhD, did a small study of autistic children. They found strabismus in 21% and acuity problems in nearly 44%.[31] Other studies show that eye movement problems are also common.[32]

An eye may turn in (*esotropia*) or out (*exotropia*). A strabismus is sometimes seen after a traumatic birth, an infection, a high fever, surgery, heavy metal toxicity, sensory deprivation, or any assault on the growing nervous system. Recently a group of American ophthalmologists searched a commercial insurance claims database of over 12 million children and found a moderate association between each strabismus type (esotropia and exotropia) and anxiety disorder, schizophrenia, bipolar disorder, and depressive disorder.[33] Yikes! How about visual screenings *before* the psychiatric diagnosis?

In strabismus, one eye accurately aims at the object of regard, while the other eye misses it by aiming above, below, or to the left or right of it. A strabismus affects everything, especially learning and behavior. Double vision (*diplopia*) results and the brain must learn how to cope and adapt. The misalignment may be constant or intermittent, and thus not always noticeable. Disorganization and confusion follow as the brain struggles to integrate competing messages.

In order to minimize the disorganization and confusion, sometimes the unconscious mind adapts to strabismus by suppressing signals from the faulty-aiming eye. Eventually, visual suppression leads to *amblyopia*, or *lazy eye*, in which the nerves that transport and interpret visual information lose some of their ability. The result is poor vision in one eye due to an interference in the neurological interpretive mechanism. Optometrists view strabismus not as a structural deficit of the eye, like a muscle weakness, but rather as a functional motor-sensory misalignment caused by a person's unconscious adaptive response to neural dysfunction.

Strabismus always requires attention. A common recommendation for treating a strabismus is patching and, if this intervention is unsuccessful, surgery to "straighten" the eye(s). While this approach may create a cosmetically straight eye, it rarely achieves binocularity.[34] This is not surprising, given that Nobel Prize-winning neurophysiologists David Hubel, MD, and Torsten Wiesel, MD, concluded that patching eventually leads to the elimination of the visual cells that respond to both eyes simultaneously, resulting in an individual attending to the input of only one eye at a time.[35] Major stress!

In many instances the reduced eyesight cannot be corrected with glasses or surgery. With the eyes functioning at less than 100% efficiency, any sustained visual activity such as reading adds the need for extra effort and more stress. In about half of cases, amblyopia occurs with strabismus, but in the remaining half doctors can find no sign that one eye sees more poorly than the other. However, some people with amblyopia may turn their heads to see certain things or close one eye when reading.[36]

Mel Kaplan believed that the body and brain's adaptations to a strabismus are wide-ranging. Individuals with strabismus experience three basic losses, which overlap:

- loss of control of the environment and of the self in relationship to the environment,
- loss of mobility, and
- limitation of the range and variety of visual concepts.

All of these losses result from the eyes not working together as a team. When perception does not occur as an end product of a dynamic process involving the use and integration of *all* sensory receptors, including the eyes, the best outcome that a strabismic individual can attain is a construct of space, not true perception of space.[37]

Since a good number of the eye's neural fibers bring information to the body's balance system, it follows that a person's sense of where they are in space is compromised if these fibers deliver inaccurate information. Anyone who is disoriented in space experiences himself and his environment as unstable and unpredictable. He may demonstrate sensory defensiveness, grow increasingly inward, or even become belligerent. Because space perception also affects movement, language development, social skills, reading, writing, mathematics, and many untold other areas, these end-product skills all suffer as well. Strabismic behavior, like mercury poisoning, looks a lot like the diagnoses targeted by this book. See Strabismus.org.

Vision Exams

Choosing an Eye Doctor

Everyone with a neurodevelopmental diagnosis should have a complete vision evaluation as soon as the condition is suspected. Different professions view vision differently, due to their widely diverse training.

Ophthalmologists, who are MDs, define vision largely in terms of structural pathology. They are the best option if an eye disease like glaucoma is present.

Optometrists, or ODs, take a functional approach to vision. Optometrists are the best choice, in my opinion, because for them function—not structure or eye health—is their primary concern. Their training emphasizes that vision operates in relationship to the rest of the body. These professionals look at how someone's vision functions in all aspects of life.

A small subset of optometrists has advanced training and passed rigorous tests qualifying them as Fellows of the Optometric Vision Development and Rehabilitation Association (FOVDRA), formerly called The College of Optometrists in Vision Development (COVD). In their evaluations, in addition to checking for eye health and structural pathology, these doctors also run a full set of tests measuring the essential visual skills.

Most also provide remediation to guide the patient so he can learn how to use vision more efficiently (see below). I *strongly* recommend getting a second and even third opinion by one of them before making any decision about treatment, especially surgery and patching. To find a qualified doctor near you, go to COVD. org.

Anyone debating their options for vision evaluations, testing, and treatment should join the Facebook group Vision Therapy Parents Unite. Cofounded in 2013 and run by vision and autism advocate Michele Hillman, that group can get you immediate answers to your questions from more than 35,000 member professionals and parents worldwide who have "been there, done that."

Optometrists interested in learning more about working with patients on the autism spectrum can read "Autism Spectrum Disorders: A Primer for the Optometrist"[38] and "Serving the Needs of the Patient with Autism."[39] To become proficient, optometrists may take postgraduate courses from the Optometric Extension Program (OEP), the continuing-education arm of optometry.

What Does a Complete Vision Exam Include?

Depending upon the developmental age and cooperation of the patient, a thorough exam consists of both informal observations and formal testing. Identifying and treating underlying visual issues includes acuity and eye health, as well as all aspects of visual function. The skill and training of the examining doctor are key to obtaining important information about a particular individual's visual skills.

For those at the low end of functioning, the optometrist might simply watch the patient move and play, first with large motor skills and then with fine motor activities. They may present the individual with objects such as puzzles, balls, or floating and other attention-getting objects such as puppets, balloons, or bubbles. Testing may resemble an occupational therapy evaluation more than a vision test if motor skills still dominate using the eyes meaningfully.

Many behavioral optometrists are administering reflex evaluations with increasing frequency. They may ask the patient to lie on the floor to see how strong their core is and whether they can follow verbal commands to move the whole body or isolate component parts, such as the right index finger. They observe whether a patient looks with the eyes alone or with head and body for assistance.

If the individual is able to comply with demands, the eye doctor may use some special equipment to look at the essential visual skills, such as binocularity and convergence. He or she may also try different lenses and prisms, noting carefully a patient's change in focus, posture, breathing patterns, and maybe language. Even if the patient cannot tolerate testing glasses on the face, a doctor can measure a response using handheld lenses known as "flippers" in front of the eyes.

In patients with higher-level skills, vision testing will be more typical. The goal is to see how well an individual uses vision to direct behavior, how well vision is integrating with the other senses, and to determine if and how dysfunctional vision is affecting general development, language, learning, social skills, and behavior. Then, specific visual interventions can be prescribed.

One brilliant Chicago-area optometrist I have known for over 25 years, Deborah Zelinsky, OD, FNORA, FCOVD, has developed a unique, patented, easy-to-administer evaluation called the Z-Bell[SM] Test. This test measures the efficiency of integration between visual processing and listening. It is based on the scientific concept that the retina is made of brain tissue and collects information through its more than 120 million photoreceptors and specialized cells. This sensory information processing is continuous day and night.

A 2014 study at Vanderbilt University found that many children with autism do not synchronize their seeing and hearing.[40] Yet, before now, the linkage between the eyes and the ears has gone untested. What these individuals might be experiencing is a constant mismatch between what they are seeing and hearing, just like what happens when you listen to and watch a video where the sound and images do not match.

Test administration is simple: the doctor rings a bell outside the patient's peripheral vision, and the patient localizes it by pointing in the direction he thinks the sound is coming from. Patients with good eye-ear integration can do this without difficulty. Then the patient is blindfolded (no eyesight), and the test is repeated. Again, patients with good eye-ear coordination can succeed while those with a mismatch rarely point in the right direction. I have watched Dr. Zelinsky administer this test to disbelieving colleagues, who were astounded by its accuracy and results.

Based on the Z-Bell results, the optometrist can prescribe eyeglasses. Prescriptions developed through the Z-Bell Test have been called nothing short of "amazing," even "miraculous" by patients, because they have brought relief for a range of symptoms, including headaches, dizziness, difficulty reading, learning and behavioral problems, attention deficit disorders, and anxiety.

These lenses can often remediate the discrepancy between the two sensory systems, even if the patient has 20/20 eyesight! Furthermore, if the patient wears previously prescribed lenses under the blindfold, the test can determine whether the prescription is helping or hindering the ability to integrate the two senses.

The Z-Bell Test has become internationally recognized by the scientific community. Dr. Zelinsky now offers seminars to certify optometrists and screeners. Visit MindEye.com.

Change Behavior by Changing Vision

Following the examination, the eye doctor will determine if vision therapy is appropriate and the timing right. Some patients with major health issues need a referral to deal with biomedical issues first. Those for whom reflexes are problematic must work on them. Occasionally, an optometrist will prescribe lenses and a home program for the patient while other therapies are pursued and schedule a progress evaluation in three to six months.

If the optometrist believes that vision therapy is warranted, they will pinpoint problem areas and develop a vision therapy program that includes activities to help the patient learn to use vision more efficiently. This program may include reflex integration.

Vision Therapy

Vision therapy (VT), also known as *vision training*, can be an astounding missing piece! It is available around the globe. The ultimate goal is to teach the eyes, body, and brain to work together. It is never too early or too late for VT.

For anyone with a strabismus, vision therapy can be life changing. VT recodes how the brain interprets the new information in a new visual-spatial framework. As the patient's mind recognizes the difference between the "old" way of processing and the "new" way, neural pathways develop that result in higher-level visual performance.

Because it builds new neural pathways, VT is far superior to patching or surgery, which often do not improve—and may in fact impair—visual processing. That was the experience of Sue Barry, PhD, a neuroscience professor at Mount Holyoke College, whose life changed forever when she learned to use her eyes together and saw in three dimensions for the first time at 50 years of age!

Barry, nicknamed "Stereo Sue" by the late neurologist Oliver Sacks, MD (1933–2015), was born with severe strabismus and underwent several unsuccessful surgeries as a child to straighten her crossed eyes and give her depth perception. She believed, as do most of her neuroscientist colleagues, that it was impossible for her eyes to learn to work as a team with the brain to see in three dimensions at her age.[41]

Enter behavioral optometrist Theresa Ruggiero, OD, FCOVD, who told her that with determined effort for a long time, Sue might be able to become binocular through vision therapy. Sue liked challenges (she is married to an astronaut!), so she decided, *Why not try?*

Her amazing book, *Fixing My Gaze: A Scientist's Journey into Seeing in Three Dimensions*[42] is the story of how she learned to see the shapes of flowers in a vase, individual snowflakes falling instead of a white sheet, and the steering wheel of her car jump out from the dashboard.

Vision Therapy Activities

VT generally includes motor, motor-visual, visual-motor, and visual activities, with or without the use of lenses and prisms (see below). It typically takes place once or twice a week in a doctor's office, supplemented with 15–30 minutes per day of home therapy to reinforce skills. While improvement is often seen in a month or so, therapy frequently continues for an average of three to nine months, and sometimes for as long as a couple of years, to stabilize and solidify learned skills, depending on the severity of the visual dysfunction.

For the lowest-functioning patients, a skilled optometrist might begin VT with simple, gentle visual arousal activities, such as blowing bubbles, playing flashlight tag, beanbag games, or Lite Brite, with or without lenses. For those with significant cognitive, language, and social-emotional challenges, these simple VT activities can sometimes result in turning on attention.

Early treatment might also include whole-body movement procedures like walking and marching, eventually integrating movement with listening and seeing by adding rhythm and balance challenges, such as walking on a rail to the beat of a metronome. Enhancing these visual skills often results in improved consistency and frequency of eye contact and social interaction outside of the therapy room.

As a person learns to use the two eyes more efficiently, therapy moves to more visually directed motor tasks such as throwing a beanbag at a target or putting pegs in holes on a rotating board. These activities are often done both with and without lenses and prisms to break up old habits and establish new ones.

For higher-functioning patients, VT can enhance organizational skills, visual thinking (see below) and language, and decrease behavioral and emotional issues, such as fears, and self-stimulatory and disruptive behaviors. Incremental gains in these areas could be sufficient to allow a young patient to move outside of a self-contained special education class and work independently for at least a part of the day in a mainstream setting.

For adult patients, at any level, vision therapy could move functioning up the ladder. Those requiring 24/7 monitoring might be able to function in assisted living. Those needing help with daily skills might be able to live independently. Higher-level education may become a realistic option. Holding a job, finding a life partner, having children may all be within reach. At the highest level, managing nervous system reactions to feeling unsafe, anxious, panicky, and unable to function may gradually diminish without direct therapy focused on those emotions and behaviors.

At the heart of vision therapy are activities individually designed to teach a person's eyes to move, align, fixate, and focus as a team. The ultimate purpose of VT is to increase visual and spatial awareness and to stimulate and develop using both eyes together. As the brain learns to coordinate new messages from the eyes, improved perception and cognition emerge. As in any treatment, VT providers need to be cognizant of a patient's sensory issues and possible overload.

Optometrists usually schedule periodic progress evaluations about every twelve sessions, or at least every six months, to evaluate visual skills, to review improvements in related areas, and to determine whether further therapy is needed. I recommend having two optometrists on the team: one to do therapy, and one to measure progress, because two opinions are better than one. Some patients improve quickly, others only gradually.

Lenses and Prisms

Counter to what many believe, lens prescribing is an art, not a science. While a machine might indicate one Rx, that prescription might make a patient dizzy or nauseous. Lenses have particular purposes, to do a specific job such as reading or driving a truck in the rain at night. Unfortunately, for most people, one lens cannot solve both problems. Different prescriptions are thus necessary, depending upon the task.

Ophthalmologists use compensatory lenses to improve eyesight for specific purposes, such as those just described. Many people reading this chapter are wearing compensatory lenses. Optometrists, however, often use lenses not only to correct *eyesight* but also therapeutically to correct *vision*.

Some prescribe *learning lenses* or *training lenses* that offer an individual a "new point of view." These lenses may be used temporarily to give the brain a preview of what to expect, and eventually the mind can perceive differently without the lens. Motor experiences are necessary for consolidation and permanent changes in perception.

All lenses displace light. Single lenses address focal vision and help us see *What is it?* Ambient prism lenses operate on the *Where is it?* function. They deflect the light rays differently, influencing how the brain interprets where the body is in space. Prisms are available in various magnitudes from weak to strong, measured in diopters. Optometrists can alter both the magnitude and the direction of the prism.

Prisms stimulate spatial awareness, redirect visual focus, increase visual attention, and facilitate visual change, all by working on the ambient or peripheral visual system. They can be powerful temporary tools because they create an unconscious change in posture or attention, thus changing a patient's perception.

Yoked prisms. When the prisms' bases face in the same direction in both eyes, they are called "yoked." These prisms can cause the environment to appear to be shifted up, down, left, or right. Objects may thus appear closer, farther away, or sloped.

Yoked prisms can have a very dramatic impact because they address a patient's ability to organize space and create a coherent body schema—the *Where is it?* and *Where am I?* functions. Most importantly, yoked prisms can make those who have tuned out their visual surroundings begin to notice their environment, thus leading to generally increased attention and awareness.

These lenses are used therapeutically; after rehabilitation occurs, they are no longer needed.[43] As the eyes reorient, the motor system adjusts, sending information to the brain, which comprehends that the individual must adapt his or her body to this new position in space. This readjustment causes a reorganization of the motor and sensory data in the cortex. In time, and with vision therapy, this reorganization becomes permanent.

According to prism guru Melvin Kaplan, success in making the shift in motor orientation occurs not simply because the eye muscles change in alignment, but also because the brain matches the visual information, while at the same time connecting vision, the motor components of vision, the vestibular process, the kinesthetic process, and proprioception. The prisms thus serve to establish balance among all the senses.[44]

The use of prism lenses for behavioral and learning problems goes back only to the 1970s when optometric pioneers began using these tools at the Gesell Institute. In the 1990s, Melvin Kaplan and psychologist Stephen M. Edelson, PhD, of the Autism Research Institute found that prism lenses improved behavior, attention, posture, coordination, and spatial orientation.[45 46]

Kaplan used prism lenses with great success on many populations, including those with anxiety and other psychiatric disorders. One of his most dramatic cases, and one that permanently altered the course of his work more than 30 years ago, was Rickie, the daughter of a prominent psychiatrist. At age 13 she developed symptoms of schizophrenia, requiring hospitalization. After 10 years of failed attempts at medications, electroshock treatments, and psychotherapy, she visited Kaplan's office.

Kaplan found that Rickie had severely impaired visual processing skills, including tunnel vision, lack of binocularity, and lack of depth perception. Her history showed visual dysfunction at age three, which continued with significant learning disabilities. Eventually, Rickie's highly stressed-out visual system collapsed and relapsed into a psychiatrically diagnosed state.

After a year of intensive vision therapy, combined with nutritional support and counseling, Rickie blossomed. Today, she is a trained practical nurse, married, and has children.[47]

Rickie's father, Frederic Flach, MD, and Kaplan conducted joint research that revealed that two-thirds of psychiatric patients suffer from some form of visual dysfunction. Nearly 85% of patients with severe, chronic "mental" illness exhibited marked impairments in spatial organization. Those with the most severe visual problems exhibited the highest levels of social withdrawal, academic failure, and employment difficulties.[48]

This research, now over 40 years old, has been ignored by the psychiatric community. My hope is that like the ketogenic diet, some metabolic psychiatrists will discover it soon!

Colored lenses. The medical use of colored light to treat conditions such as jaundiced newborns and seasonal affective disorder (SAD) is well-known. Optometrists use color to treat visual dysfunctions of all types, including strabismus, amblyopia, focusing and convergence problems, learning disorders, and effects of stress and trauma. The name of this method is *syntonics*, also known as *optometric phototherapy*.

Syntonics uses many frequencies of light to enhance visual information processing and overall visual performance. Each frequency is a different color. The colors can be delivered to a patient's eyes by having him stare at a colored filter placed over a light bulb or in a special light box. Some patients prefer to wear temporary cardboard glasses frames with the colored filter in them for 5–10 minutes a day.

Reported results are increased binocularity and focusing, and improved use of peripheral vision. Visit CSOVision.org.

Eye doctors and other professionals sometimes tint or color lenses to help patients deal with glare or visual sensitivities. More than 30 years ago, Helen L. Irlen, a California psychologist interested in visual sensitivities, named the underlying problem *scotopic sensitivity syndrome*; today, the condition is simply called *Irlen syndrome*. Irlen found that a subgroup of individuals who were sensitive to light, especially light from fluorescent bulbs, showed a marked improvement in their reading ability when they covered what they were reading with colored acetate sheets.

Irlen has trained hundreds of screeners and diagnosticians around the world to use her method, which filters out the wavelengths of light that create visual stress, allowing the brain to make the normal adjustments for various lighting conditions, glare, and brightness.

While colored overlays may indeed make it easier for some to read, these overlays may just be Ritalin for the eyes, in that they are treating symptoms but not getting to the underlying problem. Research by optometrists at the Pennsylvania College of Optometry found that 90% of 39 subjects ages 10 to 49 had significant underlying vision problems.[49] A review of 13 research papers on this subject by eminent optometrists at the Learning Disabilities Unit of the State University of New York College of Optometry corroborated that those with Irlen syndrome had various refractive, binocular, and accommodative disorders.[50]

Bifocal lenses. Optometrists frequently prescribe bifocal lenses, which can have an enormous immediate impact on functioning. Bifocals allow the eyes and the brain to readjust and respond differently as they move together into a new visual field (about every eight inches). Some people, like me, wear bifocals for the convenience of having two different prescriptions in one frame, one for near work and one for distance. The purpose of the prescriptions can either be correcting an eyesight problem or relieving the stress of the eyes and brain working together.

The Vision-Vestibular Connection

The previous chapter described the relationship between vestibular function and vision. Many with impaired vestibular systems depend upon vision for balance. With eyes closed, the balance system works even less efficiently.[51] Not surprisingly, research shows that children who received therapy for both vestibular and visual function made the best progress. While vestibular intervention alone improved balance, sensory integration, and socialization, adding visual therapy and including the use of lenses and prisms also improved binocular function.[52]

Melvin Kaplan was particularly interested in the effects of yoked prisms on the visual-vestibular feedback loop. He compared this loop to the telephone, in that incoming and outgoing "calls" move from one sensory system to another. Visual input affects vestibular function, and vestibular input affects visual function. Using prisms to stabilize eye movements also resulted in decreased motion sickness and anxiety.[53]

Many developmental optometrists collaborate with occupational therapists (OTs) today to improve the vestibular and motor coordination of their patients. Optometry stresses the role of vision as primary and movement skills as foundational. Occupational therapy focuses on movement and balance, viewing vision as one of the most important sensory systems.

In 1990 and again in 1999, Lynn Hellerstein, OD, FCOVD, and Beth Fishman-McCaffrey, OTR, wrote about the synergistic relationship that can develop between behavioral ODs and OTs.[54] Workshops on OT/OD collaboration have spawned numerous OT/OD pairs whose collaborative work may be familiar to the reader.

Vision's Role in Language and Social Skills

Pragmatic language and social skill development are the most complex outcomes of tactile-motor-visual integration. Speech-language pathologists and mental health professionals sometimes collaborate with optometrists. At the optometrists' suggestions, many even add breathing exercises to their therapy programs because they recognize that natural and spontaneous breathing brings increased amounts of oxygen to the cells and the brain, leading to improved movement, language, vision, cognition, learning, and behavior.

Many years ago, I wrote an article entitled "Say Hi to Patty." In it I elaborate upon how efficient vision allows a child to focus on someone who greets them; give the obligatory handshake, hug, or kiss; listen to what the person is saying, even if seeing double; give those words meaning; and respond—all in a matter of seconds!

Underlying visual dysfunction is a frequently missed cause of ADHD. If one examines the DSM behavioral criteria necessary to diagnose this condition, a majority of them are visual: trouble focusing, planning, organizing, etcetera. A developmental vision exam should be required before any child or adult receives that diagnosis or is put on medication. Same is true of anyone with anxiety. If your eyes don't work together efficiently, you are going to be anxious!

Most professionals working with ODs find that lenses, prisms, and vision therapy complement and enhance their own interventions by treating underlying dysfunctions of the visual system that are load factors in an individual's Total Load. They thus see positive changes in muscle tone, posture, movement, and cognitive function, as well as visual-motor, visual-spatial, visual-perceptual, language, and social-emotional skills that cannot be attributed to their therapy alone.

Vision in the Schools

An educational model defines *vision* only as *acuity* and offers vision services only for students with "low vision" or eyesight problems. School-based occupational therapists (OTs), special educators, and school psychologists all note visual perception issues and sometimes even write IEP goals to enhance visual perception as it relates to academic achievement. However, if the eyes are not working together or if there are other underlying functional vision issues, school-based professionals cannot evaluate or remediate these problems, but rather only look at and work on the end-product skills they impede. They need to recognize the importance of making an optometric referral.

School Vision Screenings

In 1999, the National PTA passed a resolution stating that the screening tests that most schools offer are merely measures of eyesight at distance and are insufficient to identify students with underlying vision problems. This leads to mistaken conclusions that academic deficiencies are a result of dyslexia, dyscalculia, dysgraphia, or poor motivation, rather than caused by serious vision issues such as convergence insufficiency.

How then can schools know if they are doing a good job of picking up students with underlying vision problems? By testing students' acuity at both distance *and* near point and by looking at essential visual skills such as focus, accommodation, and convergence in addition to acuity at 20 feet.

School nurses or teachers can use a screening machine that costs about $1,000, called a Keystone Telebinocular, to do thorough vision screenings. This instrument is portable and can move from school to school. It is well worth the price to ensure that all students are working with efficient visual skills. See KeystoneView.com.

Kindergarten readiness tests can also include items that screen vision. One of my favorite copying tests is "cat's whiskers," a horizontal line intersected by an X. Developmentally, the ability to reproduce this design occurs at about age five and one-half, the age at which most children enter kindergarten. A child should copy the design in three strokes, preferably from left to right and top to bottom. Most important is that the rays of the design are not fragmented and that the three lines intersect at a single point. Any variation, such as six rays, the letter "V", or right-to-left drawing shows lack of visual readiness for reading and writing, and the need for intervention. Being able to copy this correctly is a sign that a student is probably using both eyes together and crossing the midline of his or her body.

Environmental Changes that Enhance Vision

Many modifications of the classroom environment can enhance vision. Consider the following:

- *Ensure quality lighting*—Natural light is best; full spectrum bulbs help. Fluorescent lighting is the least desirable.[55]
- *Decrease visual "busyness" of room and written work*—Avoid mobiles, venetian blinds, wallpaper, and clutter; they are all distractors. Make sure photocopied materials are fully legible.

- *Assure good ergonomics*—Invest in a good computer chair that supports the back. Feet should be planted on the floor with knees and elbows at a 90° angle, and neutral wrist (< 15°). The head should not be tilted either backwards or forward, but rather balanced on neck. The upper arms are relaxed and close to the body, not abducted to the side or flexed forward. For young students, cushions and beanbag chairs are more forgiving than hard, ill-fitting plastic chairs.
- *Allow breaks*—Moving around gives the eyes a rest.

Classroom Visual Activities

Some of the motor, motor-visual, visual-motor, and visualization activities done in vision therapy with special lenses and prisms can easily be incorporated into the curriculum, especially at the preschool and elementary levels. Game-like activities with beanbags, balls, balloons, and other items help vision develop, solidify emerging visual skills, and consolidate vision for everyone. A number of books directed at teachers and occupational therapists offer suggestions: *Begin Where They Are*,[56] *Classroom Visual Activities: A Manual to Enhance the Development of Visual Skills*,[57] *Developing Your Child for Success*,[58] and others are all available from OEPF.org.

Visual Thinking

Visual thinking, the ability to generate and use imagery, is the culmination of efficient vision development. The skill of running movies in the mind's eye is important for reading comprehension, written language, understanding mathematics concepts, following directions, and more.

Temple Grandin is masterful at "seeing in pictures." However, many do not have her gifts and are attracted to video games, television, and computers to fill the empty space in their minds' eye where imagination should be. Far better than electronic images are those that come out of the sensory experiences of moving the body through space.

The late Harry Wachs, OD (1924–2016), an optometrist who devoted his long life to combining knowledge of vision with Piagetian theory, is one of the pioneers of visual thinking. Over 50 years ago, he collaborated with Swiss developmental psychologist Hans Furth to produce *Thinking Goes to School*,[59] a 1975 classic still in print. It is a truly remarkable book of over 200 activities that pair vision with every other sense for those of all ages. Chapter headings include "Movement Thinking," "Hand Thinking," "Logical Thinking," and "Tongue and Lip Thinking." Anyone doing intervention with those on the spectrum will find this a valuable addition to their therapy libraries.

At over 80 years old, Wachs produced another extraordinary manual of visual/ spatial/cognitive exercises with psychologist Serena Wieder, PhD, cofounder with the late Stanley Greenspan, MD (1941–2010), of the DIR method, described later in the book. With the cumbersome but all-inclusive title of *Visual/Spatial Portals to Thinking/Feeling and Movement: Advancing Competencies and Emotional Development in Children with Learning and Autism Spectrum Disorders*, it is another gem. Categorically similar to *Thinking Goes to School*, it includes comprehensive photos and drawings explaining Wachs's model.

Wachs used classic vision therapy tools such as a Marsden ball (a small ball hanging on a string, with letters and numbers on it) as well as more familiar items such as parquetry blocks. He starts with simple patterns and eventually presents constructions from different points of view because taking another's perspective is an abstract concept that starts visually and concretely and then develops into abstract thought.

Another professional with many years' experience enhancing and improving visual thinking is Colorado optometrist Lynn Hellerstein, OD, FCOVD. Hellerstein has packaged her ideas under the brand "See It, Say It, Do It." She offers books and workbooks. Her website is LynnHellerstein.com.

For insight into the relationship between visual thinking and academics, go to Chapter 20. Note the importance of this foundational skill, especially for reading comprehension and putting one's thoughts on paper. Reading and writing are the culmination of strong visualization skills.

Take-Home Points

Focusing on vision is the next step in removing important load factors. Addressing and correcting the visual dysfunction that underlies so many disorders is the missing link for many, especially those with autism, ADHD, OCD, and anxiety.

Lenses and prisms are very powerful tools for increasing visual awareness and redirecting and extending the range of vision. Prescribing and using them therapeutically is an art that can be life changing, as it was for neuroscientist Sue Barry.

Vision therapy must be part of an overall treatment plan that includes biomedical intervention and other sensory and motor therapies. Everyone is unique, and the therapies for an individual must be appropriate for their developmental age and history.

Parents, educators, therapists, physicians, counselors, and others not familiar with vision therapy may wish to consult additional references for the basics of vision. Some good resources are OEPF.org, COVD.org, and VisionTherapy.org.

Oh, yes, to answer those questions at the beginning of the chapter: Steve Jobs had vision. "I see" is in your mind's eye, and people who are myopic in their glasses prescriptions are often also myopic in their thinking.

CHAPTER 18

IMPROVING
COMMUNICATION SKILLS

"One of my deepest wishes is that I could hear my son say, 'Mommy, I love you.'" I have heard the mothers of children with autism say those words a gazillion times! If only they could hear their child's voice; if only they could know what their child was thinking.

And it goes both ways: it's not just the parents who are suffering. At age nine, Elizabeth Bonker explained her locked-in world by tediously typing the following poem:

> Me
> I sometimes fear
> That people cannot understand
> That I hear
> And I know
> That they don't believe I go
> To every extreme
> To try to express
> My need to talk
> If only they could walk
> In my shoes
> They would share my news:
> I AM HERE
> And trying to speak every day
> In some kind of way.[1]

Finally! Three-quarters of the way through this book, we get to what many parents value most: verbal communication. A close second on their wish list is that their child have friends, with the option to get married and have children of their own someday if they wish. This chapter focuses on the former, while social-emotional skills and relationship-building are in Chapter 19.

Unreliable, unusual, and absent communication skills are the hallmarks of autism spectrum disorders. That is why this chapter pertains mostly to those on the spectrum. Review the symptoms of autism presented in Chapter 1. They include

- Delayed, regressed, or nonexistent language development,
- Inappropriate eye contact or facial expression, and/or
- Poor conversational abilities if language is present.

The chapter begins with an overview of possible root causes for these symptoms, and how behavior *always* represents meaningful ways of trying to compensate for deficits. It then moves into the difference between speech and language.

Next is a short history of augmentative and alternative communication methods, computer programs, and apps for a range of tablets and phones, such as the iPad and Windows-based Android devices. These are constantly improving with the use of artificial intelligence and are helping non- and unreliable-speaking individuals make their needs known.

Then comes the biggest event in autism intervention that I have witnessed since we discovered the microbiome: the letterboard revolution! And if that isn't earth-shattering enough by itself, I discuss the possibility that in addition to many nonspeakers being cognitively intact, some may also have telepathic ability!

The chapter ends with a fun topic: building language and communication through nonhuman sources. Dolphins, dogs, and robots are some of the playful partners that help elicit those words parents so desperately want to hear. Let's start with the basic ingredients necessary for this complicated subject.

Speech Versus Language

Speech and language are closely related but different aspects of communication. Just as we make distinctions between eyesight and vision and hearing and listening, it is important to clarify the differences between speech and language.

Speech is the **physical** ability to produce sounds. Components include

- *Articulation*—moving the tongue, lips, teeth, and palate to produce sound;
- *Voice*—adding the vocal cords and breath give the sound pitch, volume, and quality; and
- *Fluency*—rhythm, flow, and smoothness of speech are called *fluency*.

Initially, babies imitate the sounds they hear in the language of caregivers and eventually put these all together to make meaningful words. This is another instance of individual instruments combining to play a beautiful symphony.

Some individuals with autism use "scripted" speech, with understandable words and sentences, which appears to lack meaning and appropriate context. *Echolalia* and repetition of sentences, songs, and adult commands—such as "no stimming"—are common, but real communication is much more than human "noise."

Language involves understanding and using gestures, signs, words, and sentences to convey meaning. The main components of language are

- *Phonology*—The sound system of a language, including rules about how sounds are combined and used;
- *Morphology*—The structure of words and how they are formed with units of meaning like *un-* or *-ing*;
- *Syntax*—The rules that govern how words are arranged to form meaningful sentences, like subject, verb, object;
- *Semantics*—The relationships between and meaning of words in the sentences: "Let's eat, Grandma," versus "Let's eat Grandma!"; and
- *Pragmatics*—The subtle social give-and-take rules of language, including turn-taking in conversation, what to say, when, and to whom, and interpreting nonverbal cues such as a raised eyebrow.
- *Prosody*—an essential aspect of both human speech and language that includes variations in pitch, loudness, and duration of speech sounds conveying information about structure, meaning, and emotion of an utterance. Are the sounds loving and supportive, or are they angry?

Numerous other components are necessary to communicate effectively, both verbally and nonverbally. They include vocabulary, grammar, and the ability to express thoughts, ask questions, or follow instructions. Irregularities in any of these can interfere with learning and relationships at home, school, and in the workplace, because they lead to misunderstandings.

Language is conceptual and involves the brain. Language pathologists discriminate between *receptive language*—or invisible, internalized comprehension—and *expressive language*, which is externalized. An individual can have speech without good language or intact receptive language without being able to speak. Assuming that because someone does not talk they cannot understand what is said is an enormous mistake that even "experts" make frequently.

Any component of the Total Load can affect any aspect of speech and language development. Unfortunately, specialists frequently send youngsters for speech-language therapy without considering *why* speech is delayed, regressed, absent, misarticulated, echolalic, or scripted. It's equivalent to assuming that behaviors such as stimming need to be "extinguished," when we should be looking at root causes.

Prior to discussing therapies that act directly upon speech and language, let's review how humans develop all the skills necessary to talk, engage, relate, and converse.

Foundations for Language

By now, you should have gotten the point: the emergence of speech and language depends upon *everything* discussed in the previous chapters. First and foremost is taking the nervous system out of fight-or-flight, a state in which development and healing cannot take place. Lifestyle adjustments further remove impediments. Biomedical treatments serve to reduce oxidative stress and subtract immune and inflammatory stressors from the body by focusing on the microbiome. As digestion, detoxification, and other bodily functions become more efficient and normalize, the body has more energy for sensory processing and reflex integration. Visual, auditory, and occupational therapies can then be most effective and produce secondary benefits in communication.

Before proceeding further reconsider whether:

- ***Sensory and oral-motor development are normal.*** First and foremost, is hearing intact? Infants are given screenings soon after birth. Hearing spoken words is essential to developing speech and language. Think Helen Keller. Low tone throughout the body may be causing oral-motor issues, making it difficult to form words. Vestibular problems can also interfere with the emergence of speech and language, because balance is processed in the inner ear. Chronic ear infections and their treatments can affect the emergence and development of both speech and language.

- ***Structural, reflex, and/or dental abnormalities are present.*** If a child was born by C-section or the birth was otherwise traumatic, these areas should be addressed. (See Chapters 10, 11, and 15.) Are tongue, lip, or cheek ties interfering? Has dys- or apraxia been added to the heap of diagnoses?

- *The individual or family has experienced significant trauma recently or there are stressors that go back several generations.* Everything from deaths of loved ones or pets to deep psychological wounds from previous generations are common in families with autoimmune and other disorders. In rare cases not speaking is selective mutism. Return to Chapter 6 to review healing traumas.

- *There are other factors in an individual's Total Load sucking up significant energy, leaving little available for talking.* Remember: the body's top priority is staying well. If the physical body is battling chemicals, heavy metals, gut bugs, viruses, bacteria, mold, or any other unwelcome toxins, it is difficult to focus outward. If you have not already done so, consider running laboratory testing to evaluate these areas. In the long run, testing will save considerable time and money.

- *The body is reacting to a food, beverage, or something in the air.* This possibility is an extension of the previous one. Allergies to food or environmental pests, such as bee stings and pollen, indicate an imbalanced immune system. If behavior changes markedly with the seasons of the year, consider this possibility, go back and improve immune functioning with additional dietary modification and supplementation.

- *Vision difficulties such as lack of binocularity have not been resolved.* A developmental vision exam is an essential prerequisite to addressing difficulties with language. Recall that surgery for a strabismus does not assure consistent binocularity. Quirky behavior and/or anything but smooth, fluent communication skills may indicate that the eyes are not working together properly. Refer back to Chapter 17 for more on this subject.

Speech-Language Therapy

A referral to a speech-language pathologist (SLP) is often the first recommendation a pediatrician makes when a child has delayed or unusual communication. SLPs are a part of every early intervention team. While well-meaning, most have little training in any of the material in the previous chapters. Occasionally, you will find one who works collaboratively with an occupational therapist, but they usually do their own thing.

If you work with one, suggest they take some continuing education courses in adjunct subjects like nutrition and oral-motor skills. Encourage them to collaborate with an OT. It can be a magical experience for a child to do language therapy while on a swing or therapy ball that stimulates the vestibular system.

Augmentative and Alternative Communication (AAC) Devices

For many years, therapists and caregivers have developed various ways to assist those who don't speak. These all fall under the huge umbrella of *alternative and augmentative communication (AAC)*, the term for any strategy other than speech used to express thoughts, needs, wants, and ideas. In the past, those who were nonverbal, non-interactive, or otherwise locked into their bodies could communicate only with those who could understand the meaning of their nonverbal behaviors and gestures. Friends were few and opportunities to socialize negligible.

Then someone got the idea to use laminated pictures and photos so that our nonspeaking friends could point to what they wanted. Even with the option to personalize visuals, it was a trial-and-error interaction that, though sometimes successful, often resulted in frustration.

Technology changed all that about 30 years ago as engineers developed devices and programs that could allow those who were not able to speak to communicate at home and school as well as outside their small circle of family and friends. This *assistive technology* offered not only a way to speak, but also some new exciting methods to increase the quality and frequency of interactions.

At first, some parents worried that if their kids used technology to communicate that they would never talk themselves. However, the opposite is true. Veteran SLP Susan Berkowitz busts this myth. She says that in her experience, 89% demonstrated gains in speech with AAC intervention.[2]

Early low-tech AAC devices, with ports for switches, head pointers, joysticks, and other access methods, allowed anyone to use common word-processing programs. They could even surf the Internet, e-mail, and link to the telephone, CD and DVD players, television, and light switches.

The change was profound because the new technology of tablets and phones gave nonspeakers a voice. By using a pointer attached to a body part, an adapted computer mouse, or even eye tracking or scanning to select target pictures or symbols, many could communicate for the first time in their lives. Sometimes these tools were used on a permanent basis; for others, they were only temporary. Originally, these products were miraculous; today they are an anachronism.

Tobii Dynavox

Dynavox, the granddaddy of assistive technology devices, is one of the few early devices that has kept up with the times. The company is now in its sixth generation of computer-based communication. Founded in 1983 and still based in my hometown, Pittsburgh, Pennsylvania, Dynavox's passion is the belief that technology's greatest benefits are realized when it empowers individuals with disabilities.

In the beginning, the Dynavox was a clunky box, mostly used by those in wheelchairs. Eventually, it evolved into a sleek, elegant, and lightweight programmable computer with excellent voice quality and ease of use, offering many choices of devices, all of which can be customized.

The first step in obtaining a Tobii Dynavox device is getting a speech-language pathologist (SLP) to assess the individual's assistive technology needs. If an AAC is deemed appropriate, a local Tobii Dynavox sales consultant speaks with the SLP and meets with the patient to determine the best match for a device.

Next comes funding the device, which can cost from $2,500–$20,000. Sometimes Medicaid or insurance will cover the cost. If either of these options does not work out, in many cases school systems can include the device in the Individualized Education Plan (IEP), so the cost does not fall on the family.

A Tobii Dynavox device can give a voice to many children and adults who need help communicating with friends, family, teachers and coworkers. TobiiDynavox.com.

The iPad, Android, and Windows Tablets

Software developers quickly recognized that computers were not just for word processing and video games. Hundreds of programs emerged in the 1990s aimed at teaching everything from reading, writing, and arithmetic to skiing and tennis and, of course, remedial language and social skills.

In June 2007 the first iPhone hit the market, and in January 2010 Apple introduced the iPad, smaller than a computer and bigger than a phone. In September 2012, another invention, the iPod Touch, arrived. Now extinct, it combined the ease of augmentative and alternative communication with the magic of technology. Soon people abandoned their non-portable PCs, Macs, and augmentative devices in favor of tablets and phones for communicating.

With those inventions, the way people communicate has forever changed. Our preferred method of communication is no longer a telephone—rather we e-mail or text. With tablets and phones, those with disabilities now have convenient, portable pieces of equipment that match what everyone else is using.

Many families now have one or more members of the iPad family; most are used for playing games, watching and making movies, listening to music, and taking pictures. However, the iPad can also be a powerful communication and assistive technology tool. Loading it with some special apps turn it into an AAC device.

If a student is already facile using the iPad as an entertainment device, learning to use it as a communication tool may be tricky and will require some changes in thinking. Jonathan Campbell, a Minneapolis expert in assistive technology suggests using a different iPad to communicate. It's like having two pairs of shoes: one for hiking and one for dancing. The color of the cover or the size of the iPad can allow individuals to identify its function easily.

Simply owning an iPad, like owning a smartphone or a computer, is not enough. The ultimate goal is to use it consistently, regularly, and eventually, automatically, as a communication tool, instead of grunting, pointing, and dragging an adult toward the desired object or having them guess. Training and consistent practice are the two most important factors in success with the use of an iPad as an AAC. Having the tablet in a case with a strap lessens the chance of loss.

Here is an early success story: Ido Kedar spent his early years trapped in silence, misunderstood by adults who presumed that he lacked understanding because he could not reply verbally. First using a communication board, and now an iPad, Ido was finally able to explain what was going on in his mind. He is advocating for thousands of others with autism who are locked in, as he was, and thus unable to show their true capacities. Read his story in *Ido in Autismland: Climbing Out of Autism's Silent Prison.*[3]

Obviously, in order to be successful at using an iPad app, not only must the individual learn its many functions and how to access them, but so must all those in his world: his family members, teachers, friends, and anyone else he has regular contact with. That too takes time and training.

Two-Dimensional Versus Three-Dimensional Learning

In Chapter 3, too much screen time was listed as a load factor in neurodevelopment. Touching, moving, and all the sensory experiences involved in good old-fashioned play were touted as far superior to playing games two-dimensionally on a screen.

In a paper presented at the 2017 Canadian Pediatric Academic Societies (PAS) Meeting, scientists reported on speech/language development of 894 children ages 6 months to 2 years:

> The more handheld screen time a child's parent reported, the more likely the child was to have delays in expressive speech. For each 30-minute increase in handheld screen time, researchers found a 49% increased risk of expressive speech delay. There was no apparent link between handheld device screen time and other communications delays, such as social interactions, body language or gestures.[4]

Bottom line: using a device as a voice for an individual who is learning to speak can be a life-changing tool. But so can improved nutrition, diet, and health, as well as shoring up foundational skills in the sensory, visual, motor, and reflex areas. Before signing on to using apps, consider the child's age and review earlier chapters on remediating possible underlying causes. Young children need to touch and experience in three dimensions, not only on two-dimensional screens.

Computer Programs

Text to Speech (TTS)

Soon technology improved to the point that nonspeakers could type words into a device and software could convert them into a voice. Text-to-speech (TTS), the process of converting written text into spoken audio was born. This may bring to mind Stephen Hawking, the brilliant physicist with the neurological disease amyotrophic lateral sclerosis (ALS), which robbed him of purposeful motor skills, including the ability to talk for the last 30 years of his life. Yet, no one questioned his intelligence, because he continued to share his brilliance through this methodology.

Early TTS consisted of synthetic robotic and unnatural sounds, but as technology evolved it became more human sounding, like "Alexa," full of emotion and intonation. Those who could use a keyboard got a new lease on life, but many of those with apraxia found even this motor-planning task tedious—or even impossible. They still had no voice.

Again, computer engineers were able to solve this problem. A new word entered our vocabulary: *app*, short for *application*, a software program designed to perform a specific function or task. Just as we replaced typewriters with word processing programs, we could now search for information without going to the library, pay our bills without writing a check, or talk to friends without a telephone! Communication apps were born!

Communication Apps

The disability community jumped all over this development. Millions of apps are available, and many of them are free. These apps used strictly as communication devices are today's augmentative and alternative communication tools, or AAC apps. Many of the early apps have been updated and are still in use. New and increasingly sophisticated ones have entered the market.

Proloquo2Go™. In 2009, Proloquo2Go brought AAC to phones and tablets. Today, it is the world's most popular AAC app. Proloquo2Go includes 100 natural-sounding text-to-speech voices, up-to-date symbols, powerful automatic constructions, a variety of languages, a huge vocabulary, and ease of use for the iPhone and iPad.

In 2022 they released Proloquo, based on over a decade of research insights and the latest technology. With 16,500 words, over four times more than any other AAC system, it is consistent for home, school, and other environments. Both apps are available and are designed to support communication and language growth. They vary in complexity and design and take motor-planning issues and levels of development into account. Research study subjects demonstrated language growth using both apps.

This extremely well-conceived and designed app is free to download and requires a monthly subscription. Watch the videos at AssistiveWare.com before deciding.

Scene & Heard Pro uses standard scenes or your own photos to which you add touchpoints. Each "hotspot" can be programmed to play audio and video media when it's tapped on.

Scene & Heard Pro caters to a range of needs, starting with simple single-message-based communication to sophisticated interlinking scenes with a range of interactive features. It is a good, inexpensive option to consider for younger children or when introducing AAC. Check out Therapybox.co.uk/scene-and-heard-pro.

SceneSpeak is another scene-based AAC app for iPhone and iPad. It includes some scenes with hotspots, and you can add your own photos and customize them with hotspots that speak. This app can also be used to build interactive story books. See GoodKarmaApplications.com.

TouchChat 2.0, released in May 2018, is a full-featured communication solution for the iPhone. It includes basic words, phrases, and messages spoken with a built-in voice synthesizer or a recorded message. Additional vocabulary words, language packs, and voices can be purchased. Text from other applications can be copied to TouchChat so it can be spoken aloud. Text generated in TouchChat can be copied to other apps. See TouchChatApp.com

CoughDrop, unlike other apps, can be used on any Android, Apple, or Windows device. Visit CoughDrop.com.

Speak for Yourself was developed by speech pathologists for young children. The icons are always in the same location on the screen. This is a great app for beginning talkers with motor-planning issues. See SpeakForYourself.org.

Determining Which Device and Apps Are Best

The best way to choose an appropriate device and the most useful apps for an individual is to work with an assistive technology professional, often a speech-language pathologist, who knows many options. First comes an assessment of needs, for the user and for the various settings in which the communication device will be used.

This evaluation will determine whether the student is capable of using an AAC and whether it will facilitate learning. Can the individual isolate a finger to point, the eye to focus, or learn how to move to the next page? How many buttons per page can they handle? Can they combine two or more icons into sentences? Is focus consistent? Is the home, school, and/or workplace conducive to the use of a communication device? Who are the key players in making sure the device is used properly?

Only after completing an assistive technology assessment is an appropriate device chosen. An AAC device can be written into an IEP. If an iPad or other tablet is the choice, then appropriate apps can be evaluated by an expert. Needs change as the student grows in skills. Annual evaluations of choices should be written into the program.

Letterboards and Spelling to Communicate

Facilitated Communication

Many AAC methods, by necessity, include a communication partner to assist the dyspraxic student in pointing to letters and pictures by supporting an arm, elbow, or hand. This is called *facilitated communication* or *supported typing*.

Enter Soma Mukhopadhyay, a highly educated chemist from India, who was convinced that her nonspeaking son Tito, diagnosed at age three with severe autism, had strong intellectual aptitude, despite his limited motor abilities. During the 1990s, she tutored him tirelessly, prompting him to point to symbols in books, while gently supporting his body parts as they went through the motions. *Supporting*, not *guiding*, is the important concept.

Soma developed what she called the *Rapid Prompting Method (RPM)* by working with Tito. As is inherent in the name, RPM is a fast-paced way to engage attention and elicit responses in a number of ways: verbally, auditorily, visually, or tactilely. The goal is to bypass motor and sensory issues that prevent effective communication. As Tito became more and more focused, the prompts were faded, and he could respond independently.

Now an adult, Tito wrote *How Can I Talk If My Lips Don't Move?*[5] when he was only a preteen. In 1999, experts from the National Autistic Society of Great Britain determined that Soma was correct: Tito was intellectually gifted. Soma and Tito later came to the United States to share their expertise with American children and have settled in Austin, Texas. To learn more, read *Developing Expressive Language in Verbal Students with Autism Using Rapid Prompting Method.* [6]

Tito was one of the first in a wave of people with autism who have since found their voices. Another was Jeremy Sicile-Kira. When Jeremy's mother, Chantal, one of the most respected voices in the autism community, heard about Soma and Tito in 2002, she was impressed by her method and her high expectations for these severely autistic, nonspeaking students. Having tried different well-known strategies with limited success for her son born in 1989, she decided to try RPM. It worked!

Jeremy wrote, "Autism is an important influence in my life. The hardest part is not being able to talk. God must have been out of voices when he made me. Having autism has hindered my ability to speak, but not my ability to communicate." [7] Jeremy's extraordinary attempts to communicate landed him a prize from MTV as the second-most inspirational moment out of nearly 300 entries in 2007. Jeremy is a successful visionary artist; he and his mother now lecture together, giving hope and a voice to those who have neither. Learn more at JeremyVision.com.

Despite the fact that facilitated communication has proven to be a miracle for thousands of nonspeaking individuals, it has fallen under severe scrutiny. Some experts in the field of speech and language believe that the communication partner manipulates the nonspeaker's hand and arm to elicit the desired response. Yet, the fact that some students successfully move beyond having a communication partner makes it clear that cannot be universally true.

Spelling to Communicate

On July 1, 2019, Judy Chinitz's life and that of her 27-year-old nonspeaking son, Alex, drastically changed. At the insistence of a friend, Judy drove Alex from their home in New York to Virginia. The purpose of this trip was for Alex to be evaluated by Elizabeth Vosseller, a seasoned speech-language pathologist, who had learned RPM while training with Soma around 2015–2016.

Judy was one of the first "warrior moms." She holds a master's degree in special education, had been to hundreds of professionals and spent hundreds of thousands of dollars on interventions, some of which helped Alex. She had truly tried everything. She felt she had nothing to lose by adding one more crazy-sounding idea to the list.

Since his diagnosis of regressive autism at the age of two, Alex had attended special education classes, where he was taught only basic skills. His IQ measured around 40, and experts concurred that he was both profoundly autistic and cognitively impaired. Alex's inability to communicate led everyone to believe he had limited ability to learn.

Vosseller is the Founder of Growing Kids Therapy Center (GKTC), GrowingKidsTherapy.com, and the Executive Director of The International Association for Spelling as Communication (I-ASC), pronounced "I Ask," a 501(c)(3) nonprofit. I-ASC's mission is to advance communication access for nonspeaking individuals globally, through training, education, advocacy, and research. See I-ASC.org.

Vosseller offers both in-person and virtual services assisting families near and far. Her many years of experience include using assistive technology to help her clients communicate. Although many were successful, others struggled. In 2017 she began using letterboards, which look like stencils. She named what she was doing *Spelling to Communicate* (S2C). Word of Vosseller's innovative adaptation of RPM spread like wildfire in the autism community.

After working with Alex for only a short time, Elizabeth shocked Judy when she declared, "Your son is not cognitively impaired; he has perfect language in his brain." Even though Judy had suspected it, no one had ever told her that her son had potential.

Once Alex began to communicate, Judy discovered he had been sneaking into his brother's room and reading his college texts. Not only could he read, think, and reason, he was able to comprehend French, do higher-level mathematics, and discourse about history and geography! Judy could barely contain herself. She shared her joy on Facebook and with me on *The Autism Detectives* podcast. Listen to it on Spotify.

A few months later in Portland, Oregon, J.B. Handley, a warrior dad and father of then 17-year-old Jamie, had a similar experience. J.B. had been active in the autism community for many years and knew families all over the world who had nonspeaking children.

Handley cofounded Generation Rescue, one of the first parent organizations that believed autism could be healed, in 2005. A text from his friend Honey Rinicella in Philadelphia, the mother of identical twin sons with autism, related a remarkable story. Her nonspeaking son Vince, slightly older than Jamie, had over the course of only a few months emerged from his silent prison through the use of a letterboard.

Honey begged J.B. to take Jamie to be evaluated by Dawnmarie Gaivin (a.k.a. "DM"), an RN with two nonspeaking autistic sons who lived near San Diego. DM and Dana Johnson, an occupational therapist in Tampa, Florida, had worked with Elizabeth Vosseller and helped define the standards and scope of S2C before they left to form Spellers Method, a slightly different application. (See below.)

J.B., like Judy, was blown away when, after a short time, Jamie began communicating. "We didn't even know that Jamie could spell," confessed Handley.

Apraxia

Both Judy and J.B. were told their sons were unable to speak because they had apraxia. You read about apraxia and its cousin dyspraxia in Chapter 16. They are impairments that affect the coordination of both gross- and fine-motor skills and are common in autism.

Because the brain and body are "out of sync," an individual with even mild dyspraxia has difficulty initiating action, motor planning, and maintaining momentum to complete an action. Often, the person they are trying to communicate with does not allow sufficient time for their minds and bodies to coordinate, so lack of understanding is assumed. "My body is not my friend," claims an anonymous speller. "I cannot count on it to do what my brain tells it to do in a reliable fashion."

Judy and J.B. had both heard of apraxia and dyspraxia over the years from their speech-language pathologists and occupational therapists. They understood how their sons' motor impairments had affected their ability to initiate play, ride a bike, write, and participate in sports. What they didn't "get" is that *speech* is motor, while *language* is cognitive. No one had ever explained that speaking is the most complex motor skill we humans perform.

Unlike Stephen Hawking mentioned above, Alex, Jamie, and other nonspeakers like them, had not spoken, or spoken only minimally, before regressing, so they had not proven their cognition was intact. No one had "assumed competence." Today, *assume competence* is a basic tenet of all methodologies that use spelling as communication.

Bypassing Apraxia and Dyspraxia

Individuals who appear cognitively impaired because of the motor-planning challenges inherent in apraxia and dyspraxia have demonstrated by using S2C that they have a great deal to say when taught to use a letterboard. Skilled and rigorously trained helpers called "communication and regulation partners" (CRPs) teach purposeful motor skills using a hierarchy of verbal and gestural prompts. With repeated practice, new neural pathways myelinate and motor skills improve.

Those with autism, Down syndrome, cerebral palsy, and other diagnoses can communicate by pointing to letters, one by one, and spell as an alternative means of communication. Those with learning disabilities can also show significantly better expressive language when communicating with a letterboard than with pencil and paper, or even on a computer. In the words of a 20-year-old young man I know, "I think that I'm challenged by communication in a way that is isolating, because I seem like I can talk, but mostly what I say only scratches the surface." The goal is to gradually achieve synchrony between the brain and body.

RPM Versus S2C

While similar, RPM and S2C are actually inversely related. What these two methods have in common is that they use "low tech" assistive technology tools with different approaches.

RPM is an *educational* technique first; communication is a byproduct, not the primary objective. S2C, is a *communication* method first, and a byproduct is that students can then access academics. The distinction matters because it explains why some things, like touching the student, are acceptable in RPM but are not even part of the practice standards in S2C, and vice versa.

Spellers Method

Visual Skills and Dyspraxia

The relationship between motor and vision skills is discussed in Chapter 17. Because inefficient eye movements can interfere with all aspects of learning and behavior, it should be no surprise that not using the eyes together well make using a letterboard difficult. Recall that many individuals with autism have a strabismus, or eye turn, that prevents them from having efficient binocular vision and depth perception. Understand that vision therapy can often remediate this condition and teach the eyes and brain to work together consistently and effectively.

DM and Dana recognized not only the necessity of efficient, purposeful motor skills for pointing, but also the role of underlying visual and visual-motor skills in scanning the letters. They were both familiar with developmental optometry, and coincidentally, DM's office was in the same town as neurodevelopmental optometrist Dr. Susan Daniel, who also has an adult nonspeaking son with autism. They began collaborating.

When Dr. Daniel started using a letterboard with her son, she already fully understood that apraxia can affect many aspects of vision, including eye movements and fixation. She observed firsthand how difficult it was for her son and some other students to scan all 26 letters on one board. So, she suggested that the alphabet be divided onto three separate boards, making letter selection easier, at least in the first stages of learning to spell to communicate. At her suggestion, DM and Dana made this and other modifications to the more traditional letterboards used in RPM and S2C.

Modified letterboards that could accommodate the visual-motor differences unique to each student eventually became a key to success for struggling students by eliciting increasingly complex communication. Dr. Daniel's evaluation of foundational visual skills in DM's and Dana's clients are an important addition to success in spelling with letterboards, because the supplementary vision therapy programs she prescribes can improve and remediate deficient visual skills.

Typical students can hold focus on a target for at least ten seconds; many nonspeakers can fixate for only a second or two. With customized vision therapy designed to meet each individual's needs, DM, Dana, and Susan have watched with awe as nonspeakers with various degrees of dyspraxia learned to control their strabismus and express themselves effectively.

Students progress from pointing to letters on a single letterboard, containing only seven or eight letters each, to pointing to letters on three letterboards with all 26 letters of the English alphabet. Eventually, as motor and visual skills progress, many students learn to type on a traditional keyboard. Responses move from concrete, one-word answers to specific questions to spontaneous writing, abstract thinking, and even poetry. Check out a collection of poetry, songs, and stories produced by nonspeakers, *Spellbound: The Voices of the Silent,*[8] edited by Judy Chinitz.

DM and Dana are committed to incorporating occupational and vision therapy from developmental optometrists to improve intentional motor and visual skills. They also offer letterboard modifications to help certain students access spelling for communication and focus on parent coaching to round out a supportive program. In 2022, they rebranded the expanded program as Spellers Method (SM).

The Spellers Method's goals have grown beyond being just a communication method—it is an open doorway to a life full of agency and autonomy. DM and Dana believe firmly that it is a parent-driven model that is guided by a provider, not the other way around. They also believe that if they aren't working themselves out of a job with every speller, then they aren't doing their job correctly.

S2C Versus SM

So, for clarity, S2C and SM are related. Spellers Method includes occupational and vision therapy as inherent components. The founders coach parents and caregivers to become fluent Communication Partners and confident "motor coaches" for their children with apraxia. They teach the intentional motor skills that each student needs, including initiation and inhibition. The end result is that many previously "locked in" nonspeakers are not only able to communicate autonomously, but to also hold jobs, attend college, and eventually, live independently. DM and Dana have a Facebook group called Spellers Community that parents and spellers using any method can join for information and support.

Anyone interested in letterboards and using spelling to communicate needs to do their research and ask lots of questions to find a provider they are comfortable with. To quote DM, "You are in charge of this journey with your child. . . . If you believe in neuroplasticity, then you already know a little bit about how the brain can rewire."

Spellers the Movie

"What if everything you thought you knew about autism was wrong?" That is the opening line of the full-length documentary, *Spellers the Movie*, produced by Jenny McCarthy-Wahlberg and her husband Donnie and showcasing Elizabeth, DM, and Dana. The film answers that question through the stories of eight nonspeakers—Aydan, Evan, Sid, Maddie, Jamie, Vince, Cade, and Elizabeth—who all found their voices through the miraculous process of S2C and/or RPM. As Jamie explains, "we think, feel, and learn just like everyone else."

Spellers won Best Documentary at the Phoenix film festival in April 2023 and was funded by The Spellers Freedom Foundation, a 501(c)(3), not-for-profit organization whose board of directors is composed of nonspeaking individuals, passionate professionals, and dedicated parents advocating for the presumption of competence and the empowerment of nonspeakers everywhere. They believe that "communication is a fundamental right, essential for claiming meaningful education, employment, and relationships."[9] This film is available for viewing online and at selected locations around the globe. See SpellersFreedomFoundation.org.

Empowered Voices

Vince Rinicella, son of Honey, mentioned above, and one of the featured spellers in the movie, has become such a proficient speller that he now offers consulting services to families interested in learning more about spelling for their loved ones. His group of educators and therapists trained in using letterboards supports individuals of all ages with motor and sensory differences. Visit EmpoweredVoices.net.

Spellers Information

Spellers.com

855–326-3727

admin@spellers.com

This website offers Spellers Method training programs, Spellers Connect Conference, Spellers University, Spellers Freedom Foundation, and the Spellers Center network.

Spellers Center San Diego

405 Oceanside Blvd. Suite B

Oceanside, CA 92056

760–536-3896

sandiego@spellers.com

Speller Center Tampa

1816 Health Care Drive

Trinity, FL 34655

727–275-1155

Tampa@spellers.com

Dawnmarie and Dana have written *The Spellers Guidebook*[10] for anyone desiring to pursue Spellers Method for a non- or unreliable speaker. This terrific book explains the theory and steps behind this approach. With time and practice, the authors say, nonspeakers can develop accuracy and achieve open, fluent communication. Each individual is unique; the time from introduction to fluency depends on many factors. Dawnmarie explains the process in a 2023 interview she did for my podcast, *The Autism Detective*. Listen in on Spotify.

Mouth-to-Hand Learning Center (M2H)

Since Alex began spelling, Judy has devoted herself to bringing the gift of spelling using letterboards to other nonspeakers by opening the Mouth-to-Hand Learning Center in Mt. Kisco, NY. See MouthToHandLearning.com

There she offers individual sessions and community classes in an integrated environment that provides learning, friendship, and fun. Students can study science, history, literature, creative writing, music appreciation, and other subjects they were denied in their special education placements. In addition to the enrichment of the course contents, these previously dumbed-down young adults are forming deep friendships with both speaking and nonspeaking peers. Their happiness and excitement in learning is palpable. Judy updates us in a second *Autism Detective* episode, also on Spotify.

Spelling to Communicate in School

While many of the successful spellers are young adults who have aged out of school, some are still in high school or even in middle or elementary school. Eager parents, attempting to write the use of letterboards into their children's IEPs, face resistance. Why?

One reason is that school systems do not yet have trained communication partners. Another is that in 2017 the American Speech-Language Hearing Association (ASHA) issued a position statement declaring Facilitated Communication (including RPM, S2C, and SM) a discredited technique that should not be used.[11] This outdated opinion clearly needs to be reevaluated and changed.

All those in the Spellers community urge parents to work closely with their school district's special education departments to educate them on the benefits of spelling communication as an efficacious tool to maximize their children's potential. Like many new ideas and techniques, this one may need research and time to convince skeptics of its value. As DM says, "Once you have seen your own child spell out their thoughts and ideas, you cannot unsee it! The truth becomes self-evident.

And what has happened to Alex and Jamie since beginning to spell? Both are works in progress. In 2021, Alex passed the New York State high school equivalency exam with distinction and was accepted into the playwriting and screenwriting program at the State University of New York at Purchase. In 2025, at age 31, he accomplished a longtime dream of celebrating his bar mitzvah, the Jewish ceremony marking the transition to manhood which typically occurs at age 13.

Both Alex and Jamie began taking college-level courses fewer than two years after beginning to spell. In 2021 the Handleys, father and son, published a book about their experience, *Underestimated: An Autism Miracle*.[12] Jamie stars in episodes of the five-part made-for-TV docuseries *Underestimated: The Heroic Rise of Nonspeaking Spellers*. The importance of ocular-motor skills in spelling is featured in Episode 4. Access it at Underestimated.tv.

Spelling to Communicate and the Spellers Method have absolutely leveled the playing field for nonspeakers. Today, more than 500 providers of several methods have been trained. This is a game-changer for sure!

SpellRBoard App

The obvious next step is an app that puts the letterboard onto a phone or tablet. That is exactly what SpellRBoard is. Developed for the iPhone and iPad by Dylan Paul Kazak, the sibling of a nonspeaker, it is a new AAC app paired with text-to-speech technology.

SpellRboard uses digital customizable alphanumeric and QWERTY letterboards in a variety of colors that include all the letters of the alphabet, numbers, emojis, and commonly used symbols. You can choose from a variety of voices and adjust speech speed, pitch, and volume to ensure the voice output meets individual preferences and needs. It's available in English, Spanish, and French at the Apple App store for under $50. See SpellRBoard.com.

Telepathy Tapes

Learning that many nonspeakers are cognitively intact was a huge eye-opener to families all over the world. Yet, the scenario gets even "curiouser and curiouser," to quote Alice after she fell down the rabbit hole and found herself in Wonderland.

Some nonspeakers may have the ability to read the minds of those they are close to! Many parents have reported this "out-there" skill to disbelieving friends and family. Yet, neuroscientist and savant expert Diane Hennacy Powell, MD, has witnessed it and run rigorous tests that have verified the phenomenon. A parent can *think* a letter, number, or word, and a nonspeaking speller will type it! How do they do it? Watch the 10-part series online to see and decide for yourself: TheTelepathyTapes.com.

Animals as Communication Partners

Dolphin-Assisted Therapy (DAT)

Speech-language pathologist Janet Flowers, CCC-SLP, EdD, works with children in dolphin-assisted therapy (DAT) sessions at Florida's Gulf World Marine Park in Panama City, Florida. Flowers's aim is to achieve greater results than traditional therapy in decreasing response time and increasing expressive communication skills. She looks at participants' attention span, communication, speech, language, gross- and fine-motor skills, and academics.

Children in her studies exhibit

- increases from 25% to 250% in time on-task,
- significant increases in mean length of utterance as well as in the complexity of expressive skills, and
- greater long-term retention than six months of conventional therapy.

DAT is also more cost-effective than conventional therapy.[13]

Flowers believes that having a positive emotional experience during therapy increases motivation and confidence and enhances a child's long-term memory. Her experience is that a dolphin's unconditional acceptance of a child with a disability in a safe environment provides exceptional motivation to overcome obstacles and increase confidence.

The program motivates and jump-starts participants, complements and reinforces therapy, and provides a stimulating reward. The achievements made with DAT also assist other professionals who work with the child. See GulfWorldMarinePark.com.

Dolphins are not the only partners that provide comfort and communication opportunities. Read about dogs, horses, and robots in the next chapter and how they encourage both speech/language and social-emotional development.

Take-Home Points

Communication's not just the words that you speak,
It's the signs, the looks, the ways that you reach;
Non-speakers got the power
It's time to shine
Open up your ears, let the message unwind.
A board, a screen, a glance or a tap
It's time for them to bridge the gap
—"No More Silence," a rap song by Julia Kazak, mother of a nonspeaker

Now is an amazing time because technology has opened doors to new ways of communicating. Those who do not speak, and who previously had no voice, can now use augmentative and alternative devices, apps, text-to-speech, and letter-boards to interact, communicate, make friends, and learn.

Those robotic, synthesized voices on computers, tablets, and iPads are gone. Text-to-speech voices are true-to-life and express the needs, feelings, and desires of those who lack their own. Many doors previously shut tight have opened to allow those who have no voice of their own to integrate into society. Socialization online has allowed them to make friends and stay in touch.

We can do almost anything on a screen we can do in real life. (Well, maybe not *anything*!) Learning on a screen has some advantages, but is not a substitute for in-person interacting and three-dimensional learning that engages all the senses and the body. When using technology, remember it is NOT a substitute for communicating, moving, touching, and experiencing life. That said, technology is here to stay! It is the future. Embrace it with heart and soul . . . and with caution!

It is *never* too late! Look at this testimonial from a 32-year-old woman after only a year at Mouth-to-Hand:

> I had no way to express myself. Judy Chinitz taught me to type to communicate . . . and I am a new person. To learn to do this is a long and difficult process. It takes months of work and dedicated practice by parents and caregivers to let us acquire this skill. It is not easy and it is not quick. But the people who make up the staff here are the most wonderful teachers on earth. It is the most incredible place. If you want your ing child to talk to you, this is where you come. You will never regret it.

The bottom line: as the narrator in advises, we must "Wake up to the *Spellers* reality that we may have underestimated the abilities of more than an estimated 30 million nonspeaking people worldwide." Nonspeaking's not the same as nonthinking.

From this time forth: Assume Competence.

CHAPTER 19

SOCIAL-EMOTIONAL SKILLS

Emotional stressors—including small worries, big fears, and major traumas—are a part of just about everyone's Total Load. Some worries and fears, like going to a new place for the first time, are predictable and mostly manageable. Others, especially if preceded by an unpleasant experience such as a minor car accident, can weigh heavily—every time the car enters a major highway at 70 miles per hour, for example. This type of reaction can usually be managed cognitively with some learned techniques.

But some traumas are deeper and longer lasting. Post-traumatic stress, like soldiers experience because of their indelible memories of wartime experiences, can be debilitating. Sensory triggers that evoke the original trauma can occur unpredictably at almost any time. The COVID lockdowns were traumatic for a large segment of the population, especially teenagers who missed events they'd expected to participate in. So, in addition to an epidemic of childhood developmental disorders, we now have an additional epidemic of teenage and adult loneliness, depression, anxiety, and suicide risk.

How the brain processes experiences determines how much energy is sucked up for bodily functions, including digestion and respiration, and how much is left for new experiences like learning. And how safe someone feels determines their responses. If we are not comfortable in our own skin, reaching outside ourselves to interact with others can be extremely problematic.

For years, speech-language pathologists, mental health professionals, educators, and others have designed programs to "improve" the social skills of those whose behavior does not fit the norms of society. Despite their widespread use, a 2007 review of the literature showed that these costly programs were only minimally effective.[1]

That finding may come as a surprise to those who work so hard to teach appropriate social behaviors to their clients. Even though the study referenced above is almost 20 years old, it is consistent with other previous studies on the subject. As society changes and guidelines for "appropriate" behavior become fuzzy, programs change, and new ones arrive. Although most treat only symptoms, these approaches remain popular.

Why do so many social skills training programs fail? Probably because the load factors in previous chapters are not addressed. If we train social skills from the outside-in, rather than let them emerge from the inside-out, we teach kids to distrust their own sensory processing.

But *some* programs must work, or there would not be a chapter on the subject in this book. What *are* the components of those that make a difference? A group of researchers at the University of Utah evaluated several programs and came up with some guidelines. Effective programs

- break down complex social behaviors into concrete steps and rules that can be practiced in a variety of settings;
- concretize abstract concepts through a variety of visual, tangible, "hands-on" activities that make socialization fun; and
- provide a variety of learning opportunities that allow skills to integrate and generalize.

Therapies in this chapter can be added to or combined with other therapies in this book, especially those focusing on biological foundations for behavior. However, the best approach is often to make sure the brain is not in fight-or-flight and the body is well-nourished. At the risk of sounding like a broken record, once nervous and digestive foundations are in place, relationships often improve spontaneously, rendering additional therapy unnecessary.

Foundations for Social-Emotional Development

Just as we considered the biological foundations for speech and language to emerge, we must look at the necessary prerequisites for emotional stability and social interactions. Note that some of these are the same as for communication.

- *Is the nervous system in fight-or-flight?* Remember the vagus nerve and Porges' Polyvagal Theory introduced in Chapter 6? As the neurophysiological foundation to feeling safe and secure emotionally, it is probably the most important concept in understanding emotions and behavior.
- Polyvagal Theory posits that the structures regulating social behaviors are compromised in individuals with neurodevelopmental conditions, putting them in a constant state of high anxiety.[2] *Reread about how the Safe and Sound Program stimulates the middle ear with a variety of sound frequencies*[3]

and how structural interventions can lessen anxiety and improve social and emotional availability, all in Chapter 6.

- **Have biomedical interventions reduced oxidative stress?** What about immune and inflammatory stressors on the body? Does the child have constipation, rashes, or respiratory issues? Remember that the body's number one priority is staying well. Available energy is always prioritized towards nervous system, digestive, respiratory, and circulatory functioning.

- **Are oxytocin and vasopressin levels normal?** These hormones play key roles in emotional and social behaviors and bonding. Refer back to Chapter 12 for more on this subject.

- **Are there any lifestyle adjustments that need to be made?** Recall the social-emotional deficits seen in babies left to languish in orphanages in China and Romania. Infants need frequent, positive adult interactions for bonding.

- **Are all sensory systems working well?** Are some hyperactive? Sensory integration is vitally important for appropriate social interactions.

Sensory Integration

Three senses are primarily responsible for good social-emotional development: touch, balance, and vision. Touch is grounding, and the ability to tolerate it is essential for feeling secure and confident. According to out-of-sync guru Carol Kranowitz, a child with tactile defensiveness sends out signals that he is unfriendly and prefers to be left alone.[4]

An efficient balance system also allows children to feel grounded and know where their body is in space. When one is gravitationally insecure, feeling emotionally secure is difficult, if not impossible.

Children with efficient vision not only see another person, they are also able to find appropriate meaning in subtle facial expressions and gestures. They focus *on* their mother's face, not *in front of* her or over her shoulder, and can judge appropriate social distance for conversation. Dysfunctions such as double vision, lack of visual flexibility, and difficulty focusing can interfere significantly with social interaction and inhibit vision's emergence as the dominant sense.

As the senses integrate through endless three-dimensional experiences, appropriate social interactions emerge. Eventually, as these sensory processing systems mature and integrate, language and social interactions deepen.

Efficient sensory integration is imperative for social-emotional development. "Mindsight" expert Dr. Daniel J. Siegel states, "Our brains use sensory information to create representations of others' minds. Based on these sensory inputs, we can mirror not only the behavioral intentions of others, but also their emotional states."[5]

Mirror Neurons

In the mid-1990s Italian neuroscientists discovered the mechanism that allows the brain to perform this remarkable feat: mirror neurons, cells in the premotor cortex that fire both when performing an action and when observing another living creature perform that same action. These "monkey-see, monkey-do" parts of the brain appear to be vitally important in understanding social skills and empathy. While a 2005 study showed deficits in the mirror neuron system in those with autism,[6] a 2013 systematic review of 25 neuroscientific studies showed mixed acceptance of this theory.[7]

The Importance of Early Play

According to Rebecca Klaw, MS, MEd, an autism specialist in Pittsburgh, Pennsylvania, for typically developing children social development begins with "small and simple" play. The child at play is not idle or aimless, boring or bored, or wasting time. The child is, in fact, engaged in complex activities that develop skills and build, bit by bit, concepts as complicated as physics and as essential as empathy.

As parents, teachers, siblings, and friends provide input, simple play gradually expands over time, becoming rich and very complex. The elaboration of play is fueled by positive social interactions and serves as the basis for early learning.[8]

Children who have dysregulated sensory systems tend to play alone in unusual and repetitive ways. Children with autism spectrum disorder tend to get stuck in their play, and they need to learn, through interaction with patient and skilled adults, how to play.

Figure 19-1 Ways to Engage Children with Autism in Play
- Heighten interest
- Be persistent
- Include repetition
- Establish routines
- Add sensory stimulation
- Minimize language
- Have fun

Figure 19–1 from Rebecca Klaw suggests ways to play with a child who wants to be left alone. Insist that they play with you, using persistence, intelligence, flexibility, and humor. Children with ASD need to be guided in memorable ways to explore all aspects of their world, not just how to manipulate objects, but how to share, build, pretend, elaborate, invent, describe, and create. By joining them in what they are focused on and doing what they are doing, you are meeting them where they are. This is the crucial tool used by the Kaufmans in the Son-Rise Program described later in this chapter.

Behavior as an Adaptive Response

Remember occupational therapist A. Jean Ayres, introduced in Chapter 16? I love her statement that "every behavior is an adaptive response." What that means is that when someone smiles, it is generally accepted that they are happy. If they put their hands over their ears, blink repeatedly, or put something in their mouth, their behavior is a probably a response to a stress.

As we know, stressors can be a visible threat such as an animal; an obvious annoyance such as loud music; a not-so-obvious irritant such as perfume; or an invisible reaction such as an allergic response. Depending upon how intensely and in what environment someone reacts to a stressor determines whether we can tolerate the behavior or not. For instance, kicking one's chair may be acceptable during a meal at home, but not while in church.

If you accept this definition of behavior, the next step is to decide what to do about disturbing behavior: ignore it, try to understand its cause, or try to "extinguish" it. To me, the first two are acceptable options. Except for dangerous situations, like running into a busy street, extinguishing socially unacceptable behaviors is disrespectful of the body's attempts to self-regulate.

That is the premise on which the rest of this chapter is based. Without understanding the origin of behavior, we are not helping an individual grow and thrive. We are only treating symptoms and, worse, causing more trauma.

This chapter divides interventions for developing social-emotional skills and encouraging positive relationships into three sections: behavioral therapies, cognitive behavioral therapies, and relationships-based approaches.

Social Symptoms

Autism, ADHD, and OCD are probably the most common diagnoses with social-emotional symptoms. Criteria include several impairments in social interaction:

- Little eye contact and/or blank facial expression
- Failure to develop peer relationships
- Lack of make-believe or social imitative play
- Repetitive and stereotyped behavior, interests, and activities
- Abnormally intense preoccupation with one or more interests
- Mannerisms such as finger flapping, twisting, or whole-body movements
- Preoccupations with object parts
- Hyperactivity, fidgeting, and touching things

Many of those with later-in-life diagnoses, especially ASD or ADHD, recall child-hoods that included perceived trauma and resultant PTSD. Their sensory differences often exceeded their abilities to cope with life's experiences, resulting in frequent feelings of anxiety and even panic. They were often perceived as being "shy" or "seen-and-not-heard," or the opposite, throwing tantrums or having unexpected meltdowns.

The above traits and more were described eloquently by functional medicine-trained nutritional therapist and researcher Claire Sehinson at the Optimum Health Clinic in London. Diagnosed with autism and ADHD along with other conditions as an adult, she enumerates the long list of health issues, including chronic fatigue and Ehlers-Danlos, that are rampant in late-diagnosed individuals. Unfortunately, she sees these health issues as comorbid conditions, rather than possible causes of autism symptoms. It is akin to telling the parent of a child with chronic diarrhea that the digestive problems are because the child has autism, rather than vice versa.

Important, however, are Ms. Sehinson's description of two communication issues that affect the lives of adults diagnosed with Level 1 or 2 ASD.

Masking, also known as *camouflaging* or *impression management*, is the conscious or unconscious effort an individual with autism or ADHD makes to change to the way they behave to cover up symptoms of their diagnosed disorder. More common in females than males, masking includes memorizing scripts, and excessive "people-pleasing." Some common overcompensating behaviors are arriving extremely early for an appointment to avoid being late, staying silent during a conversation to avoid interrupting, or taking copious notes during a meeting to focus attention and avoid forgetting something important. All take their toll as they can be very stressful and emotionally tiring.[9]

The *double empathy* problem is the idea that, just as autistic people find it difficult to understand neurotypical people, neurotypical people find it difficult to empathize with autistic people. This could cause disagreements between a diagnosed patient and a well-meaning doctor as both parties struggle to understand each other's perspective. For instance, a nutritional consult might be rejected by either party, resulting in vitamin and mineral deficiencies or an eating disorder, because of a lack of appropriate testing, care, and support.[10]

Behavioral Therapies

The use of consequences to modify behavior, also known as operant conditioning, is a tried-and-true training method popularized by B. F. Skinner in the mid-20th century. Currently, many individuals, clinics, and even entire school systems are offering behavioral therapies. They focus on the reduction of maladaptive behaviors and on the development of adaptive, socially significant behaviors. Like many other interventions, behavioral therapies are most effective with children at a young age.

Applied Behavior Analysis

Applied behavior analysis, or ABA for short, is presently the most popular form of operant conditioning, although it is losing favor. It is the systematic application of what scientists have learned by training laboratory animals to human beings in real-life situations.[11] A comprehensive understanding of ABA requires many years of practice and study.[12]

The umbrella term ABA includes many interventions based on direct observations of behavior that make up a functional behavioral assessment or a functional analysis of behavior. All ABA programs utilize the manipulation of antecedents and consequences of behavior to

- reduce inappropriate, self-injurious, and stereotypical behaviors;
- modify conditions under which interfering behaviors occur;
- teach new skills; and
- transfer and generalize behavior from one situation to another.

Clinical psychologist Ivar Lovaas, PhD, is considered the father of ABA therapy for autism. In 1987, as a professor of psychology at the University of California at Los Angeles (UCLA), Lovaas demonstrated scientifically what no one believed possible: the behavior of children with autism could be modified through teaching. Today the term *Lovaas therapy*, or simply *Lovaas*, is often used synonymously with applied behavior analysis.

Discrete trial training (DTT) is a specific one-on-one intensive ABA teaching strategy that breaks down a complex set of behavioral skills into small, manageable tasks. Basic communication, play, motor, and daily living skills complement and build upon each other. Some consider DTT or Lovaas to be the only true ABA.

Studies documented the effectiveness of ABA for treating autism across a wide range of

- *behaviors*, including those that are self-injurious,[13] stereotypical,[14] communicative,[15] social, academic, leisurely, and functional;
- *populations*, including children and adults with a variety of behavioral, learning, and developmental disorders, as well as ADHD;[16]

- *therapists and educators*, including parents, teachers, and remedial specialists; and
- *settings*, including schools,[17] homes, institutions, residential placements, hospitals,[18] and businesses.[19]

ABA therapy became the therapy of choice in the 1990s. In 1999, the late Bernard Rimland, PhD, founder and director of the Autism Research Institute (ARI), wrote a passionate letter in his newsletter that he entitled "The ABA Controversy."[20] He believed it was a mistake to think of ABA as competitive with, rather than complementary to, many other interventions and urged therapists, schools, and others recommending ABA as "the *only* scientifically based therapy for autism" to consider biomedical treatments as well. By that point in time, much research on the benefits of vitamin B-6, magnesium, and DMG had been published.

Centers where patients with autism could receive biomedical interventions combined with ABA was one of Dr. Rimland's dying wishes. The Rimland Center for Integrative Medicine in Lynchburg, Virginia, was named in his honor.

ABA is certainly compatible with other therapies. Integrating it with biomedical interventions can be quite effective. Some ABA therapists and other professionals are finding ways of accommodating a child with sensory issues by incorporating sensory integration techniques into ABA therapy.[21] Return to Chapter 16 for some ideas about how to modify environmental stimuli, encourage proper seating posture, and provide opportunities for individualized sensory input.

Pivotal Response Treatment (PRT)®

A variation of ABA that emerged in the 1970s, *PRT*® was developed by Robert and Lynn Koegel, PhD. It addresses the acquisition of complex skills that go beyond simply teaching basic behaviors. The word *pivotal* refers to important areas of development such as motivation, initiating responses, self-management, and knowing which cues to respond to and which to ignore. The T can stand for *therapy*, *training*, or *teaching*, in addition to *treatment*.

By working with each child's natural motivations in natural learning opportunities, PRT® stresses functional communication over rote learning. It targets and modifies communication and language, behavior, and social skills by capitalizing upon an individual child's idiosyncratic interests.

PRT® rewards children by creating less-structured, playful opportunities to do more of what they already enjoy. For instance, in one study, therapists played a game of tag on a huge map. Because the child was a geography expert, he was motivated to move from place to place in the game. The increase in social interactions that happened during the game of tag continued after the game ended.[22]

The Drs. Koegel are currently conducting research at Stanford University on applying PRT to teens and adults. Their legacy is the Koegel Autism Center, a part of the Gevirtz Graduate School of Education at the University of California, Santa Barbara. Read *The PRT Pocket Guide: Pivotal Response Treatment for Autism Spectrum Disorders*.[23] See AutismPRTHelp.com.

TEACCH

The late Eric Schopler, PhD (1927–2006), a psychology professor at the University of North Carolina, Chapel Hill, developed *TEACCH*. The acronym stands for ***Treatment and Education of Autistic and Related Communication Handicapped Children***. A less-intensive form of ABA, the TEACCH philosophy embraces and accommodates, rather than remediates, the "culture of autism." The TEACCH program relies on small, laminated squares with picture symbols on them depicting everything from locations in the classroom to curriculum materials and self-help skills. These "prompts" guide students through their days.

The long-term goals of the TEACCH approach include both skill development and fulfillment of fundamental human needs such as dignity, engagement in productive and personally meaningful activities, and feelings of security, self-efficacy, and self-confidence.

Because those on the autism spectrum do well with these visual materials, the TEACCH conclusion is that vision is a strength in autism. This profound misunderstanding of what vision is frequently prevents students working in a TEACCH environment from receiving the essential visual interventions they require.

For an in-depth look at this method, read *The TEACCH Approach to Autism Spectrum Disorders*.[24] Visit TEACCH.com.

Verbal Behavior

Another popular behavioral approach is a methodology designed by Vincent Carbone, EdD, that places an emphasis on day-to-day activities and learning information in context. In Verbal Behavior, learning is functional; the methodology utilizes a systematic, highly structured teaching approach that moves away from the traditional ABA and DTT by combining behavioral principles with functionality and generalization to emphasize language development. Verbal Behavior uses children's own motivations for reinforcement and relies on a fast tempo to increase spontaneous language and encourage conversational skills. Like S2C, Verbal Behavior assumes competence. See AllPointsABA.com.

Cognitive Behavioral Therapies

A second class of therapies that applies operant conditioning techniques is *cognitive behavioral therapies*. These therapies differ from behavioral therapies in that they also take an individual's feelings into account. Many view the approaches included in this section as a more humane form of ABA. The use of words such as "affective" and "fun" to describe cognitive behavioral approaches hints that these approaches are also somewhat less mechanical than those using stricter operant conditioning.

Cognitive behavioral programs are now considered first-line therapies for teens and adults with anxiety and/or obsessive behavior that interferes with them fulfilling obligations at home, school, and work.[25] In my opinion, although they can be temporarily successful, they do not get to the traumas and dysregulated nervous system underlying the behavior.

The Social Skills Training Project

Clinical psychologist Jeb Baker, PhD, represents the next generation of cognitive behavioral training. Entertaining and irreverent, Baker has developed a unique, practical, and very popular approach to social-communication difficulties. Three principal goals guide Baker's method:

- To provide social skill instruction that generalizes into daily routines;
- To make socializing fun;
- To help "typical" peers and professionals become more understanding, accepting and engaging of those with social difficulties.

Baker's flexible five-part model includes the following:

1. *Assessment:* prioritizing three or four relevant skill goals to work on for a specified period.
2. *Motivation:* establishing a purpose to learn and use these skills.
3. *Initial skill acquisition:* determining strategies appropriate for developmental levels.
4. *Generalization:* coaching to use newfound skills in natural settings and to broaden interests and preferences.
5. *Peer sensitivity training:* interacting with typical peers to increase generalization, reduce isolation, increase opportunities for friendship, and decrease bullying. Visit SocialSkillsTrainingProject.com.

Cognitive Affective Training

Cognitive affective training (CAT) is a method developed by British autistic adult Tony Attwood, PhD, for inspiring and structuring conversation on thoughts, emotions and behavior. CAT teaches about nine feeling categories: joy, sorrow, fear, love, anger, pride, shame, surprise, and safety, with 10 subgroups under each, making 90 emotions available for discussion. It uses a set of carefully designed tools collectively referred to as the CAT-kit.

Special tools help clients develop understanding of different emotions. They learn how a person can be very happy and feel comfortable in one situation and then a second later become angry or sad in another.

Four different types of color-coded behavior are presented: red (outright aggressive), yellow (passive aggressive), grey (submissive), and green (assertive). These tools promote understanding and help develop the student's ability to self-regulate. See CAT-Kit.com.

Programs for Relationship Building

Relationship-based approaches are appropriate at any developmental level or age. It is never too early or too late to use them because they all respect an individual's interests and build on the belief that all behavior is meaningful.

Most importantly, they recognize the role of efficient sensory processing in social-emotional development. Interactions focus on building trusting relationships.

The programs for relationship building covered here are

- DIR/Floortime;
- Profectum and FCD™;
- The Affect-Based Language Curriculum;
- RDI®;
- The Son-Rise Program®;
- Social Stories™;
- Social Thinking® and The Zones of Regulation®; and
- SCERTS®.

Greenspan's DIR® and Floortime™

The **Developmental, Individual Difference, Relationship-based Approach (DIR)** is a comprehensive, intensive interdisciplinary program developed by the late child psychiatrist Stanley Greenspan, MD (1941–2010), and psychologist Serena

Wieder, PhD. Greenspan and Wieder are the coauthors of *The Child with Special Needs*[26] and *Engaging Autism*,[27] from which I have drawn much of the information in this section.

The components of DIR˚ include the following:

1. ***Developmental.*** The ***D*** component of the model describes the developmental building blocks. Understanding where the child is developmentally is critical to planning a treatment program. DIR takes into account six sequential stages of development that are mobilized simultaneously.

Therapists start with shared attention, move to the use of gestures to communicate, then purposeful two-way interaction and problem-solving, and finally emotional ideation and thinking. The final goal is answering "wh" questions.

2. ***Individual differences.*** The ***I*** in DIR stands for individual differences and describes the unique ways each child takes in, regulates, responds to, and comprehends sensations and the planning and sequencing of actions and ideas.
3. ***Relationship-based.*** The ***R*** component refers to the interactions and relationships with caregivers, educators, therapists, peers, and others who tailor their behavior to the child's individual differences and developmental capacities to enable progress. Understanding each child's unique set is crucial in planning emotion-based interactions, which are at the heart of intervention.

The first goal in the DIR approach is to help the child work around sensory processing difficulties to establish a meaningful relationship with parents. For a child who is in his own world and not relating to others, the emphasis is on enticing him into the world by giving him a greater degree of pleasure in relating. Having adult/child relationships with all caregivers allows a child to develop meaningful relationships with peers and siblings.

DIR builds healthy foundations for social, emotional, and intellectual capacities rather than focusing on skills and isolated behaviors. It incorporates and is fully compatible with many of the therapeutic programs described in previous chapters, including biomedical intervention, occupational therapy, sound-based therapies, and vision therapy.

***Floortime*™.** At the heart of the DIR approach is *Floortime*™, a specific technique to both follow the child's lead and at the same time challenge the child toward greater and greater mastery of social, emotional, and intellectual capacities.

By joining in at a child's developmental level, an adult uses exaggerated affect and action to woo a child into interacting.

Greenspan and Wieder created the concept of "opening and closing circles of communication," built upon a child's natural interests. Motor planning and sensory processing differences are gently overcome. Greenspan and Wieder found that the degree of autism does not necessarily determine prognosis with DIR. See Floortime.org.

The Home Program. Unlike many in-office therapy programs, most DIR sessions take place as part of intensive home programs of at least 20 minutes, eight or more times a day. DIR's goal is to spark a response to every utterance or gesture. When a child is ready, the program can be integrated into the classroom. Once a child begins to interact, peer play is introduced.

Structured activities at least three times a day address motor and sensory skills. Occupational therapists and vision experts, including the previously mentioned optometrist Harry Wachs, were longtime Greenspan collaborators.

Prior to his death, Greenspan began consulting with the Rebecca School in New York City. The program at Rebecca School still fully integrates the DIRFloortime® model. In 2011, Rebecca School clinical director, Gil Tippy, PsyD, collaborated with Greenspan and published *Respecting Autism*, describing the successes of students there.[28] The official organization of DIRFloortime is the Interdisciplinary Council on Development and Learning (ICDL). Visit ICDL.com.

Profectum and FCD™

In 2011 an outgrowth of DIRFloortime was founded by over 60 DIR-trained therapists and professionals. Called *Profectum*, from the Latin for *advancement* or *progress*, the new model, *Foundational Capacities for Development™ (FCD™)*, is the result of further recognition and elaboration of the significant role of vision, space, and movement in emotional and symbolic development. It also includes understanding of how anxiety and behavior play a role in the sequential and organizational capacities needed for functional competence. Affect and cognition are two sides of the coin.[29]

In the 1970s Wachs combined optometry with Piagetian psychology in his groundbreaking book, *Thinking Goes to School*.[30] Greenspan, Wieder, and Wachs came from the fields of psychiatry, psychology, and optometry, respectively. As they examined the relationship between the DIR model and Wachs's unique model of vision therapy, they concluded that their seemingly different approaches were all going down the same developmental paths to cognition.

Emotions are an inherent part of cognition; emotional interactions lead the way in virtually every stage of development. As typically developing children learn where they are in space, their understanding is accompanied by an emotional need for safety. For those who are "lost in space," like many on the autism spectrum, their poor visual-spatial thinking undermines their sense of safety, resulting in anxiety.

After Greenspan's death, Wieder and Wachs expanded on the intersection of their fields. In their book *Spatial Portals to Thinking, Feeling and Movement*, they describe the developmental processes that activate, organize, and integrate all experiences with some specific exercises that follow the DIR six stages of development. The final stage is hybridized from "emotional thinking" to "logical/abstract thinking." The six stages could take anywhere from a week to a year, depending upon the beginning developmental level of the individual.[31]

"The Five Cs" is used as a guideline to capture the components of experiences that advance development: comfort, competence, confidence, control, and communication. Each individual's uniqueness holds true in emotional development as well, as each one's experiences have a unique developmental progression. See Profectum.org.

Solving the Relationship Puzzle: Relationship Development Intervention© (RDI)

RDI shares many features with **DIR**. In 1995 psychologist Steven E. Gutstein, PhD, developed *Relationship Development Intervention* or RDI©, a program to help those with autism derive pleasure from social encounters. He called this *experience sharing*. It focuses on the developmental step of "you-me" thinking and emotional attunement that Gutstein believes children with autism missed sometime during the first year of life. Three major principles lay the foundation for RDI:

- *Social Referencing:* a highly specialized form of perception and information processing that allows a person to evaluate the state of a relationship.
- *Functions Precede Means:* a stage at which children develop a desire for deeper emotional experiences, leading them to pursue and spend many hours mastering new skills.
- *Co-Regulation:* the ability to understand how one partner's actions impact the actions of another in order to maintain the relationship.

The RDI Program Protocol

Nine essential elements must be in place for the intervention to be classified as an RDI Program. It begins with comprehensive medical evaluation and developmental assessments of language, cognitive, neurological, perceptual, and motor skills. Parents must *attend a series of workshops about how to incorporate* RDI into their daily lives.

Then a customized intervention plan with lengthy recommendations is drawn up. *Parents are encouraged to spend three to six hours per week interacting with the child using RDI methods.* Most participants in RDI Programs also receive other interventions such as dietary modification, occupational therapy, vision therapy, and/or speech and language intervention.

The RDI curriculum consists of hundreds of specific objectives and customized activities developed after a child undergoes the evaluations. Activities and objectives at each of six levels represent a dramatic developmental shift in the central focus of relationships. As a reference, Gutstein has labeled the six levels in the curriculum as Novice, Apprentice, Challenger, Voyager, Explorer, and Partner.

RDI provides a structured path for people on the autism spectrum to learn friendship, empathy, and a love of sharing their world with others. The program begins at an individual's level of capability and carefully, systematically teaches them the skills they need for competence and fulfillment in a complex world. Eventually, they learn not only to tolerate, but to enjoy change, transition, and going with the flow.

The Connections Center in Houston, Texas, the home of RDI Program©, is an international consultation and training center. Hundreds of professionals in child development are certified as RDI Program Consultants. Gutstein describes RDI in his book, *Autism/Asperger's: Solving the Relationship Puzzle.*[32] See RDIConnect.com.

The Son-Rise Program®

Like other parents and professionals in this book, Barry Neil Kaufman and Samahria Lyte Kaufman refused to listen in the 1970s when, at 18 months old, their son Raun was diagnosed as severely and incurably autistic. Experts advised the Kaufmans to institutionalize Raun because of his "hopeless, lifelong condition."

Instead, they designed an innovative, unique, home-based, child-centered program to reach their son. It transformed Raun from a mute, withdrawn child with a low measurable IQ into a highly verbal, socially interactive youngster with a near-genius IQ. Bearing no traces of his former condition, Raun graduated from an Ivy League college and today is married, lectures internationally, and works to help families like his. His book *Autism Breakthrough* presents the groundbreaking principles behind the program that helped him and thousands of other families.[33]

The Son-Rise Program is based on the idea that we should meet the child in their world rather than asking the child to enter ours. It encourages parents to follow the child's lead or actions to establish strong connection before offering opportunities to safely engage with the wider world. By enthusiastically mirroring repetitive and ritualistic behaviors and interacting with the child through play, loving caregivers can establish trust and gradually lead a child toward greater and greater interaction.

The unique feature of this program is the commitment to happiness. Parents are encouraged to explore their own belief system and to question judgments that limit them. When parents' happiness isn't dependent upon their child's "success," the child feels safe enough to progress at their own pace.

Responding to the demand to teach others their program, The Kaufman family established The Option Institute and the Autism Treatment Center of America™, where they have been offering The Son-Rise Program® since 1983. Families can either come and stay at the Option Institute for a week or more at a time or learn the program online. The skills they acquire allow them to accept their child and become their teacher.

Read Raun's remarkable story in *Son-Rise: The Miracle Continues* and watch the award-winning NBC-TV movie Son-Rise: *A Miracle of Love*. Visit AutismTreatmentCenter.org.

Social Stories™

Carol Gray, director of The Gray Center for Social Learning and Understanding in Michigan, became interested in the difficulty some individuals have assuming the perspective of another person. In 1991 she devised a technique called Social Stories™ to help them learn how to understand others' behaviors.

Social Stories teaches people how to "read" and understand social situations by answering *who*, *what*, *when*, *where*, and *why* questions for a variety of situations presented in the form of stories. Each story describes a scenario, skill, or concept in terms of relevant social cues, perspectives, and common responses in a specifically defined style and format.

The goal of a Social Story™ is to impart accurate social information in a patient and reassuring manner, not to change an individual's behavior. However, heightened understanding of social events and expectations often leads to more mature behavior.

Although Gray developed Social Stories for use with children with ASD, her approach has also been successful with others, including adolescents and adults, who have social issues. Social Stories are applicable for both readers and nonreaders.

Gray and her associates have written three books with collections of Social Stories. Therapists can refer to these for working on common potentially anxiety-producing situations such as going to a new place. In addition to teaching routines, stories can be individualized to teach how to do specific activities, ask for assistance, or even respond in a socially appropriate way to feelings such as anger and frustration.

My Social Stories Book includes topics such as grooming, dealing with unexpected noises such as barking dogs and ringing telephones, and going places like the movies and the grocery store.[34] The *New Social Story Book* was revised and expanded in 2010 to celebrate the tenth anniversary of Social Stories.[35] It covers topics such as chewing gum, giving hugs, using the telephone, sharing, knowing when to say "thank you" and "excuse me," caring for pets or oneself, cooking, helping around the house, picking flowers, going to church, getting a haircut, shopping, understanding weather, and special topics like new shoes, school issues (such as fire drills, recess, and homework), escalators, seat belts, holidays, vacations, and others. Gray also teaches the reader how to construct a Social Story.

Carol celebrated the 15th anniversary of Social Stories in 2015 with another book including more sophisticated topics for older individuals, such as giving and receiving gifts, making mistakes, respect for others and the earth, and understanding adults.[36]

The Gray Center offers workshops, support groups, DVDs, webinars, and other educational materials. Anyone can join Carol's Club for free and have an ongoing relationship with the Social Story movement. Carol Gray travels extensively to teach her method. See CarolGraySocialStories.com.

Social Thinking® and The Zones of Regulation

Leah Kuypers, MAEd, OTR/L, began her professional career working in public schools as an occupational therapist (OT) and Autism Resource Specialist, where she saw her students struggling with both sensory and emotional regulation. She came up with an idea: What if there was a program that combined her passions, sensory integration and social-emotional success?

Zones of Regulation was influenced by some tried-and-true OT and autism interventions. Leah incorporated Michelle Garcia Winner's *Social Thinking®* concepts to help students gain a deeper understanding of the impact their behavior has on their relationships and become more aware of how others perceive them when they were in regulated states versus less-regulated states.

Leah and Michelle began collaborating, and Michelle offered to publish Leah's materials. The Zones team has grown with the development of more materials, including apps for the computer and phone.

The Zones program allows students to move away from depending upon adult prompts to assume personal responsibility for self-regulation. It provides a system of classifying states of arousal, feelings, and emotions into four easily identifiable distinct color-coded Zones, similar to traffic signs with an added color of blue.

The Red Zone describes heightened states of alertness and intense emotions, like elation and rage. **The Yellow Zone** describes a state of elevated stress, frustration, or anxiety where one still has control. **The Green Zone** is the desirable one, a calm state of happiness, focus, or contentment. **The Blue Zone** describes a state of low alertness and/or energy such as when one feels sad, tired, sick, or bored.

Currently Leah provides trainings and consultations to parents, schools, and professionals and offers workshops on the Zones curriculum in the US and abroad. She also works with businesses to guide their development of social-emotional learning supports. Learn more from *The Zones of Regulation*: *A Curriculum Designed to Foster Self-Regulation and Emotional Control*[37], SocialThinking.com and TheZonesofRegulation.com.

The SCERTS® Model

SCERTS® is a comprehensive, multidisciplinary, integrative team approach that was developed out of 25 years of research and clinical/educational practice by speech-language pathologist Barry Prizant, PhD, CCC-SLP, and a multidisciplinary team of professionals trained in communication disorders, special education, occupational therapy, and psychology. SCERTS is an acronym for *Social Communication, Emotional Regulation, and Transactional Support.*

SCERTS promotes social communication and emotional competence by building meaning into daily experiences from the preverbal to conversational level. The strengths of this program are its recognition of individual differences and its focus on the family. It capitalizes on forming a solid foundation of mutual respect, meaning, logic, and predictability.

All activities and strategies are designed to enhance the development of emotional self- and mutual-regulatory capacities, and to modify attentional, arousal, and emotional states. SCERTS borrows from other sensory, motor, vision, and language models, always looking at an individual child's strengths and needs.

Prizant may thus recommend that one child jump on a trampoline for arousal, and that another listen to soothing music for calming.

SCERTS provides support to ensure that behaviors generalize across settings, thus fostering successful interpersonal interactions, relationships, and productive learning experiences at school, in the community, and elsewhere.[38] The SCERTS Model encourages professionals from different disciplines to collaborate with each other and is thus totally compatible with other treatment approaches in this book. Families measure progress by noting the number of functional activities an individual can do with a variety of partners.

For a complete overview of SCERTS, watch the three video or DVD set, *The SCERTS™ Model: A Comprehensive Educational Approach for Children with Autism Spectrum Disorders*,[39] or read *Autism Spectrum Disorders: A Transactional Developmental Perspective*.[40]

Prizant describes his approach in *Uniquely Human: A Different Way of Seeing Autism*, updated and expanded in 2022.[41] He has also developed workshops, seminars, and a podcast. See BarryPrizant.com.

Building Social Skills through Animals

Animals were introduced as sensory-integration therapy helpers in Chapter 16. Whoever the human and whatever the animal, research shows that both typically developing children[42] and children with autism[43] benefit from interaction with animals. In the study referenced above, children with autism engaged in 55% more social behaviors when they were with animals, compared to toys.

Service Dogs

Service dogs can provide a physical and emotional anchor, thus enhancing opportunities for those who lack independence to access a variety of environments safely. With their child tethered to a service dog, families feel free to engage in activities as simple as going out to eat.

In many cases, the service dog accompanies the child to school, where its calming presence can minimize and often eliminate emotional outbursts, enabling the child to more fully participate in his or her school day. Transitioning among school activities is easier, and the service dog provides a focus through which the child can interact with other children. These opportunities increase a child's ability to develop better social and language skills, mobility, independence, and autonomy.

Research on animal-assisted intervention for ASD has doubled in size from 42 studies prior to 2015 to 85 studies in 2020. Horses remain the most commonly researched animal, followed by dogs.[44] Improved social interaction is the most commonly cited outcome.

Robots and Technology as Therapists

Speech-language pathologists and psychologists have long used puppets to catch the attention of their patients and get them talking and interacting. Why not use today's technology and modernize this idea by replacing puppets with robots?

That idea occurred to researchers at Notre Dame, who did just that. Nineteen children ages six to 13 completed 12 one-hour sessions with a trained human therapist, who was joined by a robotic co-therapist for half of the sessions. The robot spoke in a computer-like voice and moved mechanically. Observers used a checklist to record the subject's interactive social behaviors during both typical sessions and those accompanied by the robotic cotherapist.

Results showed that robots are clearly beneficial to those with autism. Seventeen of the 19 subjects showed increased social skills and language when the robot was present. Furthermore, skills learned with the robot were generalizable both with the therapist and at home.[45]

Just as moving from puppets to robots is a short step, so is moving from robots to technology. Let's look at the powerful potential of technology for improving social skills and reducing problematic behaviors.

Video Modeling

Watching other people interact and then mimicking what they do is the basis for many social skills training programs. Do you think you could ski after watching a skiing video? Doubtful. Give a speech after watching a graduation address? Probably not!

Likewise, those with dyspraxia and other motor planning problems, oral-motor issues, and unintegrated reflexes are probably going to struggle to copy behaviors they see on videos. Yet, programs promoting this approach persist. A better option is to remediate the foundations for these coveted abilities first, and watch these higher-level skills emerge spontaneously.

That said, some social training software is still popular. Motivating interactive games that depict social scenarios challenge the player(s) to determine appropriate responses, thus learning the rules of social communication in a variety of unstructured environments, such as in the grocery store and at school. One program allows an individual to practice everything from the right amount of social touch in the locker room to appropriate lunchtime interaction.

Apps for Social Interaction

Because of adult safety concerns leading to the tendency to overprotect, many vulnerable kids don't develop an understanding of cause-and-effect that neuro-typical kids learn. They miss experiencing positive cause-and effect outcomes like the wind in your hair from zooming down a slide or the anticipation of winding the crank on a Jack-in-the-box because the outcomes are not positive to them. Sometimes the negative effect teaches a positive lesson, such as forgetting to put on gloves and getting cold, or the proverbial touching a hot stove and getting burned. Frequently, lack of understanding is not a cognitive issue, but rather a lack of experience. At other times unreliable motor skills interfere, and results are inconsistent.

No worries. We have an app for that! The market for social skills apps has exploded in recent years, offering a wide array of options for users of all ages and backgrounds.

For young children, apps for teaching cause and effect and can be a tool for them to learn social and other skills safely. Cause and effect apps can be an adjunct to learning why things happen, but they are never a substitute for real, hands-on learning experiences. Visit Neurolaunch.com/social-skills-app/

Combining Assistive Technology with Artificial Intelligence

In 2019, Vanderbilt University opened the Frist Center for Autism and Innovation, which offers many resources including virtual reality (VR) to improve social skills and emotional recognition at any age. Virtual reality is a computer-generated simulation of a three-dimensional image or environment that can be interacted with in a seemingly real or physical way by a person using special electronic equipment, such as glasses or a helmet with an internal screen. It is applying today's new technology to social stories!

VR therapy creates structured, predictable, realistic life scenarios in a safe, controlled environment, offering opportunities to practice new skills and apply them to real-life situations. By controlling the level of sensory stimulation, VR helps our supersensitive kids develop tolerance to lights, sounds, and sensory overload, thus reducing anxiety.[46]

New developments are currently underway that intersect AI with robotics.[47] The invention of increasingly complex wearable devices is one application that can provide enhanced communication, interaction, and social engagement.[48]

Take-Home Points

The emergence of appropriate interpersonal interaction is the final step on the path to resolving neurodevelopmental issues. While it is tempting to address interpersonal skill deficits aggressively at a young age, that is probably not the best plan. Consulting with an optometrist, occupational therapist, biomedical doctor, and/or audiologist should usually take precedence over making an appointment with a speech-language pathologist or mental health professional. Even though some of the therapies in this chapter can begin in the first year of life, social skills therapies should almost always wait until a child's immune, gastrointestinal, and respiratory systems are working efficiently and motor and sensory skills are sound.

For adults, especially parents, trying to interact with a young child who doesn't look at, laugh with, talk to, or imitate them can be very frustrating. Observing a botched interaction with grandma can be heartbreaking. Acknowledging that a child has few friends can be extremely painful. Imagining how a teenager's social awkwardness might affect future personal relationships and job performance is downright frightening. Children feel more confident and able as parental anxiety is reduced by working with them as well.

This chapter summarizes some extremely innovative programs based on a solid understanding of sensory processing. Most importantly, they respect an individual's inefficient sensory systems, and view "aberrant" behaviors as attempts at coping with a confusing world. Joining a child in what might appear to be purposeless play, following DIR, RDI, or the Son-Rise Program, often yields remarkable results. While intensive, demanding, and costly, they all demonstrate that parents are the key to success.

Animals, especially service dogs and horses, can be life-changing therapy partners. Technology is opening even more incredible doors to socializing with apps, online, and virtually.

According to Daniel Goleman, author of *Emotional Intelligence*,[49] social intelligence is a better predictor of success than SAT scores. Most people on the autism spectrum or with other neurodevelopmental conditions lose jobs because of social-emotional, rather than skill, deficits.

Applying some of the brilliant methodologies described in this chapter empowers those who are less interactive to join others, interact in socially acceptable ways, have meaningful relationships, and hold good jobs. Best of all, they can realize whatever potential they have to become confident, happy, productive individuals in a demanding society.

CHAPTER 20

ACADEMICS: FINALLY LEARNING TO READ, WRITE, SPELL, AND CALCULATE

So far, the emphasis in this book has been on differences in behaviors, language development, and socialization that lead to diagnoses of autism spectrum disorder (ASD), obsessive compulsive disorder (OCD), anxiety, mood and sensory processing disorders (SPD), and "mental" illness. There has been little mention of the diagnoses of learning disabilities (LD), nonverbal learning disabilities (NVLD), attention deficits (ADHD), dyslexia, dysgraphia, and dyscalculia that all reflect difficulties in learning.

I started my career as an "educational diagnostician," a profession I made up to describe what I did every day: testing to determine whether a child truly had one of these diagnoses. As you might guess, I rarely agreed to a learning disability or attention deficit diagnosis because in almost 100 percent of the cases I found weaknesses in one or more foundational skills.

I then referred my clients to a nutritionist, OT, developmental optometrist or all three for evaluations which inevitably showed vitamin deficits, an unbalanced gut, lack of reflex integration, motor delays, sensory processing differences, and a myriad of visual problems. Once these were identified and remediated, the child's academic skills spontaneously improved, and their testing profile no longer indicated a learning disability.

One of the most challenging areas was the diagnosis of NVLD, defined in Chapter 1 as "a set of strengths in verbal memory and vocabulary, accompanied by visual-spatial, fine motor, and social difficulties that include decoding body language and understanding inference and humor." To me, this disorder is clearly a developmental motor-visual delay. Once intervention with both occupational and vision therapy was undertaken, academic skills came together beautifully.

So now we have finally reached the point in this book to address these higher-level concerns that show up as trouble with academics in the elementary grades and continue to plague some adults to a degree that their work and personal lives are affected. I know several adults with severe reading problems who have painfully hidden their struggles all their lives by avoiding reading and writing. Thank goodness for today's technology that allows them to listen to get information and use speech-to-text and spell-check to get through communication necessities with family and colleagues.

This chapter begins with addressing the Total Load stressors that interfere with the acquisition of early academic skills and functional literacy. Are reading and writing primarily visual or auditory? What role do reflexes play?

Some enhancement and remediation programs that can benefit everyone at any age to build a strong foundation for literacy are included in this chapter. They all go beyond traditional tutoring to help inefficient readers, writers, and mathematicians get unstuck. It is never too late to become good at—and even enjoy—reading for pleasure, writing in a journal, or keeping a budget!

Ensuring Readiness for Academic Success

Academic Readiness

When *are* children really ready to read, write, and do mathematics? When they have strong sensory and motor foundations and show interest in and begin to participate in those activities spontaneously. Reading specialist Debra Em Wilson targets six specific supporting foundations for academics:

- good postural control,
- efficient tactile, proprioceptive, and kinesthetic processing,
- solid midline skills,
- appropriate vision and visual processing,
- normal hearing and processing of sound, and
- strong motivation and focus.[1]

Emphasizing academics before a majority of these skills are present can result in compensatory strategies, frustration, heightened anxiety, and failure. Furthermore, we end up slapping on more labels unnecessarily.

A few students with ASD read very young, some as early as two. Extremely early reading, called *hyperlexia*, is a sign of uneven development. These children's remarkable ability to recognize patterns of letters and "read" words they don't even know the meaning of is not a sign of a budding genius, but rather an indication that their vision is too focal.

▶

Believe it or not, early reading should be discouraged! The preschool child needs to develop control over the body first, and when the body is "still," the mind is ready to learn.

Visual Readiness

Considerable research, much of it ignored by today's educators, is available on the role of vision in learning disabilities. In a now-dated review of over 300 references, the role of inefficient vision in learning disabilities and dyslexia is documented. Studies affirm a positive relationship between learning problems and the following: poor eye movement skills, convergence insufficiency, faulty binocular vision, farsightedness, lazy eye, poor visual processing, weak visual motor skills, and suppression of one eye.[2]

Ready readers and writers use their eyes together as a team, can change focus easily from the book or paper to the teacher, and can perceive just-noticeable differences among visually similar objects and words. They can move their eyes across the page[3] without upper body or head movements using rapid, small, simultaneous movements of both eyes in the same direction from one point of fixation to the next, called *saccades*.[4] They can recognize whole words by sight, sound out words phonetically, and read with understanding.

Refer back to Chapter 17 for some of the ways that vision affects reading and writing. As a review, be aware of the following:

- *Visual acuity at near point.* Readers must be able to see the printed page clearly. If the page is not clear, the use of lenses for magnification or correction of an astigmatism might be indicated.
- *Convergence.* If the eyes do not converge with ease they may drift out, causing blurring, headaches, and the tendency to get sleepy when reading.
- *Eye turn or strabismus.* If either eye turns in, out, up, or down, a student may demonstrate a head or upper body turn to compensate and use only one eye. This condition could slow down reading and writing and possibly affect comprehension.
- *Holding the breath.* If reading or writing is stressful, students may hold their breath. This symptom could be present with almost any visual stress.
- *Visual thinking.* The ability to run pictures or movies in the mind's eye is an essential skill for reading with comprehension and writing with clarity and description. Those with poor visualization often need more body-in-space experiences to become good visual thinkers.

Auditory Readiness

English words are made up of sounds, or phonemes, the components in a word that affect its meaning. Readiness for language on paper includes the ability to hear, identify, discriminate, blend, and sequence phonemes in words, a skill called *phonemic awareness.*

Because English is not a fully phonetic language, however, phonemic awareness and the associated skill of using phonics to sound out unknown words is somewhat limited in its usefulness. Furthermore, the ability to blend sounds requires good visualization, as a student must keep the sounds and associated symbols in the mind's eye to blend the sounds in the right order successfully.

Some signs that a student lacks auditory readiness are mispronunciation of words, such as "pasketti" for "spaghetti," difficulties following sequential directions, such as "go upstairs and bring me a roll of toilet paper from the hall closet," and frequent interruptions during conversations or reading.

Just as eyesight and vision are different, with the former related to clarity and the latter a result of the brain processing what it sees, so do hearing and listening differ. Hearing refers to clarity of sound, while good listening skills require the brain to give meaning to what it hears. Audiologists and speech-language pathologists may thus be members of the team working with a student who has difficulty with auditory processing.

Vision and Audition Must Integrate

While visual and auditory readiness share many components, such as *sequencing, reversals, directionality,* and *figure-ground,* the late Harry Wachs, OD, noted some important differences. In his landmark book, *Thinking Goes to School,*[5] his goal is to integrate vision with every other sense, which he accomplishes with more than 150 games and activities. Though written over 50 years ago, this resource is still a gem.

For visual-auditory integration Wachs begins with sequencing concrete objects such as colored tokens, each representing a sound. For instance, the nonsense word "pap" could be represented by red-green-red. To change "pap" to "pip," a child replaces the green token with any other color, say blue, to show that the middle vowel sound has changed. The next stage is to change "pip" into "pit" by replacing the ending red token with yet another color, say black. So, red-blue-red is "pip" and red-blue-black is "pit." The final stage is to change "pit" into "tip." Yes! Black-blue-red represents "tip."

Eventually, printed letters replace colored tokens. Letters introduce laterality concepts, as a child must be able to distinguish a *b* from a *d* and *p* from *q*, directional differences that are bypassed by using colors alone.

Finally, Wachs removes the visual clues, asking a child to repeat a word such as "carpenter" without the middle syllable, "pen." In order to do this correctly, the student must visualize the whole word and remove the targeted part to get "carter." Not easy, is it?

Wachs's book is full of activities that connect vision to audition. Integrating what is seen and what one hears is foundational to so many aspects of learning and life. That is exactly what we do when we follow directions, watch a play, and listen to someone telling us about their summer vacation.

Reflexes Must Integrate

Understanding the role of infant and postural reflexes in the emergence of foundational skills for academic development is essential. Integration of lower-level reflexes, such as the Moro discussed in Chapter 6 and the Tonic Labyrinthine Reflex (TLR) described in Chapter 16, are key to the integration of higher-level reflexes.

Integrating reflexes in developmental order is mandatory! Programs that bypass lower-level reflexes and focus on higher ones end up spending an excessive amount of time, because remedial exercises could be fighting against lower-level reflexes. The best reflex integration programs and academic readiness programs understand the importance of integrating reflexes in order.

Reflexes for Learning

Following are two reflexes that are key to learning to read and write with ease.

The Asymmetrical Tonic Neck Reflex (ATNR), which develops in utero about halfway through pregnancy, is fully present at birth. It is the familiar "warrior's pose" seen in many infant photographs.

Movement of the baby's head to one side elicits extension of the arm and leg on the same side and flexion of opposite limbs. During uterine life the ATNR facilitates continuous motion, thus stimulating the vestibular system and developing muscle tone. During labor, the ATNR helps the infant "unscrew" itself and move along the birth canal. This twisting movement is the first experience of the infant in coordinating both sides of the body.

Babies born by either forceps delivery or cesarean section do not have the opportunity to experience the ATNR. Deprived of the twisting action of the birth process, they may not go through the process of early bilateral integration necessary for later development of auditory processing and cross-lateral skills such as crawling, walking, and skipping. Results of this deprivation can include later problems with balance, confusion with crossing the midline, mixed laterality, and poor language development. An active ATNR is also present in cases of torticollis or any condition that includes asymmetry or twist in the body, favoring one side.

Symptoms of a hyperactive ATNR can often be seen when writing, as the unintegrated reflex causes a person to tighten the dominant arm and shoulder muscles in an effort to prevent the arm from straightening out when writing on the right side of a paper. It also creates difficulties in crossing the midline and organizing work in visual or auditory midfield. When the ATNR is not integrated, the presence of new auditory information can cause stress, resulting in confusion or shutting down of auditory processing.

Suspect an active ATNR in individuals who tend to flop sideways on or even fall off their chairs. Students who turn their heads to look at the blackboard, or read with their heads to one side, also exhibit active ATNRs. Any irregular walking or crawling pattern, when a person "lists" or falls to one side, is also ATNR-driven. An active ATNR can also trigger insomnia when the thoughts cannot "turn off" for sleep.

The Symmetrical Tonic Neck Reflex (STNR), which emerges around 13 weeks in utero, is active from 6 to 10 months of life, and integrates between 9 and 12 months. It's a very short lifespan! A major purpose of the STNR is to help the infant to defy gravity and get up on hands and knees, rock, creep, and eventually crawl. When a child is on all fours, dropping the head down causes the arms to bend and legs extend; on the other hand, lifting the head up causes the arms to straighten and the legs to flex. It is this series of coordinated movements that makes crawling possible.

Almost 50 years ago, Miriam Bender recognized that a retained STNR was present in a majority of those diagnosed with learning disabilities and attention deficits.[6]

Pavlides found that a high percentage of children later diagnosed with learning disabilities or ASD omitted stages of crawling and creeping in infancy and retained the STNR.[7] Their histories show that they rarely crawled on hands and knees; they scooted, shuffled along, or "bear walked" on their hands and feet alone. If they did crawl, they might have done so in an unsynchronized, unusual fashion, with rotated hands, locked elbows, and raised feet.

Creeping and crawling are essential movement patterns for sensory development. The vestibular, proprioceptive, and visual systems connect for the first time through creeping. Without going through this stage, babies end up with a poorly developed sense of balance, poor understanding of spatial relationships, and poor ability to use their eyes together, resulting in lack of depth perception.

Reading specialists O'Dell and Cook show a relationship between a retained STNR and

- academic issues such as reading, writing, mathematics, spelling, art, and music;
- coordination problems in athletics;
- difficulties acclimating to new settings;
- psychological consequences such as poor peer relationships, low self-esteem, frustration, avoidance, aggression, and inflexibility.

Furthermore, they found that both academic and focus skills improved markedly with specific activities focused on diminishing STNR reactivity.

Two other rarely recognized signs of a retained STNR are the tendency to wrap the ankle around the chair leg to prevent the knees from straightening when the arms are bent, and sitting in a "W position," with the legs splayed out behind the body. Sitting in a W inhibits trunk rotation, making it impossible for a child to cross the body's midline, an important developmental milestone. Perhaps the extended legs give them better postural control. Perhaps they are trying to stabilize a body with low muscle tone.

Australian movement consultant Brendan O'Hara believes that W-sitters process the world slowly because they cannot distinguish whether their own bodies or the world is moving. So, often they sit and watch, unable to move forward, lowering their center of gravity to improve their balance and feel more secure.[8]

This short-lived and important reflex is one of the most pesky as it has the potential to interfere with many aspects of development. Recognizing the signs and integrating it can be life changing. Red flags indicating an unintegrated STNR must be attended to for prevention of possible later developmental delays. How did that important knowledge get lost? The classic book *Stopping ADHD: A Unique and Proven Drug-free Program for Treating ADHD in Children and Adults* offers activities and exercises for remediation.[9]

Programs for Enhancing Academic Readiness

Many excellent prepackaged programs are available to develop academic readiness. They work equally well in small groups of preschool students as they do in later grades, and I believe they work better than academic tutoring for those in need of extra help.

Here are a few of my favorites. I have known many of these program developers personally for over 30 years. Strongly consider using them before going directly to the alphabet, reading, and writing.

S'Cool Moves™ for Learning

Reading specialist Debra Em Wilson was given a dire prognosis for her infant daughter, Shalea: "She has microcephaly, cognitive delays, visual and other sensory processing disorders and a rare genetic syndrome. She will never, walk, talk or function in a meaningful way."

That was more than 20 years ago. Debra began researching and, along with occupational therapist Margot Heiniger-White, recognized that many children with special needs don't have the motor, visual, and auditory foundations necessary for success in reading and writing. So, they developed a program that includes engaging reflex integration and vision therapy exercises that transition students to reading and writing and called it *S'Cool Moves™*. Shalea and others bloomed.

By focusing on improving motor planning, rhythm, timing, core posture, vestibular activation, brain integration, and sensory-motor integration in a fun setting, reading skills emerge. See SchoolMoves.com.

Fast forward to one day about five years ago, a social worker who had attended a S'Cool Moves training workshop introduced Debra to the vagus nerve and Polyvagal Theory. It was a watershed moment. Return to Chapter 6 to refresh your memory on this vital foundation for feeling safe. Debra studied with Deb Dana, the translator of Steve Porges's complicated ideas, and realized that having a safe-feeling nervous system was the most important variable in resolving learning challenges. She then developed *The Polyvagal Path to Joyful Learning*.[10]

Before the ink was even dry on this first book, Debra charged ahead with a companion volume, *The Resilient Learner's Backpack: Mind-Body Activities for Focus, Regulation, and Engagement*.[11] The backpack offers tools for parents and teachers coupled with activities for students based on befriending the nervous system. This unique pair of programs should be a mandatory part of every pre-academic program. With safety and connection at the core of her parenting and teaching, Debra's daughter and students now feel safe enough to stretch beyond their limitations.

Equipping Minds

For many years, educational consultant Kathy Johnson specialized in helping people struggling with academics to achieve individual success by focusing on laying foundations through integrating key reflexes. Her *Growing Brains* program combined specialized auditory, visual, reflex, motor, and cognitive activities. Dozens of videos where she explains how to evaluate and remediate reflex anomalies are available on YouTube.

Now retired, Kathy's program is a part of *Equipping Minds*, the brainchild of Carol Brown, PhD, a longtime advocate for students with dyslexia and other learning disabilities. Her struggling son was her primary motivation. While he finally conquered learning to read and write, processing speed and working memory remained difficult. Her Equipping Minds Cognitive Development Curriculum developed from her in-depth study of neuroscience and focuses on exercises that build higher-level processing such as working memory, executive functioning, reasoning, and comprehension. Visit EquippingMinds.com.

Multisensory Programs to Enhance Learning from Occupational Therapists

A number of occupational therapists have developed packaged programs that address the sensory needs in academics. Here are two programs that have proven their worth for 35+ years.

The Alert Program for Self-Regulation, also known as *How Does Your Engine Run?*,[12] was developed by OTs Mary Sue Williams and Sherry Shellenberger. It empowers everyone to learn how to regulate their own arousal systems. Comparing their bodies to cars, kids rev up their engines for active times of day, such as recess, and rev down for reentering the classroom for reading.

The techniques can also be used to prepare for anxiety-producing situations, such as going to the doctor, and overstimulating environments like restaurants and shopping malls. Adults, including veterans and first responders, can learn how to be calmer, more productive, and more alert, think more clearly, and work better with others.

In their book, *Take Five! Staying Alert at Home and School*, these innovative OTs provide a myriad of activities for the hypo- and hypersensitive child in five areas: touch, movement, vision, audition, and oral-motor.[13] The manual also includes an extremely valuable checklist for adults with sensory issues. To learn more about use of this program, go to AlertProgram.com.

Tool Chest for Teachers, Parents and Students[14] developed by Diana Henry, OT, includes comprehensive handbooks and DVDs offering many activities for home and school. Her focus is on mitigating undesirable behaviors by addressing their sensory roots. See ATeachAbout.com.

HANDLE, an acronym for the *Holistic Approach to NeuroDevelopment and Learning Efficiency*, grew out of the need for its late founder, Judith Bluestone

(1945–2009), to understand and heal her own challenges. Since 1994 the HANDLE Institute has certified hundreds of trainers and treated thousands of individuals with learning disabilities, attention deficits, ASD, dyspraxia, language disorders, Tourette syndrome, and a plethora of perceptual, learning, and behavioral disorders.

Strengthening the vestibular system, enhancing muscle tone, and increasing differentiation are the goals of HANDLE. These are all accomplished through deceptively simplistic activities that are sensitive to the client's physical and social-emotional needs. Small, measured doses of specific activities are incorporated into daily activities in the home, daycare, or school setting. Therapy time typically requires a total of approximately a half hour, preferably interspersed throughout the day. Some of the more frequently suggested activities involve

- drinking from a crazy straw,
- playing follow-the-leader with a flashlight,
- rhythmic ball bouncing,
- copying designs by only feeling them,
- catching a suspended ball, and
- stepping through a hula hoop maze.

Read *The Fabric of Autism: Weaving the Threads into a Cogent Theory*.[15] Visit HANDLE.org.

Educational Kinesiology and Brain Gym®

Kinesiology is the study of movement, the mechanics of how muscles and bones interact to enable us to move. *Educational Kinesiology (Edu-K)* is a comprehensive whole-body program using specific movements, postures, and balancing procedures.

The relationship between dysfunctional behavior and stress is central to Educational Kinesiology. Edu-K's goal is to reduce stress and integrate the brain for enhanced learning and performance. In the early 1980s Paul Dennison, PhD, began teaching 26 specific movements designed to counter the stress response by moving attention and energy away from the survival centers of the brain. These 26 movements are the introduction to Edu-K and are referred to as *Brain Gym®*. The Edu-K umbrella covers Brain Gym as well as numerous advanced techniques for enhancing learning and performance.

Brain Gym activities work extremely well for those experiencing anxiety about performing in competitive situations and in those where they believe others are judging them. I have seen star athletes use a Brain Gym move before serving in tennis, prior to giving an important speech, or as an extra confidence measure when entering into an important conversation or confrontation.

Edu-K comprises techniques from the fields of motor development, applied kinesiology, developmental optometry, Neuro-Linguistic Programming, acupuncture, yoga, martial arts, language development, psychology, academics, and brain research. One can debate whether this program belongs in this chapter or another covering one of any of the above areas. It is included here because of its broad applications.

Dennison first became interested in the role of movement in learning through his own personal struggles, and later through his observations of the children at the Valley Remedial Group Learning Center in Southern California, which he directed in the 1960s and 1970s.

As Dennison observed the direct connections between subtle dysfunctional movement patterns his students displayed and their learning difficulties, he became convinced that when physical skills are inefficient, the brain cannot attend to the demands of higher-level cognitive processing. He found that more energy was then available to activate the cortex and promote integrated visual, auditory, and kinesthetic functioning.

The Three Dimensions of Brain Function

Edu-K recognizes three primary dimensions of brain function.

- *Focus Dimension* refers to the integration of the front and back of the body and the front (frontal lobes) and back (brainstem) of the brain. Children labeled ADHD or OCD often have imbalances in the Focus Dimension. Reading with comprehension is the quintessential Focus Dimension task.
- *Centering Dimension* refers to integration of the top and bottom halves of the body, and the rational top (cortex) and emotional bottom (limbic system) of the brain. This integration arises from the interrelationships among proprioception, balance, and vision. These systems work together to provide a sense of the center of one's body as a point of reference for the directions up, down, back, front, left, right, in, and out.
- *Laterality Dimension* is concerned with the coordination of the right and left sides of the body and the right and left hemispheres of the cortex. Joined by the *corpus callosum*, the right hemisphere controls the left side of the body, and the left hemisphere controls the right side. Integration of

the two hemispheres is essential for the development of all bilateral skills, including binocular vision and binaural hearing.

Binocularity and lateral integration are the foundations for reading, writing, and communicating. A student with deficient bilateral skills may have difficulty crossing the midline and be labeled "learning disabled" or "dyslexic." Learning handwriting, an important bilateral integration skill, is frequently missing in educational curricula today.

PACE

Four of the 26 Brain Gym movements are used frequently as a warm-up. Called *PACE* for *Positive, Active, Clear, Energetic*, this sequence is an easy, efficient way to achieve readiness for any activity in fewer than four minutes. Some teachers I know begin every class doing PACE with their students.

- *ENERGETIC:* Drink some water.
- *CLEAR:* **Brain Buttons.** While holding one hand over the navel, use the rub the thumb and forefinger of the other hand to rub the hollow areas on either side of the sternum just below the collarbone for 30 seconds. Switch hands and massage for another 30 seconds.
- *ACTIVE:* **Cross Crawl.** Touch hand to opposite knee, alternately moving one arm and opposite leg. Even better, touch elbow to knee. Continue for one minute. Done standing, the movement is more like marching in place than crawling.
- *POSITIVE:* **Hook-ups.** Part I: Cross ankles, hold arms out with hands back to back, and cross one arm over the other so that palms touch. Clasp hands and bring them to the chest. Touch the roof of your mouth with your tongue. Sit this way for one-half to a full minute, listening to your breath. Part II: Uncross legs, place feet squarely on the ground, release arms, and hold fingertips together. Sit this way for another half to one minute.

Resources

In 1987, Dennison and his wife Gail founded the Educational Kinesiology Foundation (EKF). Today Brain Gym® is used worldwide and is taught in thousands of public and private schools as well as in corporate, performing arts, and athletic training programs. In 2012, the board of directors voted to expand the organization to include other movement-based programs under the name Breakthroughs International. Learn more at BreakthroughsInternational.org.

Books:
* *Hands On: How to Use Brain Gym® in the Classr*oom[16]
* *The Brain Gym Teacher's Edition*[17]
* *Smart Moves: Why Learning Is Not All in Your Head*[18]

Figure Eights. Occupational therapists, developmental optometrists, and others have long recognized the value of the *lazy eight*: the number eight lying on its side instead of standing on end ∞. Its formal name is a *lemniscate*. Today, therapists of many disciplines believe that moving the whole body, or one or more of its parts, in the shape of a lazy eight integrates the left and right sides of the brain, leading to increased focus and attention, spatial awareness, lateralization, and bilateral integration of eyes, ears, and limbs.

Debra Em Wilson claims to have improved students' reading fluency rates 20–30 words per minute by having them do figure eights before testing.[19] Australian movement expert Brendan O'Hara has created several figure eights exercises using beanbags.[20]

Infinity Walk. A variation on the figure eight pattern was developed in the late 1980s by psychologist Deborah Sunbeck, PhD. The *Infinity Walk* elaborates upon moving through the simple lemniscate pattern by adding the visual component and other multitasking skills to the basic movements, requiring the eyes to constantly cross the midline. This action stimulates the vestibulo-oculo-cervical (VOC) triad described in Chapter 16. Refer back for more information or go to InfinityWalk.org.

Enhancing Academic Skills

Reading
Many parents introduce storytelling and reading at birth. Children naturally become aware of sounds when listening to oral language, whether spoken or read. Rhyming games further enrich their vocabularies by pairing letters with sounds, words with pictures, and objects with concepts.

Children who have strong visual memories can recall whole-word configurations and develop a sight vocabulary. Others learn through phonemic awareness and instruction using phonics. As the eyes move across the page, fluency increases, and children increase their reading vocabularies.

In the early elementary grades, students "learn to read" by combining visual, auditory, and language skills. By third grade, when they have broken the code, they "read to learn" subjects such as science and social studies.

When we read, many pieces come together as the visual, auditory, language, and memory centers integrate, and comprehension takes place automatically. Eventually, pictures to accompany the words are no longer necessary because students derive meaning by running movies in their minds' eye.

The best reading programs for students contain subject matter that is developmentally appropriate. Just as kindergarten and first grade readers focus on content of interest to five- and six-year-olds, the "just right" reading program for older children who are slow to read should include more mature interests, while not being too difficult to decode. Many publishers specialize in these "high interest, low level" readers for older students who are struggling. We now know we made a terrible mistake with many of our nonspeaking students, who were "dumbed down" because we did not assume competence.

Here are some tried-and-true programs:

Fast ForWord® is a family of interactive computer-based listening programs developed in the mid-1990s by four research scientists: Michael Merzenich, PhD, William Jenkins, PhD, Paula Tallal, PhD, and Steven L. Miller, PhD. Their collaboration resulted in the formation of Scientific Learning®, a subsidiary of Carnegie Learning.

Fast ForWord is designed to improve language, reading, and learning skills from preschool through adulthood by listening through headphones. Based on over 30 years of brain research, it includes many different levels.

Using an intensive series of adaptive, interactive exercises of acoustically modified speech and speech sounds, six different training programs target temporal sequencing, sound discrimination, phonological awareness, decoding, sustained focus, attention and listening comprehension, vocabulary development, word recognition, and other important reading skills.

An individual plays a series of appealing computer games that automatically adjust to a student's improving competence level. Scientific Learning Reading Assistant Plus™ combines advanced speech recognition technology with scientifically based courseware to help students strengthen fluency, vocabulary, and comprehension to become proficient, lifelong readers. See SciLearn.com.

Lindamood-Bell® Learning Processes has been developing reading programs that improve students' sensory-cognitive processing of language for over 30 years. The programs are intended for students in the middle elementary grades with all types of learning issues. *Seeing Stars® for Symbol Imagery (SI™)* develops phonemic awareness for reading, spelling, and fluency, while *Visualizing and Verbalizing® (V/V®)* develops concept imagery for language comprehension, vocabulary, and critical thinking. Read further on in this chapter under "Written Language."

Lindamood-Bell also has a mathematics remediation program called *On Cloud Nine® (OCN™)* which applies symbol and concept imagery to arithmetic and mathematical reasoning, with the goal of improving computational skills and the ability to solve word problems.

About 50 Lindamood-Bell® centers are located throughout the United States, the United Kingdom, and Australia. Visit LindamoodBell.com.

Handwriting

Handwriting is a complex process that requires the integration of touch, proprioception, kinesthesia, vision, motor coordination, and language. Subtle feedback from the muscles to the brain, called *proprioception*, monitors the pressure and speed of writing. Eventually, the muscles develop memory, and writing becomes an automatic skill, like walking. The brain actually depends more on tactile and proprioception feedback than on vision to write.

Forming letters is a bilateral integration task that prepares the brain for lateralization and specialization. Cursive writing activates areas of the brain that do not light up when keyboarding or touching a screen. In fact, notetakers consistently outperformed laptop users in retention of material.[21] Yet, with the trend toward tablets, touch screens, and voice activation, handwriting may be on its way to becoming extinct.

Anyone having difficulty with handwriting should have developmental evaluations by both an occupational therapist and a behavioral optometrist. Specific remediation delivered as occupational or vision therapy can result in improved letter formation. Difficulties could be based in weakness in motor and/or visual functions or, possibly, in the lack of integration of key reflexes.

For anyone who is struggling with any aspect of handwriting, occupational therapists can provide aids such as special papers, grippers, slant boards, cushions, and writing implements. The need for these, however, usually means that the demands exceed a person's physical maturity levels. A better approach is to fall back and work at a level at which the writer is comfortable, while remediating visual and motor weaknesses and reflex issues.

Handwriting Warm-ups

Dr. Debra Em Wilson, the developer of S'Cool Moves (see above), recommends *One-Minute Moves* developed by physical therapist Freddie Ann Chandler to improve foundational handwriting skills. These include the following:

- *Deep pressure stimulation.* Press the thumb deeply and firmly into all parts of the palm of the opposite hand, about 10 pushes. Next, cross the arms over each other, and squeeze the forearm, upper arm and shoulder of the opposite side five or six times.

- *Skin sensation.* Rub the palms, then the backs of the hands together. Next, rub in between the fingers. Clap the hands together five times. Lastly, pat the forearms, then shoulders, then the back.

- *Muscle sensation.* Pretend to put a long glove on the writing hand, using the nondominant hand. Pull the glove all the way up to the shoulder, using firm pressure. Repeat several times.

- *Resistive pressure.* Push the palms together with fingers straight. Then holding the hands at chest level, bend the fingers and intertwine them, keeping the palms together. While grasping tightly, try to pull the palms apart.

- *Joint compression.* Press both hands onto the desk and rub the desk several times. Then press the hands into the thighs as if to push the feet into the floor.

- *Strengthen core postural muscles.* Get in position to write. Sit up tall, planting both feet firmly onto the floor. Place the paper at a slight angle to the midline of the body. For right-handed students, tilt the paper and head to the left; left-handed students tilt to the right.

Remedial Programs for Handwriting

A number of programs for handwriting focus on helping those with developmental delays to write legibly.

Handwriting Without Tears is the granddaddy of handwriting programs. Developed in 1977 by Maryland occupational therapist Jan Olsen, this guide expanded into a full, developmentally based multisensory K–5 curriculum—including assessments, music, and engaging manipulatives—that incorporates visual, auditory, manipulative, tactile, and kinesthetic teaching strategies. In 2018 it was rebranded into *Learning Without Tears* and includes *Keyboarding Without Tears*. See LWTears.com.

Loops & Other Groups: A Kinesthetic Writing System, by Mary D. Benbow, MS, OTR/L, is a widely used handwriting curriculum based on studies of neurology and developmental hand anatomy. Students grades 2 and up learn to form letters in groups that share common movement patterns. Available online.

Callirobics (from *CALLIgraphy* and *aeROBICS*) is a series of simple exercises in five levels from prewriting to adult set to music, and devised by handwriting expert Liora Laufer. The multisensory exercises are designed to improve eye-hand coordination, fine-motor skills, self-esteem, and work habits. Exercises are set to "old-fashioned" songs such as "Danny Boy" and "London Bridge," as well as to original melodies. Each program comes with a workbook and CD. Learn more at Callirobics.com.

Let's Do It—Write! is a series of workbooks designed to teach how to write legibly and quickly. Exercises include ways for improving sitting posture, sensory awareness, cutting skills, pencil grasp, spatial orientation, shape formation, and problem-solving skills. Available from Therapro.com.

Barchowsky Fluent Handwriting (**BFH**) is an inexpensive app for the iPad geared to all ages. See BFHHandwriting.com.

Handwriting Repair. Created by Kate Gladstone, a woman diagnosed with ASD. Visit HandwritingRepair.info.

Handwriting for Lefties

The incidence of left-handedness is slightly above 10% in the general population, and as high as 65% in those with ASDs.[22] "Left-handedness is connected to a lot of neurodevelopmental disorders," says Daniel Geschwind, MD, a UCLA expert in neurobehavioral genetics. "People with autism and schizophrenia are more likely to be left-handed," he adds.[23] Many are ambidextrous, preferring one hand for strength-related tasks and the other for fine-motor skills. This is often thought to indicate poor brain lateralization[24] and could also be due to vision issues.

Being a "southpaw" in a right-handed world is yet another challenging comorbidity. Without proper instruction, left-handers often develop unusual pencil grasps and body postures that hide and smudge what they've just written. Their motor systems can tire easily from their compensatory efforts. With the early introduction of correct fine-motor skills, experts believe that left-handers can write just as well as their right-handed friends.

A British company has a subspecialty in handwriting for lefties. A series of four books with CDs give lefties a chance. The printed books are spiral-bound at the top to allow free left-handed movement from the left side and across the page. Every page carries icons reminding users exactly how to hold their writing implements and how to position their paper in front of them on their work surface. Find them at RobinswoodPress.com.

Written Language

Putting one's thoughts down on paper is the most complex form of communication. Yet, as early as kindergarten, today's students are being asked to write about their experiences, often before strong foundations are laid.

Written language is to academics as social skills are to communication. For success, all the senses—especially audition, vision, and proprioception—must be well-developed and integrated with each other, the brain, and the body. This arduous task requires memory, motor coordination, perception, visualization, and focus. Coherent writing, along with social skills, is among the last areas to develop.

Back to vision! The ability to visualize is the highest form of visual thinking. Visualization allows us to read with understanding, write with clarity, and organize our lives. Refer back "Visual Thinking" in Chapter 17 for more on this important subject including resources for developing this skill.

Today communicating in writing is complicated by texting, where we abbreviate and use shorthand such as "LOL" ("laughing out loud") to express ourselves. No doubt electronics have complicated written language so much that many students do not know what is proper English and what is not.

Several of the programs mentioned above, such as Lindamood-Bell, include writing as a natural extension of reading. According to Nanci Bell, the cofounder of Lindamood-Bell, "You must be able to think before you can write. Without the ability to think—rational organized lucid thoughts—writing is unorganized, disjointed, non-specific, and rambling."[25]

Bell's Visualizing and Verbalizing program (V/V) is a much-needed bridge between oral and written language. It develops foundations for writing by helping students create images in their minds' eye. It encourages them to write, edit, write, edit, write, and edit some more.

Here is a summary of the steps Bell uses in her V/V program:

1. Stimulate gestalt (big picture) processing.
2. Teach sentence structure, basic punctuation, and spelling.
3. Move to paragraph writing, sequencing from copying to editing and finally writing from cue cards.
4. Write expository paragraphs following a specific organizational format.
5. Write narrative paragraphs.

Consistent with the V/V model is the request some teachers make to draw a picture to accompany a child's writing if time permits. For struggling students, completing the picture first, rather than as an afterthought, is a better way to go. Drawing a specific scene or object forces visualization of color, size, setting, and more.

The ability to write a sentence with a subject, verb, and object with correct grammar, correct spelling, and appropriate punctuation and capitalization is still a reasonable goal. Despite the use of electronics, students *must* learn this essential skill.

Spelling

Have you ever watched the televised Scripps National Spelling Bee? How do the best spellers recall words? Some look up or close their eyes, as if searching their brain for the word. Others write the word with their finger in the air or on their palm or forearm. All are using their visualization skills to recall the "look" of the whole-word gestalt.

Yes, spelling is a visual memory skill, and because vision is so intimately tied to our motor system, pairing a spelling word with a motor activity is a good way to seal it into our memories. Vision therapist Donna Wendelburg created this program using balls to teach spelling:

Grade	Type Of Ball	Bounce/Dribble 1x For Each Letter	Spelling Out Loud
K	Playground ball	Bounce and catch ball with 2 hands. (Parent writes the word on large piece of paper.)	Spell word forward. Spell word backward. Ask, "What letter comes before/after the letter _?"
1	Tennis ball	Bounce ball with right hand, catch ball with left hand. Bounce ball with left hand, catch ball with right hand. (Parent writes the word on large piece of paper.)	Spell word forward. Spell word backward. Ask, What letter comes before/after the letter_?"
2	Tennis ball	Dribble ball with alternating hands. Visualize the word.	Spell word forward. Spell word backward. Ask, "What letter comes before/after the letter _?"
3	Tennis ball	Dribble ball in pattern: 2x with right hand, 1x with left hand. Visualize the word.	Spell word forward. Spell word backward. Ask, "What letter comes before/after the letter _?"
4	1½" diameter	Dribble ball in pattern: 3x with left hand, 2x with right hand. Visualize the word.	Spell word forward. Spell word backward. Ask, "What letter comes before/after the letter _?"
5-6	1" super bouncer	Dribble in patterns of 3s: 1x right, 2x left, 3x right. 1x left, 2x right, 3x left. Visualize the word.	Spell word forward. Spell word backward. Ask, "What letter comes before/after the letter _?"
7-8	1" super bouncer	Dribble in patterns of 3s, USING ONLY TWO FINGERS. Visualize the word.	Spell word forward. Spell word backward. Ask, "What letter comes before/after the letter _?"

Chart courtesy of Donna Wendelburg, Achievers Wisconsin www.achieverswisconsin.org

Even though most writers have an automatic spell check available and often text without regard to correct spelling, knowing how to spell is a life skill that serves a student well. This especially includes the commonly confused homonyms, such as *two, to,* and *too* and *their, there,* and *they're.*

As far as I am concerned, teaching spelling to those with learning disabilities is no different than teaching it to anyone else. Thus, I don't list any "special" programs here.

Mathematics

Like language, the foundation for understanding numbers is based in sensory experiences. According to optometrist Harry Wachs, mentioned several times, a young child learns concepts such as same/different, greater/smaller, and first/last through touch, movement, pressure, vision, and even taste and smell by creating images in his mind's eye. Rote learning of number facts or rote counting is not the same.

Wachs cites three foundational skills based on Piagetian conservation principles, all essential for successful understanding of mathematics:

1. *Visual thinking concepts* include *part-whole, figure-ground, perspective,* and *time perception.* Activities using parquetry blocks and pegboards can enhance visual thinking.
2. *Numerical literacy* includes the ability to read numerals, position commas, and understand place value. Base ten blocks, available commercially, can assist in building skills in this area.
3. *Visual logic* adds the component of logical reasoning to visual thinking. Young children need to understand one-to-one correspondence, order, conservation of mass, weight, volume, number, and linear length, as well as classification, seriation, and probability, to learn traditional academic arithmetic or geometry. Children with poorly developed visual logic might master basic computations but will struggle to solve word problems.[26]

Good programs focusing on teaching mathematics adhere to these constructs to get children—with or without issues—to fully comprehend number concepts. Using manipulative materials and number lines are the best choices. Activities that include estimation of time, money, and measurement are preferable to drill with paper and pencil. The Lindamood-Bell program On Cloud Nine, mentioned above, is an example.

Organizational and Study Skills

If good organizational and study skills are goals for secondary school students with ADHD, ASD or LD, then working with mathematics concepts is the way to go. Comprehending mathematics is closely aligned with the understanding of time and space, both visual concepts. This is complicated because some people who have excellent visual-spatial skills are "lost in space." They can be superb at moving things around in their minds' eye but miss something that is right in front of their nose. The key is flexibility, to be able to move from one to the other with ease.

Students who are well organized are usually good in math and are able to develop their own study methods. While courses in study skills can offer tools to get organized, most disorganized students have underlying visual-spatial issues, and thus cannot conform to someone else's timelines and structuring of space.

Provide students with the tools to organize themselves, such as a wristwatch with a built-in stopwatch, alarm clock, calendar (weekly and monthly), organizer, measuring tape, allowance (in a denomination that is easily broken down, such as one dollar), and pedometer.

Apps for Learning

Using apps to help with organization is another option. Organization apps provide tools for being successful at school and in life. They can be helpful for anyone who needs assistance in note-taking; making lists; keeping a calendar; managing time; scheduling; organizing medical information, emails, or their thoughts; or keeping track of finances, messaging, and even meditating!

This book does not have ample space to include the multitude of apps for these purposes. Ask around to learn which ones work for people you trust.

Other Considerations

Tutoring

In the same way that parents are very concerned when their young children are not talking, they become anxious about late readers and writers. The chosen panacea is often hiring a tutor. When applying a developmental model, tutoring is inappropriate until at least age eight, or when foundational skills are at a five- to six-year level. Spend the time and money on foundational therapies instead. For older children, find a tutor who uses a combination of methods, rather than one who recommends only one approach.

Positioning and Lighting

The quality of seating and lighting should be taken into account for all academics. Ensure that desks and chairs are the right size for every body; one size does not fit all. Each should be adjustable, so that the student's feet are fully on the floor at right angles to the waist and the desk meets the body slightly above the waist. Desks and chairs that do not match in size or are too high or low can make learning more difficult.

Use of ergonomically validated lighting systems can increase the speed with which students perform reading and writing tasks.[27] Lights must not have glare or buzz. Refer back to Chapter 3 for more on this subject.

Take-Home Points

Teaching students with learning disabilities how to read, write, and understand number concepts occurs in Chapter 20, not earlier, because many sensory, motor, language, and cognitive foundations are necessary for success. *No*, we are not "dumbing down" our special kids; we are protecting them from undue failure. Until students master control over and become competent in visual, auditory, language, and gross- and fine-motor skills, academics will be challenging and frustrating for them. And we know they don't need more stressors in their lives!

Parents, teachers, and therapists must be patient and focus on improving lower-level foundational skills rather than moving prematurely to higher-level skills. Pushing academic tasks such as writing and reading prematurely can result in diagnosed learning disabilities, attentional issues, or decompensatory behavior later on.

At any point in the above developmental sequence, if the demands for production exceed motor and visual maturity or if a reflex has not integrated, the child's body must physically compensate, and aberrant postures will emerge. This is often what happens in autism. A teacher may then observe a child slumping at the desk, wrapping a foot around the table leg, laying her head on an arm to write, or demonstrating other symptoms indicative of the lack of motor and visual readiness.

That said, many programs exist that focus on building and strengthening of foundations, thus allowing the natural progression of reading, writing, and counting to take place spontaneously. These programs are well-conceived, fun, and readily available, both as tangible materials and electronically. Therapists of many types can access them and supplement their remedial programs. Furthermore, schools using some of the programs described in this chapter will find that students of all ages take to them easily and happily.

If, in your impatience for academics, you are reading this chapter out of sequence and have skipped earlier ones on more foundational skills, please refer back to ensure that all systems are "go" before proceeding onward, no matter the individual's age. Even teens with visual and auditory processing deficits can benefit from a review of the basics. Our older students and young adults are the subject of the next chapter. Keep reading to learn about transitioning.

CHAPTER 21

TRANSITIONING OUT
OF SCHOOL INTO A LIFE

In April 2012, I received an unexpected email: "Hi Patricia. We are parents from Kuwait with an autistic son. We met you in 1996 in Washington, DC. Our son was five at that time; he is 21 now. We are one of over 3000 families who petitioned the Kuwaiti government to fund a model center for young adults with disabilities. We have now received approval, and are requesting your expert advice. Awaiting your kind reply."

Six weeks later my team of three jetted off to Kuwait, a country I had only heard of and could barely locate on a map. We knew not what awaited us: a life-altering experience for me.

Over the next two years, I visited Kuwait four times with international teams of speech-language and occupational therapists, optometrists, special educators, and other professionals. Together, we designed a beautiful center and advised Kuwaiti officials, professionals, and parents on all aspects of diet, movement, vision, curriculum, remediation, recreation, architecture, and engineering. I consulted with dozens of families, put on conferences, and visited every government agency and nonprofit organization having anything to do with autism, cerebral palsy, Down syndrome, genetic disorders, and other developmental disabilities. All in 110-degree heat, with a mandatory siesta each afternoon.

My team and I heard universal questions: "Will my child ever lead a 'normal' life?" "How can I calm my two nonverbal adult sons with autism sufficiently so they can fly out of the country?" "How can I stop my son from masturbating in public?" "Two of my five children have autism, and my wife is pregnant. How can I prevent my new baby from becoming autistic?" "What will happen to my son when I am gone?"

More than a decade later, Center 21, named for the age of maturity, is thriving. I am proud of this life-changing project that indeed is the model they desired, compatible with Kuwaiti culture, religious beliefs, and family values.

A Universal Problem

The need for young adult programs is widespread and prodigious. Children born in the 1990s to the early 2000s are now young adults and aging out of school. An increasing number of parents around the world are frantic about what their kids will do all day once they have no school, no program, nothing to get up for in the morning.

In the poignant 2010 Chinese film *Ocean Heaven*,[1] we see the universal nature of this problem. Wang, a single father of a severely autistic, nonspeaking young adult male struggles to find a place for his son to live after discovering that he is dying of liver cancer. Each place he calls and visits is inappropriate; his boy doesn't function well enough for a post-secondary school and is too old or too "normal" for the asylum of the mentally ill. Now, fifteen years later, have we made any progress?

In his desperation, Wang teaches his son to boil eggs, counting rotely until they are cooked. He travel-trains him to take the bus to his beloved aquarium, where he swims with the dolphins, and playfully teaches him how to dress and undress himself. Despite feeling deep sadness as the father passes away, the audience is gladdened that his efforts pay off as the boy becomes more independent.

An estimated 50,000 students with autism age out of schools in the United States each year.[2] For about 20 years of mandated education they have been supported in their studies and relationships. While their typical peers have choices for "what's next," they have few. According to a 2018 report of the A.J. Drexel Autism Institute in New Jersey, many received insufficient transition services in high school, leaving families struggling to navigate what's out there on their own. The result: many youth fail to launch into any program and, at best, are placed on long, slow-moving waitlists.[3]

Increasing awareness of the critical need for young adult services is vital because after awareness comes programs and support. This chapter provides an overview of transitioning, finances, self-advocacy, presently available resources, and sources of funding. The next one covers continuing treatment, further education, living arrangements, employment, and personal relationships.

Preparing for Transition

From birth, parents around the world strive to support their children to become independent, successful, happy humans. As with planning for aging parents, experts recommend that you cannot start too soon. The groundwork for transition actually starts before anyone has even thought about a transition plan. Can he use the toilet on his own? Brush his own teeth? Make a sandwich? Do laundry? Swim without a life jacket?

From the beginning of the IEP process, fighting for some degree of inclusion in the "least restrictive environment," which the law mandates, is the best way to challenge all students to gradually become more independent. Early decisions that underestimate a student's abilities and place him in a too-restrictive environment can be extremely difficult to overcome, especially in an inflexible school system that offers only "academic" and "nonacademic" tracks.

Even minimal inclusion early on makes learning and gaining new skills easier during transition. For instance, if a nonspeaking student with autism has been placed in skills-based classes rather than academic ones, by sixth grade it may be very difficult for them to catch up in content subjects like social studies and the sciences.

During elementary and middle school, parents must be extremely vigilant to keep teachers and related service providers accountable for meeting IEP goals and objectives, thus assuring that their child continues to move forward, rather than fails to meet the same unreachable goals year after year. Sometimes this process can lead to irreversible decisions that put students on an educational "track" that is impossible to veer from.

When transitioning to middle school, IEP students may already be too far behind to be placed in a mainstream class. Before you know it, they are qualified only to receive a certificate of attendance and not a high school diploma. This possibility is especially relevant in mathematics where students are entering accelerated math classes earlier and earlier.

Transition Services

The transition period of an individual's life (ages 18–29) is now being considered a new developmental stage called "emerging adulthood." For many young people, the transition to independence or a monitored life can be fraught with anxiety centered around adult expectations, decision-making, and moving forward. Some students experience mild, manageable anxiety; others are possessed by severe, sometimes uncontrollable panic attacks. Transition is a time to reassess intervention strategies to avoid potential mental health crises.

The late Andrew S. Halpern (1939–2008), a professor at the University of Oregon, is considered the grandfather of transition. His definition, written more than 30 years ago, takes into account both the individual and his environment. At the time of his death in 2008, Halpern could not have imagined how important this well-conceived idea would be to so many millions of students today.

Transition refers to a change in status from behaving primarily as a student to assuming emergent adult roles in the community. These roles include employment, participating in post-secondary education, maintaining a home, becoming appropriately involved in the community, and experiencing satisfactory personal and social relationships. The process of enhancing transition involves the participation and coordination of school programs, adult agency services, and natural supports within the community. The foundations for transition should be laid during the elementary and middle school years, guided by the broad concept of career development. Transition planning should begin no later than age 14, and students should be encouraged, to the full extent of their capabilities, to assume a maximum amount of responsibility for such planning.[4]

Transition services are based on all of an individual student's needs, taking into account many aspects of a student's future life in different environments. These include, but are not limited to, transportation training, diet, medical needs, recreation, leisure, friendships, and more. At some point, teachers and parents realize that they must change their primary focus from remediating weaknesses to capitalizing on strengths.

Examples of transition services are adult services, post-secondary education, continuing and adult education, vocational education, integrated employment (including supported employment), independent living, supported living, and any type of community participation. Services might also involve specific types of instruction, related services, community experiences, the development of employment and other post-school, adult-living objectives, and, when appropriate, acquisition of daily-living skills.

Assessment

Before writing a student's annual transition plan, school systems are required to perform a thorough functional vocational evaluation. Transition assessment is an ongoing process that continues throughout the transition years and is updated each year as part of the IEP process. As young adults get closer to graduation, educational needs change and goals become more specific.

The assessment process must include anecdotal and formal observations in a variety of settings that can include on-the-job tryouts, classroom performance examples, formal and informal tests, work samples, inventories, and any other sources of information that lead to an understanding of the following:

- aptitudes,
- oral and written communication skills,
- interests,
- strengths,
- weaknesses,
- learning style,
- medical issues,
- work habits,
- unusual behaviors,
- hygiene,
- social skills, and
- values.

Unlike diagnostic testing completed to qualify a student for special education, functional assessments provide a description of what an individual with a disability *can* do, learn, and achieve, rather than focusing on deficits. Unlike previous assessments, during which the student may participate passively, the functional assessment involves the student's *active* participation.

Students whose goals include attending post-secondary educational institutions of various types must determine which ones can accommodate them at their level of academic competency in reading, writing, and mathematics. What are the requirements for graduation? Do they need to pass a competency test to enter? Is there a foreign language requirement?

Many post-secondary schools provide only accommodations, not curriculum modifications. More and more mainstream colleges and universities are now offering special supportive "college-like" certificate programs for students with learning disabilities. Don't despair! If a student graduates with an IEP diploma, indicating that the high school curriculum has been modified, they can still have a successful transition to further education and potential job opportunities, even if the educational focus has been on life skills and functional reading, writing, and arithmetic. See some options in the next chapter.

Transition Is a Team Sport

From the first multidisciplinary team evaluation that precedes diagnosis through adulthood, a multitude of professionals comprise a decision-making group that plans for the future. Members of the transition team are as varied as the people doing the transitioning.

In addition to the transitioning teen, parents, and school personnel, consider including advocates and experts in the following areas:

- assistive technology,
- community living,
- employment,
- exercise and physical health,
- financial planning,
- mental health,
- nutrition,
- sensory integration,
- transportation training,
- vision, and
- vocational rehabilitation.

The young adults themselves, no matter how limited, are important members of the transition team; their interaction, to the best of their ability, is essential to increase their chances of achieving their dreams successfully. Parents and teachers are obviously also crucial, and their goals and objectives for their kids must be "just right," neither too low nor unrealistic.

According to the 2004 reauthorization of the Individuals with Disabilities Education Act (IDEA),[5] by age 16 the Individualized Education Plan (IEP) must provide for "transition services" to move a student from under the umbrella of the school system to the "real world." State laws vary markedly on transition services. Parents must learn what services are available in their resident state, as well as how to access them and how they are funded.

IDEA guarantees the right to an appropriate, free education for all students with disabilities from birth through age 21. Many parents choose to continue in the school system until their offspring's twenty-second birthday to take advantage of those extra four years of education and vocational training. I know one parent who sued the system for four more years after her school system graduated her minimally educated son at 18 with only a certificate of attendance and limited academic skills. She won, and the town was required to pay for one-on-one remedial services to the tune of over $100,000, which allowed the young adult to pass the GED (a high school equivalency exam) and attend college.

Homeschooling

Homeschooling has become increasingly popular, with the population of those choosing this option growing at an estimated 2% to 8% per year over the past several years. As would be expected, it increased drastically from 2019–2021, during the pandemic. Some families of kids with IEPs found that it was more productive for their kids than sitting in front of a computer all day during lockdowns. In fact, they liked it so much they continued homeschooling even after others returned to the brick-and-mortar school buildings when they reopened.

According to the National Home Education Research Institute, an estimated three million plus American kids are homeschooled. Many of these homeschoolers are kids with significant learning and behavioral issues. At some point in a student's schooling, many parents either run out of steam fighting for their student's rights or run up against a wall of a nonnegotiable mandate, such as wearing a mask during the pandemic. At that point, pulling the child out of the school and looking at homeschool programs becomes the only option.

Religion is another reason many parents decide to homeschool. Some of the values of public schools are incompatible with their personal and religious beliefs, and schools run by their local churches may not have the staff or knowledge to undertake the education of students with autism, ADHD, or severe anxiety.

Before endeavoring to jump on the homeschool bandwagon, the first step is to check with the local school to see what requirements must be met. Some districts require only an occasional "check-in," while others are more demanding and must approve the curriculum you are following. All hold annual IEP meetings to measure progress and reevaluate goals and objectives.

Homeschooling a child with ADHD, OCD, or autism can be a challenging task. Finding the right curriculum that caters to their unique learning needs is crucial. Traditional curricula that rely on textbooks, worksheets, and quizzes may not be the most effective approach.

How about cyber school? Local programs abound. Check around with other homeschooling families and see about joining an already formed group. Many have very flexible scheduling that allows students to learn "in the field" with daily trips to museums, factories, or other real-life places for older students to have some on-the-job experiences that could lead to further training.

Community-Based Instruction

For those students with lower skill sets, most schools introduce community-based instruction (CBI) at some point. Many private programs in the community, such as Goodwill Industries, collaborate with the school system, local universities, and technical schools. Programs outside the school system usually combine educational and academic components with job training and life skills instruction. These on-the-job training programs provide opportunities for students to learn not only what they like, but also what they don't like to do.

Starting as trips to local sites which offer job-training opportunities in real-life settings and moving on to assisted internships and "trial" jobs, each unique program is crafted to meet the particular needs of a student and to utilize strengths while continuing to implement IEP goals and objectives. Focus is generally on communication and social goals, such as

- advocating for oneself,
- purchasing food,
- balancing a checkbook,
- doing laundry,
- using the library,
- locating, carrying, and purchasing items in a store,
- using public transportation,
- attending community events,
- ordering food in a restaurant, and
- identifying potential employers in the community.

I believe many "experts" are in too big a hurry to prepare students for the workplace. Some schools start placing them outside the school as early as age 14. Many kids are "late bloomers." Despite everyone's best efforts, most students, especially those with autism, are not prepared for the working world even at age 22 or beyond. Premature transition to a "job" could result in placements that are hardly more than babysitting services or are settings for which the students are neither cognitively nor emotionally ready.

Work on personal life skills—such as being able to communicate either orally or using assistive technology, grooming, personal management, relationship building, and merging home and work life—is all preparation for transition. Instruction on important skills, including managing one's emotions, reading body language, taking criticism, suppressing reactions, and generally responding appropriately, all need to be ongoing. These are the skills that are vitally important to be successful in further education and to keep a job at any level in the future. Waiting for a couple of years of physical, social, and emotional maturation makes more sense to me.

Successful transition also requires some degree of literacy, such as reading signs, writing one's name, address, and telephone number, and comprehending basic concepts of time, money, and measurement through schedules, budgeting, cooking, and building. Extra time to solidify foundational skills can make the difference between restocking shelves in a grocery warehouse and keeping inventory at point-of-sale. While these two jobs may not seem dissimilar, their contrasting demands may account for increased job satisfaction and moving from minimum wage to a decent salary.

The Americans with Disabilities Act (ADA)

The Americans with Disabilities Act of 1990 (ADA) extends services beyond secondary school into college and the workplace. The ADA requires publicly funded colleges and universities to remove barriers that keep out disabled students. Those with disabilities can arrange to take college entrance exams with accommodations. Once it accepts a disabled student, a college must accommodate the student's needs, whether it's with recorded books and lectures, a quiet, isolated area to take tests, or permission to dictate and record, rather than write, reports. Today most colleges have a Students with Disabilities Office.

The ADA also applies to the workplace, where it guarantees equal employment opportunities for people with disabilities and protects disabled workers against job discrimination. Employers may not consider the disability when selecting among job applicants. They must also make "reasonable accommodations" to help workers who have handicaps do their job. Such accommodations could include shifting job responsibilities, modifying equipment, or adjusting work schedules. During transition, parents, educators, employers, and the student himself need to be aware of the provisions ADA mandates so that everyone can advocate for and respect the rights of the employee or student with a disability.

Finances

Who assumes the cost of care for an individual with a disability who cannot take care of himself? Every family is different. The stress on the family of having a child with special needs can be unfathomable. RFK Jr. was so bold as to say that "autism destroys families." I agree; it often does. The divorce rate is high, and both financial and emotional resources are stretched inequitably.

Who is responsible once the parents are gone? If there are multiple children, the obvious solution is to engage siblings in the long-term care plan. Depending on their ages and relationship status, this can be challenging. Some become natural caretakers as their dependent sibling matures. Others are wrapped up in their own lives. I know families with both situations.

From day of diagnosis, this is one of the biggest concerns facing families. While parents' health insurance covers young adults up to the age of 26, when costs exceed insurance caps, this question looms large. Various options are available, none is perfect.

Guardianship and Special Needs Trusts

Guardianship

One option is *guardianship*; the process begins once a child is 17½. Delaying could cause a gap with the potential to jeopardize an emergency medical situation. Parents are usually the legal guardians of all their children until they turn 18, at which time the children legally become adults, responsible for making their own decisions about money, healthcare, residence, and more.

Many parents of children with profound special needs choose to apply for guardianship because they recognize that at 18 their children are not able to make these decisions independently. Consult with an attorney specializing in this area before making this momentous decision.

Both *full* and *limited* guardianship are options. Obtaining full guardianship presumes that a child is "legally incompetent." Another adult takes complete responsibility for a child's welfare, stripping him of some legal rights, such as the right to choose a home, vote, or marry without consent.

Alternatives to full guardianship include

- Obtaining limited guardianship, which makes a parent or someone else responsible for money or medical care and allows an individual to retain their remaining rights;
- Obtaining power of attorney, which gives someone else authority to make legal, financial, and medical decisions on behalf of the individual; and
- Appointing a temporary guardian or conservator in an emergency, when certain decisions must be made immediately and the individual cannot make them independently.

Guardianship Capacity Questionnaires[6] are available from each state for help in deciding which option to choose. Whichever option works during parents' lifetimes, make arrangements for a replacement guardian after death and put this decision in parents' wills.

Special Needs Trusts

Since 1993, the federal government has authorized several specific trusts that can fund supplemental care and quality of life for individuals with disabilities. The term *Special Needs Trust* usually refers to (d)(4)(A) trusts; the letters and number allude to where they can be found in the statute that authorizes them.

Special Needs Trusts are solely for the benefit of individuals under the age of 65 years and disabled according to Social Security standards. Also called *Disability Trusts* or *Supplemental Needs Trusts*, these funds can be established by an individual's parent, grandparent, legal guardian, or the Court for the sole benefit of an individual with a disability.

Special Needs Trusts allow a person with a disability to earn or be gifted money, thus protecting him from losing certain government benefits that are denied people who have personal bank accounts that exceed $2,000 at any given point in time. Benefits may include Supplemental Security Income (SSI), Medicaid, vocational rehabilitation, subsidized housing, food stamps, and other benefits based upon need. A Special Needs Trust provides for supplemental and extra care over and above that which the government provides.

A properly drafted trust includes provisions for termination or dissolution under certain circumstances and explicit directions for amendment when necessary. All assets in the trust belong to the beneficiary. Each Supplemental Needs Trust is its own "entity" with its own EIN ("employer identification number") issued by the Internal Revenue Service. A number of types of trusts require legal help to understand and draft. Variations include *Third-Party Irrevocable Trusts*, *Sole-Party Trusts*, *First-Party Trusts*, and *Pool Trusts*. The Special Needs Alliance offers a handbook on this subject. Learn more at SpecialNeedsAlliance.org.

Special Needs Trusts are a tax benefit for a person with a disability and his family. Unlike a traditional trust, a properly drafted and administered special needs trust is not counted as an asset, and trust disbursements are not included as income under the rules that apply to SSI and Medicaid.[7]

Families interested in establishing a Special Needs Trust must find an attorney with expertise in trusts and discuss their unique situation. Some insurance companies and pro bono lawyers also provide such services.

Government Funding

Allocation of funds for those with disabilities varies markedly from state to state and year to year. The law requires that during transition school districts put families in touch with local programs and benefits as a part of the transition process. However, they do so with rather wide-ranging success or lack thereof.

Many school systems offer a free annual open house where all providers—such as lawyers, housing providers, and other local resources in the area—can exhibit. The school is only obligated to provide the information; whether a program is appropriate is up to the parent to discover.

Here are some of the revenue streams that are available to pay for housing, rehabilitation, nursing, medications, and other continuous needs once an individual turns 21.

- *Supplemental Security Income (SSI) and Medicaid* are called *means-tested* benefits because to qualify an applicant must be both disabled and have extremely limited assets and income. For SSI an applicant can have no more than $2,000 in assets.

  Those with severe and profound issues generally qualify for Supplemental Security Income (SSI), a payment made jointly by the federal government and the state of residence to cover some of the living costs for individuals with disabilities. If a young adult lives at home and pays rent, he may qualify for higher-income SSI.

  People who qualify for SSI and live outside their family home usually receive *Personal and Incidental Money (P&I)* every month as part of their SSI payment. This money is for clothing, entertainment, or other "incidental" expenses. If the person earns an income, then this money is deducted from the P&I and must be reported to Social Security.

- *Medicaid Waiver*—A *Home and Community-Based Services Waiver* is open to those with several specific disabilities, including autism. The waiver provides both Medicaid and services (potentially including *personal care assistants*, or *PCAs*) for behavior management, respite, transportation, etcetera. There are several types of waivers. A good source of information is FamiliesUSA.org.

- *Social Security Disability Insurance (SSDI)*—Some individuals may also be receiving Social Security disability income under their own claim or Social Security *Disabled Adult Child* (DAC) benefits under a parent's claim. The amount of money received by the adult child depends upon the income and the number of years of Social Security contributions paid by the parent. After 24 months on SSDI, an individual is eligible for Medicare.

- *Health Insurance*—The Affordable Care Act of 2010 made many provisions expanding access to insurance options through Affordable Insurance

Exchanges and improvements in Medicaid and Medicare. This may be changing now with the Trump administration.

New health plans sold in the individual and small group markets, including Exchanges, cover "essential health benefits" to help make sure that health insurance is comprehensive. Health insurers also offer annual out-of-pocket limits to protect families' incomes against the high cost of healthcare services.

These provisions can change quickly as the government passes new laws. Check with your insurance agent and school system to see what is available locally today because this is a moving target.

- *In-Home Supportive Services (IHSS)*—Money from this source can be used to pay someone to come to the home to help with personal tasks such as cooking, shopping, cleaning, paramedical services, medical transportation, and protective supervision. Parents may also be the paid provider for an adult with disabilities.

- *Shift Nursing*—Several government programs offer funding for nursing, if needed, under the *Nursing Facility Waiver* for persons over age 21. Shift nursing may cover nursing expenses up to what it would cost if the individual were to be placed in residential care. This care is available for adults residing in their own homes, licensed homes, foster homes, and for those in independent or supported living.

Letting Go

One of the most difficult parts of the developmental process for all parents and caregivers is knowing when and how much to let go. For parents of children with special needs, a continuous "push-me/pull-you" occurs between the need to protect and the need to allow more independence.

If a child has had chronic medical or behavioral issues, the need to protect can be so strong that letting go is even more difficult. Maybe he will have a seizure, get stung by a bee without an EpiPen, unknowingly eat something with peanut oil, wander away, or be bullied or taken advantage of in unthinkable ways. But as philosopher Friedrich Nietzsche noted, maybe "that which does not kill us makes us stronger."[8]

Most people can cite a mistake from which they learned a lesson that changed their life. Making mistakes is crucial to understanding cause-and-effect relationships. Making "little" mistakes, like being late and missing an event are important steps in preventing "big" ones, like having unprotected sex or driving drunk, that can have dire consequences. With proper planning, letting go becomes easier.

Thomas W. Welch, PsyD, has become an expert on transitioning, based on his own upbringing as a student with ADHD and dyslexia. Now a child and family psychologist who specializes in work with those on the autism spectrum near Denver, CO, he founded Humanex Academy, a middle and high school dedicated to the growth and development of this population.

Welch's book *The Breakaway: A Parent's Guide to Transitioning the Autistic and Twice Exceptional Adolescent into Young Adulthood*[9] guides families and professionals through the challenges of helping teens gradually take greater initiative and responsibility for themselves. It includes such important subjects as timing purposeful nudges forward, staying engaged and flexible, when to intervene, and how to avoid power struggles, set achievable goals, create game plans, reassess, and resist the impulse to panic.

Self-Advocacy

This chapter would be incomplete without a discussion of advocacy. For their entire lives most individuals with special needs have depended upon parents, teachers, and others to advocate for them. No matter how disabled a person is, at some point during transition, they must learn some self-advocacy skills to survive. These may be as simple as indicating the need for help when lost, a more complex skill such as learning to use assistive technology to communicate, or as complicated as going to court to access desired services.

Learning how to advocate for oneself starts young. We teach preschoolers to stand up for themselves on the playground and in the block corner. Likewise, we should work from the start with our students with special needs to assure that their needs, hopes, and desires are being met. The latest technology and assistive aids, like letterboards, give a voice to those who lack the skills to "speak for themselves."

Tom Iland can help. A unique Certified Human Potential Coach, TEDx Speaker, and a Toastmasters International Accredited Speaker, Tom was diagnosed on the autism spectrum at age 13. Today he helps those like him find *their* meaning of L-I-F-E: *L* for *love and relationships*, *I* for *independent living*, *F* for *further education*, and *E* for *employment*. See ComeToLifeCoaching.com.

For those with special needs, self-advocacy starts with knowing what the law says about rights and responsibilities. Self-advocacy courses should be a part of every transition program. Any student who wants to be included in the mainstream, whether as a part of a sports team, a club member, or a participant in a dance contest, must learn that inclusion is a right. After graduation, pamphlets, books, workshops, webinars, and seminars from many groups can help young adults understand the various ways in which they can stand up for themselves.

A common thread running through successful transition programs is a balance between accommodating an individual's personal sensory needs and learning to navigate his/her sensory environment. To me, this is a missing link in many of those diagnosed late with ASD.

These late-diagnosed individuals often cite their sensory differences as simply "unique wiring" instead of recognizing that hyperalert senses are how the nervous system tells the brain that you are not safe. The premise of this book from the beginning is that not feeing safe is a poor foundation for future social, behavioral, communication, and educational functioning.

Accepting that an out-of-balance nervous system has diagnosable causes and, most importantly, efficacious treatments is crucial. Improving nervous system function, when possible, is better than requiring the world to stand on its collective head to accommodate "quirkiness." I just attended a webinar given by an adult, who took a 10-minute break for a needed breather only 15 minutes into their presentation. While the speaker had learned good self-advocacy, they might be also be served by learning calming techniques that could obviate the need to inconvenience their listeners.

The area of sensory needs is an especially important one for self-advocacy. Revisit Chapter 16 for more information on how to develop the "just right" sensory diet. Only the individual can determine whether the environment is too loud, his clothes too tight, or his world too bright. Providing young adults with sensory barometers and ways to communicate discomfort assures a more peaceful transition at home and in the workplace.

Read more about this important topic in the next chapter, which extends self-advocacy to the workplace and relationships.

Additional Resources

Books, websites, organizations' materials, webinars, conferences, and other sources abound for those working with teens transitioning to adulthood. Grassroots organizations are popping up in Japan, the United Kingdom, Australia, and the U.S. In addition to those mentioned in this chapter, here are a couple of others that are excellent:

- Grigal, Madaus, Dukes, and Hart, *Navigating the Transition from High School to College for Students with Disabilities, 1st Edition*, 2018.
- Murin, Hellriegel, and Mandy, *Autism Spectrum Disorder and the Transition into Secondary School: A Handbook for Implementing Strategies in the Mainstream School Setting*, 2016.

- Laviano and Swanson, *IEP Guide for All: What Parents and Teachers Need to Know About Individualized Education Programs*, 2025.
- ADayInOurShoes.org offers printable lists of IEP goals and objectives for specific behaviors, including anxiety.
- Homeschool.com has suggestions for homeschooling those with special needs, as well as state laws on homeschooling.
- National Home Education Research Institute: NHERI.org is a 501(c)(3) nonprofit organization specializing in research, facts, statistics, scholarly articles, and information on home education.

Take-Home Points

The time to start planning for the future of our loved ones at all levels is now! Educational, prevocational, and financial decisions should look beyond the day after tomorrow. Work with experts in transitioning to learn what options are available in your geographical area. Get acquainted with key people so you know who to go to when the time comes. Assume competency! Make every effort to encourage independence even when it seems like a real reach for your family member with a neurodevelopmental condition to accomplish a goal without support.

Living in the world of special needs can be full of surprises like my email almost 15 years ago. Center 21 and other innovative programs around the world can be models for others. Keep reading to learn about them.

CHAPTER 22

LIVING, EMPLOYMENT, AND BEYOND

When the school bus stops coming, life halts for thousands of students graduating with IEPs who exit the only schools they have known without sufficient skills to be independent. They have nowhere to go every day . . . no school, no program, no life. Few alternatives exist, although more are showing up each year.

Unlike their siblings who go off to college or another post-secondary placement, 18- to 22-year-olds with special needs are frequently stuck at home with Mommy and Daddy. Their parents, who might otherwise have become empty nesters, are also stuck, tethered to an adult child who still needs support for the activities of daily living. Many cannot imagine their offspring living anywhere but with them, and most live in fear of what will happen to their kids when they are gone.

In this chapter, the subject of transitioning to adulthood continues. We start with more on self-advocacy. Whatever communication, social, and academic skills an individual brings out of transition will determine next steps. Though a free and appropriate education has ended, learning has not. Continuing to advocate for oneself and increase knowledge in content areas, in addition to finding a vocation, is as crucial to those with IEPs as it is for those without.

The chapter moves to a discussion of day programs and temporary residential transition options as well as some permanent living arrangements. Then, I tackle the importance of evaluating career choices and employment possibilities and the need for periodic reassessment. We'll then explore the ins and outs of neurodiversity in relationships and the loaded subject of gender identity. We end with a discussion of the benefits of the martial arts.

More Self-Advocacy

The previous chapter introduced this important subject through the end of high school. When someone with a disability is still a minor, their parents usually advocate for them while, at the same time, assisting them to develop self-advocacy skills. During transition, self-advocacy should be an IEP goal. After leaving high school, teaching self-advocacy falls on the shoulders of parents, post-secondary mentors, and caregivers.

During emerging adulthood, self-advocacy is mandatory to asserting one's disability rights, especially for "invisible" disorders like learning disabilities and anxiety. Self-advocacy can be as simple as asking for a momentary need to be fulfilled, such as requesting a glass of water or that loud music be turned down. Or it could be more strategic, such as requesting headphones to dull noise at work. At its extreme, it can be life changing and/or lifesaving, such as recognizing the need for essential medical care or an assistant at work or, in personal relationships, recognizing and addressing abusive or sexually inappropriate behavior.

One's level of self-advocacy depends upon one's level of communication and social skills. Self-advocacy *ability* varies tremendously, and many folks are going to require advocacy help for their entire lives. As expressive language and socialization improve, the ability to self-advocate will continue to improve as well. This is an ongoing lifelong process for all human beings.

The Autistic Self Advocacy Network (ASAN), a 501(c)(3) nonprofit organization founded in 2006, is there to help those on the spectrum. ASAN, run by and for Autistic people, or *Autists*, promotes a culture of inclusion and respect for all; however, its focus is primarily on those with Level 1 autism.

This group advocates at both federal and local levels to enforce the rights of Autists to equal opportunities at school, in restaurants, public places, and most importantly, in the workplace. An *autism-friendly* environment accommodates and accepts neurodiversity in all aspects of society. This includes creating spaces that minimize sensory overload, provide flexibility, communicate clearly and consistently, and foster an inclusive culture.

ASAN members seek to improve funding for community services and support and provide research on the above subjects. They believe self-advocacy is essential to this process. Their slogan is "Nothing About Us Without Us." See AutisticAdvocacy.org.

Continued Treatment

As bodies get bigger and heavier and hormones rage, needs of all kinds change. Working with healthcare practitioners to revisit dosages and frequencies of medications and supplements is essential. Small adjustments can make big differences in behavior, focus, language, and learning. Interventions that previously failed to result in gains may now be just what is needed.

Instead of discontinuing therapies as students age out of school, consider applying them to the workplace. For instance, the emphasis in vision therapy might move from tracking and eye teaming to visual thinking, planning, and organization. Maybe motor and visual skills have matured enough to use a letterboard, which was not generally available five years ago.

Once communication is more articulate, speech and language support can focus on enriching vocabulary and improving pragmatic language skills. Knowing when it is your turn to talk, making small talk, asking for what you need, and knowing how to phrase what you want to say are all complex language skills that can enhance self-advocacy.

In the motor areas, clumsy children and teens have the potential to become young adults who enjoy simple individual, noncompetitive sports such as skating, swimming, and biking. Equipment is now available for almost any size and body.

Post-Secondary Education and Living Arrangements

Many students with IEPs graduate with certificates of attendance rather than high school diplomas. They are often "dumbed down" early in their education and never exposed to math beyond simple calculations, science, social studies, or literature because no one assumes competence. It is truly shameful!

What about a GED? This abbreviation can stand for *general education degree* or a similar phrase. It is a high school equivalency diploma that requires an eighth grade reading level, writing competence, knowledge of some history, geography, science, and a few other basics. A high school diploma or GED equivalent is usually necessary to attend higher levels of education.

Remember Alex, one of the first Spellers whose mother Judy has become an amazing advocate in the use of letterboards? Alex, like many of his peers did not receive a high school diploma, but rather a certificate. Yet, unbeknownst to any adult, his knowledge was rich because he snuck into his brother's room and read his textbooks! He took the GED test and passed with flying colors.

If a student does not have the basics to pass the GED, courses are available in virtually every community and online. Check them out at GED.com.

Day Programs

Most young adults who have finished high school without a degree start their post-secondary education by commuting to day programs in their community. These vary markedly, from understaffed adult day care centers to some that offer precollege workshops and GED prep. Day habilitation programs emphasize keeping safe and busy, medical monitoring, using available community resources, making friends, and maybe developing some interests and skills. The ultimate goals for all young adults are the same: to have friends, to have something to get up for in the morning, and to lead productive, meaningful lives. Those are the goals in the

United States, Japan, Europe, and for the members of Center 21 in Kuwait; they are universal in other words, whether the new adult lives at home, in a supported residential setting, and/or attends a post-secondary educational facility.

States differ markedly in their offerings, and financial arrangements vary according to a young adult's needs. Parents must take the time to investigate options along with the public school system during transition.

Community colleges are becoming increasingly flexible in their admissions policies and programming. Some now allow students to enter without a GED and take courses pass/fail; others offer support, such as mentors, readers, and even tutors. Options change frequently, usually depending on how money is allocated.

Don't like any of the programs in your area? Consider Autism College, founded by Chantal Sicile-Kira in 2011 to share with others what she had learned about advocating for her son Jeremy, who is described in Chapter 18. It is not an actual brick-and-mortar college, but rather an online resource for parents and professionals to make important life decisions regarding education and therapy for young adults with severe autism. Although nonspeaking from apraxia, and requiring daily personal support, Jeremy graduated from high school in California with a diploma. Today, he is a successful artist. Visit AutismCollege.com.

Temporary Residential Post-Secondary Programs

How many of our neurotypical kids have ever done their own laundry before going off to college? Our kids with neurodevelopmental differences can learn too. Whether the placement is at home, in an adult day center, or a group home, increasing independence is key to success.

Experts strongly recommend beginning the transition to supported adult living early on, when a student is still in high school. Instead of going away to college, consider some of these college-like "bridge" programs:

Chapel Haven was established in 1972 (by parents, of course!), making it one of the oldest programs of its kind in the nation. Today, it is an award-winning, nationally accredited school and transition program serving 250-plus adults with a variety of abilities and needs.

Chapel Haven offers several two-year residential programs for those age 18 and older with Level 1 or 2 ASD and other learning and social disabilities that could make them vulnerable in other settings. Four- and eight-week summer programs are also available in New Haven, Connecticut, and at a satellite campus called Chapel Haven West in Tucson, Arizona.

The Residential Education at Chapel Haven (REACH) program provides life skills and functional academics along with social skills training, communication competency, and recreation opportunities. Participants prepare for supported employment and life outside the home. Residential students live in staffed apartments where they learn housekeeping, meal preparation, time management, and social skills.

The Asperger's Syndrome Adult Transition (ASAT) program is for students who could benefit from transitional housing with individualized support. ASAT offers opportunities to attend college-level classes at local schools and work in appropriate internships.

In 2017, the program was expanded to offer day programming for local students needing support for college and/or vocational development. For the first time, services became available for older adults with previous failed experiences in independent living. Chapel Haven offers targeted instruction in executive functioning, time management, college survival skills, and social communication.

Chapel Haven is approved for funding in several New England states as well as in Arizona. See ChapelHaven.org.

The College Internship Program (CIP) is one of the most comprehensive transition programs in the world, helping adolescents and young adults with autism, ADHD, and other learning differences find success in college, employment, and independent living. It was founded in 1984 by Michael McManmon, EdD, who was diagnosed with Asperger syndrome in his early fifties, CIP's mission is to help emerging adults make successful transitions from adolescence to adulthood.

This multidimensional program teaches independent living skills such as organizing, cooking, and cleaning, while students either attend college or gain employment experience. Residents live in furnished apartments in California, Florida, Indiana, and Massachusetts where they are monitored carefully. Well-trained staff provide whatever supports are necessary for a student to be successful socially, academically, and personally. Services can include developing plans for personal fitness, nutrition, stress reduction, a sensory diet, and sleep.

The College/Certification Track is a yearlong post-secondary transition program offering individualized academic, social, career, and life skills support for those with autism, attention deficits, and learning disabilities while they attend a local college or university. A unique community-living model providing an integrated transition to the next level of independence follows.

All CIP students eventually gain appropriate employment skills through extensive training, counseling, community service, resume development, and internship and job placements. Visit CIPWorldwide.org.

Center for Discovery is a continuously growing 1,000+ acres in the Catskill Mountains, 90 minutes northwest of New York City. Among other services, it offers residential-living options to individuals over 21. The program educates and supports the whole person through the HealthE6 Model™, six pillars designed to optimize health, functioning, and learning and to decrease stress on body and brain for individuals with complex forms of autism and medical frailties:

Environment: physical, academic, and social;
Eating: organic, optimized nutrition;
Energy Regulation: focus on movement, exercise, and restorative sleep;
Emotional Regulation: self-regulation, empathy, and self-efficacy;
Evidence-Based: technology and individualized assessments;
Education: scientific research, focused on family and other collaboration.

The result is decreased maladaptive behaviors as well as increased focus, attention, and positive interaction.

The Center for Discovery provides a wide variety of healthy vocational and experiential activities such as exercise, yoga, swimming, concerts, sporting events, and enrichment classes on a working farm, with an on-site bakery, theater, art studios, and other venues. See TheCenterForDiscovery.org.

Spectrum Tech Trade School, Village, and Training Center in Monmouth, NJ, was conceived and founded by Lisa Sosnowski in 2015. It's more than a school—it's a movement, one that embraces individuals with language delays, sensory issues, and learning differences often overlooked in conventional education. This amazing mother-driven project prepares students for a lifetime of supported independence. This is an evolving program that assumes competence and embraces neurodiversity by capitalizing on strengths and challenging stigmas and stereotypes. Visit SpectrumTechTradeSchool.com.

Maple Mountain is a fairly new 35- to 60-day program located near Salt Lake City, Utah, designed as an integrative program for those with mental health issues and chemical dependency. All-inclusive services start with a full panel of metabolic and genetic testing followed by individualized nutraceutical protocols, diets focused on gut-brain health, IV therapies, yoga, acupressure, and sound therapies. All are designed to address physical and mental health concurrently. See MapleMountainRecovery.com.

Permanent Living Options

In Kuwait, where families are large and extended and help is readily available, taking care of an adult who is unable to take of himself is hardly problematic. Support is everywhere, from beloved nannies to siblings and cousins.

In countries where families are scattered and everyone is busy with work and their own lives, a time eventually comes to explore different living arrangement options. Who can shoulder the responsibilities of caring for a child who might never grow up: parents, a sibling, another relative, a friend, a paid aide in a rental apartment, a licensed or group home, or can this individual live alone?

Camphill

Camphill is an international movement consisting of more than 100 communities in 22 countries. Inspired by the works of philosopher, educator, and Renaissance man Rudolf Steiner—collectively called "anthroposophy"—Camphill was founded in 1939 in Scotland by Karl Koenig, MD, an Austrian pediatrician and educator, and others. At Camphill, children, youth, and adults, with and without special needs, learn and work together while living the philosophy that those with special needs should be included and valued, not warehoused away from the rest of the world.

Camphill's goals are

1. providing education, advocacy, therapeutic care, and other services to support people with disabilities and helping them participate fully in the world as contributing citizens;
2. caring for and healing the earth through sustainable and healthy methods of consumption, agriculture, and natural resource use; and
3. creating social arrangements designed to nurture the growth and development of individuals, so that all community members contribute their time and skills according to their capabilities.

Camphill was established in North America in 1959. Today, 15 independent communities serve over 800 people. Camphill is interested in expanding its programs to include individuals with every type of brain. Four communities specifically serve young adults.

Heartbeet Lifesharing, located in Hardwick, VT, was founded circa 2005 by Hannah Schwartz and her husband, Jonathan Gilbert. It is a thriving intentional biodynamic farming community where individuals of all abilities live together in beautiful extended-family, life-sharing homes. It includes residents with and without special needs, volunteers, coworkers, families, dogs, cats, cows, goats, chickens, and pigs.

Schwartz grew up at Camphill Village Kimberton Hills, PA, near Philadelphia, where both of her parents worked. She had always dreamed of starting her own

Camphill and found the perfect spot with 150 acres in rural Vermont. She built this grassroots farm from scratch.

At Heartbeet each person participates fully in life and is able to experience a sense of purpose and accomplishment. In addition to the ongoing rhythm of farming, residents learn the traditional crafts of weaving, papermaking, felting, and culinary arts. It also includes Camphill Academy, a practice-integrated post-secondary program in Inclusive Social Development, the art of building inclusive communities. If you visit, I bet you will fall in love with the place, as I did. See Heartbeet.org.

Riverflow Community, also in Vermont, is the newest addition. It started with tenacious local parents of young adults with no suitable place to go, an idea, and the renovation of an abandoned eight-bedroom house. The founders approached Hannah, above, and their first residence opened in 2024. Check out progress at RiverflowCommunity.org.

Soltane has been dedicated to their mission of helping young adults with a variety of differences curate a vibrant, purpose-driven life after high school for 37 years. It houses approximately 90 people, about half of whom have disabilities, on a 50-acre site about an hour west of Philadelphia. Students have access to classes offered through an extensive education program. Visit CamphillSoltane.org.

Triform is a dynamic residential and day community of over 100 people spanning several generations, including about 40 young adults with developmental disabilities as residential or day students, who are at the beginning of transitioning to adulthood. Located in the Berkshire Mountains near Hudson, New York, this 410-acre biodynamic/organic farm is home to those with autism, Down syndrome, and other special needs who live and work side by side with staff. See Triform.org.

Go to Camphill.org to learn about other Camphill programs throughout the world.

Other Farming Communities

Being outdoors, growing and eating organic food, working with animals and bonding with the earth are natural options for anyone trying to maintain good mental health. That's why farming communities are forming all over the world to offer living and employment opportunities for those with all levels of abilities. Two near where I live:

- *Bittersweet Farms*, established in 1983 as the first farmstead-based program for adults with autism in North America. Initially located near Toledo, OH, Bittersweet has expanded across northwest Ohio, opening a second farmstead in Lima in 2005 and a third in Pemberville in 2006. Today, they provide residential, vocational, transitional, and recreational services to over 100 individuals at these three locations. Learn about it in the 2023 book *Creating Quality of Life for Adults on the Autism Spectrum: The Story of Bittersweet Farms*[1] or at BittersweetFarms.org.

- **Good Works Farm**, located in southwest Ohio, started in 2015 by Nancy Bernotaitis and her family, when her son with autism was fifteen. It offers both a summer respite program and individualized transition programs. See GoodWorksFarm.org.

Careers and Employment

About 50 years ago, I started out my own career as a vocational counselor. My choice of that profession went back to my teens when my mother, then a volunteer in an adult day care center, brought home some completed interest inventories she had given to the center's residents. Instead of leaving the participants to stare at the TV all day, someone got the brilliant idea to offer classes and investigated what might be of interest. I was fascinated.

I had always loved math, and these questionnaires quantified interests and looked at how to apply them to something meaningful. Not only was I taken with this idea, I decided then and there that I wanted to help people find their life's work as my own life's work. My dream job appeared at a local university when I was in my thirties. There, I assisted undergraduate students in finding a satisfying major, helped empty nesters determine "what next," and brainstormed with disgruntled lawyers and burned-out teachers on new ways to use their hard-earned skills in environments that were less stressful than courtrooms and inner-city schools.

I still enjoy exploring options with anyone who is struggling to find meaningful work. In the absence of a career counselor, parents and teachers must take on this role of turning talents, interests, and obsessions into paid work and a career. For those with neurodevelopmental issues, choosing a vocation is a good time to turn perceived negatives into positives. The flip side of inflexible thinking is loving routine; the positive outcome of difficulty making transitions is a strong ability to concentrate for lengthy periods of time. In fact, qualities often viewed as weaknesses could turn out to be important, unique strengths, especially when applied to the latest technology and artificial intelligence.

Obviously, individuals with anxiety, ADHD, or ASD are as different in their career choices as they are in every other aspect of their being. Twenty years ago, when Temple Grandin started counseling families of those with autism, she wrote

Developing Talents: Career Planning, Including Higher Education, for Students with Autism and Asperger Syndrome[2] about some fields she thought were particularly well-suited to those individuals. These include

- animal handler,
- artist,
- auto mechanic,
- computer programmer or technician,
- crafts,
- drafting,
- engineering,
- graphic design,
- landscaping,
- theatre lighting and sound,
- music, and
- web design.

While I think a wide variety of fields are potentially open to everyone, interest in and love of computer technology, music videos, car parts, collecting, classifying, and categorizing are common among those on the spectrum. Furthermore, they can all be translated into meaningful careers.

OCD is another diagnosis that makes it possible to predict jobs that could bring pleasure and some to avoid. A love for order, detail, routine, cleanliness, and a preference for working alone makes it easy to rank job categories. OCD-friendly jobs usually have clear rules and let you manage your work. The trend toward working virtually can be a boon to job seekers with OCD.

Web designer, proofreader, medical coder, photographer, housekeeper, and accountant are just a few professions in which those with an OCD diagnosis have found success. Jobs like customer service, sales, childcare, and nursing could be stressful.

Not every person with autism or another neurometabolic diagnosis shows strong interests, talents, or even obsessions. Like those without diagnoses, many need guidance and encouragement to find something they can be passionate about.

Employment

Finding employment that is meaningful, satisfying, and pays the bills is difficult enough for new graduates without nervous-system dysregulation. Add a little anxiety, an occasional panic attack, poor communication skills, and some procrastination, and landing that first job could be a nightmare. Fortunately, more and more companies realize that attracting neurologically atypical potential employees requires some innovative methods. Some even view neurodiversity as a competitive advantage and are tailoring the application process to find individuals with niche skills they need.

Parent-Founded Programs

Not surprisingly, fathers and mothers have been at the forefront of developing innovative post-secondary programs for their offspring whose needs are unmet by available offerings. For those still developing communication and social skills, dozens of innovative opportunities have arisen in the last couple of years. Here are a few amazing ones I have learned about.

Rising Tide Car Wash. In 2011 when his son Andrew aged out of school, John D'Eri came up with the novel idea of buying a car wash to employ Andrew and others like him. The family started Rising Tide Car Wash in Florida, along with D'Eri's other son, Tom, now the COO. See RisingTideCarWash.com.

Today, this thriving business is one of the largest employers of people with autism in the United States! Four unexpected wins:

- every employee felt safe;
- accountability became a tool for growth;
- everyone's work gave them a sense of purpose; and
- customers loved their experience.

And they now offer an online course with step-by-step instructions for creating a business that can sustainably employ your loved one with autism. Visit TheAutismAdvantage.com.

The Chocolate Spectrum. Valerie Herskowitz created The Chocolate Spectrum, an artisan culinary outlet in Jupiter, Florida, where her son and about a dozen young adults with autism produce delicious fudge, truffles, and other goodies. Marco, a 26-year-old with autism, including severe language and auditory processing deficits, began working at The Chocolate Spectrum for only two hours a week. He gradually worked his way up to nine hours. Under Valerie's loving guidance and much supervision at first, he learned to independently combine ingredients, decorate finished products, and wrap them for distribution. Visit TheChocolateSpectrum.com.

The Teaching Hotel. Jeff Huffman partnered with Marriott and his local ARC to build The Teaching Hotel, a six-story, 150-bed Courtyard in Muncie, Indiana. Luckily for Huffman, Muncie's mayor had a grandson with autism and was interested in helping him find funding: a five-million-dollar grant from the state. Opened in 2016, the hotel provides 130 jobs, 20 percent of which are for citizens with disabilities who are educated by the Erskine Green Training Institute for positions in management, catering, culinary, front desk, laundry, and room service. This project is a model for similar programs nationally.[3]

Specialisterne. The Specialist People Foundation. Looking ahead to a possible career for his son with autism, Dane Thorkil Sonne left his job of 15 years, refinanced his house, and founded Specialisterne, aimed at creating jobs for adults with autism. Today, Specialisterne employs techies who solve computer problems for leading telecommunications companies in over a dozen countries around the world. Consultants provide a range of services, from software management to data testing and logistics and consultancy and design, to meet today's business needs.

Specialisterne USA hired its first group of consultants in July 2013; all had ASD diagnoses. When they sent consultants to EY (formerly Ernst & Young),[4] everyone was surprised that some employees with autism grew beyond the company's expectations and defied autism stereotypes.

A dandelion seed is the company's logo "to remind everyone that we all have the potential to realize our potential if welcomed and understood. It is the settings we end up in that determine if we will be able to grow or wither."

The Specialist offers an individual assessment where participants clarify their strengths, weaknesses, special aptitudes, capabilities, and interests. This is followed by a training program during which they map out their needs for support, guidance, and environmental adjustments in order to perform successfully as IT consultants.

Specialisterne's four-week Pre-employment Program is designed for autistic high school and college students. Participants explore opportunities in the labor market and how they can prepare for active careers by assessing and building their skill sets, motivation, and workability. Specialisterne also runs a school in Denmark with a three-year program for neurodivergent young adults aged 16 to 24.

The Specialist People Foundation has set the ambitious goal of providing meaningful and productive jobs for one million people with autism and other "invisible" disorders. To do this, they will replicate Specialisterne operations around the world.

Sonne recognized that the likelihood of an individual with autism completing the required steps of applying for a job by filling out an application, sending in a resume, enduring a round of interviews, and visiting the human resources department of a major company was slim. Through Specialisterne, the German software company SAP[5] and Microsoft[6] became interested in leveraging the strengths of those on the spectrum; they now modify application procedures, provide mentoring and training beyond a typical orientation, and support those with autism as long as is necessary. They have found their efforts worth it, as their employees with autism are generally loyal, hard-working, and skilled. Moreover, they feel valued, enjoy what they do, and are even growing in social and other skills that might have prevented them from achieving prior to these opportunities.

SAP's Autism at Work program hopes that employees on the autism spectrum will reach one percent of SAP's workforce. Not only are many good at testing software, but some also become graphic designers and even supervisors, a job requiring "people skills" often considered lacking in those on the spectrum.

Bottom line: innovative businesses are popping up all over the globe. All it takes is an idea, some serious fundraising, and determination. Read Tom D'Eri's book[7] about how you can emulate the Rising Tide Car Wash and start a meaningful social enterprise like his family did. Step out of your comfort zone, speak to those at work, your church, synagogue, or book club. Not-for-profits, foundations, and philanthropists are all interested in supporting those with potential.

Underemployment

Poor mental health and a lack of job satisfaction make a vicious circle. If someone is depressed, anxious, or obsessive, managing to find a suitable position and go through an application process can feel like climbing Mt. Everest. Once a job has been secured, showing up daily, on time, and getting through the required tasks of a position takes perseverance and support.

Despite efforts of companies large and small, only a small percentage of those with serious neurodevelopmental disorders hold full-time jobs for the very reasons they are diagnosed in the first place. They may be unfocused, inflexible, and/or unable to follow the unwritten rules of the workplace. Many show up late, have hygiene issues, don't comply with the nuances of social "niceties," or have a myriad of other potential problems. They often feel misunderstood, judged, and powerless.[8]

In her frequent keynote addresses to autism groups, Temple Grandin is adamant that parents of those with autism help them learn the "rules" by guiding them as her own mother did. She expands upon this concept in her 2016 book, *The Loving Push*.[9]

A guidebook, authored by two adults on the spectrum, that covers the unwritten rules of employment is *The Hidden Curriculum of Getting and Keeping a Job: Navigating the Social Landscape of Employment: A Guide for Individuals with Autism Spectrum and Other Social-Cognitive Challenges.*[10]

In 2012 the Ohio Center for Autism and Low Incidence (OCALI) published a booklet on "Employment during Transition to Adulthood."[11] It states what everyone already knows: employment of those with serious disabilities is woefully low, and significant supports must be in play for employment to be successful.

Supports for Employment

Working virtually has been a wonderful option for those who do things at their own speed and who march to their own drummer. They can sleep late, work in pajamas, never shower, and still be conscientious about making deadlines. Technology is their friend. They can attend meetings on Zoom and make brilliant presentations while leaving their video off.

Thank goodness for apps! If the boss insists they show up in person once in a while, an app can get them out of bed and to work on time as well as tell them how to take public transportation to a distant office.

Jason worked at a local grocery store; the adults supervising him assumed competence and believed he could learn how to travel independently by public bus from his home to his job. Using an app, his parents took photos of each step required to get Jason from home to work. After almost a month Jason learned how to use his iPad and follow the 15-step sequence of events from standing at the bus stop to getting off at the proper stop and walking to the store. He was then able to travel alone without using the iPad.

Government and Equal Employment

The Americans with Disabilities Act (ADA), mentioned in previous chapters, applies very specifically to the workplace. This federal law guarantees equal employment opportunities for people with disabilities and protects disabled workers against job discrimination. Employers may not consider the disability when selecting among job applicants. They must also make "reasonable accommodations" to help workers do their job. Such accommodations could include shifting job responsibilities, modifying equipment, or adjusting work schedules.

Our complex governmental system with its many departments and subdepartments has a multitude of agencies that are supposed to support individuals with disabilities. For instance, in the US Dept. of Labor, the Office of Disability

Employment Policy exists to assure customized support for individuals with mental health challenges and autism. Obviously, more is needed.

The Autism Full Employment Act

During the pandemic of 2020, seasoned California employment attorney Michael Bernick and Louis Vismara, MD, a founding member of the UC Davis MIND Institute, gathered a group of adults with autism, practitioners, and advocates and set out to develop an Autism Full Employment Act. Their goal: to show that there can be a place in the job world for a wide range of adults with autism, ADHD, and other learning and mental health differences—many of whom were unemployed. Read about the six broad strategies they designed to explore professionalizing support services to help with transitions and comorbidities in their 2021 book.[12]

For a decade prior, Berick had been working on ideas for projects that would support full employment of those on the spectrum. He conceived the idea of job clubs, where anyone could help those with autism and other conditions network, prepare for interviews, and land jobs. An afterword in the above book covers the many employment initiatives for adults on the autism spectrum launched in the three years after the book was originally published.[13] Another dream of Berick's is the establishment, by 2030, of Autism City, a community with housing, public safety, mental health support, and employment for those with ASD.

Self-Employment

Many quirky people (including me!) find it much easier to work for themselves than for someone else. Set your own hours, follow your own rules, work in your own space, and do your own thing. It requires discipline, passion, and support from family and friends, and, for some, it is the best answer.

You may have heard of John Elder Robison, a multitalented man who did not receive an autism spectrum diagnosis until well into adulthood. His lifelong love of tinkering with cars led to a very successful high-end automobile repair and detailing business, described in his delightful book *Look Me in the Eye*.[14]

Mark Rimland, autistic son of Bernie Rimland, the late founder of the Autism Research Institute (ARI), and his wife Gloria, and Jeremy Sicle-Kira are both well-known self-employed professional artists, whose paintings sell for large sums. They both turned their love of painting into careers from which they derive pleasure and income.

British autist Robyn Steward has mentored those with ASD and other conditions, and her 2021 book is a great little guide for turning passions, ideas, and obsessions into a satisfying, lucrative self-employed life. She covers such topics as weighing the risks and benefits, turning ideas into reality, marketing, and dealing with finances and pesky details.[15]

Relationships

Everyone (some more than others) needs friends and companions with whom to share interests, activities, and even intimacy. The new friendships that have blossomed among Spellers, who prior to using letterboards mostly hung out with only family members, is quite remarkable. These minimal speakers are now communicating with each other frequently and have bonded.

Online

The Internet has opened the floodgates to global social network and dating sites. It is now possible to find a job, a friend, and even a mate with the click of a mouse or a touch screen.

The Web offers the opportunity for many types of friendships, and getting to know someone (assuming they are telling the truth!) without face time, reading social cues, and the awkwardness of making small talk: all skills some find difficult. Thus, the Internet can also offer the opportunity to be targeted for abuse and bullying. It can be difficult to sort out who is truly a friend and who isn't. Clearly, having a friend on Facebook or an online relationship is not the same as having one in person, though one can lead to the other.

I have almost 5,000 friends from around the world on Facebook. Some are people I've known since childhood; others are from my professional life. In my travels I have met some in faraway places, like the autism mom from Sicily who saw from my daily posts that I was visiting nearby and made the effort to meet me. I *love* the opportunities social media offers!

A variety of sites offer social communities for specialized groups, albeit with some vulnerability and even one-on-one guidance through the entire relationship process. Wrong Planet is a Web community for individuals with autism, ADHD, and other neurological differences. It includes many discussion forums, articles, how-to guides, a blog, and more.

Many chapters of national organizations offer workshops and support groups that bring together people with similar diagnoses. These are people who understand you because they have walked in your shoes. They also have social functions and recreational activities, ranging from bowling to barbeques, that can foster friendships and build relationships.

Dating and Marriage

For those with neurodevelopmental disorders and language, social skills, sensory processing, and other issues, moving from friendship to dating to sexual contact can be especially challenging. Those who cannot communicate well must rely on caretakers to channel the physical needs that arise with hormonal changes.

Two famous couples have taken on dating and marriage on the autism spectrum. Australians Sarah and Tony Attwood have written extensively about sexuality and intimacy, and their books are available from Jessica Kingsley Publishers. They often lecture in the United States and offer advice through various types of media.

Americans Jerry and Mary Newport met at an Asperger support meeting, married, divorced, and then remarried each other. Together, they wrote their story in *Mozart and the Whale*, which was fictionalized in the movie of the same name. They have both passed away, but their knowledge is accessible through their books and conferences from Future Horizons.

For those interested in more specifics, *The Asperger Love Guide: A Practical Guide for Adults with Asperger's Syndrome to Seeking, Establishing and Maintaining Successful Relationships*[16] is a free download online. It includes subjects such as assessing your readiness for a relationship, how and where to seek and find someone, whether or not to disclose, dating safely, maintaining a relationship, cohabiting, and marriage. Autism Victoria (Australia), now known as Amaze (because "life with autism is a maze in which our members constantly amaze us!"), has a brilliant piece called "Romantic Relationships and Autism Spectrum Disorder," by psychologist Kristy Kerr, detailing the many aspects of dating and relationship building.[17]

Want to watch some reality TV stories? Tune into *Love on the Spectrum* on Netflix.

Gender Identity

The beginning of the autism epidemic in the late 1980s and early 1990s parallels the increase in a universally more open discussion of gender differences, as well as the increased visibility of the LGBTQ+ movement. While the media and academia have treated this phenomenon sociologically, there is strong evidence that gender differences, like the neurological disorders discussed in this book, are biologically based.

This book takes the stance that the same endocrine disruptors that interfere with detoxification, digestion, and development are likely accountable for *gender dysphoria* (GD). This disorder, defined by the DSM-5 as "a mismatch between a person's assigned sex at birth and that person's identity," was previously called *gender identity disorder*.[18]

Research strongly supports this premise. Only a hypothesis in 2011,[19] after additional research in Sweden, France, and Japan, it now appears to be a near certainty. In one study, prenatal exposure to phthalates (see Chapter 2) was associated with a less typically "masculine" play behavior in boys.[20] In another, males with ASD were found to look younger than their chronological age, have sparse body hair, and high-pitched voices. Many were androgynous, not only in appearance, but also in their self-concepts and sexual preferences.[21]

Females with ASD were often similarly androgynous and found to have elevated testosterone levels. Furthermore, individuals with ASD presented a higher rate of asexuality. Temple Grandin, for instance, is open about "lacking an interest in romantic relationships."

In addition, results indicated that females with ASD showed a significantly lower degree of heterosexuality when compared to males with ASD. The results also suggested a higher degree of homosexuality among females with ASD although this effect did not reach significance.[22]

Few case reports or studies in English-speaking countries have dared to broach the subject of the co-occurrence of ASD and gender identity questioning, even though they seem to be closely related. In his 2024 book, *Raising Healthy Kids*, mentioned in Chapter 2, environmentalist David Steinman tells the fascinating story of the alligators in Lake Apopka in Florida. Since 1985, these animals have had an 80–95 percent failure to reproduce. Some are intersex, born with both testes and ovaries. Why? On the shores of the lake, behind a chain-link fence with an ominous sign reading "CONTAMINATED SITE," are the remnants of the abandoned Tower Chemical Company, a designated federal Superfund site.

Tower Chemical manufactured the pesticide *dicofol* for years, with several major spills into the lake. Dicofol is up to 20 percent DDT, which is highly estrogenic. Dr. Louis Guillette, the reproductive biologist who studied the alligators for more than 30 years was able to prove that their reproductive systems were in a state of gender chaos and that it was conclusively a result of the dicofol turning the lake's water into a sea of estrogen. These alligators had morphed into a third gender when compared to those untouched by DDT.[23]

Steinman takes this subject an important step further and discusses the biology of vertebrates, from alligators to humans. Quoting the work of neurobiologist Dick Swaab, he explains that sexual differentiation of the genitals takes place in the first two months of pregnancy, whereas the sexual differentiation of the brain comes later, during the second half of pregnancy. These two processes are influenced independently, and, if the fetus is exposed to different hormones during these key times, transsexuality may result. In other words, "the degree of masculinization of the genitals may not reflect the degree of masculinization of the brain." Dr. Swaab and his colleagues adamantly believe that biology, not society, determine gender identity.[24]

Why is this germane to this book? Because it supports, in part, the idea that Total Load, including the increased toxins in our environment, is causing the rise of autism, ADHD, and other learning and mental health differences. Furthermore, it highlights the need to both reduce exposure to toxins and detoxify those who have been affected. It also implies that transgender people deserve inclusion and accommodation as much as neurodivergent people do.

According to a 2013 French study,[25] gender identity clinics are now reporting an overrepresentation of individuals with ASD among their patients. In a Dutch study, the prevalence of ASD was 10-fold higher among patients with gender identity disorder than in general population.[26] Another study in Japan[27] found similar results.

Want to learn more about gender dysphoria from an amazing resource who is a positive voice on this subject? Read and listen to Wenn Lawson, a brilliant expert on gender, sexuality, and autism, at BuildSomethingPositive.com.

Obviously, this subject is very loaded with broad implications, especially regarding treatment options. Further research is clearly indicated.

Martial Arts

Several experts have explored ancient and Eastern movement and mind-body techniques for improving sensory, balance, and coordination issues, as well as the resultant anxiety and behaviors.[28] These interventions focus on posture and the breath, and since deep breathing is known to lessen anxiety, their success is not surprising.

While this section could have been included earlier in this book, placement in this chapter makes sense because mind-body techniques help with body control and promote "mindfulness."[29] Sports and activities such as yoga, aikido, karate, tae kwon do, tai chi, and judo can all benefit anyone at any age.

A 2012 study by New York occupational therapists[30] utilizing a classroom yoga program called *Get Ready to Learn (GRTL)* also showed significantly fewer maladaptive behaviors in the treated group. Aikido is a Japanese martial art that is both a practical and effective form of nonviolent self-defense, and a means of compassionate conflict prevention. It is also a way to relax and get physical exercise. On its deepest level, aikido is a path of mind/body/spirit awareness and integration.

Paul Linden, PhD, is a somatic educator and martial artist in Columbus, Ohio, and the developer of *Being In Movement*, a mind-body training program he has used successfully for over 50 years. His specific body awareness training methods help with self-monitoring and self-regulation. He sees improvement in managing aversive stimuli, as well as better focus, even in overstimulating, distracting environments. In addition, his students have learned more graceful, effective styles of movement.

The newest resource is *The Mindful Warrior: Martial Arts as a Path to Mental Wellbeing*[31] by Buddhist clinical psychologist Michael Jones, PhD. He shares lessons on how to face mental health challenges with calm, confidence, and clarity that he has learned from his spiritual practice, his own physical pursuits, and the care of hundreds of clients for more than 35 years.

Look for local resources that have experience with individuals with special needs and the martial arts. These wholesome activities can help anyone learn control over multiple facets of their body and become a source of ongoing healing from the inside out.

Annual Evaluations

After a student has left the school system, periodic reviews of physical, emotional. and mental status are still necessary. Reviews can take place quarterly, semi-annually, or annually at the least. Team members can meet in person or use technology. Each should be very familiar with the latest services offered, status of the adult with special needs, and reports of progress.

Additional Resources

- Chantal and Jeremy Sicle-Kira: *A Full Life with Autism: From Learning to Forming Relationships to Achieving Independence*, 2012.
- Francis Tabone, PhD: *The ASD Independence Workbook: Transition Skills for Teens and Young Adults with Autism*, 2018.
- Jason Jones: *Adult Autism Essentials: A Step-By-Step Approach to Navigating Relationships, Professional Life and Finding Resources While Celebrating Our Strengths*, 2024.
- Cammie McGovern: *Hard Landings: LOOKING into the FUTURE for a CHILD with AUTISM*, 2021.

Take-Home Points

During transition, parents are required to make decisions that might seem irreversible. However, life after the school bus stops coming has many options and facets as well as a multitude of ups, downs, and surprises. Deciding on where and how a young adult with challenges should live, spend the day, and be active are all looming choices that have an increasing number of possibilities. Even as this book is published, new ones will have arisen, requiring the diligent parent or guardian to do ongoing research.

Whether a loved one should stay home, move in with a relative, attend a day program, pursue post-secondary education, live on a farm, learn a vocation, or be a part of a community depends upon individual circumstances. How about a social life? Friends make life fun. Online friends are different from friends in person. While they work differently, both have advantages. Moving from friendship to romance to marriage are all real possibilities that offer both pitfalls and pleasure for those on the spectrum, and help is available.

Martial arts of various types are a terrific resource for those of all ages to improve physical well-being, as well as unite body, mind and spirit. Aikido, tai chi, and judo all show efficacy for those with mental health differences.

Finally, remember that reaching adulthood is a milestone, but not a stopping point. Our kids, like their neurotypical peers, keep growing and changing. As they self-advocate and become more independent, different options become available to them. Many are late bloomers, with needs that are constantly changing and must be reassessed annually.

CHAPTER 23

HAVING HEALTHY BABIES

If neurodevelopmental disorders are treatable when we know the risk factors, then the obvious next question is, are these conditions preventable?

Maureen McDonnell, RN, who for over 10 years planned programs for parents of children diagnosed with autism and their providers, states that "we now have a pool of wisdom, good science and common sense from which to draw safe, effective and practical recommendations for preventing autism."[1]

Why has the fertility rate dropped to an all-time low?[2] Why does the United States have one of the highest maternal and infant mortality rates among developed countries?[3] Why is the cause of autism still a "mystery"? Hopefully, before this book is published, RFK Jr. will have the data to provide us with the definitive answer: "It's Total Load."

Here is why no one dared to give that response sooner: getting pregnant, staying pregnant, giving birth, and raising babies is big business. Jennifer Margulis, PhD, tells it all in *Your Baby, Your Way*, originally titled *The Business of Baby: What Doctors Don't Tell You, What Corporations Try to Sell You, and How to Put Your Pregnancy, Childbirth, and Baby BEFORE Their Bottom Line.*[4]

"Life can only be understood backwards; but it must be lived forwards."[5] This profound quote from Søren Kierkegaard most assuredly describes what many parents of affected kids must feel. So many have confided, "If only I knew then what I know now. I would have made very different choices."

In August 2014, "Mountain Mama," one of the founders of the Thinking Moms' Revolution (TMR), bared her heart and soul to the world by bravely posting "How I Gave My Son Autism."[6] In it, she takes full responsibility for her son's regression.

I don't know if she read my first book,[7] but clearly Mountain Mama subscribes to Total Load Theory. She believes her son's autism resulted not from a single trigger, but rather from the accumulation of assaults on his nervous, sensory, and other systems. Some of the culprits—ultrasounds, acetaminophen, and vaccines—are included in this chapter. The actions she blames were based in fear and recommended by well-meaning healthcare professionals. She strongly believes that autism and other neurodevelopmental disorders are preventable.

For those already diagnosed, it's too late to cancel the exterminator, retract the needle from the second MMR dose, and eat only organic food during pregnancy. But it's not too late to educate the next generation of parents, and we have the responsibility to do so. The goal of this chapter is to fulfill Mountain Mama's hope of saving other children by educating prospective parents.

Education is the goal of parent Beth Lambert, author of *A Compromised Generation*[8] and executive director of Documenting Hope, formerly Epidemic Answers, with which I merged my early nonprofit Developmental Delay Resources in 2013. It is also the goal of environmental advocate Justin Valley, author of *Healthy from Day 1*, a tiny e-book written for expectant parents,[9] and neurodevelopmental optometrist Sarah Lane, OD, founder of Natural Beginnings NJ, which offers autism prevention workshops in New Jersey.

Young people should understand the importance of good physical and mental health by adolescence. I would love to see a mandatory course in personal health in every high school in America.

Pediatrician David Berger, MD, FAAP, counsels parents who already have a child with autism. He believes they can take many precautionary steps *before* conception to increase their chances of having a healthy baby. Since starting his Preconception to Infancy Project more than 15 years ago, not a single child born into his medical practice has developed autism spectrum disorder, not even one sibling of a child previously diagnosed. That is remarkable, since the incidence of autism in subsequent children is estimated to be one in five.[10] Read his recommendations in an article published in 2012.[11]

Preparation should start as much as a year preconception. Both partners should limit exposure to toxins while maximizing nutrition and health. Toxins in fathers can lower testosterone levels, possibly damaging sperm, limiting fertility, and causing birth defects. Poor sperm motility is often related to toxicity.[12]

Learning when to trust the body's healing wisdom, recognizing when a medical professional is needed, and understanding how good lifestyle choices support good health should also start then. Why? Because all these steps improve fertility and help avert miscarriage, pregnancy complications, and problems during delivery.

A full year out? Yes, because that's how long it takes to replace the bad stuff, to learn about the good stuff, and for the body to detoxify safely. If you are that prospective parent and are reading this, congratulations! Keep track of your family load factors and think about how you might reduce them. Healthy parents tend to have healthy babies. Now, let's get started!

Preconception

Run Laboratory Tests

The information tests provide empowers you to make informed decisions. Consider running the following tests, described in detail in Chapter 5, *before* becoming pregnant. None are routine; you must request them from your doctor.

Genetics and Nutrigenomics. Know your single-nucleotide polymorphisms (SNPs), those genetic glitches that can cause possible difficulties with getting and staying pregnant. Work with a healthcare professional to identify ways to mitigate them.

Toxic elements. The earlier in gestation toxic exposures occur, the more detrimental they can be to development. According to Philippe Grandjean, MD, author of *Only One Chance: How Environmental Pollution Impairs Brain Development— and How to Protect the Brains of the Next Generation*,[13] every woman should know what toxins her body is holding. Then she can detox preconception to ensure that her baby isn't exposed in those early weeks before a pregnancy test is positive. Ask your doctor to run one of the toxic elements tests introduced in Chapter 5. These tests can pinpoint what's in the toxic load you could dump into your unborn baby.

Thyroid function. Testing for and treatment of thyroid dysfunction is complicated and controversial. See Chapter 12 to learn about the vital importance of the thyroid gland and the hormones it secretes. Environmental toxins are often endocrine disrupters, affecting both egg and sperm.[14]

If a woman is toxic, her thyroid is probably not working properly. A low-functioning thyroid (hypothyroidism) in the mother is a known cause of infertility,[15] miscarriage, increased risk of a cesarean birth, and increased bleeding after delivery.[16]

During early critical stages of brain development, often before a woman is even aware she is pregnant, a couple of hours of insufficient T4 or marginally elevated levels of thyroid-stimulating hormone (TSH) can affect the baby's cognitive function.[17] Endocrine damage may also occur to the hypothalamus, impeding its ability to produce oxytocin and vasopressin, the bonding hormones.[18]

All tests and supplements are not created equal. Routine blood tests for thyroid function, which measure TSH, T4, and T3, frequently fail to detect a problem. New York thyroid specialist Raphael Kellman, MD, believes the TSH reference range is too wide, with the upper boundary of "normal" being too high; such tests thus fail to detect marginally low thyroid in many patients.

Kellman recommends an alternative: the TRH stimulation test, which he has used in over 15,000 patients. This test employs a hormone called TRH to stimulate the pituitary gland, which, in turn, produces TSH to stimulate the thyroid to produce thyroid hormones. When the thyroid is sluggish, the pituitary must produce more TSH. However, in those with hypothyroidism high levels of TSH frequently do not show up in the blood, rendering the routine thyroid blood test inadequate in a significant percentage of patients.

Even when blood levels of TSH are normal in hypothyroidism, TSH is always high in the pituitary gland. Upon stimulation with TRH, TSH is released on the spot, causing levels to rise and allowing physicians to make a proper diagnosis and treat the patient accordingly (KellmanMD.com). Israeli researchers have shown that a significant percentage of women with fertility issues whose routine blood testing showed normal TSH levels had abnormal results when evaluated using the TRH stimulation test.[19]

Vitamin D levels. Every day we are learning more about the relationship between vitamin D and health (see Chapter 11). The importance of prenatal, neonatal, and postnatal vitamin D supplementation cannot be overestimated. Vitamin D during gestation and early infancy is essential for normal brain functioning. In 2009 researchers concluded that vitamin D deficiencies in pregnant women should be considered a risk factor for neurodevelopmental disorders.[20]

You want your number to be at least 50. If your level is low, your healthcare professional will probably suggest supplementation. Recheck in three months. High doses are sometimes necessary for a short time to elevate levels.

A food sensitivities test. Look for gluten, casein, soy, egg, garlic, and other intolerances. Discomfort after eating, such as gas or bloating, minor skin irritations, or more serious issues such as constipation or asthma could be a result of food sensitivities. Once you know the problematic foods, you can rotate and eliminate those with moderate to severe reactions.

Antibody titers. Ask your doctor to run blood titers to find out which diseases you are immune to. Make sure you are not a hepatitis carrier. If you are hep B-free, show test results to the obstetrician and put it in writing to prevent your baby from getting the hep B shot at birth.

Remove Mercury-Containing Amalgams

If you have any mercury-containing amalgams in your mouth, strongly consider having them removed safely by a biological dentist at least six months before getting pregnant. Follow up with a mercury detoxification program for at least a couple of months.

Mercury amalgams off-gas into a mother's blood, and mercury crosses the placenta, landing in the liver and kidneys of the fetus. It even shows up in her breast milk.[21] Infants' mercury levels correlate with the number of amalgams in the mother.[22] There is no "safe" level of mercury. As few as one or two "silver" fillings off-gassing into the mouth with brushing, chewing, and drinking hot or cold liquids can cause problems in the child.

Heal the Gut

Read Chapter 7 to understand the importance of our gut and microbiome, the variety of organisms that live inside us. Being aware of how your gut works, caring for it, and noticing triggers for improper function are keys to establishing a healthy gut.

Prior to conception, consider normalizing digestion by eliminating problematic foods and taking digestive enzymes and probiotics. Test for candida, parasites, and other gut bugs. Use natural products to kill them if present. Make sure you poop daily. Consider colon hydrotherapy.

Detoxify the Body

No baby is born today without toxic exposure in the womb. Mothers dump a good part of their body burden into their unborn babies through the umbilical cord and into the placenta. The lower a mother's toxic load, the lower the baby's.

The Environmental Working Group (EWG) and Commonweal performed a now classic study of 10 children born in August and September of 2004 in US hospitals. Researchers found an average of 200 industrial chemicals and pollutants in umbilical cord blood. Tests revealed a total of 287 pesticides, consumer product ingredients, and wastes from burning coal, gasoline, and garbage, 180 of which are carcinogenic, 217 of which are toxic to the brain and nervous system, and 208 of which can cause birth defects or abnormal development.[23]

Once tests have identified what toxins make up a potential mother's body burden, mercury amalgams are removed, and the gut is working well, it is time to detoxify (see Chapter 13). In preparation for pregnancy, detoxification can increase circulation to the reproductive system, thus supporting the production of healthy sperm and eggs and increasing fertility.[24] Work with a healthcare professional to find a detox protocol you like.

Consider using a sauna to sweat out the toxins. This was the method of choice for first responders from 9/11 to help rid their bodies of asbestos and other poisons.[25] Another is a homeopathic detox program that clears out chemicals, metals, parasites, bacteria, viruses, and radiation (see Chapter 14).

Detoxification without removing mercury amalgams and with abnormal digestion is dangerous!

Detoxify the Mind and Spirit

Preconception is a good time to repair broken relationships and clean the skeletons out of your relationship closet. If you have a history of family problems, consider a family constellation, also in Chapter 14.

Improve Your Diet

Eat organic, in-season, vegetables galore, with few white foods and no artificial colors, flavors, or preservatives. Consider going gluten-free. Eliminate or greatly reduce tobacco, alcohol, and caffeine. Minimize consumption of large fish. Use EWG's handy seafood calculator to check mercury levels of your favorite fish.[26] Buy a juicer, blender, or mixer to juice organic vegetables and make nutrient-rich smoothies. Consume fermented foods like sauerkraut and kimchi daily.

Decrease Exposure to Electrosmog

Return to Chapter 2 to refresh your memory about the dangers of electromagnetic fields and electromagnetic radiation, especially radio frequency and microwave radiation. Damage falls into three categories: cellular growth, DNA replication, and neurological functioning.[27] Why take the risk?

Your bedroom should be for sleeping and sex. Consider removing all electronics, including computers, TVs, wireless phones, iPads, Bluetooth devices and cell phones from this sacred place. When out and about, use cell phones with wired headsets; keep your cell in a purse or briefcase; avoid the bra and pockets. Turn off Wi-Fi at night.

Exercise and Breathe Good Air

Strengthen muscles and increase stamina by exercising. Stretch to increase flexibility. Walk—outside whenever possible. Open the windows; outdoor air is cleaner than indoor air. Consider filtering the air at home and in the office.

Green Your Environment

Gradually switch to nontoxic cleaners (no dry cleaning), laundry soaps, detergents, pest removal, gardening, and personal care products. If you are doing any renovations, be mindful about which products you are choosing for paint (no VOCs), flooring (consider cork or bamboo), insulation, and other building materials. Cabinetry often contains toxins such as formaldehyde. Any extra money you spend on being "green" is worth it for your family's long-term health. See GreenBuildingSupply.com

Take Quality Supplements

Contraceptives and other medications can deplete minerals. Find a good comprehensive, natural, and easily absorbed multivitamin, without fillers, colors or other additives. Make sure you get ample vitamin D3 with K2, mercury-free omega-3 oils, B vitamins, and probiotics. A healthcare professional can help determine the right products, ingredients, and dosages.

Stay Hydrated

Carry water wherever you go in a glass, stainless steel, or safe plastic water bottle. Drink only filtered water.

Say "No Thank You" to Vaccines

It's best to avoid flu, COVID, or any other shots at least a year prior to conception. By design, most vaccines contain toxins, with which your body must contend, to increase the body's immune response. Excipients such as aluminum and propylene glycol are not ingredients that belong in the environment where you are going to grow a healthy baby. Instead, build your immune system by following the guidelines in Chapter 9 so your body can fight off any bugs that come your way.

Start Reading

My two favorites:

- *The Natural Pregnancy Book, Third Edition: Your Complete Guide to a Safe, Organic Pregnancy and Childbirth with Herbs, Nutrition, and Other Holistic Choices.*[28] The author started out as an herbalist and midwife and became a physician midlife.
- *Your Baby, Your Way: Taking Charge of your Pregnancy, Childbirth, and Parenting Decisions for a Happier, Healthier Family*[29]

Pregnancy

Many decisions a mother-to-be makes prior to her baby's birth can be crucial to the baby's healthy future. Spend at least as much time and energy on this major event as you did planning your wedding.

Diet

Seek out nutrient-rich, not high-caloric choices. Eat 75–100 grams of protein, organic fruits, vegetables, legumes, nuts, and free-range, antibiotic-free, grass-fed animals. Avoid sugar and its substitutes, conventionally produced wheat and dairy, and hydrogenated fats. Limit fish and soy products. Take the time to sit down and eat slowly, chewing well. The more you chew your food, the less work the rest of your digestive system has to do and the more access your body has to the nutrients. Consider five or six small meals instead of three large ones. Snack on protein.

Good quality water is still the best drink. Minimize alcohol, soda, and diet or caffeinated beverages. Consider consulting with a herbologist or Traditional Chinese Medicine practitioner regarding herbal teas and infusions. Many have medicinal qualities and nutrients that can strengthen you.

Supplements

While the right, high-quality foods can provide much-needed nutrition, eating adequate amounts of some nutrients is simply impossible. Research prenatal vitamins; not all are created equal. Be sure yours includes omega-6 fats and folinic acid, not folic acid, which is synthetic.

If you have an MTHFR mutation, this is vitally important as your body cannot convert folic acid into a usable form. Folate protects against neural tube defects such as spina bifida. A trusted brand is Seeking Health, created by MTHFR guru Ben Lynch.

Exercise

Movement during pregnancy is beneficial to both mother and baby. Take walks in nature. Finding a balance between exercise you enjoy and movement you can manage may evolve as you get closer to the baby's birth.

Reflexes

Simple reflex integration activities during pregnancy could help the birth go more smoothly. Integrated Spinal Galant and Asymmetric Tonic Neck (ATNR) reflexes in the mother and emerging ones in the baby assist passage through the birth canal. Mothers who retain either or both these reflexes may have difficulty giving birth vaginally.

Yoga

Prenatal yoga is a unique and oft-recommended exercise choice; it can be challenging physically, however. Some poses are energetic, others restorative. The physical focus is on stretching and strengthening muscles used for birthing, as well as the upper back, shoulders, and arms to prepare the body for holding a growing infant. It also improves digestion and circulation and keeps the bowels moving.

Controlling the breath, a fundamental skill for birthing, is an inherent part of yoga. It also offers emotional support by preparing body, mind, and spirit for the big event. Research shows the multiple benefits of this ancient practice.[30]

Find a class for pregnant women or an instructor familiar with modifying poses during pregnancy. Avoid hot yoga and vigorous vinyasa yoga due to the excess heat.

Prenatal yoga classes often include a "check-in" to share experiences as well as activities focused on centering, breathing, visualization, and relaxation. They can also provide a community of like-minded women who can be supportive throughout pregnancy and potentially evolve into a postpartum "mommies" group. If you cannot find a local class, look for videos on DVD, YouTube, or exercise apps for doing prenatal yoga at home.

Relaxation and Sleep

Practice deep breathing daily. Oxygenation of cells enhances their function. Meditation can help to release stress, allowing the body to put its energy into growing a healthy baby.

Sleep is restorative. Try to turn in before 10:00 p.m. and sleep at least nine hours every night during pregnancy. Get your zzzs while you can!

Dental Care

If possible, avoid dental work while pregnant. This includes cleaning, root canals, and the removal or insertion of fillings.

Chiropractic Care

Spinal alignment is essential to an easy pregnancy and birth, as well as a quick postpartum recovery. If you can, schedule weekly chiropractic care throughout pregnancy to keep your structure balanced in preparation for birth. The earlier these adjustments take place, the more likely they can prevent concerns arising later. The birth process is intense, and working with a chiropractor might help prevent structural stress during birth. [31]

Find a doctor who works with pregnant women. Those who know the Webster® technique have advanced training in techniques of gentle body and muscle manipulation to properly position a baby for birth.

Continue EMF Protection

Consider using Belly Armor products to protect you and your baby. They offer tank tops, blankets, and other soft, cotton products which shield the dangerous rays. See BellyArmor.com.

Ultrasounds

Just because we know how to look into the womb doesn't necessarily mean we should. Routine ultrasounds are in vogue, even though a research study of over 15,000 pregnant women published in the *New England Journal of Medicine* concluded that "ultrasound screening does not improve perinatal outcome in current U.S. practice." Many doctors prescribe them monthly, and even more often if a pregnancy is "high risk" because of the mother's age or multiples. Some pregnant women even get them weekly. British consumer activist Beverly Beech called repeated ultrasounds in pregnancy "the biggest uncontrolled experiment in history."[32]

Ultrasound is energy in the form of sound waves vibrating at approximately a hundred times the frequency of normal sound. When transmitted through the amniotic fluid, this vibration constitutes an assault on the cells of the baby's body, particularly the developing brain. In addition to sound waves up to 100 decibels, ultrasounds emit heat. Both sound and heat can stress the fetus. Recent research supports neurologist Manuel Casanova, MD, who contends that ultrasound exposure is a load factor contributing to the rise in autism.[33]

Medications

The safety of virtually all prescription *and* nonprescription drugs has not been fully tested during pregnancy. This includes antidepressants.[34] If you get sick, consult with your doctor, drink more water, rest, and take vitamin C up to bowel tolerance; try to avoid antibiotics and other prescription and over-the-counter medications. Look for natural approaches such as herbs, acupuncture, or homeopathy, but be careful. Some herbs and homeopathic remedies can increase risk of preterm labor.

Vaccines

Doctors are now advising their patients to get vaccinated during pregnancy. According to the American College of Obstetricians and Gynecologists (ACOG), the shots most often recommended are flu, Tdap, COVID-19, and RSV (respiratory syncytial virus),[35] but that is constantly changing.

One justification for this radical departure from the precautionary principle is that the immune system naturally slacks off during pregnancy so the body won't reject the growing fetus. This means pregnant women can be subject to increased complications from infection. Another reason is that mothers pass antibodies to their newborns through breast milk. Thus, vaccinating a pregnant woman is believed to protect her baby. But every medication carries some risk, and the risks are higher during pregnancy. The CDC insists there's no relationship between vaccination and miscarriage. However, foreign studies show otherwise.[36]

The CDC has also advocated "cocooning" in recent years: vaccinating everyone, including parents, grandparents, and caregivers, in the newborn's proximity with Tdap to prevent pertussis infection.[37] Unfortunately, there is no good scientific evidence backing up this practice. In fact, the Tdap shot does not prevent pertussis infection and transmission; it only reduces symptoms. Thus, the recently vaccinated are less likely to exhibit symptoms when infected and less likely to stay away from a vulnerable newborn.[38]

Again, why take the risk? If you are healthy, consider avoiding any vaccinations while pregnant. Vaccines contain potentially harmful toxins (see Chapter 10).

Prepare Your Nipples for Nursing
Two to four weeks before you are to give birth, begin conditioning your nipples with oil or an over-the-counter ointment such as Lansinoh. You will be happy you did.

Surround Yourself with Supportive People
Pregnancy is a time to focus on only positive, supportive interactions and relationships. Spend time with those who are encouraging of your choices. Minimize contacts with those who are negative.

Shop around for a Holistic Childbirth Education Class
Many of the classes associated with hospitals focus on preparation for a medicalized birth, not for one that emphasizes the involvement of birthing parents. Consider options such as HypnoBirthing® and the Bradley Method®, both of which focus on cultivating trust in a woman's body to birth normally. Consider home birth, which has a lower rate of medical interventions for mother and child than hospital birth as long as it occurs with a certified practitioner.[39]

Put Together Your Birthing Team

Choose calm, educated professionals for a positive birth experience. Speak with friends who have recently had children to learn which practitioners in your area have philosophies consistent with yours. Interview doulas, midwives, and obstetricians (OBs).

A doula can be a valuable support for the birthing mother while the OB or midwife attends to the well-being of the baby. An experienced doula who has seen hundreds of births can help interpret your sensations, evaluate how labor is progressing, and help determine whether it is necessary to go to the hospital.

Ask a gazillion questions until you find people you like. If, midstream, you feel you made the wrong choice, switch—it's never too late!

Interview Pediatricians

Shop around for a pediatrician with whom you are philosophically compatible. A family doctor may be a better choice. Sometimes doctors of osteopathy (DOs) are more flexible, especially about vaccination schedules. Choose someone who supports health rather than treats illness.

Make a Birthing Plan

A birthing plan anticipates and states your choices regarding intrusive procedures—such as induction, an epidural, use of forceps or vacuum extraction, C-section, and clamping the cord prematurely—so you do not have to make decisions about them in a crisis mode. These and other medical interventions could affect future function for both mother and child. Labor induction and C-sections are load factors for developmental delays. In one study, cesarean delivery was associated with a 26% increased risk of autism.[40]

Talk to your team about positioning during labor and delivery. Being on your back may not be the best choice. Sitting upright or crouching on all fours for pushing and delivery could enhance your ability to birth with greater ease.

Stock Your Library

Invest in books for later, such as the classic *How to Raise a Healthy Child in Spite of Your Doctor*,[41] *Smart Medicine for a Healthier Child*,[42] and *Holistic Baby Guide: Alternative Care for Common Health Problems*.[43]

These great references advise parents on home treatments for routine childhood illnesses, allergies, and fevers. *Most* importantly, read, read, read, so that you recognize when you need expert medical help. Then you could go to the nearest urgent care center, a brilliant modern convenience. Consider whether you *really* need a pediatrician to tell you your baby is thriving.

Prepare the Baby's Room

Spend more money on an organic mattress than on a fancy crib. If you can find a used crib in good shape, that's even better because it has already off-gassed. Be sure the slats are close together, so there is no danger of stuck body parts! Use non-toxic products, such as no-VOC paints and natural flooring instead of carpeting, which can harbor bugs and mold. Consider alternative sleeping options for the newborn, such as a safe co-sleeper.

Watch those baby monitors; some emit EMFs, which are not good for baby. Ideally, the baby monitor base-station transmitter should be voice activated, meaning that it transmits only when it senses a sound from the baby. This would reduce the microwave exposure of both infant and parent as well as everyone else in the home.[44]

Labor and Delivery

Labor is an intense process—like nothing you have ever experienced. No two pregnancies are the same. Fear of pain, labor, and complications is the basis of today's medicalized birth model. Fear is a terrible motivator! Even though it may be tempting to consider avoiding labor altogether, moving through the stages of labor and delivery are important learning experiences for both mother and baby.

For the baby, passing through the birth canal exercises reflex patterns that function automatically, without conscious effort. A baby who has not used these reflexes to get through the birth canal may need help later in development to organize the reflex system.

A natural vaginal birth not only enhances an infant's neurological function[45] it also exposes a newborn to the mother's vaginal bacteria.[46] This process "seeds" the baby's unique microbiome. For those born by C-section, doctors are now swabbing the baby with secretions from the mother's vagina to make up for this disadvantage.[47]

While most doctors consider a baby "full-term" as early as 37 weeks, and some consider inducing labor at any sign of stress, Pediatrician David Berger, MD, instructs parents that a baby is not even considered past due until after 42 weeks. Furthermore, he sees no reason to artificially rush the delivery of a baby, as induction has the potential to have negative effects on a child's birth and development.

Antidote fear with knowledge. Knowledge is power. Immerse yourself in learning about natural alternatives to use during labor and delivery.[48]

Approach birthing as you would studying for the most important final exam you have ever taken, times ten! Know your options. A stressless, fearless, and trauma-free birth is a wonderful gift you can give your child. For you, birthing can be joyful and potentially even orgasmic.

Newborn and First Months of Life

Take Care of Yourself First

Sleep when the baby sleeps. When baby is awake, touch, move, talk, laugh, emote, and exaggerate interactions. Maintain good dietary habits, including ample fruit, vegetables, high-quality protein, and fiber to keep your bowels moving. Make bone broth to replenish minerals and soups and stews with root vegetables. Use warming spices like cinnamon, cloves, turmeric, and curry. Get a massage once a month, or more often, if finances permit.

If you are feeling depressed, work with a nutritionist or healthcare practitioner familiar with postpartum issues to adjust diet and supplements. An interesting method of preventing postpartum depression is placental encapsulation, which involves saving and preparing tissue from the placenta as a supplement. Watch a video on this fascinating option.[49]

More on Chiropractic Care

Both mother and baby should be evaluated by a qualified chiropractor shortly after birth. This step is imperative. Some chiropractors even want to be in the delivery room to mold a newborn's head within hours of delivery to ensure proper blood flow to the brain. Proper spinal alignment is essential for mom's healthy recovery and continued stamina. The baby's future cranial and spinal development depends on it.

Consider ongoing chiropractic care for your newborn baby. A baby with acid reflux, colic, torticollis, or ear infections should be seen by a chiropractor before resorting to medication. In one study, over 90 percent of babies adjusted by chiropractors showed improvement in function, 25 percent after the first adjustment.[50] Many of these common problems can be healed with routine chiropractic care and other lifestyle adjustments. To learn more, read *Kids First: Health with No Interference*.[51]

Feeding

Breast or bottle? Research is overwhelmingly supportive of mother's milk. Nursing is one of the best ways to provide optimal nutrition and to ensure quality bonding time between mother and child. Babies who are breastfed have lower incidence of sudden infant death syndrome (SIDS), fewer illnesses, and healthier gastrointestinal systems.[52]

Milk takes up to four days to come in. Immediately following birth, nursing babies get *colostrum*, a clear substance that is very beneficial for their immune systems. If you decide not to breastfeed, purchase colostrum as *transfer factor* and give it to your baby. Read about this remarkable product in Chapter 11.

While breastfeeding, keep lubricating your breasts. Remember to stay well hydrated. Keep water with you at all times. Nurse often, as a baby's stomach is small and can hold only a few tablespoons of fluid at a time.

Having trouble nursing? Support is key to success. Hire a lactation consultant, join a group, or call La Leche League. A breast pump can be helpful, especially if your baby is sick or hospitalized; breast milk is good medicine. Use a nipple shield to help baby latch on. Read books. Go online. Lactation consultant Nancy Mohrbacher offers a website, courses, books, and more. Visit NancyMohrbacher.com and Lactalearning.com.

How long should a mother nurse? It depends on so many factors. The American Academy of Pediatrics recommends exclusive breastfeeding for the first six months and continued breastfeeding for at least the rest of the first year, or longer, to support a baby's developing brain, immune, and digestive systems.[53] The World Health Organization recommends a minimum of two years.[54]

When is the right time to introduce solid food, and what foods are best? Only you and your healthcare provider can determine that. Most doctors are now recommending postponing solid food until six months.[55]

Feeding your baby solid food too early can compromise gut health and the immune system and trigger food allergies. Some signs of developmental readiness for solid foods are that the baby

- can sit up without support;
- does not push solids out of his mouth with his tongue;
- has teeth or is able to chew;
- can pick up food and other objects between thumb and forefinger;
- is interested in mealtime, may try to grab food and put it in his mouth.[56]

Avocado is a popular "first food" because it is already mushy and contains excellent nutrition. Steam organic vegetables and puree them, or buy an organic brand of prepared vegetables without sugar or other additives. Introduce one food at a time to minimize allergic reactions. Red cheeks, dark circles under the eyes, tummy problems, irritability, a cough, or runny nose can all indicate food reactions. No early cow's milk, soy, or grains. If you insist on rice cereal, make sure it is organic, although even organic rice could contain arsenic.

Diapers, Bottles, and Other Baby Products

Another area that requires research: cloth or disposable diapers? With chemicals or without? Glass or plastic? All diapers, powders, lotions, creams, and bottles are not created equal.

Did you know that what makes disposable diapers so absorbent, lotions so smooth, and bottles so colorful are toxic chemicals and plastics? Choosing to diaper with organic diapers, slather your baby with organic oils, and buy glass instead of plastic is similar to choosing to eat, clean, and decorate with nontoxic products.

Sarah Lane, OD, of Natural Beginnings NJ asks

- Is this product free of volatile organic compounds (VOCs), Bisphenol-A (BPA), and other toxic chemicals?
- Do I feel 100% comfortable with all the ingredients in this product?
- Am I choosing this product because it's cute or because it's safe?
- Is this product going to off-gas?

Do your homework and make decisions that are consistent with your values, philosophy, and finances.

Bonding with Baby

You cannot spoil a newborn! Hold, talk to, and wear your baby. Even though a baby does not yet understand words, tone and facial expressions speak volumes. Keep baby close by utilizing an on-the-body baby carrier. Wearing baby while you clean, vacuum, and move around the house offers the vestibular system much-needed stimulation. As you and the baby move in space, neurons myelinate and grow. Babies *must* move for their brains to develop.

So many different types of baby carriers exist. Make sure that newborn legs are not dangling; baby's hips should be in a supported sitting position. Try on many carriers and find one that is comfortable for both you and baby. If money allows, purchase a different one for each parent, allowing baby to be held in different positions.

Movement is food for a baby's nervous system. Limit time in popular seats, walkers, carriers, jumpers, and buckets that inhibit movements, put a baby's reflexes "in jail," or keep the baby on its back at a 45-degree angle. Lifting the head from this position is extremely difficult. These devices allow your child to attempt higher-level movements before the body is ready, potentially leading to atypical movement patterns and poorly controlled movements.[57] One very popular seat that has gotten bad press from physical therapists for its inhibition of an infant's movement is the Bumbo Baby Seat.[58]

Avoid electronic devices of all kinds. Babies do not benefit from videos, television, phones, iPads, or other electronic games. Their eyes and ears are not ready for that type of stimulation, and using these technological marvels could interfere with development.[59]

Sleeping Position

In the past 25 years, doctors have recommended back sleeping to lower the risk of *sudden infant death syndrome (SIDS)*.[60] Before SIDS emerged as a phenomenon in 1969 and the "Back to Sleep" campaign launched in 1994, most babies slept on their tummies.

A new study on SIDS supports Total Load Theory, including genetic predisposition and vaccines as load factors that can increase liver enzymes beyond the level an infant's immature liver is able to tolerate.[61]

In the 1990s, Barry Richardson, a British expert in materials degradation, and T. James Sprott, a New Zealand chemist and forensic scientist, independently suggested that SIDS is the result of accidental poisoning due to toxic gases released from baby mattresses.[62][63] Gases are produced by the interaction of common household fungi with phosphorus, arsenic, and antimony, chemicals which are either present naturally in the mattresses or which have been added as flame retardants. These toxins are heavier than air, so a baby sleeping face-up is less likely to inhale a lethal dose.

Another plausible explanation is that SIDs is a vaccine reaction. In 2021, Neil Z. Miller, a longtime critic of vaccination, analyzed data from the Vaccine Adverse Event Reporting System (VAERS) to investigate the timing of infant deaths after vaccination. He found that a significant proportion of reported infant deaths occurred within a short period post-vaccination: 58 percent happened within three days and 78.3 percent occurred within seven days. The author concluded, "While the findings are not proof of an association between infant vaccines and infant deaths, they are highly suggestive of a causal relationship."[64]

If either of these possibilities is true, SIDS is preventable with choices made in bedding and sleeping options.

Babies on their backs are neurologically upside-down. From that position they can barely raise their heads, much less put weight on their arms and hands or begin to crawl and creep. Many of them have flattened heads with bald spots. Although back-sleeping babies usually appear to develop normally, lack of early lower-brain development from spending too much time on their backs could translate into later learning and behavior challenges.

Tummy Time

Antidote back sleeping with tummy time. Tummy time promotes the development of strong head and neck muscles by allowing the baby to hold up the head against gravity. During tummy time, babies bear weight on their arms and hands, learn to reach with the eyes and the hands together, strengthen their cores, and eventually move through space to explore.

Place newborn babies on their tummies several times a day after naps, eating, and diaper changes, when they are awake and content. Wedge a rolled-up towel under baby's chest and armpits, allowing them to be slightly inclined. Most will probably reward you with a smile and reach out for stimulating objects.

Once a baby has head control, all waking hours should be spent prone. This position is extremely important for hand development, the emergence and integration of primitive reflexes, and bilateral and binocular integration.

As babies gain muscle strength in about the third month, tummy time becomes more fun. Infants' first random movements enhance posture and coordination, eye tracking, arm rotation, and hand strength. The result is a firm foundation for crawling, creeping, manipulating toys, and later using a pencil.[65] Four- and five-month-old babies are more purposeful, scooting and pivoting to reach toys.

Crawling and creeping organize important parts of the central nervous system that provide the foundation for all future growth and learning. When the lower brain develops appropriately, higher-level cognitive skills emerge naturally and easily. Disorders involving self-regulation, sensory integration, and learning could signal a lack of appropriate development in these key areas.

Babies who do not experience adequate tummy time sometimes show a propensity to turn or hold their heads predominately to one side. If neck muscles contract too much, medical help may be necessary to restore the full motion needed for normal development. Untreated, this condition, called *torticollis,* can require extensive therapy or even surgery. To avoid torticollis,

- switch arms often when feeding;
- frequently reverse the baby's position on the diaper table;
- carry the baby in a variety of ways, allowing them to experience different head positions;

- in a baby carrier or car seat, keep the neck straight with head supports;
- avoid equipment that holds the baby in a semi-reclining position;
- provide baby with adequate vestibular stimulation through (slow and controlled) swaying, dancing, and dipping upside-down to encourage eye movement development.

More on Vaccinations

Of all the subjects to research, this one is the *most important*. Read *everything* you can to educate yourself. Learn about vaccine ingredients, including aluminum, formaldehyde, and MSG.

Even though mainstream medicine denies a cause-and-effect relationship between vaccines and autism,[66] many believe that vaccination is a load factor in some children. Research indicates that those with MTHFR polymorphisms are genetically more vulnerable.[67]

Learn what is mandated in your state for daycare, school, and participation in other activities. Learn about the various types of exemptions available in your state. The 50 states differ markedly. Go to the website of the National Vaccine Information Center (NVIC.org) to find out how your state government interprets vaccine law.

Parents are required to make their first vaccination decision within 24 hours of a baby's birth. The hepatitis B shot is usually given at birth, despite the fact that more than 99% of newborns will derive no benefit, only risk. Unless a mother is positive for hepatitis B, a newborn probably does not need to receive this vaccine because the disease is transmitted primarily through sexual activity.

Your choice of pediatrician will determine how much flexibility you have in vaccine decision-making. Some are requiring that you subscribe to the CDC's one-size-fits-all vaccination schedule to be in their practices. This means that a tiny preemie who is struggling for life could get the same shots as a robust, full-term baby. One pediatrician I know says she wants to know "who is in there" first. Now, that makes sense! Infants with immune defects should not get live vaccines; however, few pediatricians test for immune deficiency before giving shots.

The NVIC has established guidelines to help parents make informed vaccine decisions.

1. Is my child sick right now?
2. Has my child had a bad reaction to a vaccination before?
3. Does my child have a personal or family history of
   - previous vaccine reactions,

- convulsions or neurological disorders,
- severe allergies, or
- immune system disorders?

4. Do I know if my child is at high risk of reacting? How to identify and report a vaccine reaction? Do I know the vaccine manufacturer's name and lot number? That I have a choice?

A "yes" to any of the questions one through three is reason to reconsider vaccination carefully. In order to make an informed decision about vaccination, parents must consider a child's birth history and genetic background. How many load factors does a child already have? Could a vaccination today be "the straw that breaks the camel's back"? Everyone should be prepared to answer "yes" to all of the questions in number four.

If you have *any* reservations about vaccinating a child, postpone, spread out, or avoid vaccines altogether. *No* long-term studies show that giving multiple vaccines at once is safe. *Never* allow a doctor to administer more than one shot on a given day. Some, like the MMR or DTaP, already contain three pathogens.

Never vaccinate a sick child or one who is on or just coming off an antibiotic. The child's immune system may not be able to handle the insult of the vaccination while trying to fight off the infection.

Did you know that many doctors and health departments use multidose vials? Protocol requires that the healthcare provider shake the vial vigorously to ensure even distribution of pathogens and adjuvants. Is it even possible that each and every one of the ten doses is identical? Unlikely.

Never give acetaminophen (Tylenol®) before or after a vaccine. William Shaw, PhD, Founder of Great Plains Lab, now Mosaic, has shown a clear relationship between acetaminophen use and autism.[68] Instead, use vitamin C or echinacea drops and cod liver oil to boost a baby's immune system for a few days prior to and after vaccinating. Also consider using homeopathic vaccination before and after the first shot of any pathogen to introduce the pathogen to the immune system, thus lessening the likelihood of a vaccine reaction later.

Vaccine exemptions are available in every state if you decide not to vaccinate. However, vaccine laws are changing, so check and recheck your state. At the time of publication, the following are true:

- ***Medical Exemption.*** All 50 states allow medical exemptions to vaccination.[69] Proof of medical exemption is a signed document from an MD or DO stating that administering one or more vaccines would be detrimental to the health of an individual. Fewer and fewer doctors are willing to write these because of retribution. Some states accept a private physician's written exemption without question. Others ask the state health department to review the doctor's statement; accepting or revoking it are options.

- **Religious Exemption.** All states allow religious exemptions to vaccination except California, New York, Maine, and Connecticut. West Virginia's new governor recently signed an executive order establishing one in the state, but the legislature has not yet put it into law. Religious exemptions are intended for people who hold sincere religious beliefs opposing vaccination; forcing vaccination would infringe on their right to exercise their religious beliefs.

    Some state laws define religious exemptions broadly to include personal religious beliefs, similar to personal philosophical beliefs (see below). Other states require an individual who claims a religious exemption to be a member of a state-recognized religion with official tenets that prohibit invasive medical procedures such as vaccination. Some laws require a signed affidavit from the pastor or spiritual advisor of a parent exercising religious exemption that affirms the parent's sincere religious belief about vaccination, while others allow the parent to sign a notarized waiver. Bizarrely, Washington and Oregon require religious exemptions to be signed by a doctor.

- **Philosophical, Personal, and Conscientious Belief Exemption.** The following 16 states allow exemption to vaccination based on philosophical, personal, or conscientiously held beliefs: Arizona, Arkansas, Colorado, Idaho, Louisiana, Michigan, Minnesota, North Dakota, Ohio, Oklahoma, Oregon, Pennsylvania, Texas, Utah, Washington, and Wisconsin. In some of these states, individuals must object to all vaccines, not just a particular vaccine, in order to use the philosophical or personal belief exemption.[70]

If parents believe a child is at high risk for suffering vaccine-induced injury or death, they have the moral right to protect their child from harm. If they choose to selectively vaccinate a child or use no vaccines at all, they should be aware there is a small possibility they could be charged with medical neglect for failing to vaccinate a child with all state-required vaccines.

Vaccine-induced immunity is temporary and they don't always "take," which is why more than one dose and boosters are recommended. Some states accept proof of "immunity" in lieu of subsequent doses of a vaccine. A blood test from a private laboratory, costing about $60, can determine whether a child has sufficiently high antibody titers to demonstrate immunity to a specific disease. Some children produce sufficient antibodies after a single shot, and others cannot show proof of immunity even after many boosters.[71]

Duration of vaccine-induced immunity varies from person to person, just as the risk for vaccine reactions varies. Consider asking your doctor to draw a titer before administering a booster. They probably won't like this and will tell you it's likely your insurance won't cover the cost. Do it anyway! Inoculating your child with a dose of a pathogen to which their body already has immunity unnecessarily activates their immune system and exposes them to other toxins included in the vaccine.

Vision Development

Babies have predictable visual milestones, just as they do in the areas of movement and language. By six weeks, a baby should look at you and smile. Eyes should work together, as if a string were attached to them: if one eye looks left, the other eye should also look left. Any variation of this is of concern and requires an immediate evaluation by an eye care professional.

For more than 20 years, neuroscientist Karen Pierce, PhD, at UC San Diego Autism Center of Excellence has investigated gaze patterns in infants. She discovered that babies at risk for autism spend greater time visually examining geometric patterns than they do social patterns as young as 12 months.[72]

Even if no eye or vision problems are apparent, the American Optometric Association (AOA) recommends scheduling your baby's first eye assessment between six and twelve months. InfantSEE®, a public health program managed by Optometry Cares®—The AOA Foundation, is designed to ensure that eye and vision care become an essential part of infant wellness care to improve children's quality of life. Under this program, participating optometrists provide a comprehensive infant eye assessment as a no-cost public service. The doctor looks for excessive or unequal amounts of nearsightedness, farsightedness, or astigmatism, as well as eye movement ability and eye health problems. Find a doctor near you at InfantSee.org.

Take-Home Points

Neurodevelopmental conditions are preventable! Can we guarantee that by following the guidelines in this chapter we will wipe them out? Of course not. However, we know the risk factors, and by being mindful of decisions we make along the way, we can ensure healthier babies by reducing load factors. For more information, several excellent books, some of which were mentioned earlier, cover the topics in this chapter in greater depth:

- *Brighton Baby: The Complete Guide to Preconception and Conception* by Roy Dittmann, OMD, MH
- *Healthy from Day 1* (an e-book) by Justin Valley
- *How to Prevent Autism: Expert Advice from Medical Professionals* by Dara Berger[73]
- *Naturally Healthy Babies and Children: A Commonsense Guide to Herbal Remedies, Nutrition, and Health* by Aviva Romm
- *The Nourishing Traditions Book of Baby & Child Care* by Sally Fallon Morell and Tom Cowan of The Weston A. Price Foundation
- *Preventing Autism & ADHD: Controlling Risk Factors Before, During and After Pregnancy* by Debby Hamilton, MD
- *Preventing Autism: What You Can Do to Protect Your Children Before and After Birth* by Jay Gordon, MD
- *Turtles All the Way Down: Vaccine Science and Myth*

AFTERWORD

Now it is more important than ever to advocate for the next generation of children, their children, and their children's children.

To quote Sayer Ji, founder of GreenMedInfo: "This is just the beginning. Join the work. We are building the future together . . . from the soil up . . . from the heart out."

- Read *The MAHA Report*.[1] It is a map of healing for the next generation.
- Support your local farmers and regenerative food systems.
- Speak the truth, even when it's costly.
- Get involved locally—run for the school board, join health councils and community coalitions.
- Reclaim your own health.
- Campaign to improve education and awareness of the lifestyle choices we can make to minimize toxicity and therefore reduce the risk of developing neurodevelopmental delays.
- Join organizations and fund education and research on the importance of a toxin-free, less stressful world for conception and birth.
- Support organizations that promote health freedom, such as Children's Health Defense, the World Council of Health, REACT19, and the Global Wellness Forum.
- "The children are watching. The ancestors are listening. The future is calling."[2]

Future generations of kids cannot wait for tomorrow's research. We must act *now*. We can do it together!

ENDNOTES

Chapter 1

1   GBD 2019 Mental Disorders Collaborators, "Global, Regional, and National Burden of 12 Mental Disorders in 204 Countries and Territories, 1990–2019: A Systematic Analysis for the Global Burden of Disease Study 2019." *Lancet Psych*, 2022;9:137–50.

2   B. Zablotsky et al., "Prevalence and Trends of Developmental Disabilities among Children in the U.S: 2009–2017," *Pediatrics*, Sept. 26, 2019.

3   Autism Speaks, www.AutismSpeaks.org, accessed Nov. 8, 2024.

4   "General Prevalence of ADHD in Children," CHADD, https://chadd.org/about-adhd /general-prevalence-children/, accessed Nov. 8, 2024.

5   Learning Disabilities Association of America, www.ldaAmerica.org/, accessed Nov. 8, 2024.

6   National Alliance on Mental Illness, www.NAMI.org/, accessed Nov. 8, 2024.

7   Katherine Schaeffer, "What Federal Education Data Shows about Students with Disabilities in the U.S., Pew Research Center, July 24, 2023, https://www.pewresearch .org/short-reads/2023/07/24/what-federal-education-data-shows-about-students-with -disabilities-in-the-us/.

8   "Fact Sheet: The World's Nearly 240 Million Children Living with Disabilities," UNICEF, Dec. 2, 2021, https://www.unicef.org/press-releases/fact-sheet-worlds-nearly-240-million -children-living-disabilities-are-being-denied/.

9   B. Zablotsky et al. "Prevalence and Trends of Developmental Disabilities among Children in the United States: 2009–2017," *Pediatrics*: Oct. 2019, 144, 4.

10 "Epidemiology Glossary," Centers for Disease Control, https://www.cdc.gov/reproductive -health/glossary/, accessed Dec. 3, 2024.

11 Hans Selye, *The Stress of Life* (New York, NY: McGraw-Hill, 1956).

12 S. Y. Tan and A. Yip, "Hans Selye (1907–1982): Founder of the Stress Theory," *Singapore Medical Journal*, 2018 April, 59(4):170–171.

13 G. Goodman and M. J. Poillion, "ADD: Acronym for Any Dysfunction or Difficulty," *Journal Spec Educ*, 1992; 26(1), 37–56.

14 A. Mathews, "The Good-Enough Parent," *Psychology Today*, Sept. 19, 2024.

15 G. W. Miller, *The Exposome: A New Paradigm for the Environment and Health, Second Edition* (Cambridge, MA: Academic Press, 2020).

16 American Psychiatric Association, *Diagnostic and Statistical Manual of Mental Disorders, 5th ed. text revision*, (Washington, DC: American Psychiatric Association, 2022).

17 K. Leadbitter et al., "Autistic Self-Advocacy and the Neurodiversity Movement: Implications for Autism Early Intervention Research and Practice," *Front Psychology*, Apr. 12, 2021;12:635690.

18 "About NAMI," National Alliance on Mental Illness, https://www.nami.org/about-nami/, accessed Nov. 10, 2024.

19 B. Meyer, "NVLD: What Is It and Why Is It Not in the DSM?," NVLD Project, Feb. 26, 2020, https://nvld.org/nvld-what-is-it-why-not-in-dsm/.

20 A. Marshall, "Pathological Demand Avoidance in Autism and Beyond," verywellmind, Dec. 10, 2024 https://www.verywellmind.com/pathological-demand-avoidance.

21   A. S. F. Lutz, "An Interview with Neurodiversity Originator Judy Singer," *Psychology Today*, June 26, 2023, https://www.psychologytoday.com/us/blog/inspectrum/202306/an-interview-with-neurodiversity-originator-judy-singer.

22   Steve Silberman, *Neurotribes: The Legacy of Autism and the Future of Neurodiversity* (Garden City Park, NY: Avery Books, 2016), 16.

23   Jim Sinclair, "Why I Dislike 'Person First' Language," 1999. https://autismmythbusters.com/general-public/autistic-vs-people-with-autism/jim-sinclair-why-i-dislike-person-first-language, accessed July 25, 2025.

24   D. Muzikar, "Person First Language—Autistic, Person with Autism, Aspergian, Aspienaut?" The Art of Autism, Nov. 1, 2018, http://the-art-of-autism.com/person-first-language-autistic-person-with-autism-aspienaut/.

25   J. E. Robison, "Neurodiversity and Autism in College," Aug. 30, 2017, https://www.psychologytoday.com/us/blog/my-life-aspergers/201708/neurodiversity-and-autism-in-college.

26   P. Dwyer, "The Neurodiversity Approach(es): What Are They and What Do They Mean for Researchers?" *Human Development*, May, 2022; 66, 2: 73–92.

27   Dara Berger, *How to Prevent Autism* (NY: Skyhorse, 2017), 34.

28   C. Nevison, "In New Autism Report, CDC Again Fails to Address Root Causes," Children's Health Defense, Mar. 24, 2023, https://childrenshealthdefense.org/defender/autism-report-cdc-failure/

29   Dan Olmstead and Mark Blaxill, *Denial: How Refusing to Face the Facts about our Autism Epidemic Hurts Children, Families and our Future* (New York: Skyhorse Publishing, 2017).

30   Autism Society, https://autismsociety.org/, accessed Nov. 2024.

31   G. Xu et al., "Prevalence of Autism Spectrum Disorder Among US Children and Adolescents, 2014–2016," *JAMA*, 2018;319:11, 81–82.

32   "Autism CARES Act of 2024," Interagency Autism Coordinating Committee, https://www.iacc.hhs.gov/about-iacc/legislation/autism/cares-act-2024/, accessed Nov. 8, 2024.

33   Individuals with Disabilities Education Act (IDEA). 20 U.S.C. § 1401 (3) (26) *§300, 1997.*

34   Americans with Disabilities Act. (ADA) 42 U.S.C. *§§12101–12213, 1990.*

35   C.T. Gordon, "Pharmacological Treatment Options for Autism: Part 1," *J National Alliance for Autism Research*, Spring 2003, 8.

36   S. Bhandari, "Drugs to Treat Mental Illness," WebMD, https://www.webmd.com/mental-health/medications-treat-disorders/, accessed Nov. 10, 2024.

37   Peter R. Breggin, MD, www.Breggin.org, accessed Nov. 10, 2024.

38   Adams Autism Research, http://www.adamsautismresearch.com/, accessed Nov. 10, 2024.

39   Leo Galland, *The Four Pillars of Healing* (New York, NY: Random House, 1997) 54–5, 85–6.

40   Martha Herbert, *The Autism Revolution* (New York: Ballantine Books, 2012).

41   M.J. Goldberg and E. Goldberg, *The Myth of Autism* (New York: Skyhorse Publishing, 2011).

42   Neil Nathan, *The Sensitive Patient's Healing Guide* (Fort Bragg, CA: Cypress House, 2024).

43   C.P. Johnson and S. Myers, "Identification and Evaluation of Children with Autism Spectrum Disorders," *Pediatrics*, Oct. 29, 2007.

44   A. Einstein and L. Infeld, *The Evolution of Physics*, CP Snow, Editor (Cambridge, UK: Cambridge Univ Press, 1938).

45   Leo Galland, *The Four Pillars of Healing* (New York: Random House, 1997), xiii–xvi.

46   C. M. Palmer, *Brain Energy: A Revolutionary Breakthrough in Understanding Mental Health—And Improving Treatment for Anxiety, Depression, OCD, PTSD, and More* (Dallas, TX: BenBella Books, 2022), 77.

Chapter 2

1   "Exposome and Exposomics," Centers for Disease Control, https://archive.cdc.gov/# /details?url=https://www.cdc.gov/niosh/topics/exposome/default.html, accessed Nov. 12, 2024.

2   Candela K. "Genetics and Health: What Percentage of Diseases Are Genetic?," Parsley Health, Aug. 29, 2024, https://www.parsleyhealth.com/blog/is-health-genetic/.

3   S. S. Negah and F. Forouzanfar, "Oxidative Stress is a New Avenue for Treatment of Neuropsychiatric Disorders: Hype of Hope?," *Curr Mol Med.* 2024;24(12):1494–1505.

4   W. McGinnis, "Oxidative Stress in Autism," *Alternative Therapy Health Medicine*, Nov–Dec 2004,10:6, 22–36.

5   P. Grandjean, *Only One Chance: How Environmental Pollution Impairs Brain Development —and How to Protect the Brains of the Next Generation* (New York: Oxford Univ Press, 2013).

6   "Off the Books II: More Secret Chemicals," Environmental Working Group, May 9, 2016, https://www.ewg.org/research/books-ii-more-secret-chemicals/.

7   "Toxics Release Inventory (TRI) Program," US Environmental Protection Agency, https://www.epa.gov/toxics-release-inventory-tri-program, accessed Nov. 12, 2024.

8   Rachel Carson, *Silent Spring* (Boston: Houghton Mifflin, 1962).

9   T. Colburn, D. Dumanoski, and J. P. Myers, *Our Stolen Future* (New York: Dutton, 1996).

10 David Steinman, *Diet for a Poisoned Planet: How to Choose Safe Foods for Your Family* (New York, NY: Harmony Books, 1990).

11 David Steinman, *Raising Healthy Kids: Protecting Your Children from Hidden Chemical Toxins*, (New York, NY: Skyhorse Publishing, 2024), 78–105.

12 S. A. Abrams and P. J. Landrigan, "Are GMO Foods Safe for My Child? AAP Policy Explained," American Academy of Pediatrics, Dec. 18, 2023, www.HealthyChildren.org.

13 Stephanie Seneff, *Toxic Legacy: How the Weedkiller Glyphosate Is Destroying Our Health and the Environment* (White River Junction, VT: Chelsea Green Publishing, 2021).

14 "Glyphosate is Everywhere," Detox Project, https://detoxproject.org/glyphosate/glyphosate-is-everywhere/, accessed Nov. 6, 2024.

15 Zen Honeycutt, *Unstoppable: Transforming Sickness and Struggle into Triumph, Empowerment, and a Celebration of Community* (Mission Viejo, CA: Moms Across America Publishing, 2018).

16 A. Samsel and S. Seneff, "Glyphosate's Suppression of Cytochrome P450 Enzymes and Amino Acid Biosynthesis by the Gut Microbiome: Pathways to Modern Diseases," *Entropy* 2013, 15, 1416–63.

17 Sarah Pope, "70+ nonGMO Crops Sprayed with Roundup Just before Harvest," The Healthy Home Economist, https://www.thehealthyhomeeconomist.com/pre-harvest-roundup-crops-not-just-wheat/, accessed Nov. 6, 2024.

18 Greater Boston Physicians for Social Responsibility, *In Harm's Way: Toxic Threats to Child Development*. (Cambridge, MA: GBPSR, 2000).

19 Russell Blaylock, *Excitotoxins: The Taste that Kills* (Santa Fe, NM: Health Press, 1997).

20 A. L. Choi, G. Sun G, Y. Zhang, and P. Grandjean P, "Developmental Fluoride Neurotoxicity: A Systematic Review and Meta-Analysis," *Environ Health Perspect*, 2012: 120, 1362–8.

21 National Toxicology Program, Public Health Service U.S. Dept of Health and Human Services, *NTP Monograph on the State of the Science Concerning Fluoride Exposure and Neurodevelopment and Cognition: A Systematic Review* (Research Triangle Park, NC: August 2024).

22   A. J. Malin et al. "Maternal Urinary Fluoride and Child Neurobehavior at Age 36 Months." *JAMA Netw Open*, 2024;7(5).

23   J. Shulman and L. M. Wells, "Acute Fluoride Toxicity from Ingesting Home-Use Dental Products in Children, Birth to 6 Years of Age," *J Public Health Dent.* 1997 Summer; 57:3, 150–8.

24   Gary Null, "Interview with EPA's William Marcus on NTP's Fluoride/Cancer Study," Fluoride Action Network, Mar. 10, 1995, https://fluoridealert.org/content/marcus -interview/.

25   J. C. May, *My House Is Killing Me* (Baltimore: Johns Hopkins University Press, 2001).

26   D. Broom, "Our Poisonous Air Is Harming Our Children's Brains," World Economic Forum, Mar. 20, 2019, World Economic Forum. https://www.weforum.org/stories /2019/03/our-poisonous-air-is-harming-our-children-s-brains/.

27   U.S. Department of Health and Human Services (USDHHS). *A Report of the Surgeon General: How Tobacco Smoke Causes Disease: What It Means to You* (Consumer Booklet) (Atlanta, GA: U.S. Department of Health and Human Services, Centers for Disease Control and Prevention, National Center for Chronic Disease Prevention and Health Promotion, Office on Smoking and Health; 2010).

28   "How Is Secondhand Smoke Linked to Asthma and Allergies?," Asthma and Allergy Foundation of America, "https://aafa.org/asthma/asthma-triggers-causes/secondhand -smoke-environmental-tobacco-asthma/, accessed Nov. 18, 2024.

29   N. M. Johnson et al., "Air Pollution and Children's Health—A Review of Adverse Effects Associated with Prenatal Exposure from Fine to Ultrafine Particulate Matter," *Environ Health Prev Med*, 2021 Jul 12;26(1):72.

30   T. Sieber, "Are Your Headphones or Earbuds a Health Risk?," Sound Guys, July 17, 2023, https://www.soundguys.com/headphone-safety-earbuds-health-risk-63011/.

31   T. Jafari et al., "The Association between Mercury Levels and Autism Spectrum Disorders: A Systematic Review and Meta-Analysis," *J Trace Elements in Medicine and Biology*, 44: Dec. 2017, 289–D7.

32   D. Klinghardt, Heavy Metals Summit, http://TheHeavyMetalsSummit.com, Feb. 4, 2018.

33   G. Filippilli, "How Poisonous Mercury Gets from Coal-fired Power Plants into the Fish You Eat," *The Conversation*, Feb. 15, 2022, https://theconversation.com/how-poisonous -mercury-can-get-from-coal-fired-power-plants-into-the-fish-you-eat-trumps-epa-plans -to-weaken-emissions-rules-meant-to-lower-the-risk-176434.

34   H. L. Needleman et al., "Deficits in Psychologic and Classroom Performance of Children with Elevated Dentine Lead Levels," *New England Journal of Medicine*, Mar. 29, 1979, 300:13, 689–95.

35   B. P. Lanphear et al., "Cognitive Deficits Associated with Blood Lead Concentrations <10 µg/dL in US Children and Adolescents," *Public Health Reports*, Nov 2000, 115, 521–9.

36   "Reducing Your Child's Lead Levels," Consumer Reports, Sept. 2013, https://www.con-sumerreports.org/cro/2012/03/cdc-advisers-call-for-less-allowable-lead/index.html/.

37   R. Liang et al., "Aluminium-Maltolate-Induced Impairment of Learning, Memory and Hippocampal Long-Term Potentiation in Rats, *Ind Health*, 2012; 50:5, 428–36

38   M. Mold, D. Umar, A. King, and C. Exley, "Aluminum in Brain Tissue in Autism," *Journal Trace Elements in Medicine & Biology*, 2018; 46:76–82.

39   "Aluminum's Impact on Brain Health: Exploring Potential Risks and Effects,"NeuroLaunch, Sept. 30, 2024, https://neurolaunch.com/effects-of-aluminum-on-the-brain/.

40 "Arsenic and Your Health," National Institute of Environmental Health Sciences, https://www.niehs.nih.gov/sites/default/files/health/materials/arsenic_and_your_health_508.pdf, accessed Nov. 12, 2024.

41 J. Pizzorno, "Time to Recognize and Address the Serious Arsenic Problem," *Integr Med* (Encinitas). 2024 Mar;23(1):6–9.

42 P. J. Chedrese, M. Piasek, and M. C. Henson, "Cadmium as an Endocrine Disruptor in the Reproductive System," *Immunology, Endocrine & Metabolic Agents in Medicinal Chemistry*, 6:1, 2006, Page: [27–35].

43 S. Bernard et al., "Autism: A Novel Form of Mercury Poisoning," *Medical Hypotheses*, 2001, 56:4, 462–71.

44 J. L. Jacobson and S. W. Jacobson, "Intellectual Impairment in Children Exposed to Polychlorinated Biphenyls in Utero," *New England Journal of Medicine*, Sept. 1996; 335:11.

45 G. M. Lehmann et al., "Environmental Chemicals in Breast Milk and Formula: Exposure and Risk Assessment Implications." *Environ Health Perspect*. 2018 Sept.;126(9).

46 J. G. Vos et al. "Brominated Flame Retardants and Endocrine Disruption," *Pure and Applied Chemistry*, 2003, 75:11–12, 2039–46.

47 P. Grandjean, *Only One Chance: How Environmental Pollution Impairs Brain Development—and How to Protect the Brains of the Next Generation* (New York: Oxford University Press, 2013).

48 D. L. Dadd, *Toxic Free*, (New York, New York: Penguin, 2011) 44–5.

49 A. C. Steinemann et al., "Chemical Emissions from Residential Dryer Vents During Use of Fragranced Laundry Products," *Air Quality, Atmosphere and Health*, 2011.

50 "Are Your Art Supplies Toxic?," Green America, https://www.greenamerica.org/green-living/are-art-supplies-toxic, accessed Nov. 18, 2024.

51 "Phthalates," Campaign for Safe Cosmetics, "http://www.safecosmetics.org/get-the-facts/chemicals-of-concern/phthalates/, accessed Nov. 18, 2024.

52 T. Schettler, "Human Exposure to Phthalates via Consumer Products," *Int J Androl*, Feb 2006, 29:1, 134–39.

53 S. Swan, "Environmental Phthalate Exposure in Relation to Reproductive Outcomes and Other Health Endpoints in Humans," *Environmental Health Perspectives*, 2008.

54 S. Asarch, "Can Plastic Go in the Microwave? Here's What the Experts Say," Tom's Guide, Dec. 11, 2023, https://www.tomsguide.com/features/can-plastic-go-in-the-microwave-heres-what-the-experts-say.

55 J. Rano, "Sen. Lautenberg Introduces Safe Chemicals Act of 2011," *Environmental Working Group*, Apr. 14, 2011, https://www.ewg.org/news-insights/news/sen-lautenberg-introduces-safe-chemicals-act-2011.

56 Toxic-FreeFuture.org, www.Toxic-FreeFuture.org, accessed Nov. 13, 2024.

57 Neil Nathan, *The Sensitive Patient's Healing Guide*, (Fort Bragg, CA: Cypress House, 2024).

58 ElectromagneticHealth.org, https://electromagnetichealth.org/, accessed Nov. 2, 2017.

59 "Understanding How Electromagnetic Fields (EMFs) Can Damage Your Health," Joseph Mercola interview with Dietrich Klinghardt, https://www.youtube.com/watch?v=XYUNC2uA9nw, accessed Nov. 17, 2024.

60 "Geopathic Stress—The Ultimate Guide," Helios3, http://www.helios3.com/geopathic-stress.html, accessed Nov.17, 2024.

61 "Health Impacts & Reporting," Environmental Health Project, https://www.environmentalhealthproject.org/health-impacts, accessed Nov. 17, 2024.

62 J. Thurnell-Read, *Geopathic Stress: How Earth Energies Affect Our Lives* (Boston, MA: Element Books, 1995).

63   EnergyStar.gov, https://www.energystar.gov/, accessed Nov. 2, 2017.

64   J. Cloud, "Why Your DNA is not Your Destiny," *Time Magazine*, Jan. 6, 2010.

65   R. Rusting, "Epigenetics Explained," *Scientific American*, Nov. 22, 2011.

66   S. Ebrahim, "Epigenetics: The Next Big Thing," *Int. J. Epidemiol*, Feb. 2012 41:1, 1–3.

67   S. Mukherjee, "Same but Different: How Epigenetics Can Blur the Line Between Nature and Nurture," *The New Yorker*, May 2, 2016.

68   L. Dall'Aglio et al., "The Role of Epigenetic Modifications in Neurodevelopmental Disorders: A Systematic Review," *Neurosci Biobehav Rev*, 2018 Nov.;94:17–30.

69   A. Baccarelli and V. Bollati, "Epigenetics and Environmental Chemicals," *Curr Opin Pediatr*, 2009, 21:2, 243–51.

70   James Lyons-Weiler, *The Environmental and Genetic Causes of Autism* (New York: Skyhorse Publishing, 2016).

Chapter 3

1   Dietrich Klinghardt, "Klinghardt Academy Conference: Healing the Brain," New York, May 2013.

2   H. Gardener et al., "Perinatal and Neonatal Risk Factors for Autism: A Comprehensive Meta-Analysis," *Pediatrics*, Aug. 2011; 128(2): 344–355.

3   V. Frymann, "The Trauma of Birth," DrFreely.com, Nov. 6, 2011, https://drfeely.com/articles/the-trauma-of-birth/.

4   D. Stein et al., "Obstetric Complications in Individuals Diagnosed with Autism and in Healthy Controls," *Compr Psychiatry*, 2006;47(1):69–75.

5   S. Kahn and P. R., *Jaws: The Story of a Hidden Epidemic* (Redwood City, CA: Stanford University Press, 2018).

6   C. Hocking, "Understanding Retained Infant Reflexes," Whole-Brain Living and Learning, Dec. 8, 2015, https://www.wholebrainliving.com/2015/12/understanding-retained-infant-reflexes.html.

7   "The 2025 Dirty Dozen," Environmental Working Group, https://www.ewg.org/foodnews/dirty-dozen.php, accessed July 27, 2025.

8   L. Clausen, "Why Hydration Is Crucial for Children," Central Oregon Pediatric Associates, Oct. 16, 2024, https://copakids.com/child-healthcare-news/why-hydration-is-crucial-for-children/.

9   C. Sissons C. "What Is the Average Percentage of Water in the Human Body?," *Medical News Today*, May 27, 2020. https://www.medicalnewstoday.com/articles/what-percentage-of-the-human-body-is-water.

10   J. Nestor, *Breath: The New Science of a Lost Art* (New York: Riverhead Books, 2020) 172–173.

11   M. G. White, "Oxygen and Human Body Metabolism," Optimal Breathing, Sept. 26, 2023, https://optimalbreathing.com/blogs/respiratory-chemistry/oxygen-and-metabolism.

12   M. Kaplan, *Seeing through New Eyes* (Philadelphia: Jessica Kingsley Publishers, 2006), 161.

13   M. Theobald and C. Chai, "What Happens When You Don't Sleep for Days," Everyday Health, Aug. 9, 2022, Accessed November 18, 2024, https://www.everydayhealth.com/conditions/what-happens-when-you-dont-sleep-days/.

14   N. Mahajan et al., "Hyperactive-Impulsive Symptoms Associated with Self-reported Sleep Quality in Non-medicated Adults with ADHD," *Journal of Attention Disorders*, Sept. 2010, 14:2, 132–37.

15   M. Walker, *Why We Sleep* (New York: Scribner, 2017) 314–315.

16 R. Mendelsohn, *How to Raise Healthy Children in Spite of Your Doctor* (New York: Ballantine Books, 1987).

17 Vaccine Schedule," Centers for Disease Control and Prevention, https://www.cdc.gov /vaccines-children/schedules/index.html, accessed Nov. 20, 2024.

18 J. Healy, *Failure to Connect: How Computers Affect Our Children's Minds—and What We Can Do About It* (New York: Simon and Schuster, 1999).

19 S. R. Johnson, "TV and Our Children's Minds," You and Your Child's Health, Aug. 2017, http://www.youandyourchildshealth.org/articles/tv-article.html, accessed Nov. 20, 2024.

20 J. Haidt, *The Anxious Generation* (New York: Penguin Press, 2024).

21 J. M. Nagata et al. "Screen Time and Mental Health: A Prospective Analysis of the Adolescent Brain Cognitive Development (ABCD) Study," *BMC Public Health*, 2024 Oct. 7;24(1).

22 V. L. Dunckley, *Reset Your Child's Brain: A Four-Week Plan to End Meltdowns, Raise Grades, and Boost Social Skills by Reversing the Effects of Electronic Screen Time* (Novato, CA: New World Library, 2015).

23 R. Louv, *The Last Child in the Woods: Saving Our Children From Nature Deficit Disorder* (Chapel Hill, NC: Algonquin Books, 2008).

24 M. Sarris, "Stress and the Autism Parent," Kennedy Krieger, Jan. 2, 2019, https://www .kennedykrieger.org/stories/interactive-autism-network-ian/stress-and-autism-parent.

25 I. Ferwana and L. R. Varshney, "The Impact of COVID-19 Lockdowns on Mental Health Patient Populations in the United States," *Sci Rep.* 2024 Mar 7;14(1):5689.

26 Y. Wang et al. "Long-term Risk of Psychiatric Disorder and Psychotropic Prescription after SARS-CoV-2 Infection Among UK General Population," *Nat Hum Behav* 8, 1076–1087, 2024.

27 M. R. Dvorsky et al., "Impacts of COVID-19 on the School Experience of Children and Adolescents with Special Educational Needs and Disabilities," *Curr Opin Psychol*, 2023 Jun. 17.

28 M. S. Herbert and K. Weintraub, *The Autism Revolution: Whole Body Strategies for Making Life All It Can Be* (New York: Ballantine Books, 2012).

Chapter 4

1   "EWG's Shopper's Guide to Pesticides in Produce," Environmental Working Group, https://www.ewg.org/foodnews/full-list.php, accessed Dec. 4, 2024.

2   J. L. Carwile, "Canned Soup Consumption and Urinary Bisphenol A: A Randomized Crossover Trial," *JAMA*, 2011, 306:20, 2218–20.

3   M. Pollan, *A Natural history of Transformation* (New York: Penguin Books, 2013).

4   J. Calmia, "Fast Food Nation: Americans Cook Less than Any Developed Country," *LiveScience*, Apr. 2, 2011, https://www.livescience.com/13930-americans-cook-obese. html.

5   R. Wrangham, *Catching Fire: How Cooking Made Us Human* (New York: Basic Books, 2009).

6   A. Abrahamson, "Everything You Need to Know About Green Cleaning," Apartment Therapy, Nov. 28, 2023, https://www.apartmenttherapy.com/everything-you-need-to -know-about-green-cleaning-37342224.

7   "How We Keep Creativity Safe," The Art and Creative Materials Institute, https://www .acmiart.org/acmi-seals, accessed June 25, 2025.

8   D. Davis, *Disconnect: A Scientist's Solutions for Safer Technology, and How to Protect your Family* (New York, NY: New Voices Press, 2024).

9   N. Pineault, *The Non-Tinfoil Guide to EMFs* (Columbia, SC: N&G Media, 2017).

10 A. L. Gittleman, *Zapped: Why Your Cell Phone Shouldn't Be Your Alarm Clock and 1,268 Ways to Outsmart the Hazards of Electronic Pollution* (New York: Harpers, 2011).

11 B. Blake Levitt, *Electromagnetic Fields: A Consumer's Guide to the Issues and How to Protect Ourselves*. (Harcourt Brace, 2011).

12 I. Karim, *BioGeometry: Back to a Future for Mankind* (Egypt, 2010).

13 E. Suni "How Much Sleep Do You Need?" Sleep Foundation, May 13, 2024, https://sleep foundation.org/press-release/national-sleep-foundation-recommends-new-sleep-times.

14 "M Water," Life Enthusiast, https://www.life-enthusiast.com/shop/hbwm-m-water -3258#attr=2465, accessed July 29, 2025.

15 R. Louv, *The Last Child in the Woods: Saving Our Children from Nature Deficit Disorder* (Chapel Hill, NC: Algonquin Books, 2008).

16 J. Zand et al., *Smart Medicine for a Healthier Child* (New York, NY: Avery Publishing, 2003).

17 L. Hellerstein, *See It, Say It, Do It* (Centennial, CO: HiClear Publishing, 2009).

18 P. J. Landrigan and M. M. Landrigan MM, *Children and Environmental Toxins: What Everyone Needs to Know** (New York, NY: Oxford University Press, 2018).

19 David Steinman, *Raising Healthy Kids: Protecting Your Children from Hidden Chemical Toxins* (New York, NY: Skyhorse Publishing, 2024).

20 V. Naumov, *Generation Sick: The Power, Politics & Propaganda Behind America's Health Crisis* (New Milford, NJ: Vincent Crow Press, 2015).

21 C. Gavigan, *Healthy Child Healthy World: Creating a Cleaner Greener Safer Home* (New York: Dutton, 2008).

22 S. Lantz and T. McIntosh, *One Bite at a Time: Reduce Toxic Exposure and Eat the World You Want* (Australia: Ink Asia, 2016).

23 E. Ryan et al., *Squeaky Green* (San Francisco: Chronicle Books, 2008).

24 D. Wentz and M. Wentz, *The Healthy Home: Simple Truths to Protect Your Family from Hidden Household Dangers* (Philadelphia: Vanguard Press, 2011).

25 M. Perro and V. Adams, *What's Making Our Children Sick? How Industrial Food Is Causing an Epidemic of Chronic Illness, and What Parents (and Doctors) Can Do About It* (White River Junction, VT: Chelsea Green, 2017).

Chapter 5

1   Ben Lynch, *Dirty Genes: A Breakthrough Program to Treat the Root Cause of Illness and Optimize Your Health* (New York: HarperOne, 2018).

2   S. Hausman-Cohen, "Utilizing Genomically Targeted Molecular Data to Improve Patient-Specific Outcomes in Autism Spectrum Disorder," *International Journal of Molecular Science* 2022, 23(4).

3   R. Ruwa, "What Is a SIBO Breath Test?" April 12, 2023, https://www.healthline.com /health/sibo-breath-test.

4   A. Holmes, M. Blaxill, and B. Haley, "Reduced Levels of Mercury in First Baby Haircuts of Autistic Children," *International Journal of Toxicology,* 22: 277–85, 2003.

5   Neil Nathan, *The Sensitive Patient's Healing Guide* (Fort Bragg, CA: Cypress House) 376–379.

6   R. M. Suen and S. Gordon, "The Clinical Relevance of IgG Food Allergy Testing Through ELISA," *Townsend Letter for Doctors & Patients,* Jan. 2004, 61–66.

7   L. Heuer et al., "Reduced Levels of Immunoglobulin in Children with Autism Correlates with Behavioral Symptoms," *Autism Research,* Oct. 2008, 1:5, 275–83.

8   M. Casanova et al., "Minicolumnar Pathology in Autism," *Neurology,* 2002, 58, 428–32.

9   H. Hazlett et al., "Early Brain Overgrowth in Autism Associated with an Increase in Cortical Surface Area before Age 2 Years," *Archives of General Psychiatry*, May 2011, 68:5, 467–76. E. Courchesne et al., "Evidence of Brain Overgrowth in the First Year of Life in Autism," *JAMA*, July 2003, 290:3, 337–44.

10 E. Anagnostou and M. J. Taylor, "Review of Neuroimaging in Autism Spectrum Disorders: What Have We Learned and Where We Go from Here," *Molecular Autism*, Apr. 2011, 2:1, 4.

11 A. Frandsen et al., "Autonomic Response Testing Compared with Immunoglobulin E Allergy Panel Test Results: Preliminary Report," *Alternative Therapies in Health and Medicine*, Jan. 15, 2018.

12 Korotkov, "Prof. Fritz-Albert Popp," International Union of Medical and Applied Bioelectrography, Aug. 10, 2018, https://www.iumab.org/prof-fritz-albert-popp/.

13 Y. W. et al., "How to Solve Clinical Challenges in Mood Disorders; Machine Learning Approaches Using Electrophysiological Markers," *Clinical Psychopharmacology and Neuroscience*, Aug. 31, 2024;22(3):416–430.

Chapter 6

1   S. W. Porges and S. Porges, *Our Polyvagal World: How Safety and Trauma Change Us* (New York, New York: WW Norton, 2023) 116–7.

2   J. Goodlatte, "How a Mother's Emotions Affect Her Unborn Child," Fit for Birth, Jan. 9, 2024, https://getfitforbirth.com/a-mothers-emotions-affect-her-unborn-child/.

3   K. Power, *How to Hack Your Vagus Nerve: Exercises to Dramatically Reduce Inflammation, Trauma and Anxiety with Vagal Stimulation* (Middletown, DE: Katrina Power, 2020).

4   R. K. Naviaux, "Metabolic Features of the Cell Danger Response," *Mitochondrion* 16 (2014): 7–17.

5   Neil Nathan, *The Sensitive Patient's Healing Guide* (Fort Bragg, CA: Cypress House, 2024) 45–52.

6   S. W. Porges, "The Vagus: A Mediator of Behavioral and Physiologic Features Associated with Autism," In Bauman and Kemper, *The Neurobiology of Autism, 2nd ed.* (Baltimore, MD: Johns Hopkins Press, 2005) 65–78.

7   "What Is Polyvagal Theory?," Polyvagal Institute, https://www.polyvagalinstitute.org /whatispolyvagaltheory, accessed Jan. 10, 2025.

8   R. Wilson and R. Chicot, "The Importance of Early Bonding on the Long-term Mental Health and Resilience of Children," *London Journal of Primary Care* Jan. 2016, 8(1):12–14.

9   D. Dana, *The Polyvagal Theory in Therapy: Engaging the Rhythm of Regulation* (New York, NY: WW Norton & Co, 2018).

10 D. Dana, *Anchored: How to Befriend Your Nervous System Using Polyvagal Theory* (Boulder, CO: Sounds True, 2021).

11 S. W. Porges and S. Porges, *Our Polyvagal World: How Safety and Trauma Change Us* (New York, NY: WW Norton, 2023) 116–7.

12 S. W. Porges, *The Polyvagal Theory: Neurophysiological Foundations of Emotions, Attachment, Communication, and Self-regulation* (New York: Norton Books, 2011).

13 S. W. Porges, *The Pocket Guide to the Polyvagal Theory: The Transformative Power of Feeling Safe*, Norton Series on Interpersonal Neurobiology (New York, NY: Norton Books, 2017).

14 Z. Molnár, "On the 400th Anniversary of the Birth of Thomas Willis," *Brain*, May 7, 2021;144(4):1033–1037.

15 I. P. Pavlov, *Conditioned Reflexes: An Investigation of the Physiological Activity of the Cerebral Cortex*, Trans and ed by Anrep GV (London: Oxford University Press, 1927).

16 I. M. Setchenov, *Physiology of Behavior*, Ed by M. G. Jaroshevky (Moscow, Russia: Scientific Works, 1995).

17 L. S. Vygotsky, *Child Psychology: The Problems of Child Development, vol 4* (Moscow, Russia: 1986).

18 Svetlana Masgutova, *Reflexes: Portal to Neurodevelopment and Learning* (Orlando, FL: Svetlana Masgutova Educational Institute, 2015).

19 Neil Nathan, *The Sensitive Patient's Healing Guide* (Fort Bragg, CA: Cypress House, 2024).

20 J. Spence, *Trauma-Informed Yoga: A Toolbox for Therapists* (Eau Claire, WI: 2021).

21 T. M. Mäkinen et al., "Autonomic Nervous Function During Whole-Body Cold Exposure before and after Cold Acclimation," *Aviat Space Environ Me,* 2008 Sept.;79(9):875–82.

22 James Nestor, *Breath: The New Science of a Lost Art* (New York, NY: Riverhead Books, 2020).

23 D. Dana, *The Nervous System Workbook; Practical Exercises to Ease Anxiety, Find Safety, and Come Home to Yourself Using Polyvagal Theory* (Boulder, CO: Sounds True, 2024) 141.

24 S. A. Severinsen, *"The Role of the Vagus Nerve in Managing Stress: How Breathing Can Calm Your Mind and Body,"* Breatheology, https://www.breatheology.com/how-breathing -can-calm-your-mind-and-body/, accessed Jan. 10, 2025.

25 D. Dana, *Polyvagal Exercises for Safety and Connection: 50 Client-Centered Practices* (New York, NY: WW Norton, 2020).

26 D. Dana, *Polyvagal Card Deck: 58 Practices for Calm and Change* (New York, NY: WW Norton, 2022).

27 D. Dana, *The Nervous System Workbook: Practical Exercises to Ease Anxiety, Find Safety, and Come Home to Yourself Using Polyvagal Theory* (Boulder, CO: Sounds True, 2024).

28 N. Habib, *Activate Your Vagus Nerve: Unleash Your Body's Natural Ability to Heal* (Berkeley, CA: Ulysses Press, 2019).

29 E. Williams, *Daily Vagus Nerve Exercise: A Self-help Guide to Stimulate Vagal Tone, Relieve Anxiety and Prevent Inflammation* (Middletown, DE: Liberty Books, 2020).

30 K. Power, *How to Hack Your Vagus Nerve: Exercises to Dramatically Reduce Inflammation, Trauma and Anxiety with Vagal Stimulation* (Middletown, DE: Liberty Books, 2020).

31 S. W. Porges, "Orienting in a Defensive World: Mammalian Modifications of our Evolutionary Heritage. A Polyvagal Theory," *Psychophysiology,* 1995, 32, 301–18.

32 S. Porges and K. Onderko, *Safe and Sound: A Polyvagal Approach for Connection, Change, and Healing* (Boulder, CO: Sounds True, 2025).

33 R. I. Miller and S. K. Clarren, "Long-term Developmental Outcomes in Patients with Deformational Plagiocephaly," *Pediatrics,* 2000: Feb.,105:2, E26.

34 R. Rosenberg, *Accessing the Healing Power of the Vagus Nerve: Self-Help Exercises for Anxiety, Depression, Trauma, and Autism* (Berkeley, CA: North Atlantic Books; 2017) 163–183.

35 S. Centers, "Osteopathy: A Philosophy and Methodology for the Effective Treatment of Children with Autism," *Autism Science Digest,* Apr. 2011: 1, 101–9.

36 M. A. Block, *No More Ritalin: Treating ADHD Without Drugs* (New York, NY: Kensington Books, 1996).

37 L. Hoang, *Osteopathy for Children* (Hobart, NY: Hatherleigh Press, 2015).

38 Eve Kodiak, "Reflexes 1: Emergence, Development, Integration," https://evekodiak .com/movement-matters-blog-entries/reflexes-1-emergence-development-integration, accessed Jan. 10, 2025.

39 T. Jovanovic et al., "Childhood Abuse Is Associated with Increased Startle Reactivity in Adulthood," *Depress Anxiety* 2009; 26(11):1018–26.

40   N. Doidge, *The Brain That Changes Itself* (New York, NY: Penguin Life, 2005).
41   A. Hopper, *Wired for Healing: Remapping the Brain to Recover from Chronic and Mysterious Illnesses* (Altona, Manitoba, Canada: Friesens, 2014).
42   A. J. Bratty, "Neuroplasticity Intervention, Amygdala and Insula Retraining (AIR), Significantly Improves Overall Health and Functioning Across Various Chronic Conditions," *Integrative Medicine*, 2024 Jan.;22(6):20–28.

Chapter 7

1   A. Collen, *10% Human: How Your Body's Microbes Hold the Key to Health and Happiness* (New York, NY: Harper, 2016).
2   F. Baquero and C. Nombela, "The Microbiome as a Human Organ," *Clin Microbiology Infection*, July 2012. 18:2–4.
3   Common Fund, "Human Microbiome Project, Program Snapshot," National Institutes of Health, https://commonfund.nih.gov/hmp, accessed Dec. 9, 2024.
4   National Human Genome Research Institute, "Human Microbiome Project: Diversity of Human Microbes Greater Than Previously Predicted." *Science News*, May 21, 2010, https://www.sciencedaily.com/releases/2010/05/100520141214.htm.
5   M. Blaser, *Missing Microbes: How the Overuse of Antibiotics Is Fueling Our Modern Plagues* (New York, NY: Picador, 2015).
6   J. G. Mulle et al., "The Gut Microbiome: A New Frontier in Autism Research," *Curr Psychiatry Rep,* 2013 Feb;15(2):337.
7   L. J. et al., "Gut Mycobiome Dysbiosis and Its Impact on Intestinal Permeability in Attention-Deficit/Hyperactivity Disorder," *Journal of Child Psychology and Psychiatry,* 64:9, 1280–91, Sept. 2023.
8   J. W. Law et al., "Looking into the Gut Microbiome of Obsessive Compulsive Disorder: Can Probiotics Help?" *Gut*, 2023;72:A135.
9   J. Lai et al., "Gut Microbial Clues to Bipolar Disorder: State-of-the-Art Review of Current Findings and Future Directions," *Clinical and Translational Medicine*, 2020 Aug.;10(4):e146.
10   A. Góralczyk-Bińkowska et al., "The Microbiota-Gut-Brain Axis in Psychiatric Disorders," *International Journal of Molecular Sciences*, 2022 Sept. 24;23(19).
11   J. P. Fadok, "The Brain-Body Connection and the Vagus Nerve," *Psychology Today*, Oct. 15, 2024.
12   J. F. Cryan et al. "The Microbiota-Gut-Brain Axis," *Physiological Reviews*, Oct. 1, 2019;99(4):1877–2013.
13   M. Clapp et al., "Gut Microbiota's Effect on Mental Health: The Gut-Brain Axis," *Clinics and Practice*, Sept. 15, 2017;7(4):987.
14   M. Carabotti et al., "The Gut-Brain Axis: Interactions Between Enteric Microbiota, Central and Enteric Nervous Systems," *Annals of Gastroenterology*, Apr-Jun 2015; 28:2, 203–9.
15   M. Gershon, *The Second Brain* (New York: Harper, 1999).
16   S. Blum, "What Is Dysbiosis—And What You Can Do About It," Blum Health MD, June 17, 2017, https://blumhealthmd.com/2017/06/17/what-is-dysbiosis/.
17   R. Krajmalnik-Brown et al., "Gut Bacteria in Children with Autism Spectrum Disorders: Challenges and Promise of Studying How a Complex Community Influences a Complex Disease, *Microbial Ecology in Health and Disease*, Mar 12, 2015;26:26914.
18   Z. R. Hill et al., "Indoxyl Sulfate and Autism Spectrum Disorder: A Literature Review," *International Journal of Molecular Sciences*, 25(23), 12973, 2024.
19   A. Goswami et al., "Role of Microbes in the Pathogenesis of Neuropsychiatric Disorders," *Frontiers in Neuroendocrinology*, Jul. 2021;62:100917.

20   K. Hashimoto, "Emerging Role of the Host Microbiome in Neuropsychiatric Disorders: Overview and Future Directions," *Molecular Psychiatry*, Sept. 2023;28(9):3625–3637.

21   *The Autism Enigma*, documentary, Cogent/Benger, http://cogentbenger.com/autism/, accessed Dec. 11, 2024.

22   P. D'Eufemia et al., "Abnormal Intestinal Permeability in Children with Autism," *Acta Paediatrica*, Sept. 1996, 85:9, 1076–79.

23   E. Lipski, *Leaky Gut Syndrome*, A Keats Good Health Guide (Los Angeles, CA: Keats Publishing, 1998) 7.

24   Q. Mu et al., "Leaky Gut as a Danger Signal for Autoimmune Diseases," *Front Immunol,* May 23, 2017;8:598.

25   A. Siebecker, "SIBO—Small Intestinal Bacterial Overgrowth," SIBO Info, https://www.siboinfo.com/overview.html, accessed Dec. 12, 2024.

26   L. Wang et al. "Hydrogen Breath Test to Detect Small Intestinal Bacterial Overgrowth: A Prevalence Case-Controlled Study in Autism," *European Child and Adolescent Psychiatry*, Feb 2018; 27:2, 233–40.

27   J. Kossewska et al., "Personality, Anxiety, and Stress in Patients with Small Intestine Bacterial Overgrowth Syndrome. The Polish Preliminary Study," *International Journal of Environmental Research and Public Health.* Dec 21, 2022;20(1):93.

28   N. Habib, *Activate Your Vagus Nerve* (Berkeley, CA: Ulysses Press, 2019) 79–80.

29   B. Feingold, *Why Your Child Is Hyperactive* (New York: Random House, 19750.

30   J. Brostoff and L. Gamlin, *Food Allergy and Intolerance* (London: Bloomsbury Press, 1989).

31   "Product Spotlight: Malvin Phenolic," Acuheart. https://www.acuheart.com/post/product-spotlight-malvin-phenolic, accessed Dec. 11, 2024.

32   R. Waring R, "Enzyme and Sulfur Oxidation Deficiencies in Autistic Children with Known Food/Chemical Intolerances," *Xenobiotica*, 1990, 20, 117–22.

33   R. M. Harris et al., "Activity of Phenylsulfotransferases in the Human Gastrointestinal tract," *Life Science*, Sept. 2000, 67:17, 2051–57.

34   J. Trebatická and Z. Ďuračková, "Psychiatric Disorders and Polyphenols: Can They Be Helpful in Therapy?" *Oxidative Medicine and Cellular Longevity*, 2015;2015:248529.

35   S. Liu et al., "Relationship between Dietary Polyphenols and Gut Microbiota: New Clues to Improve Cognitive Disorders, Mood Disorders and Circadian Rhythms," *Foods*, Mar 19, 2023;12(6):1309.

36   G. Morris et al., "Polyphenols as Adjunctive Treatments in Psychiatric and Neurodegenerative Disorders: Efficacy, Mechanisms of Action, and Factors Influencing Inter-individual Response," *Free Radical Biology and Medicine,* 172: 2021, 101–122.

37   W. G. Crook, *The Yeast Connection* (New York: Vintage Books, 1986); W. G. Crook, *The Yeast Connection and the Woman* (Jackson, TN: Professional Books, 1994).

38   B. Rimland, "Candida-Caused Autism," *Autism Research Review International,* 1985, 2, 2–3.

39   W. Shaw et al., "Increased Excretion of Analogs of Krebs Cycle Metabolites and Arabinose in Two Brothers with Autistic Features," *Clinical Chemistry*, 1995, 41, 1094–104.

40   E. Severance et al., "Candida Albicans Exposures, Sex Specificity and Cognitive Deficits in Schizophrenia and Bipolar Disorder," *npj Schizophr* 2, 16018 (2016).

41   S. Chandler et al., "Parent-Reported Gastro-intestinal Symptoms in Children with Autism Spectrum Disorders," *Journal of Autism and Developmental Disorders*, Dec. 2013, 43:12, 2737–47.

42   "Constipation and Diarrhea in Autism (The Poop Page)," The Autism Community in Action (TACA), https://www.tacanow.org/family-resources/the-poop-page/, accessed Dec. 12, 2024.

43   A. Mannion and G. Leader, "Gastrointestinal Symptoms in Autism Spectrum Disorder: A Literature Review," *Review Journal of Autism and Developmental Disorders*, Oct. 2013.

44   F. Oski, *Don't Drink Your Milk* (Syracuse, NY: Mollica Press, 1983).

45   American Academy of Pediatrics Committee on Nutrition, "Why Do Infants Need Baby Formula Instead of Cow's Milk?" HealthyChildren.org, Apr. 22, 2024.

46   A. R. Tapia, "Celiac Disease vs. Non-Celiac Gluten Sensitivity vs. Food Allergy," Cleveland Clinic, Aug. 31, 2022, https://health.clevelandclinic.org/gluten-sensitivity -celiac-disease-wheat-allergy-differences.

47   H. Brooker, "Milk Allergy vs. Lactose Intolerance," Food Allergy Research & Education, https://www.foodallergy.org/resources/milk-allergy-vs-lactose-intolerance, accessed Dec. 12, 2024.

48   K. L. Reichelt et al., "Gluten, Milk Proteins and Autism: Dietary Intervention Effects on Behavior and Peptide Secretion," *Journal of Applied Nutrition*, 1990, 42:1, 1–11.

49   K. Seroussi, *Unraveling the Mystery of Autism and Pervasive Developmental Disorder: A Mother's Story of Research & Recovery* (New York: Random House, 2000).

50   T. Turbin, "Gluten Enzymes," glutenfreehelp.info http://glutenfreehelp.info/autoimmune -disorders/gluten-enzymes, accessed Dec. 12, 2024; C. Tiruppathi et al., "Genetic Evidence for Role of DPP IV in Intestinal Hydrolysis and Assimilation of Prolyl Peptides," *American Journal of Physiology-Gastrointestinal and Liver Physiology*, 265:1, July 1, 1993.

51   J. Panksepp, "A Neurochemical Theory of Autism," *Trends Neurosci*, 1979, 2, 174–77.

52   K. L. Reichelt et al., "Biologically Active Peptide Containing Fractions in Schizophrenia and Childhood Autism," *Advances in Biochemical Psychopharmacology*, 1981, 28, 627–43.

53   Shattock P, Lowdon G. "Proteins, Peptides and Autism, Part 2: Implications for the Education and Care of People with Autism," *Brain Dys*, 1991, 4:6, 323–34.

54   J. McCrone, "Gut Reaction: Is Food to Blame for Autism?," *New Scientist*, June 1998 20, 42–5.

55   N. L. Swanson et al., "Genetically Engineered Crops, Glyphosate and the Deterioration of Health in the United States of America," *Journal of Org. Syst.* 9 (2014) 6–37.

56   S. Seneff, *Toxic Legacy: How the Weedkiller Glyphosate is Destroying Our Health and the Environment* (White River Junction, VT: Chelsea Green, 2021), 156–7.

57   R. E. Frye et al., "Approaches to Studying and Manipulating the Enteric Microbiome to Improve Autism Symptoms," *Microbial Ecology in Health and Disease*, Vol 26— Supplement 1, 2015: "The Microbiome in Autism Spectrum Disorder."

58   K. De Felice, *Enzymes for Autism and Other Neurological Conditions* (Thundersnow Interactive, 2008).

59   M. Campos, "Probiotics for Bipolar Disorder Mania," Harvard Health Publishing, June 25, 2018, https://www.health.harvard.edu/blog/probiotics-for-bipolar-disorder-mania -2018062514125.

60   A. A. Shehata et al., "Distribution of Glyphosate in Chicken Organs and its Reduction by Humic Acid Supplementation," *The Journal of Poultry Science*, 2014, 51:3, 333–7.

61   E. Gough et al., "Systematic Review of Intestinal Microbiota Transplantation (Fecal Bacteriotherapy) for Recurrent Clostridium Difficile Infection," *Clinical Infectious Diseases*, 2011, 53:10, 994–1002.

62   D. W. Kang et al., "Microbiota Transfer Therapy Alters Gut Ecosystem and Improves Gastrointestinal and Autism Symptoms: An Open-Label Study," *Microbiome* 2017, 5:10.

63   Z. R. Hill et al., "Indoxyl Sulfate and Autism Spectrum Disorder: A Literature Review," *International Journal of Molecular Sciences*, 2024; 25(23):12973.

64   D. Siniscalco and N. Antonucci, "Possible Use of Trichuris Suis Ova in Autism Spectrum Disorders Therapy, *Medical Hypotheses*, July 2013;81:1, 1–4.

65   William Parker, "Autism and Helminths: The Good, the Bad, and the Rumors," *Autism Research Review International,* 29:2, 2015.

Chapter 8

1   K. Dorfman, "A Biomedical Approach to Autism Spectrum Disorders," in P. Lemer, *Envisioning a Bright Future* (Santa Ana, CA: OEPF, 2008), 120.

2   "Parent Ratings of Behavioral Effects of Drugs and Nutrients," Autism Research Institute, Mar. 2009, accessed July 28, 2025.

3   K. Dorfman, *Cure Your Child with Food* (New York: Workman Press, 2013).

4   M. Pollan, *Food Rules* (New York: Penguin Books, 2009).

5   "The Clean Fifteen," Environmental Working Group, https://www.ewg.org/foodnews /clean-fifteen.php, accessed Dec 15, 2024.

6   M. M. Bailey, "Synthetic Food Dyes: A Rainbow of Risks," https://www.cspinet.org /cspi-news/synthetic-food-dyes-rainbow-risks, accessed Dec. 15, 2024.

7   M. Perro and V. Adams, *What's Making Our Children Sick? How Industrial Food is Causing an Epidemic of Chronic Illness, and What Parents (and Doctors) Can Do About it* (White River Junction, VT: Chelsea Green Pub, 2017).

8   Doris Rapp, *Is This Your Child?* (New York: Quill Books, 1991).

9   "Top Reasons to Implement a GFCF Diet," The Autism Community in Action (TACA), https://www.tacanow.org/family-resources/in-defense-of-the-gfcfsf-diet-for-children-with-autism/, accessed Dec. 18, 2024.

10   Nsouli, et al. "The Role of Food Allergy in Serious Otitis Media," *Annals of Allergy,* 1991, 66, 91.

11   F. Ghalichi et al., "Effect of Gluten Free Diet on Gastrointestinal and Behavioral Indices for Children with Autism Spectrum Disorders: A Randomized Clinical Trial," *World J Pediatr,* Nov. 2016;12(4):436–442.

12   A. Amirova, "A Gluten-Free Diet Could Reduce Anxiety," Anxiety.org, Nov. 7, 2014, https://www.anxiety.org/gluten-free-diet-improves-depression-and-anxiety-symptoms.

13   H. Jyonouchi et al.,"Dysregulated Innate Immune Responses in Young Children with Autism Spectrum Disorders: Their Relationship to Gastrointestinal Symptoms and Dietary Intervention," *Neuropsychobiology,* 2005, 51:2, 77–85.

14   K. Daniels, *The Whole Soy Story* (Warsaw, Indiana: Newtrends Publishing, 2005).

15   V. T. Ramaekers et al., "A Milk-Free Diet Downregulates Folate Receptor Autoimmunity in Cerebral Folate Deficiency Syndrome," *Dev Med Child Neurol,* May 2008;50(5):346–52.

16   M. Gogou and G. Kolios, "The Effect of Dietary Supplements on Clinical Aspects of Autism Spectrum Disorder: A Systematic Review of the Literature," *Brain Development,* Sept. 2017;39:8, 656–64.

17   Lisa Lewis, *Special Diets for Special Kids* (Future Horizons, 2011).

18   Pete Evans, *Healthy Food for Healthy Kids: 120 Simple, Nourishing, Gluten- and Dairy-Free Recipes Your Whole Family Will Lov,* (Franklin Lakes, NJ: Children's Health Defense, 2025).

19   "Gluten-Free Products Market Size, Share & Trend Analysis Report by Product (Bakery Products, Dairy/Dairy Alternatives, Convenience Stores), by Distribution Channel (Online, Specialty Stores), by Region, And Segment Forecasts, 2023–2030," Grand View Research, http://www.grandviewresearch.com/industry-analysis/gluten-free-products -market, accessed Dec. 8, 2024.

20   M. D. Miller et al., "Potential Impacts of Synthetic Food Dyes on Activity and Attention in Children: A Review of the Human and Animal Evidence," *Environ Health,* Apr. 29, 2022;21(1):45.

21   S. J. James et al., "Metabolic Biomarkers of Increased Oxidative Stress and Impaired Methylation Capacity in Children with Autism," *Am J Clin Nutr*, 2004, 80, 1611–17.

22   J. Hersey, *Why Can't My Child Behave? Why Can't She Cope? Why Can't He Learn?* (Alexandria, VA: Pear Tree Press, 2002).

23   K. Konikowska et al., "The Influence of Components of Diet on the Symptoms of ADHD in Children," *Rocz Panstw Zakl Hig.* 2012;63(2):127–34.

24   W. G. Crook, *Help for the Hyperactive Child* (Jackson, TN: Professional Books, 1991).

25   William Shaw et al., "Increased Excretion of Analogs of Krebs Cycle Metabolites and Arabinose in Two Brothers with Autistic Features," *Clin Chem*, 1995, 41, 1094–104.

26   E. Gottschall, *Breaking the Vicious Cycle: Intestinal Health Through Diet* (Baltimore, Ontario, Canada: The Kirkton Press, 2004).

27   S. M. Baker and J. Chinitz, *We Band of Mothers: Autism, My Son, and the Specific Carbohydrate Diet* (San Diego, CA: Autism Research Institute, 2007).

28   P. Ferro and R. Prasad, *The SCD for Autism and ADHD: A Reference and Dairy-free Cookbook for the Specific Carbohydrate Diet* (Arlington, MA, Swallowtail Press, 2015).

29   Natasha Campbell-McBride, *Gut and Psychology Syndrome Diet* (Cambridge, England: Mendinform Publishing, 2010).

30   Natasha Campbell-McBride, *Gut and Physiology Syndrome: Natural Treatment for Allergies, Autoimmune Illness, Arthritis, Gut Problems, Fatigue, Hormonal Problems, Neurological Disease and More* (Cambridge, England: Mendinform Publishing, 2020).

31   Natasha Campbell-McBride, *Gut and Psychology Syndrome Diet. Natural Treatment for Autism, Dyspraxia, A.D.D., Dyslexia, A.D.H.D., Depression, Schizophrenia*, Revised & Enlarged Edition (Cambridge, England: Mendinform Publishing, 2015).

32   J. D'Adamo, *The D'Adamo Diet* (Toronto, Ontario, Canada: McGraw-Hill Ryerson, 1989).

33   T. A. Baroody, *Alkalize or Die* (Waynesville, NC: Holographic Health Press, 2002).

34   M. R. Herbert and J. A. Buckley, "Autism and Dietary Therapy: Case Report and Review of the Literature," *J Child Neurol*, Aug. 28, 2013:8, 975–82.

35   C. M. Palmer, *Brain Energy: A Revolutionary Breakthrough in Understanding Mental Health—and Improving Treatment for Anxiety, Depression, OCD, PTSD, and More* (Dallas, TX: BenBella Press, 2022).

36   E. Aranburu et al., "Gluten and FODMAPs Relationship with Mental Disorders: Systematic Review," *Nutrients.* May 31, 2021;13(6):1894.

37   "Low Oxalate Diet," Low Oxalate Info, www.lowoxalate.info, accessed Dec. 17, 2024.

38   Julie Matthews, "Low Oxalate Diet," Nourishing Hope, https://nourishinghope.com /low-oxalate/, accessed Dec/ 17, 2024.

39   K. Reid and B. Price, *Fat, Stressed, and Sick: MSG, Processed Food, and America's Health Crisis* (Lanham, MD: Rowman & Littlefield, 2023).

40   L. Kenny and B. T. Walsh, "Avoidant/Restrictive Food Intake Disorder (ARFID): Defining ARFID," *Eating Disorders Review*, May/June 2013, 24: 3, https://edr.iaedp foundation.com/avoidantrestrictive-food-intake-disorder-arfid/.

41   J. Ingwersen et al., "A Low Glycaemic Index Breakfast Cereal Preferentially Prevents Children's Cognitive Performance from Declining throughout the Morning," *Appetite*, 2007, 49:10, 240–244.

42   D. Benton et al., "The Influence of the Glycaemic Load of Breakfast on the Behaviour of Children in School," *Physiological Behavior*, Nov. 2007, 92:4, 717–24.

Chapter 9

1　T. C. Theoharides, D. Kempuraj, and L. Redwood, "Autism: An Emerging 'Neuro-immune Disorder' in Search of a Therapy," *Expert Opinion on Pharmacotherapy*, 10, 13 (2009): 2127–43.

2　J. Dornell, "Humoral vs Cell-mediated Immunity," Technology Networks (Immunology & Microbiology), Jan. 22, 2024, https://www.technologynetworks.com/immunology/articles/humoral-vs-cell-mediated-immunity-344829.

3　"Antibodies," Cleveland Clinic, May 6, 2022, https://my.clevelandclinic.org/health/body/22971-antibodies.

4　J. R. Tisoncik et al., "Into the Eye of the Cytokine Storm," *Microbiological Molecular Biology Review*," 2012;76(1):16–32.

5　S. Cappanera et al. "When Does the Cytokine Storm Begin in COVID-19 Patients? A Quick Score to Recognize It," *Journal of Clinical Medicine*, Jan 15, 2021;10(2):297.

6　M. A. Elhelu, "The Role of Macrophages in Immunology," *Journal of National Medical Association*, March 1983, 75(3): 314–7.

7　S. S. Kirikovich et al., "The Molecular Aspects of Functional Activity of Macrophage-Activating Factor GcMAF," *Int J Mol Sci*, Dec. 12, 2023;24(24):17396.

8　J. J. Bradstreet et al., "Initial Observations of Elevated Alpha n-Acetylgalactosaminidase Activity Associated with Autism and Observed Reductions from GC Protein—Macrophage Activating Factor Injections," *Journal of Autism and Developmental Disorders*, 31:2, 2012, 175–81.

9　J. Oliveira et al., "Immune Dysfunction: A common Feature of Major Psychiatric Disorders," In A. L. Teixeira & M. E. Bauer (Eds.), *Immunopsychiatry: A Clinician's Introduction to the Immune Basis of Mental Disorders* (Oxford University Press, 2019), 141–164.

10　E. Gutierrez and J. Rosso, *The Parent's Roadmap to Autism: A Functional Medicine Approach* (Austin, TX: Lioncrest Publishing, 2018), 170–1.

11　H. Jyonouchi et al., "Innate Immunity Associated with Inflammatory Responses and Cytokine Production Against Common Dietary Proteins in Patients with Autism Spectrum Disorders," *Neuropsychobiology*, 2002, 46:2, 76–84

12　S. Bernstein, "Symptoms of Immune System Problems," WebMD, Mar. 7, 2025, https://www.webmd.com/cold-and-flu/ss/slideshow-immune-system-problems.

13　P. Ashwood et al., "Autism and the Immune Response: A New Frontier in Autism Research," *Journal of Leukocyte Biology* 2006, 80(1): 1–15.

14　R. P. Warren et al., "Immune Abnormalities in Patients with Autism," *Journal of Autism and Developmental Disorders*, 1986, 16, 189–97; R. P. Warren et al., "Immunogenetic Studies in Autism and Related Disorders," *Molecular Clinical Neuropathology*, 1996, 28, 77–81.

15　M. Chen et al., "The Prevalence of Bipolar Disorder in Autoimmune Disease: A Systematic Review and Meta-analysis," *Annals of Palliative Medicine*, 10: 1, Jan. 31, 2021.

16　S. Gupta et al., "Dysregulated Immune System in Children with Autism: Beneficial Effects of Intravenous Immune Globulin on Autistic Characteristics," *Journal of Autism and Developmental Disorders*," 1996, 26:4, 439–52.

17　D. L. Vargas et al., "Neuroglial Activation and Neuroinflammation in the Brain of Patients with Autism," *Annals Neurol*, Jan. 2005, 57:1, 67–81.

18　K. Bock, *Healing the New Childhood Epidemics: Autism, ADHD, Asthma, and Allergies: The Groundbreaking Program for the 4-A Disorders* (New York: Ballantine Books, 2008).

19　P. Ashwood et al., "In Search of Cellular Immunophenotypes in the Blood of Children with Autism," May 2011, 6:5, e19299; H. Jyonouchi et al., "Children with ASD Who

Exhibit Chronic GI Symptoms and Marked Fluctuation of Behavioral Symptoms Exhibit Distinct Innate Immune Abnormalities," *Journal of Neuroimmunology*, July 29, 2011.

20 A. Chauhan, V. Chahan, and W. T. Brown, ed. *Autism, Oxidative Stress, Inflammation and Immune Abnormalities* (Boca Raton, FL: CRC Press, 2010).

21 R. Rountree and C. Colman, *Immunoitics: Your Personal Immune-Boosting Program* (New York: Perigee, 2000), 37.

22 J. E. Libbey et al., "Autistic Disorder and Viral Infections," *Journal of Neurovirology*, Feb. 2005, 11:1, 1–10.

23 M. V. Pletnikov et al., "Rat Model of Autism Spectrum Disorders: Genetic Background Effects on Borna Disease Virus-induced Developmental Brain Damage," *Ann NY, Academy of Science*, June 2001, 939, 318–19.

24 C. N. Swisher and L. Swisher, Letter: "Congenital Rubella and Autistic Behavior," *New England Journal of Medicine*, July 1975, 293:4, 198.

25 J. M. Caruso et al., "Persistent Preceding Focal Neurologic Deficits in Children with Chronic Epstein-Barr Virus Encephalitis," *Journal of Child Neurology*, Dec. 2000, 15:12, 791–96.

26 M. Ghaziuddin et al., "Brief Report: Autism and Herpes Simplex Encephalitis," *Journal of Autism and Developmental Disorders,* 1992, 22:1, 107–113.

27 J. J. Bradstreet et al., "Detection of Measles Virus Genomic RNA in Cerebrospinal Fluid of Children with Regressive Autism, A Report of Three Cases," *Journal American Physicians and Surgeons*, Summer 2004, 9:2, 38–45; V. K. Singh et al., "Serological Association of Measles Virus and Human Herpes Virus-6 with Brain Auto-antibodies in Autism," *Clinical Immunology and Immunopathology*, Oct. 1998, 89:1, 105–8.

28 P. Ashwood et al., "Altered T-cell Responses in Children with Autism," *Brain Behavior and Immunology*, 2010;25:5, 840–9.

29 K. Heckenlively and J. Mikovits, *Plague: One Scientist's Intrepid Search for the Truth about Human Retroviruses and Chronic Fatigue Syndrome (ME/CFS), Autism, and Other Diseases* (New York, NY: Skyhorse Publishing, 2014).

30 Dietrich Klinghardt, Lyme Light Masterminds Conference, Morristown, NJ, May 10–11, 2018.

31 "COVID-19 Timeline," David J. Sencer CDC Museum: In Association with the Smithsonian Museum. https://www.cdc.gov/museum/timeline/covid19.html, accessed Jan. 19, 2025.

32 R. Del Rio et al., "Potential Role of Autonomic Dysfunction in Covid-19 Morbidity and Mortality," *Frontiers in Physiology*, 2020, 11: 561749.

33 T. M. Nsouli, "The Role of Food Allergy in Serious Otitis Media," *Annals of Allergy*, 1991, 66, 91.

34 J. L. Paradise et al., "Tympanostomy Tubes and Developmental Outcomes at 9 to 11 Years of Age," *New England Journal of Medicine*, Jan. 2007, 356:3, 248–61.

35 K. Dorfman, P. S. Lemer, and J. Nadler, "What Puts a Child at Risk for Developmental Delays?" Unpublished survey of Developmental Delay Resources (DDR), Pittsburgh, PA, 1995.

36 R. Niehaus and C. Lord, "Early Medical History of Children with Autism Spectrum Disorders in a Large Group-Model Health Plan," *Pediatrics*, Oct. 2006, 118:4, e1203–11.

37 M. Konstantareas and S. Homatidis, "Ear Infections in Autistic and Normal Children," *Journal of Autism and Developmental Disorders,"* 1987, 17, 585.

38 K. Dorfman, "Post-Traumatic Ear Infection Syndrome," *New Developments*, Spring 2004, 9:3, 7.

39 Anil Minocha, "Autism: It's Gut, Stupid," Minocha Health, http://minochahealth.typepad.com/autismadhd/propionicaciddietbacteriabrain.html, accessed Mar. 7, 2018.

40 D. F. MacFabe et al., "Effects of the Enteric Bacterial Metabolic Product Propionic Acid on Object-Directed Behavior, Social Behavior, Cognition, and Neuroinflammation in Adolescent Rats: Relevance to Autism Spectrum Disorder, Behavioral Brain Research," Feb. 2, 2011, 217:1, 47–54.

41 William Shaw et al., "Increased Urinary Excretion of Analogs of Krebs Cycle Metabolites and Arabinose in Two Brothers with Autistic Features," *Clinical Chemistry* 41 8 Pt 1 (1995): 1094–104.

42 T. K. Murphy et al., "Characterization of the Pediatric Acute-onset Neuropsychiatric Syndrome Phenotype," *Journal of Child and Adolescent Psychopharmacology*, 2015; 25:1, 14–25.

43 D. Balzer, "Babesiosis and What You Need to Know about the 2023 Tick Season," Mayo Clinic News Network, March 24, 2023, https://newsnetwork.mayoclinic.org/discussion/babesiosis-and-what-you-need-to-know-about-the-2023-tick-season/.

44 E. Greenburg, "Treating Lyme Disease," *EXPLORE for the Professional*, May 2006, 15:3.

45 "Diagnosis and Testing," Lyme Wellness Initiative, Harvard Health Publishing, https://lyme.health.harvard.edu/diagnosis-and-testing/, accessed Jan. 19, 2025.

46 D. Klinghardt and M. Ruggiero, "The Ruggiero-Klinghardt (RK) Protocol for the Diagnosis and Treatment of Chronic Conditions with Particular Focus on Lyme Disease," *American Journal of Immunology*, 2017, 13:2, 114–26.

47 Yuhang Sun et al., "Immunotoxicity of Three Environmental Mycotoxins and Their Risks of Increasing Pathogen Infections," *Toxins* 2023, 15, no. 3: 187.

48 Jill Carnahan, "Struggling with Mold Illness? How EMFs Could Be Making Your Symptoms Worse," Dr. Jill: Your Functional Medicine Expert, Sept. 16, 2021, https://www.jillcarnahan.com/2021/10/16/struggling-with-mold-illness-how-emfs-could-be-making-your-symptoms-worse.

49 J. Crista, *Break the Mold: 5 Ways to Conquer Mold and Take Back Your Health* (Wellness Ink Publishing, 2018).

50 A. Keil et al., "Parental Autoimmune Diseases Associated with Autism Spectrum Disorders in Offspring," *Epidemiology*, Nov. 2010; 21(6):805–8.

51 H. O. Atladóttir et al., "Association of Family History of Autoimmune Diseases and Autism Spectrum Disorders," *Pediatrics*, Aug. 2009, 124:2, 687–94.

52 Neurolaunch Editorial Team, "The Intricate Connection Between ADHD and Autoimmune Diseases: Unraveling the Mystery," Neurolaunch, Aug. 4, 2024, https://neurolaunch.com/adhd-and-autoimmune-disease/.

53 V. T. Ramaekers et al., "Folate Receptor Autoimmunity and Cerebral Folate Deficiency in Low-functioning Autism with Neurological Deficits," *Neuropediatrics*, Dec. 2007;38(6):276–81.

54 R. E. Frye et al., "Folinic Acid Improves Verbal Communication in Children with Autism and Language Impairment: A Randomized Double-blind Placebo-Controlled Trial," *Molecular Psychiatry* doi:10.1038/mp.2016.168.

55 R. E. Frye et al., "Cerebral Folate Receptor Autoantibodies in Autism Spectrum Disorder," *Molecular Psychiatry*, 2013 Mar; 18:3, 369–81.

56 W. G. Crook, *The Yeast Connection Handbook* (Jackson, TN: Professional Books, 1996).

57 G. Samonis et al., "Prospective Evaluation of the Impact of Broad-spectrum Antibiotics on Gastrointestinal Yeast Colonization of Humans," *Antimicrobial Agents and Chemotherapy*, 1993, 37, 51–53.

58 M. Blaser, "Antibiotic Overuse: Stop the Killing of Beneficial Bacteria," *Nature*, 2011, 476, 393–94.

59 S. B. Levy, *The Antibiotic Paradox: How the Misuse of Antibiotics Destroys Their Curative Power* (Cambridge, MA: Perseus Books, 2002).

60 M. Schmidt, *Beyond Antibiotics: Strategies for Living in a World of Emerging Infections and Antibiotic-Resistant Bacteria* (Berkeley, CA: North Atlantic Books, 2009).

61 Boyd Haley, "Toxic Overload: Assessing the Role of Mercury in Autism," *Mothering*, Nov/Dec 2002, 115, 44–6.

Chapter 10

1   V. Uhlmann et al., "Potential Viral Pathogenic Mechanism for New Variant Inflammatory Bowel Disease, *Molecular Pathology*, Apr. 2002, 55(2):84–90.

2   "Corporate and Organizational Partners," American Academy of Pediatrics, https://www.aap.org/en/philanthropy/corporate-and-organizational-partners/current-partners, accessed May 28, 2025.

3   Centers for Disease Control, *Epidemiology and Prevention of Vaccine-Preventable Diseases, 13th edition*, J. Hamborsky, A. Kroger, and S. Wolfe, Editors, (Washington, DC: Public Health Foundation, 2015), Appendix E3.

4   "Recommended Child and Adolescent Immunization Schedule, 2025," Centers for Disease Control and Prevention, https://www.cdc.gov/vaccines/hcp/imz-schedules/downloads/child/0–18yrs-child-combined-schedule.pdf, accessed July 29, 2025.

5   C. Racha, "Difference Between IgG and IgM," Biodifferences.com, Mar. 25, 2017, https://biodifferences.com/difference-between-igg-and-igm.html.

6   "Immune Function & Autism," Autism Research Institute, https://autism.org/immune-system-function-autism/, accessed May 28, 2025.

7   L. Hewitson et al., "Influence of Pediatric Vaccines on Amygdala Growth and Opioid Ligand Binding in Rhesus Macaque Infants: A Pilot Study," *Acta Neurobiologiae Experimentalis* 70:2 (2010): 147–64.

8   B. L. Fisher, "Shots in the Dark," *The Next City*, Summer 1999, 4, 4.

9   Rev. James Hindes, Eulogy for Philip Incao, Feb. 14, 1941–Feb. 28, 1922, Anthroposophical Society in America, Mar. 3, 2022, https://issuu.com/anthrousa/docs/bh29-web/s/21662572.

10 Y. Suzuki et al., "Role of Host-Encoded Proteins in Restriction of Retroviral Integration," *Front Microbiol.* 2012: 3, 227.

11 Tetyana Obukhanych, *Vaccine Illusion: How Vaccination Compromises Our Natural Immunity and What We Can Do to Regain Our Health*, www.amazon.com: Kindle book, 2012.

12 H. E. Butram, *A Commentary on Current Childhood Vaccine Programs* (Quakertown, PA: Philosophical Publishing Company, 2010).

13 Y. Shoenfeld and L. Tomjlenovic, *Vaccines and Autoimmunity* (Hoboken, NJ, Wiley-Blackwell, 2015).

14 R. Moskowitz, *Vaccines—A Reappraisal* (New York, NY: Skyhorse Publishing, 2017), 245.

15 P. Thomas and D. Hoover, *Vax Facts: What to Consider BEFORE Vaccinating at All Ages & All Stages of Life* (New York, NY: Morgan James, 2025).

16 Neil Z. Miller, "Vaccines and Sudden Infant Death: An Analysis of the VAERS Database 1990–2019 and Review of the Medical Literature," *Toxicology Reports*, June 24, 2021;8:1324–1335.

17 I. Grotto et al., "Major Adverse Reactions to Yeast-Derived Hepatitis B Vaccines—a Review," *Vaccine*, 1998, 16, 329–34.

18 A Colarosso, P. J. Johnson, and Orlando Sentinel Staff, "Dad Freed from Life Sentence in Son's Death," Orlando Sentinel, Oct. 24, 2018, https://www.orlandosentinel.com/2004/08/28/dad-freed-from-life-sentence-in-sons-death/.

19   "Vaccine Injury Compensation Program Monthly Statistics Report," Health Resources
     and Services Administration, Jan. 1, 2025, https://www.hrsa.gov/sites/default/files/hrsa
     /advisory-committees/vaccines/vicp-stats-01–01-25.pdf.

20   Bruesewitz v. Wyeth, 562 U.S. 223 (2011), https://supreme.justia.com/cases/federal
     /us/562/223/, accessed May 27, 2025.

21   Sharyl Attkisson, "Family to Receive $1.5M+ in First-Ever Vaccine-Autism Court Award,"
     CBS News, Sept. 10, 2010, https://www.cbsnews.com/news/family-to-receive-15m-plus
     -in-first-ever-vaccine-autism-court-award/.

22   "Electronic Support for Public Health—Vaccine Adverse Event Reporting System
     (ESP:VAERS)," Digital Healthcare Research, https://digital.ahrq.gov/ahrq-funded
     -projects/electronic-support-public-health-vaccine-adverse-event-reporting-system,
     accessed June 14, 2025.

23   H. Fraser, *The Peanut Allergy Epidemic: What's Causing It and How to Stop It,* Revised and
     Updated Third Edition with Foreword by Robert F. Kennedy Jr. (New York: Skyhorse
     Publishing, 2017).

24   S. H. Sicherer and H. A. Sampson, "Peanut Allergy: Emerging Concepts and Approaches
     for an Apparent Epidemic," *Journal of Allergy and Clinical Immunology*, 120:3, 491–503,
     2007.

25   I. Faulkner, "Nobel Prize 1913: Vaccines Containing Food Proteins May Trigger Deadly Food
     Allergies," *Elephant Journal*, May 4, 2019, https://www.elephantjournal.com/2019/05
     /nobel-prize-1904-vaccines-containing-food-proteins-trigger-deadly-food-allergies/.

26   P. Thomas and J. Margulis, *The Vaccine-Friendly Plan: Dr. Paul's Safe and Effective
     Approach to Immunity and Health—from Pregnancy through Your Child's Teen Years*
     (Ballantine Books, 2016).

27   J. Lyons-Weiler and P. Thomas, "Relative Incidence of Office Visits and Cumulative Rates
     of Billed Diagnosis along the Axis of Vaccination," *International Journal of Environmental
     Research and Public Health*," 2020, 17:22, 8674.

28   J. Hammond, *The War on Informed Consent: The Persecution of Dr. Paul Thomas by the
     Oregon Medical Board* (New York, NY, Skyhorse Publishing, 2021).

29   J. Lyons-Weiler and R. L. Blaylock, "Revisiting Excess Diagnoses of Illnesses and
     Conditions in Children Whose Parents Provided Informed Permission to Vaccinate
     Them," *International Journal of Vaccine Theory, Practice, and Research*, 2022, 2:2, 603–18.

30   J. Glowicz, "Multi-Dose Vial Safety Reminders for National Immunization Awareness
     Month," CDC Safe Healthcare Blog, Aug. 15, 2023, https://blogs.cdc.gov/safehealthcare
     /multi-dose-vial-safety-reminders/.

31   J. Obradovic, "Shake That Vial: Flu Shot Thimerosal Content Can Vary by Draw," Age
     of Autism, June 25, 2013, https://www.ageofautism.com/2013/06/shake-that-vial-flu
     -shot-thimerosal-content-can-vary-by-draw.html.

32   W. A. Jamieson and H. M. Powell, "Merthiolate as a Preservative for Biological Products,"
     Am J Hyg, 1931, 14, 218–24.

33   "Thimerosal and Vaccines," Centers for Disease Control and Prevention, Dec. 19, 2024,
     https://www.cdc.gov/vaccine-safety/about/thimerosal.html.

34   R. F. Kennedy Jr., Editor. *Thimerosal: Let the Science Speak: The Evidence Supporting the
     Immediate Removal of Mercury—a Known Neurotoxin—from Vaccines* (New York, NY,
     Skyhorse Publishing, 2015).

35   G. DeLong, "A Positive Association Found Between Autism Prevalence and Childhood
     Vaccination Uptake Across the US Population," *Journal of Toxicology and Environmental
     Health*, 2001, Part A, 74, 903–16.

36 P. M. Gayed, "Toward a Modern Synthesis of Immunity: Charles A. Janeway, Jr. and the Immunologist's Dirty Little Secret," *Yale Journal of Biological Medicine*, June 2011, 84:2, 131–8.

37 R. I. Blaylock, "The Danger of Excessive Vaccination during Brain Development," *Med Veritas*, 2008, 5:1, 1727–41.

38 R. A. Yokel, "The Toxicology of Aluminum in the Brain: A Review," *Neurotoxicology*, 2000, 21:5, 813–28.

39 Y. Z. Zhu et al., "Impact of Aluminum Exposure on the Immune System: A Mini Review," *Environ Toxicol Pharmacol*, Jan. 2013; 35:1, 82–7.

40 V. J. Johnson and R. P. Sharma, "Aluminum Disrupts the Pro-inflammatory Cytokine /Neurotrophin Balance in Primary Brain Rotation-Mediated Aggregate Cultures: Possible Role in Neurodegeneration," *Neurotoxicology*, 2002, 24:2, 261–68.

41 L. Tomljenovic and C. A., "Aluminum Vaccine Adjuvants: Are They Safe?," *Curr Med Chem*, 2011; 18:17, 2630–7.

42 L. Tomljenovic and C. A. Shaw, "Do Aluminum Vaccine Adjuvants Contribute to the Rising Prevalence of Autism?," *Journal of Inorganic Biochemistry*, Nov. 2011, 105:11, 1489–99.

43 T. Rosenbluth, "Yes, Some Vaccines Contain Aluminum. That's a Good Thing," The New York Times, Jan. 24, 2025.

44 A. Price, "Robert F. Kennedy Jr. and the Debate over Aluminum in Vaccines," Daily Tribune, Jan. 25, 2025.

45 J. Lyons-Weiler, "PopRat Fact Check: MSM Gets Aluminum Toxicity in Vaccines Dead Wrong—Here Are the Facts They Don't Want You to Know," https://popular rationalism.substack.com/p/poprat-fact-check-msm-gets-aluminum?utm_source=post -banner&utm_medium=web&utm_campaign=posts-open-in-app&triedRedirect=true.

46 M. Mold, D. Umar, A. King, and C. Exley, "Aluminum in Brain Tissue in Autism," *Journal of Trace Elements in Medicine & Biology*, 46, Mar. 2018, 76–82.

47 C. Exley, "An Aluminum Adjuvant in a Vaccine is an Acute Exposure to Aluminum," *Journal of Trace Elements in Medicine and Biology*," 57: Jan. 2020, 57–9.

48 C. Exley, "Autism and Aluminium: The Din of Silence," The Hippocratic Post, Jan. 14, 2018, https://www.hippocraticpost.com/ageing/autism-aluminium-din-silence/.

49 S. Das and J. Leake, "Funding Halted for Professor Chris Exley, Who Links Vaccines to Autism," The Sunday Times, Apr. 7, 2019, https://web.archive.org/web/20240927005855 /https://www.thetimes.com/uk/science/article/funding-halted-for-professor-chris -exley-linking-vaccines-to-autism-8xvwp0g8p.

50 C. Exley, *Imagine You Are an Aluminum Atom: Discussions with Mr. Aluminum* (New York, NY: Skyhorse Publishing, 2020).

51 J. B. Handley and J. Handley, *Underestimated: An Autism Miracle* (New York, NY: Skyhorse Publishing, 2021).

52 Institute of Medicine, *Adverse Effects of Vaccines: Evidence and Causality* (Washington, DC: National Academies Press, 2011).

53 D. Kirby, *Evidence of Harm: Mercury in Vaccines and The Autism Epidemic, A Medical Controversy* (New York: St. Martin's Press, 2005).

54 L. Habakus and M. Holland, *Vaccine Epidemic: How Corporate Greed, Biased Science, and Coercive Government Threaten Our Human Rights, Our Health, and Our Children* (New York: Skyhorse Publishing, 2011).

55 D. A. Geier et al., "A Two-Phase Study Evaluating the Relationship between Thimerosal-Containing Vaccine Administration and the Risk for an Autism Spectrum Disorder Diagnosis in the United States," *Translational Neurodegeneration*," Dec. 2013, 2:1, 25.

56   A. R. Mawson and B. Jacob, "Vaccination and Neurodevelopmental Disorders: A Study of Nine-Year-Old Children Enrolled in Medicaid," *Science, Public Health Policy & the Law,* Vol. 6, Jan. 23, 2025.

57   S. J. James et al., "Thimerosal Neurotoxicity Is Associated with Glutathione Depletion: Protection with Glutathione Precursors," *Neurotoxicity,* 2005, 26, 1–8.

58   D. K. Parran et al., "Effects of Thimerosal on NGF Signal Transduction and Cell Death in Neuroblastoma Cells," *Toxicological Sciences,* 2005, 86:1, 132–140.

59   M. R. Herbert, "Autism: A Brain Disorder, or a Disorder That Affects the Brain?" *Clinical Neuropsychiatry,* 2005, 2:6, 354–79.

60   L. Redwood, "More Evidence of Harm," Age of Autism, Jan. 23, 2015, https://www .ageofautism.com/2015/01/more-evidence-of-harm-thimerosal-found-to-disrupt -mitochondrial-function-in-cells-from-people-with-autism.html.

61   B. E. Haley, "Mercury Toxicity: Genetic Susceptibility and Synergistic Effects," *Medical Veritas,* 2005, 2, 535–42.

Chapter 11

1   K. Singh et al., "Sulforaphane Treatment of Autism Spectrum Disorder (ASD)," *Proceedings of the National Academy of Sciences, USA,*" Oct. 28, 2014; 111:43, 15550–55.

2   S. Rebensburg et al., "Potent in Vitro Antiviral Activity of Cistus Incanus Extract against HIV and Filoviruses Targets Viral Envelope Proteins," *Science Reports.* Feb 2, 2016;6, 20394.

3   W. A. Price, *Nutrition and Physical Degeneration* (La Mesa, CA: The Price-Pottenger Nutrition Found, 1997), 157.

4   Y. Shabo et al., "Camel Milk for Food Allergies in Children," *IMAJ,* 2005, 7, 796–98.

5   R. Yagil, "*Camel Milk and Autoimmune Diseases: Historical Medicine,* 2004, https: //bengreenfieldlife.com/wp-content/uploads/2017/02/Camel-Milk-and-Autoimmune -Diseases-Historical-Medicine.pdf, accessed July 29, 2025.

6   Y. Shabo and R. Yagil, "Etiology of Autism and Camel Milk as Therapy," *International Journal on Disability and Human Development,* 2005, 4:2, 67–70.

7   L. Y. Al-Ayadhi and N. E. Elamin, "Camel Milk as a Potential Therapy as an Antioxidant in Autism Spectrum Disorder (ASD)," *Evid Based Complement Alternat Med,* 2013, 602834.

8   Christina Adams, *Camel Crazy: A Quest for Miracles in the Mysterious World of Camels* (New World Library, 2019).

9   L. Galland, *Superimmunity for Kids: What to Feed Your Children to Keep Them Healthy Now, and Prevent Disease in Their Future,* (New York: Dell, 1989).

10   P. J. Martens et al., "Vitamin D's Effect on Immune Function," *Nutrients,* Apr. 28, 2020;12(5):1248.

11   The Vitamin D Society, www.VitaminDSociety.org, accessed Jan. 31, 2025.

12   H. A. Dissanayake et al., "Prognostic and Therapeutic Role of Vitamin D in COVID-19: Systematic Review and Meta-analysis," *Journal of Clinical Endocrinological Metabolism,* Apr. 19, 2022;107(5):1484–1502.

13   G. Glaser, "What If Vitamin D Deficiency Is a Cause of Autism," *Scientific American,* Apr. 29, 2009.

14   E. Koaovska et al., "Vitamin D and Autism: Clinical Review," *Res. Dev Disabil,* Sep–Oct 2012, 33:5, 1541–50.

15   A. A. Vinkhuyzen et al., "Gestational Vitamin D Deficiency and Autism Spectrum Disorder," *British Journal of Psychology Open,* Apr. 10, 2017;3(2):85–90.

16   "Minneapolis Somali Autism Spectrum Disorder Prevalence Project," University of Minnesota, http://rtc.umn.edu/autism/, accessed Jan. 28, 2025.

17   J. M. Greenblatt, "Mental Health in the Sun: The Role of Vitamin D Deficiency in Mental Illness," *Psychiatric Times*, Oct. 16, 2024, https://www.psychiatrictimes.com /view/mental-health-in-the-sun-the-role-of-vitamin-d-deficiency-in-mental-illness.

18   R. E. Anglin et al., "The Psychiatric Manifestations of Mitochondrial Disorders: A Case and Review of the Literature," *Journal of Clinical Psychiatry*, Apr. 2012;73(4):506–12.

19   G. T. Rezin et al., "Mitochondrial Dysfunction and Psychiatric Disorders," *Neurochemistry Research*, June 2009;34(6):1021–9.

20   B. Jepson, *Changing the Course of Autism* (Boulder, CO: Sentient Publications, 2007, 235).

21   G. S. Kelly, "Bovine Colostrums: A Review of Clinical Uses," *Altern Med Rev*, Nov. 2003, 8:4, 378–94.

22   H. H. Fudenberg, "Dialyzable Lymphocyte Extract (DLyE) in Infantile Onset Autism: A Pilot Study," *Biotherapies*, 1996, 9:1–2, 143–7.

23   Faezeh Gouhari et al., "Colostrum Supplementation Enhance Mental Health Status and Alleviate Pain in Patients with Acetabular Fracture: A Randomized, Controlled, Clinical Trial," *Journal of Functional Foods*, 119, 2024,

24   A. A. White, *A Guide to Transfer Factors and Immune System Health: Helping the Body Heal Itself by Strengthening Cell-Mediated Immunity, 2nd Ed.* (BookSurge Publishing, 2009).

25   V. C. Toreti et al., "Recent Progress of Propolis for Its Biological and Chemical Compositions and Its Botanical Origin," *Evidence Based Complementary Alternative Medicine*, 2013;2013:697390.

26   Dietrich Klinghardt, "Biological Treatment of Lyme Disease," PowerPoint presentation, https://www.theforensicnurse.org/downloads/klinghardt_biological_treatment_of _lyme_disease_protocol.pdf, accessed July 29, 2025.

27   Neil Nathan, *Toxic: Heal Your Body from Mold Toxicity, Lyme Disease, Multiple Chemical Sensitivities, and Chronic Environmental Illness* (Las Vegas, NV: Victory Belt Publishing, 2018).

28   R. Horowitz, *Why Can't I Get Better? Solving the Mystery of Lyme and Chronic Disease* (New York, NY: St. Martin's Press, 2013).

29   R. Horowitz, *How Can I Get Better? Treating Resistant Lyme & Chronic Disease* (New York, NY: St. Martin's Griffin, 2017).

30   D. Ingels, *The Lyme Solution: A 5-Part Plan to Fight the Inflammatory Auto-Immune Response and Beat Lyme Disease* (New York, NY: Penguin Random House, 2018).

31   D. A. Kinderlehrer, *Recovery from Lyme Disease: The Integrative Medicine Guide to Diagnosing and Treating Tick-Borne Illness* (New York, NY: Skyhorse Publishing, 2021).

32   S. H. Buhner, *Healing Lyme: Natural Healing of Lyme Borreliosis and the Coinfections Chlamydia and Spotted Fever Rickettsiosis, 2nd Edition* (New York, NY: Raven Press, 2015).

33   J. L. Lanciego et al., "Functional Neuroanatomy of the Basal Ganglia," *Cold Spring Harbor Perspectives in Medicine*, Dec. 1, 2012;2(12):a009621.

34   M. Thienemann et al., "Clinical Management of Pediatric Acute Onset Neuropsychiatric Syndrome: Part I–Psychiatric and Behavioral Interventions," J. Frankovich et al., "Clinical Management of Pediatric Acute Onset Neuropsychiatric Syndrome: Part II–Use of Immunomodulatory Therapies," and M. S. Cooperstock et al., "Clinical Management of Pediatric Acute Onset Neuropsychiatric Syndrome: Part III–Treatment and Prevention of Infections," *Journal of Child and Adolescent Psychopharmacoogyl*, Sept. 2017, 27:7, 566–606.

35   K. Bock, *Brain Inflamed: Uncovering the Hidden Causes of Anxiety, Depression, and Other Mood Disorders in Adolescents and Teens* (New York, NY: Harper Wave, 2021).

36   J. Crista, *A Light in the Dark for PANDAS & PANS* (2022).

37   N. O'Hara, *Demystifying PANS/PANDAS: A Functional Medicine Desktop Reference on Basal Ganglia Encephalitis* (2022).

38   B. Lambert and M. Rickert-Hong, *Brain Under Attack: A Resource for Parents and Caregivers of Children with PANS, PANDAS, and Autoimmune Encephalitis* (2018).

39   S. Swedo et al., "Pediatric Autoimmune Neuropsychiatric Disorders Associated with Streptococcal Infections (PANDAS)," July 7, 2024 [Updated Sept. 13, 2024], in J. J. Ferretti, D. L. Stevens, and V. A. Fischetti, Editors, *Streptococcus Pyogenes: Basic Biology to Clinical Manifestations, 2nd edition* (Oklahoma City, OK: University of Oklahoma Health Sciences Center, 2022), Chapter 26.

40   D. Ingels, "Allergy Desensitization: An Effective Alternative Treatment for Autism," in T. Lyons and K. Siri, *Cutting Edge therapies for Autism* (New York: Skyhorse Publishing, 2012).

41   W. A. Shrader, "LDA Therapy," http://www.drshrader.com/lda_therapy.htm, accessed Jan. 31, 2025.

42   T. Vincent, "Low Dose Immunotherapy (LDI) in the Treatment of the Sensitive Patient," in Neil Nathan, *The Sensitive Patient's Healing Guide* (Fort Bragg, CA: Cypress House, 2024), 391–402.

43   J. Teitelbaum et al., "Improving Communication Skills in Children With Allergy-Related Autism Using Nambudripad's Allergy Elimination Techniques: A Pilot Study," *Integrative Medicine*, Oct/Nov 2011, 10:5, 36–43.

44   D. Nambutripad, *Say Good-Bye to Allergy Related Autism* (Los Angeles, CA: Delta Publishers, 1999).

45   E. W. Cutler, *The Food Allergy Cure: A New Solution to Food Cravings, Obesity, Depression, Headaches, Arthritis, and Fatigue* (New York: Three Rivers Press, 2003).

46   G. M. Elchaar et al., "Efficacy and Safety of Naltrexone Use in Pediatric Patients with Autistic Disorder," *Annals of Pharmacotherapy*, June 2006, 40:6, 1086–95.

47   D. Mischoulon et al., "A Double-Blind, Randomized, Placebo-Controlled Trial of Low-Dose Naltrexone for Major Depressive Disorder," *Journal of Affective Disorders*, 2017, 208, 6–14.

48   NeuroLaunch Editorial Team, "Low Dose Naltrexone for Anxiety and Depression: A Comprehensive Guide," Neurolaunch, July 11, 2024, https://neurolaunch.com/low-dose-naltrexone-anxiety/.

49   S. Dickson, "Can You Use Low Dose Naltrexone (LDN) in Children?," LDN Research Trust, https://ldnresearchtrust.org/, accessed Apr. 24, 2025.

50   "Anti-inflammatory Activity of Curcumin," Healthy Source, http://curcumin-turmeric.net/inflammation.html, accessed Feb. 4, 2025.

51   T. C. Theoharides and B. Zhang, "Neuro-Inflammation, Blood-Brain Barrier, Seizures and Autism," *Journal of Neuro-Inflammation*, Nov. 2011, 8:1, 68.

52   E. Trevathan, "Seizures and Epilepsy among Children with Language Regression and Autistic Spectrum Disorders," *Journal of Child Neurology*," Aug. 2004, 19 Supplement 1, S49–57.

53   R. Tuchman and I. Rapin, "Epilepsy in Autism," *Lancet Neurology*, Oct. 2002, 6, 352–8.

54   M. Matsuo et al., "Characterization of Childhood-Onset Complex Partial Seizures Associated with Autism Spectrum Disorder," *Epilepsy Behaviors*, Mar. 2011, 3, 524–27.

55   P. Grinspoon, "The Endocannabinoid System: Essential and Mysterious," Harvard Health Publishing, Aug. 11, 2021, https://www.health.harvard.edu/blog/the-endocannabinoid-system-essential-and-mysterious-202108112569.

56   D. Siniscalco et al., "Cannabinoid Receptor Type 2, but Not Type 1, Is Up-regulated in Peripheral Blood Mononuclear Cells in Children Affected by Autistic Disorders," *Journal of Autism Development Discordia*, 2013:43,11: 2686–95.

57   A. Wilcox, "Autism & Cannabis," Herb, Dec. 3, 2019. https://herb.co/marijuana/news/mom-overjoyed-medical-marijuana-saved-my-sons-life.

58   "Is It Safe to Use Medical Cannabis to Treat Autism?," Conversations on Cannabis, YouTube, https://www.youtube.com/live/QdEAf30YvFg, accessed July 29, 2025.

59   R. Olivardia, "The Damaging Effects of Cannabis on the ADHD Brain," *ADDitude Magazine*, Mar. 26, 2020.

60   A. K. Davis et al., "Effects of Psilocybin-Assisted Therapy on Major Depressive Disorder: A Randomized Clinical Trial," *JAMA Psychiatry*, 2021; 78(5):481–489.

61   C. Brotheridge, "Best Mushrooms for Anxiety and Depression and How to Use Them," Calmer You, Nov. 26, 2024, https://www.calmer-you.com/best-mushrooms-for-anxiety/.

62   M. Pollan, *How to Change Your Mind: What the New Science of Psychedelics Teaches Us about Consciousness, Dying, Addiction, Depression, and Transcendence* (New York, NY, Penguin, 2019).

63   Y. T. Lv et al. "Transplantation of Human Cord Blood Mononuclear Cells and Umbilical Cord-Derived Mesenchymal Stem Cells in Autism," *Journal of Translational Medicine*, 2013:11, 196.

64   A. Sharma et al., "Autologous Bone Marrow Mononuclear Cell Therapy for Autism: An Open Label Proof of Concept Study," *Stem Cells Institute*, 2013:623875.

65   T. Slepcevic, *Mother's Journey Healing Her Son with Autism* (New York: Morgan James Publishing, 2023).

66   C. Weiss and E. Weiss, *Educating Marston: A Mother and Son's Journey Through Autism* (Changing Lives Press, 2021).

67   P. M. Nemechek and J. R. Nemechek, *The Nemechek Protocol for Autism and Developmental Disorders* (Middletown, DE: Dr. Patrick Nemechek, 2017).

68   M. M. Saeed, *The Holistic Rx: Your Guide to Healing Chronic Inflammation and Disease* (Lanham, MD: Rowman & Littlefield, 2017).

Chapter 12

1   "The Role of Hormones in Bipolar Disorder," Bay Area CBT Center, July 18, 2024, https://bayareacbtcenter.com/role-of-hormones-in-bipolar-disorder/.

2   D. Klinghardt, *Medicine 2012: New Symptoms, New Causes, New Treatments* (Warren, NJ: Klinghardt Academy, 2012).

3   C. S. Carter, "Sex Differences in Oxytocin and Vasopressin: Implications for Autism Spectrum Disorders?" *Behavioral Brain Research*, Jan. 2007, 176:1, 170–86.

4   C. Palmer, *Brain Energy* (Dallas, TX, BenBella Books, 2022), 122–125, 210–11.

5   M. Picard et al., "Mitochondrial Functions Modulate Neuroendocrine, Metabolic, Inflammatory and Transcriptional Responses to Acute Psychological Stress," *Proceedings of the National Academy of Sciences, USA*, 1122:48, 2013.

6   W. Wisner, "The Link Between Hormones and Your Mental Health," Verywell Mind, June 5, 2023, https://www.verywellmind.com/the-link-between-hormones-and-mental-health-7500077.

7   L. Lucaccioni et al., "Long Term Outcomes of Infants Born by Mothers with Thyroid Dysfunction During Pregnancy," *Acta Biomed*, Sept. 15, 2020;92(1):e202.

8   W. Thompson et al., "Maternal Thyroid Hormone Insufficiency during Pregnancy and Risk of Neurodevelopmental Disorders in Offspring: A Systematic Review and Meta-analysis," *Clinical Endocrinology (Oxf)*, Apr. 2018;88(4):575–584.

9   L. Davies and H. G. Welch, "Increasing Incidence of Thyroid Cancer in the United States, 1973–2002," *JAMA*, May 2006, 295:18, 2164–67.

10   A. Sanabria et al., "Growing Incidence of Thyroid Carcinoma in Recent Years," *Head Neck*, Apr. 2018;40(4):855–866.

11   D. Brownstein, *A Holistic Approach to Viruses* (Medical Alternatives Press, 2021).

12   D. Kharrazian, *Why Do I Still Have Thyroid Symptoms When My Lab Tests Are Normal?* (2010).

13   "What Is Cortisol," Cleveland Clinic, https://my.clevelandclinic.org/health /articles/22187-cortisol, accessed Feb. 12, 2025.

14   V. K. Patel et al., "Cortisol as a Target for Treating Mental Disorders: A Promising Avenue for Therapy," *Mini Rev Med Chem*, 2024;24(6):588–600.

15   M. Brosnan et al., "Absence of a Normal Cortisol Awakening Response (CAR) in Adolescent Males with Asperger Syndrome (AS)," *Psychoneuroendocrinology*, Aug. 2009, 34:7, 1095–100.

16   S. K. Putnam et al., "Salivary Cortisol Levels and Diurnal Patterns in Children with Autism Spectrum Disorder." *Journal of Developmental and Physical Disabilities*, 2015, 27:4, 453–65.

17   M. Pugle, "What Are Catecholamines, and What Do They Do?" VerywellHealth, Aug. 13, 2024, https://www.verywellhealth.com/catecholamines-8685190.

18   S. Watson, "Dopamine: The Pathway to Pleasure," Harvard Health Publishing, Apr. 18, 2024, https://www.health.harvard.edu/mind-and-mood/dopamine-the-pathway-to-pleasure.

19   I. Bancos, "Adrenal Hormones," Endocrine Society, Jan. 24, 2022, https://www .endocrine.org/patient-engagement/endocrine-library/hormones-and -endocrine-function/adrenal-hormones.

20   M. S. Shajib et al., "Interleukin 13 and Serotonin: Linking the Immune and Endocrine Systems in Murine Models of Intestinal Inflammation," *PLoS One*, Aug. 2013, 8, 8.

21   K. T. Daniels, *The Whole Soy Story: The Dark Side of America's Favorite Health Food* (Washington, DC: New Trends Pub, 2007).

22   E. Ingudomnukul et al., "Elevated Rates of Testosterone-Related Disorders in Women with Autism Spectrum Conditions," *Hormones and Behavior*, May 2007;51(5):597–604.

23   B. Auyeung et al., "Testosterone and Autistic Traits in 18 to 24-month-old Children," *Molecular Autism*, 2010, 1:1, 11.

24   S. Baron-Cohen et al., "Elevated Fetal Steroidogenic Activity in Autism," *Molecular Psychiatry*, 2015, 20, 369–76.

25   "Melatonin: The Master Circadian Rhythm Regulator," Chronoceuticals, https: //chronoceuticals.com/melatonin-the-master-circadian-rhythm-regulator/, accessed Feb. 13, 2025.

26   S. K. Satyanarayanan et al., "Circadian Rhythm and Melatonin in the Treatment of Depression," *Current Pharmaceutical Des*, 2018;24(22):2549–2555.

27   D. H. Skuse et al., "Common Polymorphism in the Oxytocin Receptor Gene (OXTR) Is Associated with Human Social Recognition Skills," *Proceedings of the National Academy of Sciences*, Pub online Dec. 23, 2013.

28   T. Sarachana et al., "Sex Hormones in Autism: Androgens and Estrogens Differentially and Reciprocally Regulate RORA, a Novel Candidate Gene for Autism," *PLoS ONE*, Feb. 2011, 6:2.

29   X. Xu et al., "Mothers of Autistic Children: Lower Plasma Levels of Oxytocin and Arg-Vasopressin and a Higher Level of Testosterone. PLoS One," Sept. 2013, 8:9.

30   M. Parolin et al., "Coronaviruses and Endocrine System: A Systematic Review on Evidence and Shadows," *Endocrine Metabolism Immune Disorders Drug Targets*, 2021; 21(7):1242–1251.

31   T. Colburn et al., *Our Stolen Future* (New York: Plume/Penguin, 1997).

32 M. J. Strong, "SARS-CoV-2, Aging, and Post-COVID-19 Neurodegeneration," *Journal of Neurochemistry*, Apr. 2023; 165(2):115–130.

33 D. Klinghardt, "Post-COVID Medicine," Sophia Health Institute lecture, Feb. 18. 2025.

34 J. Crista, "Mold Mycotoxins and Estrogen Levels: Dangerous Health Impact," https://drcrista.com/2019–5-29-mold-promotes-bad-estrogens/, accessed Feb. 16, 2025.

35 Grady. "The Impact of EMFs on Sleep Quality and Health," Tacoma Encounter, July 21, 2024. https://tacomaencounter.org/health/emf/the-impact-of-emfs-on-sleep-quality-and-health/.

36 S. M. Engel et al., "Prenatal Phthalate Exposure Is Associated with Childhood Behavior and Executive Functioning," *Environmental Health Perspectives*, 2010, 118, 565–71.

37 A. Miodovnik et al., "Phthalates, BPA Linked to Atypical Childhood Social Behaviors," *NeuroToxicology*, Mar. 2011, 32:2, 261–7; Y. Kim et al., "Prenatal Exposure to Phthalates and Infant Development at 6 Months: Prospective Mothers and Children's Environmental Health (MOCEH) Study," *Environ Health Perspect*, Oct. 2011, 119:10, 1495–500; Williams R. "Phthalates Affect Child Development," Environment Report, Sept 2011.

38 "TEDX List of Potential Endocrine Disruptors," The Endocrine Disruption Exchange, https://endocrinedisruption.org/interactive-tools/tedx-list-of-potential-endocrine-disruptors/search-the-tedx-list., accessed Feb. 12, 2025.

39 F. De Luca, "Endocrinological Abnormalities in Autism," *Seminars in Pediatric Neurology*, Oct. 2020;35: 100582.

40 E. G. Spratt et al., "Enhanced Cortisol Response to Stress in Children in Autism," *Journal of Autism Developmental Disorders*, Jan. 2012, 42:1, 75–81.

41 C. A. Molloy et al., "Familial Autoimmune Thyroid Disease as a Risk Factor for Regression in Children with Autism Spectrum Disorder: a CPEA Study," *Journal of Autism Developmental Disorders*, Apr. 2006, 36:3, 317–24; G. C. Roman et al. "Association of Gestational Maternal Hypothyroxinemia and Increased Autism Risk," *Annals of Neurology*, Nov. 2013, 74:5, 733–42.

42 R. E. Frye et al., "Thyroid Dysfunction in Children with Autism Spectrum Disorder Is Associated with Folate Receptor α Autoimmune Disorder," *Journal of Neuroendocrinology*, Mar.; 29:3, 2017.

43 K. L. Williams, "High Testosterone in Children," Dec. 5, 2018, www.HowtoAdult.com, Accessed February 16, 2025. K. Heckenlively, "Mercury, Testosterone and Autism—a Really Big Idea," Age of Autism, Apr. 21, 2008, https://www.ageofautism.com/2008/04/mercury-testost.html.

44 K. Heckenlively, "Mercury, Testosterone and Autism—a Really Big Idea!," Age of Autism, Apr. 21, 2008, https://www.ageofautism.com/2008/04/mercury-testost.html.

45 E. Ingudomnukul et al., "Elevated Rates of Testosterone-Related Disorders in Women with Autism Spectrum Conditions," *Hormones and Behavior*, May 2007, 51:5, 597–604.

46 Spectrum, "Puberty and Autism: An Unexplored Transition," The Transmitter, Mar. 24, 2021, https://www.thetransmitter.org/spectrum/puberty-and-autism-an-unexplored-transition/.

47 M. Cohn, "Lupron Therapy for Autism at Center of Embattled Doctor's Case: Some Parents Embrace Alternative Treatment for Autism that Scientists Don't Support," *The Baltimore Sun*, June 16, 2011.

48 L. Sykes, *Sacred Spark* (Fourth Lloyd Productions, 2009).

49 D. Gorski, "Chemical Castration of Autistic Children Leads to the Downfall of Dr. Mark Geier," Science-Based Medicine, May 9, 2011.

50 J. M. Curin et al., "Lower Cortisol and Higher ACTH levels in Individuals with Autism," *Journal of Autism Developmental Disorders*," Aug. 2003, 33:4, 443–8.

51 P. Newhouse and K. Albert, "Estrogen, Stress, and Depression: A Neurocognitive Model," *JAMA Psychiatry*, 2015, 72(7), 727–729.

52 B. Rzepka-Migut and J. Paprocka, "Efficacy and Safety of Melatonin Treatment in Children with Autism Spectrum Disorder and Attention-Deficit/Hyperactivity Disorder—A Review of the Literature," *Brain Science*, April 7, 2020;10(4):219.

53 I. M. Andersen et al., "Melatonin for Insomnia in Children with Autism Spectrum Disorders," *Journal of Child Neurology*, May 2008, 23:5, 482–5.

54 P. A. Geoffroy et al., "The Use of Melatonin in Adult Psychiatric Disorders: Expert Recommendations by the French Institute of Medical Research on Sleep (SFRMS)," *Encephale*, Nov. 2019;45(5):413–423.

55 P. Tsigaris and J. A. Teixeira da Silva, "Smoking Prevalence and COVID-19 in Europe," *Nicotine Tobacco Research*, 2020;22:1646–1649.

56 C. Curley, "New Treatment Restores Sense of Smell in Some People with Long COVID," MedicalNewsToday, https://www.medicalnewstoday.com/articles/new-treatment-restores -sense-of-smell-in-some-people-with-long-covid, accessed Feb. 16, 2025.

57 R. D. Shytle et al., "A Pilot Controlled Trial of Transdermal Nicotine in the Treatment of Attention Deficit Hyperactivity Disorder," *World Journal of Biological Psychiatry*, July 2002;3(3):150–5.

58 K. McGrane, "What Are DIM Supplements? Benefits and More," Healthline, Mar. 17, 2023, https://www.healthline.com/nutrition/dim-supplement, accessed Feb. 16, 2025.

59 A. Preti et al., "Oxytocin and Autism: A Systematic Review of Randomized Controlled Trials," *Journal of Child and Adolescent Psychopharmacology*, 2014; 24:2, 54–68.

60 A. J. Guastella and A. B. Hickie, "Oxytocin Treatment, Circuitry, and Autism: A Critical Review of the Literature Placing Oxytocin into the Autism Context," *Biological Psychiatry*, 79:3, 234–42.

61 M. G. Welch and R. J. Ludwig, "Calming Cycle Theory and the Co-Regulation of Oxytocin," *Psychodynamic Psychiatry*, 2017 Winter;45(4):519–540.

62 S. G. Gregory et al., "Genomic and Epigenetic Evidence for OXTR Deficiency in Autism," *BMC Medicine*, 2009, 7:62.

63 E. Hollander, "Social Synchrony and Oxytocin: From Behavior to Genes to Therapeutics," *American Journal of Psychiatry*, Oct. 2013;170(10):1086–9.

64 E. Clark et al., "Autism Spectrum Disorder and Induced/Augmented Labor: Epidemiologic Analysis of a Utah Cohort," *American Journal of Obstetrics & Gynecology*, Jan. 2015, 212: 1, Supp, S4–S5.

65 "Study Finds Adverse Effects of Pitocin in Newborns," press release from the American College of Obstetricians and Gynecologists regarding a paper delivered by Michael Tsimis at The Annual Clinical Meeting, May 7, 2013, https://web.archive .org/web/20130609170402/http://www.acog.org/About_ACOG/News_Room /News_Releases/2013/Study_Finds_Adverse_Effects_of_Pitocin_in_Newborns.

66 N. Schneid-Kofman et al., "Labor Augmentation with Oxytocin Decreases Glutathione Level," *Obstetrics and Gynecology International*, 2009, Article #807659.

67 E. Hollander et al., "Oxytocin Increases Retention of Social Cognition in Autism," *Biological Psychiatry*, Feb. 2007, 61:4, 498–503; A. J. Guastella et al., "Intranasal Oxytocin Improves Emotion Recognition for Youth with Autism Spectrum Disorders," *Biological Psychiatry*, Apr. 2010, 67:7, 692–94.

68 E. Andari et al., "Promoting Social Behavior with Oxytocin in High-Functioning Autism Spectrum Disorders," *Proceedings of the National Academy of Sciences*, Mar. 2, 2010;107(9):4389–94.

69 K. J. Parker et al., "Intranasal Oxytocin Treatment for Social Deficits and Biomarkers of Response in Children with Autism," *Proceedings of the National Academy of Sciences*, 114: 30: 2017, 8119–24.

70 R. S. Tareen and M. K. Kamboj, "Role of Endocrine Factors in Autistic Spectrum Disorders," *Pediatric Clinical North Americas*, Feb. 2012, 59:1, 75–88.

Chapter 13

1   A. Kalidas, "The Role of the Liver in Detoxification: Keeping Your Body Clean," The Center for Natural and Integrative Medicine Blog, Sept. 14, 2024, https://drkalidas .com/blog/the-role-of-the-liver-in-detoxification-keeping-your-body-clean/.

2   "Why You Should Care About Free Radicals," Cleveland Clinic, July 19, 2022, https: //health.clevelandclinic.org/free-radicals.

3   S. Bennet and S. Barrie, *7-Day Detox Miracle* (New York: Three Rivers Press, 2001).

4   S. Seneff, *Toxic Legacy: How the Weedkiller Glyphosate Is Destroying Our Health and the Environment,* (White River Junction, VT: Chelsea Green, 2021).

5   S. C. Owens, "Understanding the Sulfur System," *Defeat Autism Now! Conference Proceedings,* Philadelphia, PA, May 16, 2003, 65–76.

6   S. J. James et al., "Abnormal Transmethylation/Transsulfuration Metabolism and DNA Hypomethylation among Parents of Children with Autism," *Journal of Autism Developmental Disorders*, Nov. 2008 38:10, 1966–75.

7   Barbara Carletti et al., "Schizophrenia and Glutathione: A Challenging Story," *Journal of Personalized Medicine*, 202313, 11: 1526.

8   D. A. Averill-Bates, "The Antioxidant Glutathione," *Vitamins and Hormones*, 2023;121: 109–141.

9   S. A. Rogers, *Detoxify or Die* (Sarasota, FL: Sand Key Company, 2002), 48–49.

10 J. McCandless, *Children with Starving Brains*, 4th Ed. (Putney, VT: Bramble Books, 2009).

11 S. Melnyk et al., "Metabolic Imbalance Associated with Methylation Dysregulation and Oxidative Damage in Children with Autism," *Journal of Autism and Developmental Disorders*, Mar. 2012, 42:3, 367–77.

12 T. Stein T. "A Genetic Mutation That Can Affect Mental & Physical Health," *Psychology Today*, Sept. 5, 2014.

13 H. Claerhout et al., "Isolated Sulfite Oxidase Deficiency," *Journal of Inheritable Metabolic Disorders* Jan. 2018;41(1):101–108.

14 R. Pelton, "Acetaminophen Depletes Glutathione," Essential Formulas, Sept. 30, 2021, https://essentialformulas.com/acetaminophen-depletes-glutathione/.

15 W. Shaw, "Evidence That Increased Acetaminophen Use in Genetically Vulnerable Children Appears to Be a Major Cause of the Epidemics of Autism, Attention Deficit with Hyperactivity, and Asthma," *Journal of Restorative Medicine*, 2013, 2, 1–16.

16 E. Gutierrez and J. Roso, *The Parent's Roadmap to Autism* (Lioncrest Publishing, 2018).

17 S. R. Hausman-Cohen et al., "Genomics of Detoxification: How Genomics Can Be Used for Targeting Potential Intervention and Prevention Strategies Including Nutrition for Environmentally Acquired Illness," *Journal of the American College of Nutrition*, 2020, 39(2), 94–102.

18 K. Dorfman, "Nutritional Supplementation and Autism, ADHD, SPD and Other Delays," Documenting Hope, Spring, 1996, https://documentinghope.com/using -nutritional-supplementation-and-autism-adhd-spd-other-delays.

19 F. Liu et al., "Antioxidants in Neuropsychiatric Disorder Prevention: Neuroprotection, Synaptic Regulation, Microglia Modulation, and Neurotrophic Effects," *Frontal Neurosciences*, Dec. 5, 2024;18:1505153.

20   M. Ware, "Why Do We Need Magnesium?," MedicalNewsToday, Oct. 25, 2023, https: //www.medicalnewstoday.com/articles/286839.

21   "SIBO & The Connection to Sulfur," interview with D. Minich and B. Briggs, Metagenics Institute, https://www.metagenicsinstitute.com/pulse_patrol/plp-eps-5-sibo-sulfur, accessed Feb. 27, 2025.

22   W. Walsh, *Nutrient Power: Heal Your Biochemistry & Heal Your Brain* (New York: Skyhorse Publishing, 2012).

23   D. Mischoulon, "Omega-3 Fatty Acids for Mood Disorders," Harvard Health Publishing, Oct. 27, 2020, https://www.health.harvard.edu/blog/omega-3-fatty-acids -for-mood-disorders-2018080314414.

24   E. Seranova, "Can TMG Help with Autism? Understanding the Role of Methylation," NMN Bio, Mar. 9, 2024, https://nmnbio.co.uk/blogs/longevity/tmg-for-autism.

25   "Autism and DMG," EvenBetterHealth, https://www.evenbetterhealth.com/autism -and-dmg.php, accessed Feb. 28, 2025.

26   D. A. Rossignol and R. E. Frye, "The Effectiveness of Cobalamin (B12) Treatment for Autism Spectrum Disorder: A Systematic Review and Meta-Analysis," *Journal of Personal Medicine*, Aug. 11, 2021;11(8):784.

27   A. Badar, "Neuropsychiatric Disorders Associated with Vitamin B12 Deficiency: An Autobiographical Case Report," *Cureus*. Jan. 21, 2022;14(1):e21476.

28   "Methyl B12 for Autism," The Autism Community in Action (TACA), https://tacanow .org/family-resources/methyl-b12-for-autism/, accessed July 29, 2025.

29   E. Greenberg, "Practical Solutions for Autism Recovery," *Explore Magazine*, 16:4, July, 2007.

30   L. Packer et al., "Alpha-Lipoic Acid as a Biological Antioxidant," *Free Radical Biological Medicine*, Aug. 1995, 19:2, 227–50.

31   L. Patrick, "Mercury Toxicity and Antioxidants: Part 1: Role of Glutathione and Alpha-Lipoic Acid in the Treatment of Mercury Toxicity," *Alternative Medicine Review*, Dec. 2002, 7:6, 456–71.

32   J. K. Kern et al., "A Clinical Trial of Glutathione Supplementation in Autism Spectrum Disorders," *Medical Science Monitor*, Dec. 2011;17(12):CR677–82.

33   S. J. James et al., "Metabolic Biomarkers of Increased Oxidative Stress and Methylation Capacity in Children with Autism," *American Journal of Clinical Nutrition*, 2004, 80:6, 1611–17.

34   V. Chithra and S. Leelamma, "Coriandrum Sativum Changes the Levels of Lipid Peroxides and Activity of Antioxidant Enzymes in Experimental Animals," *Indian Journal of Biochemistry and Biophysics*, Feb. 1999, 36:1, 59–61.

35   S. C. Skoryna et al., "Studies on Inhibition of Intestinal Absorption of Radioactive Strontium," *Can Med Assoc J*, Aug. 1964, 91:6, 285–88.

36   T. Bito et al., "Potential of Chlorella as a Dietary Supplement to Promote Human Health," *Nutrients*, Aug. 20, 2020;12(9):2524.

37   P. Bergner, *The Healing Power of Garlic* (New York: Prima Publishing, 1996).

38   "We ALL Need to Detox from Spike Protein," World Council for Health, Nov. 1, 2024, https://worldcouncilforhealth.substack.com/p/we-all-need-to-detox-from-spike-protein.

39   A. L. Miller, "Dimercaptosuccinic Acid (DMSA), a Non-toxic, Water Soluble Treatment for Heavy Metal Toxicity," *Alternative Medical Review*, June 1998, 3:3, 199–207.

40   D. Lonsdale et al., "Treatment of Autism Spectrum Children with Thiamine Tetrahydrofurfuryl Disulfide: A Pilot Study," *Neuroendocrinology Letters*, Aug. 2002, 23:4.

41   E. M. Cranton and J. P. Frackelton, "Scientific Rationale for EDTA Chelation Therapy: Mechanism of Action," in *A Textbook on Chelation Therapy*, 2nd ed., E. M. Cranton, Editor (Newburyport, MA: Hampton Roads Publishing, 2001).

42 "Infrared Saunas: What They Do and 6 Health Benefits," Cleveland Clinic, Apr. 14, 2022, https://health.clevelandclinic.org/infrared-sauna-benefits.

Chapter 14

1   J. Kantor, *Sane Asylums: The Success of Homeopathy before Psychiatry Lost Its Mind* (Rochester, VT: Healing Arts Press, 2022).

2   A. Helmenstine, "Lymphatic System—Definition, Anatomy, Functions," Science Notes, Jan. 8, 2025, https://sciencenotes.org/lymphatic-system-definition-anatomy-functions/.

3   R. Morse, "Lymphatic System Function | Anatomy & Physiology," Dr. Morse's Healing Herbs, Apr. 28, 2023, https://web.archive.org/web/20240620002751/https://drmorses.com/blogs/body-system-webinars/lymphatic-system-function.

4   D. Thom et al., *Bioregulatory Medicine: An Innovative Holistic Approach to Self-Healing* (White River Junction, VT: Chelsea Green, 2018).

5   N. Antonucci et al., "Manual Lymphatic Drainage in Autism Treatment," *Madridge Journal of Immunology*, Dec. 12, 2018.

6   M. H. Rapaport et al., "Massage Therapy for Psychiatric Disorders," *Focus, American Psychiatric Publications*, Jan. 2018; 16(1):24–31.

7   M. Arata and C. Sternberg, "Transvascular Autonomic Modulation: Novel Venous Therapy for Autonomic Dysfunction," *Journal of Vascular Surgery: Venous and Lymphatic Disorders*, 2:1, 118–119.

8   X. Xiao et al., "Stress-Related Disorders among Young Individuals with Surgical Removal of Tonsils or Adenoids," *JAMA Network Open Network*, 2024.

9   D. Klinghardt, "The Tonsils and Their Role in Health and Chronic Illness," Klinghardt Institute, https://klinghardtinstitute.com/the-tonsils-and-their-role-in-health-and-chronic-illness/, accessed Feb. 1, 2025.

10 W. B. Jonas and E. Ernst, "The Safety of Homeopathy," in W. B. Jonas and J. S. Levin, eds., *Essentials of Complementary and Alternative Medicine* (Philadelphia: Lippincott Williams & Wilkins, 1999), 167–71.

11 J. Zand et al., *Smart Medicine for a Healthier Child* (Garden City, NY: Avery Publishing Group, 1994), 33.

12 S. Marohn, *The Natural Medicine Guide to Autism* (Charlottesville, VA: Hampton Roads, 2002), 137.

13 J. Reichenberg-Ullman et al., *A Drug-Free Approach to Asperger's and Autism: Homeopathic Medicine for Exceptional Kids* (Edmonds, WA: Picnic Point Press, 2005).

14 J. Reichenberg-Ullman and R. Ullman, *Ritalin-Free Kids: Safe and Effective Homeopathic Medicine for ADHD and Other Behavioral and Learning Problems* (Edmonds, WA: Picnic Point Press, 2013).

15 A. Lansky, *The Impossible Cure: The Promise of Homeopathy* (R.L. Portola Valley, CA: Ranch Press, 2003).

16 J. Kantor, *Autism Reversal Toolbox: Strategies, Remedies and Resources* (Haarlem, Netherlands: Emryss Publishers, 2015).

17 P. Khandelwal, "Homeopathic Remedies for Psychiatric Disorders," Lybrate, Nov. 11, 2024, https://www.lybrate.com/topic/homeopathic-remedies-for-psychiatric-disorders/7861ef3667f291905165126018f62bc0.

18 J. Reichenberg-Ullman and R. Ullman, *The Homeopathic Treatment of Depression, Anxiety, Bipolar and Other Mental and Emotional Problems: Homeopathic Alternatives to Conventional Drug Therapy* (Picnic Point Press, 2012).

19   G. Taylor, "Miasms—Understanding and Classifying Miasmatic Symptoms," Hpathy.com, Apr. 15, 2005, https://hpathy.com/organon-philosophy/miasms-understanding-and-classifying-miasmatic-symptoms/.

20   J. Kantor, *The Emotional Roots of Chronic Illness Homeopathy for Existential Stress* (Rochester, VT: Healing Arts Press, 2023).

21   Homotoxicology: Understanding Your Body's Natural Detoxification Process," Natural Therapy Pages, May 1, 2025, https://www.naturaltherapypages.com.au/article/what-is-homotoxicology.

22   E. Ernst, "Hans-Heinrich Reckeweg—Inventor of Homotoxicology," in: *Bizarre Medical Ideas and the Strange Men Who Invented Them* (New York, NY, Springer, 2024).

23   "Dr. Tinus Smits," Inspiring Homeopathy, www.tinussmits.com, accessed Mar. 4, 2025.

24   J. Elmiger, *Rediscovering Real Medicine: New Horizons of Homeopathy* (Boston, MA: Element Books, 1998).

25   T. Smits, *Autism Beyond Despair* (Haarlem, Netherlands: Emryss Publishers, 2010).

26   T. Smits, *Inspiring Homeopathy* (Haarlem, Netherlands: Emryss Publishers, 2019).

27   S. Blacher, "Emotional Freedom Technique (EFT): Tap To Relieve Stress and Burnout," *Journal of Inter-professional Education & Practice,* 30, 2023.

28   D. Klinghardt, "Neural Therapy," *Townsend Letter for Doctors and Patients,* July 1995, 96–98.

29   "Laser Detox, Basic Package I," Eagle Research LLC, https://www.eagleresearchllc.com/products/bioenergetics/laser-detox-basic-package/, accessed Mar. 6, 2025.

30   R. Tannenbaum, "Laser Energetic Detoxification," Dr. Russ Tannenbaum, https://www.drrtannenbaum.com/homeopathy/laser-energetic-detoxification-with-homeopathy, accessed Mar. 5, 2025.

31   E. Faretta and M. Dal Farra, "Efficacy of EMDR Therapy for Anxiety Disorders," *Journal of EMDR Practice and Research,* 13(4), 2018, 325–332.

32   K. Howard, "EMDR Therapy and Depression," (Season 2, No. 20) [Audio podcast episode] in *Let's Talk EMDR podcast, EMDR International Association,* Oct. 15, 2023, https://www.emdria.org/podcast/emdr-therapy-and-depression/.

33   L. Bedeschi, "EMDR for Bipolar Disorder: A Systematic Review of the Existing Studies in Literature, *Clinical Neuropsychiatry: Journal of Treatment Evaluation,* 15(3), 2018, 186–189.

34   B. Carey, "Francine Shapiro, Developer of Eye Movement Therapy, Dies at 71," The New York Times, July 11, 2019, https://www.nytimes.com/2019/07/11/science/francine-shapiro-dead.html.

35   D. Klinghardt, "APN—Applied Psycho-Neurobiology: Another Tool to Healing," https://www.consciouslivingcenter.com/wp-content/uploads/2017/02/APN.pdf, accessed July 29, 2025.

36   B. Ulsamer, *Healing the Power of the Past: The Systemic Therapy of Bert Hellinger* (Nevada City, CA: Underwood Books, 2005).

37   M. Wolynn, *It Didn't Start with You: How Inherited Family Trauma Shapes Who We Are and How to End the Cycle* (New York, NY: Penguin Publishing, 2017).

38   T. Ping et al., "The Effectiveness of Hyperbaric Oxygen Therapy in Children and Adolescents with Autism Spectrum Disorders: A Systematic Review and Meta-Analysis," *Progress in Neuro-Psychopharmacology and Biological Psychiatry,* 137:20, Mar. 2025, 111257.

39   W. Porterfield, "HBOT for Mental Health: Benefits, Risks, and How It Works," Clarity Hyperbarics, July 2024, https://desmoineshyperbarics.com/blog/hbot-for-mental-health-benefits-risks-and-how-it-works, accessed Feb. 10, 2025.

40 R. J. Rowen, "Ozone and Oxidation Therapies as a Solution to the Emerging Crisis in Infectious Disease Management: Review of Current Knowledge and Experience," *Medical Gas Research,* Oct-Dec 2019;9(4): 232–237.

41 C. L. Ross et al., "The Use of Pulsed Electromagnetic Field to Modulate Inflammation and Improve Tissue Regeneration: A Review," *Bioelectricity,* 2019; 1(4):247–259.

Chapter 15

1 W. J. Gies, "Dental Education in the United States and Canada. A Report to the Carnegie Foundation for the Advancement of Teaching," 1926. *Journal of the American College of Dentistry,* 2012 Summer; 79(2):32–49.

2 Y. Qiao et al., "Alterations of Oral Microbiota Distinguish Children with Autism Spectrum Disorders from Healthy Controls," *Scientific Reports*: 8:1597, Jan. 2018.

3 R. Baxter, *Tongue-Tied: How a Tiny String Under the Tongue Impacts Nursing, Speech, Feeding, and More,* (Alabama Tongue-Tie Center, 2018).

4 M. A. Breiner, *Whole Body Dentistry* (Fairfield, CT: Quantum Health Press, 1999), 35.

5 K. Seaverson, "A Brief History of Amalgams," Tooth by the Lake, http://www.toothby thelake.net/wellness-center/, accessed Mar. 15, 2025.

6 "Weston A. Price, DDS," The Weston A. Price Foundation, January 1, 2000, https://www.westonaprice.org/health-topics/nutrition-greats/weston-a-price-dds/#gsc.tab=0.

7 W. A. Price, *Nutrition and Physical Degeneration, 8th edition* (Washington, DC: Price-Pottenger Nutrition Foundation, 2009).

8 W. A. Price, "Dental Infections Oral & Systemic" (Vol 1), https://www.forgottenbooks.com/en/books/DentalInfectionsOralandSystemic_10913785 and "Dental Infections & the Degenerative Diseases" (Vol 2), https://www.forgottenbooks.com/en/books/DentalInfectionsandtheDegenerativeDiseases_10919178.

9 H. Huggins, *It's All in Your Head: The Link between Mercury Amalgams and Illness* (New York, NY: Avery Publishing, 1993).

10 H. A. Huggins and T. E. Levy, *Uninformed Consent: The Hidden Dangers in Dental Care* (Charlottesville, VA: Hampton Roads Pub, 1999).

11 A. H. Cutler, *Amalgam Illness: Diagnosis and Treatment. What You Can Do to Get Better, How Your Doctor Can Help* (Andy Cutler Publishing, 1999).

12 R. R. Lee and A. H. Cutler, *The Mercury Detoxification Manual: A Guide to Mercury Chelation* (Andy Cutler Publishing, 2019).

13 R. Kulacz and T. E. Levy, *The Toxic Tooth: How a Root Canal Could be Making You Sick* (Henderson, NV: MedFox Publishing, 2014).

14 T. E. Levy, *Hidden Epidemic: Silent Oral Infections Cause Most Heart Attacks and Breast Cancer* (Henderson, NV: MedFox Publishing, 2017).

15 M. Dassani, *Airway Is Life: Waking Up to Your Family's Sleep Crisis* (MeghnaDassani.com, 2021).

16 H. A. Waldron, "Did the Mad Hatter Have Mercury Poisoning?," *British Medical Journal,* 287 (6409): 1961.

17 "What Is Dental Amalgam (Silver Filling)?" Academy of General Dentistry, https://knowyourteeth.com/infobites/abc/article/?abc=w&iid=286&aid=1242, accessed Mar. 15, 2025.

18 D. W. Eggleston and M. Nylander, "Correlation of Dental Amalgam with Mercury in Brain Tissue," *Journal of Prosthetic Dentistry,* Dec. 1987, 58:6, 704–7.

19 P. Grandjean et al., "Cognitive Performance of Children Pre-natally Exposed to 'Safe' Levels of Methyl Mercury," *Environmental Research,* May 1998. 72:2, 165–72.

20   M. Breiner, "Dental Materials Bio-compatibility Testing," Breiner Whole-Body Health Center, https://wholebodymed.com/dentistry/dental-materials-bio-compatibility-testing/, accessed March 15, 2025.

21   M. Tokuyasu, "Examples of Diets for Infant's and Children's Nutritional Guidance, and Their Effects of Adding Chlorella and CGF to Food Schedule," Totori City, Japan: Conference Proceedings at the Nutritional Illness Counseling Clinic, 1983, *Japanese Journal of Nutrition*, 41:5, 275–283.

22   E. Phillips, *Kiss Your Dentist Goodbye, Second Edition* (Garden City Park, NJ: Square One Publishers, 2025), 74–76.

23   E. T. Everett, "Fluoride's Effects on the Formation of Teeth and Bones, and the Influence of Genetics," *Journal of Dental Research*, May 2011; 90:5, 552–60.

24   J. Sanders, "Tooth Decay Trends in Fluoridated vs. Unfluoridated Countries," Fluoride Action Network, July 30, 2012, https://fluoridealert.org/studies/caries01/.

25   C. Bryson, *The Fluoride Deception* (Seven Stories Press, 2006).

26   S. Violante, "The Failures of Dental Sealants: Weigh the Pros and Cons with Dental Patients," June 5, 2018. https://www.dentistryiq.com/dental-hygiene/clinical-hygiene /article/16367632/the-failures-of-dental-sealants-weigh-the-pros-and-cons-with-dental -patients.

27   W. A. Price, *Dental Infections, Oral and Systemic* (Cleveland, Ohio: Penton Publishers, 1923).

28   G. Meinig, *Root Canal Cover-up Exposed! Many Illnesses Result* (Washington, DC: Price Pottenger Nutrition, 1994).

29   S. Forsgren, "From Dis-Ease to Better Health: A Model for Recovering from Chronic Lyme Disease, Mold Illness, and Related Conditions," *Townsend Letter*, July 1, 2022.

30   S. Rankin, "Anesthesia & ASD," Autism Research Institute, https://autism.org/anesthesia -and-asd/, accessed Mar. 16, 2025.

31   P. Morgan, "When Propofol Is Problematic," Presentation at 12th annual joint winter meeting of the Society of Pediatric Anesthesia and American Academy of Pediatrics, Winter, 2007.

32   V. C. Baum, "When Nitrous Oxide Is No Laughing Matter: Nitrous Oxide and Pediatric Anesthesia," *Paediatric Anaesthesia*, Sept. 2007; 17:9, 824–30.

33   R. R. Selzer et al., "Adverse Effect of Nitrous Oxide in a Child with 5,10 -Methylenetetrahydrofolate Reductase Deficiency," *New England Journal of Medicine*, 2003; 349:45–50.

34   F. Maready, *Crooked: Man-Made Disease Explained: The Incredible Story of Metal, Microbes and Medicine—Hidden within Our Faces* (CreateSpace, 2018).

35   M. A. Breiner, *Whole Body Dentistry* (Fairfield, CT: Quantum Health Press, 2018), 22.

36   D. Klinghardt, "The Mechanics of Brain Detoxification," July 6, 2018, online webinar.

Chapter 16

1   S. Marohn, *The Natural Medicine Guide to Autism* (Charlottesville, VA: Hampton Roads Publishing Co., 2002), 194. Quote attributed to Harvard Medical School instructor Tal Kenet.

2   "Sensory Processing Disorder in Autism," The Autism Community in Action (TACA), https://tacanow.org/family-resources/sensory-processing-disorder-in-autism/, accessed Apr. 12, 2025.

3   H. M. Kuhaneck et al., "Occupational Therapy Practice Guidelines for Children and Youth with Challenges," in *Sensory Integration and Sensory Processing* (Bethesda, MD: American Occupational Therapy Assn, 2018), xi.

4   B. Rimland, *Infantile Autism: The Syndrome and Its Implications for a Neural Therapy of Behavior* (New York, NY: Appleton Century Crofts, 1964); E. M. Ornitz, "Disorders of Perception Common to Early Infantile Autism and Schizophrenia," *Comprehensive Psychiatry*, 1969, 10, 259–74.

5   L. K. Koop and P. Tadi, "Neuroanatomy, Sensory Nerves [Updated July 24, 2023]," in *StatPearls* [Internet], (Treasure Island, FL: StatPearls Publishing; 2025), www.ncbi.nlm .nih.gov/books/NBK539846/.

6   Erica, "The 12 Senses," from "Man's Twelve Senses in Their Relation to Imagination, Inspiration, and Intuition," The Steiner Connection, https://thesteinerconnection.com /2022/07/21/the-12-senses/, accessed Mar. 19, 2025.

7   A. J. Ayres, *Sensory Integration and Learning Disorders* (Los Angeles, CA: Western Psychological Services, 1972), 113–29.

8   M. Kawar, "Oculomotor Control: An Integral Part of Sensory Integration," in A. C. Bundy, S. J. Lane, and E. Murray, eds., *Sensory Integration: Theory and Practice, 2nd ed.* (Philadelphia, PA: FA Davis Company, 2002).

9   J. C. Moore, "The Functional Components of the Nervous System: Part I," *Sensory Integration Quarterly*, 1994, 22, 1–7.

10 J. R. Biggio, "Bed Rest in Pregnancy: Time to Put the Issue to Rest," *Obstetrics and Gynecology*, Jun 2013, 121:6, 1158–60.

11 "Vestibular System: Your Child's Internal GPS System for Motor Planning and Attention," Integrated Learning Strategies, https://ilslearningcorner.com/2016–04-vestibular -system-your-childs-internal-gps-system-for-motor-planning-and-attention/, accessed Mar. 19, 2025.

12 A. J. Ayres, *Sensory Integration and the Child* (Los Angeles, CA: Western Psychological Services, 1979), 184.

13 S. I. Greenspan and S. Wieder, "Developmental Patterns and Outcomes in Infants and Children with Disorders in Relating and Communicating: A Chart Review of 200 Cases of Children with Autistic Spectrum Diagnoses," *Journal of Developmental Learning Disabilities*, 1997, 1:1, 87–142.

14 A. Brown et al., "Defining Sensory Modulation: A Review of the Concept and a Contemporary Definition for Application by Occupational Therapists," *Scandinavian Journal of Occupational Therapy*, 26(7), 515–523, 2019.

15 S. Spitzer and S. Smith Roley, "Sensory Integration Revisited: A Philosophy of Practice," in S. Smith Roley, E. I. Blanche, and R. C. Schaaf, eds., *Understanding the Nature of Sensory Integration in Diverse Populations* (Tucson, AZ: Therapy Skill Builders, 2001), 5.

16 A. J. Ayres, *Sensory Integration and the Child* (Los Angeles, CA: Western Psychological Services, 1979), 124.

17 G. T. Baranek et al., "Tactile Defensiveness and Stereotyped Behaviors," *American Journal of Occupational Therapy*," 1997, 51, 91–5.

18 A. J. Ayres, Hyper-responsivity to Touch and Vestibular Stimuli as a Predictor of Positive Response to Sensory Integration Procedures in Autistic Children," *American Journal of Occupational Therapy*, 1980, 34, 375–86.

19 Z. Mailloux, "Sensory Integrative Principles in Intervention with Children with Autistic Disorder," in S. Smith Roley, E. I. Blanche, R. C. Schaaf, eds., *Understanding the Nature of Sensory Integration in Diverse Populations* (Tucson, AZ: Therapy Skill Builders, 2001), 365–84.

20 B. S. Myles et al., *Asperger Syndrome and Sensory Issues* (Shawnee Mission, KS: Autism Asperger Publishing Co., 2000).

21 P. Teitelbaum et al., "Movement Analysis in Infancy May Be Useful for Early Diagnosis of Autism," *Proceedings of the. National Academy of Sciences USA*, 1998, 95, 13982–87.

22   J. P. Piek and M. J. Dyck, "Sensory-Motor Deficits in Children with Developmental Coordination Disorder, Attention Deficit Hyperactivity Disorder and Autistic Disorder," *Human Movement Science*, 2004, 23, 475–88.

23   C. A. Molloy et al., "Postural Stability in Children with Autism Spectrum Disorder," *Journal of Autism and Developmental Disorders*, 2003, 33:6, 643–52.

24   J. Case-Smith and H. Miller, "Occupational Therapy with Children with Pervasive Developmental Disorders," *American Journal of Occupational Therapy*, 1999, 53, 506–13.

25   V. Jones and M. Prior, "Motor Imitation Abilities and Neurological Signs in Autistic Children," *Journal of Autism and Developmental Disorders*, 1985, 15, 37–46.

26   O. Bogdashina, *Sensory Perceptual Issues in Autism and Asperger Syndrome. Different Sensory Experiences, Different Perceptual Worlds* (London: Jessica Kingsley Publishers, 2003).

27   A. Trecker, "Play and Praxis in Children with an Autism Spectrum Disorder," in Miller-Kuhaneck H, ed. *Autism, A Comprehensive Occupational Therapy Approach, 2nd ed.,* (Bethesda, MD: AOTA Press, 2004).

28   A. J. Ayres, *Developmental Dyspraxia and Adult Onset Apraxia* (Torrance, CA: Sensory Integration International, 1985.

29   S. J. Rogers et. al., "Imitation and Pantomine in High Functioning Adolescents with Autism Spectrum Disorders," *Child Development*, 1996, 67, 2060–73.

30   W. F. Langley and D. Mann, "Central Nervous System Magnesium Deficiency," *Archives of Internal Medicine*, Mar. 1991;151:3, 593–6.

31   C. Peretti, "Oxytocin and Autism," interview with Dr. Luis Martinez, International Autism Summit, Cairo, Egypt, 2009.

32   K. Hessellund and J. Nutto, "Understanding Occupational Therapy's Role in Sensory Integration," in A. Barber, ed., *Vision and Sensory Integration* (Santa Ana, CA: Optometric Extension Program, 1999).

33   D. Parham et al., *The Sensory Processing Measure, Second Edition* (Los Angeles: Western Psychological Services, 2021).

34   A. J. Ayres, *Sensory Integration and Praxis Test* (Los Angeles, CA: Western Psychological Services, 1989).

35   D. Parham et al., "Sensory Processing and Praxis in High Functioning Children with Autism," paper presented at Research 2000, Feb. 4–5, 2000, Redondo Beach, CA.

36   C. Delacato, *The Ultimate Stranger: The Autistic Child* (Novato, CA: Academic Therapy Pub, 1974).

37   Z. Mailloux, "Sensory Integrative Principles," in S. Smith Roley, E. I. Blanche, and R. C. Schaaf, eds., *Understanding the Nature of Sensory Integration in Diverse Populations* (Tucson, AZ: Therapy Skill Builders, 2001), 365–84.

38   R. C. Schaaf et al., "An Intervention for Sensory Difficulties in Children with Autism: A Randomized Trial," *Journal of Autism & Developmental Disorders*, July 2014: 44:7, 1493–1506.

39   D. Davis, *Say It with Sound: Hum, Harmonize and Heal* (Dorinne Davis Publications, 2017).

40   J. Gerritsen, "A Review of Research Done on Tomatis Auditory Stimulation," Sacarin Center, 2009, https://sacarin.com/wp-content/uploads/2019/01/Review-of-Tomatis-Research.pdf, accessed Mar. 21, 2025.

41   P. Madaule, *When Listening Comes Alive* (Norval, Ontario, Canada: Moulin Publishing, 1994).

42   P. Madaule, *Terapia de Escucha* (Mexico: Editorialis Trillas, 2005).

43   N. Doidge, *The Brain's Way of Healing: Remarkable Discoveries and Recoveries from the Frontiers of Neuroplasticity* (New York, NY: Penguin Books, 2016).

44   P. Sollier, *Listening for Wellness: An Introduction to the Tomatis Method* (Walnut Creek, CA: The Mozart Center Press, 2005).

45   S. Ruben, *Awakening Ashley: Mozart Knocks Autism on its Ear* (Lincoln, NE: iUniverse, 2004).

46   G. Berard, *Hearing Equals Behavior* (Georgiana Foundation, prepublication issue, 1992), 1.

47   G. Berard and S. Brockett, *Hearing Equals Behavior, Updated and Expanded* (Manchester Center, VT: Shires Press, 2011).

48   A. Stehli A. *The Sound of a Miracle: A Child's Triumph Over Autism* (New York: Doubleday, 1991).

49   G. Thomas, *Overcoming Autism: My Story* (Kindle Book, 2015).

50   J. Panksepp et al., "Biochemical Changes as a Result of AIT-type Modulated and Unmodulated Music," *Lost & Found: Perspectives on Brain, Emotion, and Culture*, 1996/7, 2:1, 4.

51   Auditory Integration Training, AIT Institute, www.aitinstitute.org, accessed Mar. 22, 2025.

52   S. M. Edelson et al., "Behavioral and Physiological Effects of Deep Pressure on Individuals with Autism: A Pilot Study Investigating the Efficacy of Grandin's "Hug Machine," *American Journal of OT*, 1999, 53, 145–52.

53   C. Kranowitz, *The Out-of-Sync Child: Recognizing and Coping with Sensory Processing Disorder, Third edition*, (New York: Perigee Press, 2022).

54   C. Kranowitz, *The Out-of-Sync Child Grows Up: Coping with Sensory Processing Disorder in the Adolescent and Young Adult Years* (New York: Perigee Books, 2016).

55   C. Kranowitz, *Good Times with Out-of-Sync Grandkids: Activities for Grown-ups and Children with Sensory Processing Differences* (Arlington, TX: Future Horizons, 2025).

56   C. Kranowitz and J. Newman, *Growing an In-Sync Child: Simple, Fun Activities to Help Every Child Develop, Learn, and Grow* (New York: Perigee Books, 2010).

57   J. Newman and C. Kranowitz, *In-Sync Activity Cards* (Arlington, TX: Sensory World, 2012).

58   A. Baniel, *Kids Beyond Limits: The Anat Baniel Method for Awakening the Brain and Transforming the Life of Your Child with Special Needs* (New York, NY: Perigee Press, 2012).

59   A. Baniel, *Move Into Life: NeuroMovement for Lifelong Vitality* (CreateSpace, 2016).

60   R. Melillo, *Disconnected Kids: The Groundbreaking Brain Balance Program® for Children with Autism, ADHD, Dyslexia, and Other Neurological Disorders* (New York, Perigee Press, 2009).

61   J. Lara, *Autism Movement Therapy® Method: Waking up the Brain!* (London, Jessica Kingsley Publishing, 2015).

62   T. Grandin, *Animals in Translation* (New York: Scribner, 2005), 6–8.

63   A. A. Pack and L. M. Herman, "Sensory Integration in the Bottlenosed Dolphin: Immediate Recognition of Complex Shapes Across the Senses of Echolocation and Vision," *Journal of the Acoustical Society of America*, 1995, 98, 722–33.

64   A. Shkedi, "Sensory Input through Riding," *Proceedings for the 7th International Therapeutic Riding Congress* Aarhus, Denmark, 1991, 129–32.

65   M. J. Biery and N. Kauffman, "The Effects of Therapeutic Horseback Riding on Balance," *Adapted Physical Activity Quarterly*, 1989, 6, 221–29.

66   L. Jake, "Autism and the Role of Aquatic Therapy in Recreational Therapy Treatment Services," *Therapeutic Recreation Directory.*, Sept. 1, 2003.

67   Dull H. Watsu, *Freeing the Body in Water* (Middletown, CT: Watsu Publishing, 2008).

68   O. Sacks, *Musicophilia: Tales of Music and the Brain* (New York: Vintage, 2008).

69   S. Goddard, *Reflexes, Learning and Behavior* (Eugene, OR: Fern Ridge Press, 2005), 108–9.

70   "Music Therapy as a Treatment Modality for Autism Spectrum Disorders," American Music Therapy Association, June 2012. http://www.musictherapy.org/assets/1/7/MT_Autism_2012.pdf

71   J. Whipple, "Music in Intervention for Children and Adolescents with Autism: A Meta-analysis," *Journal of Music Therapy*, Summer 2004, 41:2, 90–106.

72   H. Blomberg and M. Dempsey, *Movements that Heal: Rhythmic Movement Training and Primitive Reflex Integration* (Australia: Bookpal, 2011).

73   A. Hulbert, *Off the Charts: The Hidden Lives and Lessons of American Child Prodigies* (New York, NY: Alfred A. Knopf, 2018), 216–50.

74   S. Shore, *Beyond the Wall* (Shawnee Mission, KS: Autism Asperger Publishing Co., 2003), 68–83.

75   P. A. Devnani and A. U. Hegde, "Autism and Sleep Disorders," *Journal of Pediatric Neurosciences*, Oct-Dec 2015; 10:4, 304–7.

76   E. F. Leone and S. L. Rogers, "Sensory Applications for Sleep and Toilet Training," in R. A. Huebner, ed., *Autism: A Sensorimotor Approach to Management* (Gaithersburg, MD: Aspen Publishers, 2001), 355–63.

77   M. Wheeler, *Toilet Training for Individuals with Autism and Related Disorders: A Comprehensive Guide for Parents and Teachers* (Arlington, TX: Future Horizons, 2001).

Chapter 17

1   A. Noe, *Out of Our Heads: Why You Are Not Your Brain, and Other Lessons from the Biology of Consciousness* (New York, NY: Hill and Wang, 2009).

2   G. N. Getman and A. Gesell, *Vision, Its Development in Infant and Child* (New York, NY: Harper & Brothers, 1949).

3   F. Ilg et al., *Infant and Child in the Culture of Today* (New York, NY: Harper & Brothers, 1943).

4   G. N. Getman, *How to Develop Your Child's Intelligence* (Santa Ana, CA: Optometric Extension Program, 1993).

5   Thread with Dr. Larry Palevsky and James Maskell, https://www.threads.net/@mrjamesmaskell/post/DHJkbuFtX4w, accessed July 29, 2025.

6   J. Flavell, *The Developmental Psychology of Jean Piaget* (New York, NY: D. Van Nostrand Company, 1967).

7   I. B. Suchoff, *Cognitive Development: Piaget's Theory* (Santa Ana, CA: Optometric Extension Program, 1978).

8   T. Grandin, *Thinking in Pictures and Other Reports of My Life with Autism* (New York, NY: Vintage, 1996).

9   L. W. McDonald, *Some Considerations When Developing Visual Abilities* (Santa Ana, CA: Optometric Extension Program Foundation, 1964), 2, 189–94.

10 L. J. Press, "Sensorimotor Dynamics and Two Visual Systems: Shades of Skeffington and Brock. Part 1," The Visionhelp Blog, May 22, 2011. http://visionhelp.wordpress.com/2011/05/22/sensorimotor-dynamics-and-two-visual-systems-shades-of-skeffington-brock-part-1/.

11 R. Mozlin, "Recess Matters: Skeffington's Circles in the Schoolyard," COVD blog Mindsight, Sept. 26, 2017. https://covdblog.wordpress.com/2017/09/26/skeffingtons-circles-in-the-schoolyard/.

12 S. Kandyba, "The Truth about Bumbo Chairs," Chance to Advance Home Pediatric Therapy, https://www.cta-pt.com/the-truth-about-bumbo-chairs, accessed Mar. 25, 2025.

13 S. Masgutova, *PTSD Recovery: Gentle, Rapid, and Effective Treatment with Reflex Integration* (Orlando, FL: Svetlana Masgutova Educational Institute®, 2015).

14 J. W. Streff, "Optometric Care for a Child Manifesting Qualities of Autism," *Journal of the American Optometric Association*, 1979, 46:6, 592–97.

15 "Mental Health and Vision," Beach Eye Medical Group, Apr. 5, 2023, https://beacheye .com/blog/mental-health-vision/.

16 D. M. Hornbeak and T. L. Young, "Myopia Genetics: A Review of Current Research and Emerging Trends," *Current Opinions in Ophthalmology*, Sept. 2009, 20:5, 356–62.

17 E. C. Engle, "Genetic Basis of Congenital Strabismus," *JAMA Ophthalmology*, Feb. 2007, 125, 2.

18 "Computer Vision Syndrome," American Optometric Association, https://www.aoa.org/ healthy-eyes/eye-and-vision-conditions/computer-vision-syndrome, Mar. 19, 2025.

19 "A Life-Saving Eye Exam Helped Detect Diabetes," VSP Vision Care, https://www.vsp .com/eyewear-wellness/eye-health/health-conditions/eye-exam-detects-diabetes, accessed July 29, 2025.

20 M. Megson, "The Biological Basis for Perceptual Deficits in Autism," http://megson .com/readings/BiologicalBasis.pdf, accessed Mar. 5, 2018.

21 K. B. Carman et al., "Acute Mercury Poisoning Among Children in Two Provinces of Turkey," *European Journal of Pediatrics*, June 2013, 172:6, 821–27.

22 M. Kaplan, *Seeing through New Eyes: Changing the Lives of Children with Autism, Asperger Syndrome and other Developmental Disabilities through Vision Therapy* (Philadelphia, PA: Jessica Kingsley Publishers, 2006), 126.

23 Kaplan, *Seeing through New Eyes*, 10-16.

24 E. Milne et al., "Vision in Children and Adolescents with Autistic Spectrum Disorder: Evidence for Reduced Convergence," *Journal of Autism & Developmental Disorders*, July 2009: 39:7, 965–75.

25 J. Cooper and R. Duckman, "Convergence Insufficiency: Incidence, Diagnosis and Treatment, *Journal of the American Optometric Association*, 1978, 49, 673–80.

26 M. S. Bolding et al., "Ocular Convergence Deficits in Schizophrenia," *Front Psychiatry*, Oct. 17, 2012;3:86.

27 D. Granet et al., "The Relationship between Convergence Insufficiency and ADHD," *Strabismus*, Dec. 2005, 13:4, 163–68.

28 D. Porter, "Convergence Insufficiency," American Academy of Ophthalmology, May 18, 2021, https://www.aao.org/eye-health/diseases/what-is-convergence-insufficiency.

29 M. Kaplan et al., "Strabismus in Autism Spectrum Disorder," *Focus on Autism and Other Developmental Disabilities*, Summer 1999, 14:2, 101–5.

30 D. S. Friedman et al., "Prevalence of Amblyopia and Strabismus in White and African-American Children Aged 6 through 71 Months: The Baltimore Pediatric Eye Disease Study," *Ophthalmology*, Nov. 2009, 116:11, 2128–34.

31 J. E. Scharre and M. P. Creedon, "Assessment of Visual Function in Autistic Children," *Optometric Vision Science*, June 1992, 69:6, 433–39.

32 U. Rosenhall et al., "Oculomotor Findings in Autistic Children," *Journal of Laryngological Otology*, May 1988 102:5, 435–39; C. Kemner et al., "Abnormal Saccadic Eye Movements in Autistic Children," *Journal of Autism and Developmental Disorders*, Feb. 1998, 28:1, 61–7.

33 Y. H. Lee et al., "Association of Strabismus with Mood Disorders, Schizophrenia, and Anxiety Disorders among Children," *JAMA Ophthalmology*. 2022; 140(4): 373–381.

34 L. J. Press, "Topical Review: Strabismus," *Journal of Optometry & Vision Development*, 1999, 22, 5–20.

35 P. Venkhatesh, "Do We Learn to See?," *Resonance*, Jan. 2011, 16:1, 88–99.

36 R. M. Kavner, "Strabismus and Amblyopia," *New Developments Newsletter*, Winter 2002–03, 8:2, 5.

37 Kaplan, *Seeing through New Eyes*.

38   M. Taub and R. Russell, "Autism Spectrum Disorders: A Primer for the Optometrist,"
      Review of Optometry, May 2007; R. A. Coulter, "Serving the Needs of the Patient with
      Autism," Optometric Vision Development, 2009; 40:3, 136–40.

39   R. A. Coulter, "Serving the Needs of the Patient with Autism," Optometric Vision
      Development, 2009; 40:3, 136–40.

40   R. A. Stevenson et al., "Multisensory Temporal Integration in Autism Spectrum
      Disorders," *Journal of the Neurosciences*, Jan. 15, 2014, 34:3, 691-7.

41   O. Sacks,"Stereo Sue," *New Yorker Magazine*, June 19, 2006, 64–73.

42   S. Barry S. *Fixing My Gaze: A Scientist's Journey into Seeing in Three Dimensions* (New
      York: Basic Books, 2010).

43   Kaplan, *Seeing through New Eyes*, 81.

44   Kaplan, *Seeing through New Eyes*, 47.

45   M. Kaplan et al., "Postural Orientation Modifications in Autism in Response to Ambient
      Lenses," *Child Psychiatry and Human Development*, 1996, 27:2, 81–91.

46   M. Kaplan et al., "Behavioral Changes in Autistic Individuals as a Result of Wearing
      Ambient Transitional Prism Lenses," *Child Psychiatry & Human Development*, 1998.
      29:1, 65–76.

47   Kaplan, *Seeing through New Eyes*, 37–41.

48   F. F. Flach and M. Kaplan, "Visual Perceptual Dysfunction in Psychiatric Patients,"
      Comprehensive Psychiatry, 1983, 24:4, 304–311.

49   M. Scheiman et al., "Vision Characteristics of Individuals Identified as Irlen Filter
      Candidates," *Journal of the American Optometric Association*, Aug. 1990, 61:8, 600–5.

50   H. Solan and J. Richmond, "Irlen Lenses: A Critical Appraisal," *Journal of the American
      Optometric Association*, Oct. 1990, 61:10, 789–96.

51   C. A. Molloy et al., "Postural Stability in Children with Autism Spectrum Disorder,"
      Journal of Autism and Developmental Disabilities, 2003, 33:6, 643–52.

52   H. A. Solan et al., "Vestibular Function, Sensory Integration and Balance Anomalies: A
      Brief Literature Review," *Optometry and Vision Development*, 2007, 38, 1–5.

53   Taub and Russell, "Autism Spectrum Disorders."

54   L. F. Hellerstein and B. Fishman, "Vision Therapy and Occupational Therapy:
      An Integrated Approach," *Journal of Behavioral Optometry*, 1990, 1:5, 122–26;
      L. F. Hellerstein and B. Fishman, "Collaboration Between Occupational Therapists and
      Optometrists," *Journal of Behavioral Optometry*, 1999, 10:6, 147–52.

55   R. Mumford, "Improving Visual Efficiency with Selected Lighting," *Journal of Optometric
      Vision Development*, 2002, 33:3.

56   K. Nurek and D. Wendelburg, *Begin Where They Are* (Santa Ana, CA: Optometric
      Extension Program Foundation, 1993).

57   R. Richards and K. Remick, *CVA: Classroom Visual Activities: A Manual to Enhance the
      Development of Visual Skills* (Novato, CA: Academic Therapy Press, 1988).

58   K. Lane K, *Developing Your Child for Success* (Lewisville, TX: Learning Potentials Pub,
      1991).

59   H. Furth and H. Wachs, *Thinking Goes to School* (New York: Oxford Univ Press, 1975).

Chapter 18

1   E. Bonker and V. B. Breen, *I Am in Here: The Journey of a Child with Autism Who Cannot
      Speak but Finds Her Voice* (Amazon: Kindle Books, 2011).

2   S. Berkowitz, "AAC 101: Myths and Misconceptions," Kidz Learn Language, Jan. 22,
      2017, https://kidzlearnlanguage.blogspot.com/.

3   I. Kedar, *Ido in Autismland: Climbing Out of Autism's Silent Prison* (2012).

4   "Handheld Screen Time Linked with Speech Delays in Young Children," AAP News, May 4, 2017, https://publications.aap.org/aapnews/news/10384/.

5   T. Mukhhopadhyay, *How Can I Talk if My Lips Don't Move?* (New York, NY, Arcade, 2008).

6   S. Mukhopadhyay, *Developing Expressive Language in Verbal Students with Autism Using Rapid Prompting Method* (Parker, CO: Outskirts Press, 2016).

7   Jeremy Sicile-Kira, www.jeremysvision.com, accessed Apr. 6, 2025.

8   J. H. Chinitz, *Spellbound: The Voices of the Silent* (New York, NY: The Bookbound Initiative, 2023).

9   Spellers Freedom Foundation, www.SpellersFreedomFoundation.org, accessed Apr. 2, 2024.

10 D. Gaivin and D. Johnson, *The Spellers Guidebook: Practical Advice for Parents and Students* (New York, NY: Skyhorse Publishing, 2023).

11 "Facilitated Communication," American Speech-Language-Hearing Association, https://www.asha.org/policy/PS2018-00352/, accessed Apr. 5, 2025.

12 J. B. Handley and J. Handley, *Underestimated: An Autism Miracle* (New York, NY: Skyhorse Publishing, 2021).

13 D. E. Nathanson et al., "Effectiveness of Short-Term Dolphin Assisted Therapy for Children with Severe Disabilities," *Anthrozoos*, 1997, 10:2/3, 90–100.

Chapter 19

1   S. Bellini et al., "A Meta-Analysis of School-Based Social Skills Interventions for Children with Autism Spectrum Disorders," *Remedial and Special Education*, May/Jun 2007, 28:3, 153–62.

2   S. W. Porges, "The Vagus: A Mediator of Behavioral and Physiologic Features Associated with Autism," in Bauman and Kemper, *The Neurobiology of Autism, 2nd ed.* (Baltimore, MD: Johns Hopkins Press, 2005), 65–78.

3   S. W. Porges, "Orienting in a Defensive World: Mammalian Modifications of Our Evolutionary Heritage. A Polyvagal Theory," *Psychophysiology*, 1995, 32, 301–18.

4   C. Kranowitz, *The Out-of-Sync Child, rev ed.* (New York; Perigee: 2006).

5   D. J. Siegel, *Mindsight: The New Science of Personal Transformation* (New York: Bantam Books, 2011), 60–61.

6   L. M. Oberman et al., "EEG Evidence for Mirror Neuron Dysfunction in Autism Spectrum Disorders," *Cognitive Brain Research*, 2005, 24:2, 190–98.

7   A. F. C. Hamilton, "Reflecting on the Mirror Neuron System in Autism: A Systematic Review of Current Theories," *Developmental and Cognitive Neuroscience*, Vol. 3, Jan. 2013, 91-105.

8   R. Klaw, "Thoughtful Response to Agitation, Escalation and Meltdowns in Individuals with Autism Spectrum Disorders," DVD (Pittsburgh, PA: Autism Services by Klaw, 2007).

9   F. Williams, "What to Know About ADHD Masking," MedicalNewsToday, June 19, 2023, https://www.medicalnewstoday.com/articles/adhd-masking.

10 "Milton's 'double empathy problem': A Summary for Non-academics," Reframing Autism, https://reframingautism.org.au/miltons-double-empathy-problem-a-summary-for-non-academics/, accessed June 6, 2025.

11 B. Sulzer-Azaroff and R. Mayer, *Behavior Analysis for Lasting Change* (Fort Worth, TX: Holt, Reinhart & Winston, 1991).

12   This overview of ABA is a compilation from many sources. The author is grateful to Susan Varsames, MA Ed, of the Holistic Learning Center (www.holisticlc.com) for her contributions.

13   J. Rojahn, S. R. Schroeder, and T. A. Hoch, eds., *Self-Injurious Behavior in Intellectual Disabilities* (Amsterdam, The Netherlands: Elsevier, 2007).

14   W. H. Ahearn et al., "On the Role of Preference in Response Competition," *Applied Behavioral Analysis*, 2005, 38:2, 247–50.

15   J. M. Lambert et al, "Trial-Based Functional Analysis and Functional Communication Training in an Early Childhood Setting," *Journal of Applied Behavioral Analysis*, 2012, 45, 579–83.

16   S. D. Hupp et al., "The Effects of Delayed Rewards, Tokens, and Stimulant Medication on Sportsmanlike Behavior w/ ADHD-diagnosed Children," *Behavior Modification*, 2002, 26:2, 148–62.

17   C. P. Bradshaw et al., "Examining the Effects of School-wide Positive Behavioral Interventions and Supports on Student Outcomes: Results from a Randomized Controlled Effectiveness Trial in Elementary Schools," *Journal of Positive Behavior Interventions*, July 2010, 12:3, 133–48.

18   R. H. Horner et al., "Problem Behavior Interventions for Young Children with Autism: A Research Synthesis," *Journal of Autism and Developmental Disorders*, 2002, 32:5, 423–46.

19   Center for Autism and Releated Disorders (CARD), www.CenterforAutism.com, accessed Apr. 13, 2025.

20   B. Rimland, "The ABA Controversy," *Autism Research Review International*, 1999, 13:3, 3.

21   A. Zier and K. Hoehne, "Bridging Sensory Processing Theory and Practice with Discrete Trial Teaching," *New Developments Newsletter*, 1998, 6:3, 4.

22   M. Baker et al., "Increasing Social Behavior of Young Children with Autism Using Their Obsessive Behaviors," *Journal of Association of Persons with Severe Handicaps*, 1998, 23:1, 300–8.

23   L. K. Koegel and R. L. Koegel, *The PRT Pocket Guide: Pivotal Response Treatment for Autism Spectrum Disorders, Second Edition* (Baltimore, MD: Paul H. Brookes Publishing Co, 2018).

24   G. Mesibov et al., *The TEACCH Approach to Autism Spectrum Disorders* (NY, NY: Springer, 2004).

25   J. E. Curtiss et al., "Cognitive-Behavioral Treatments for Anxiety and Stress-Related Disorders," *Focus (Am Psychiatric Publications)*, June 2021;19(2):184-189.

26   S. I. Greenspan and S. Wieder, *The Child with Special Needs: Encouraging Intellectual and Emotional Growth* (Cambridge, MA: Perseus Books, 1998).

27   S. I. Greenspan and S. Wieder, *Engaging Autism: Using the Floortime Approach to Help Children Relate, Communicate and Think* (Cambridge, MA: Perseus Books, 2006).

28   S. I. Greenspan and G. Tippy, *Respecting Autism* (New York: Vantage Press, 2011).

29   S. Wieder and H. Wachs, *Spatial Portals to Thinking, Feeling and Movement: Advancing Competencies and Emotional Development in Children with Learning and Autism Spectrum Disorders* (Mendham, NJ: Profectum Foundation, 2012), 4–5.

30   H. Furth and H. Wachs, *Thinking Goes to School* (New York, NY: Oxford Univ. Press, 1975).

31   Wieder and Wachs, *Spatial Portals to Thinking*.

32   S. Gutstein, *Autism / Asperger's: Solving the Relationship Puzzle* (Arlington, TX: Future Horizons, 2000).

33 R. Kaufman, *Autism Breakthrough: The Groundbreaking Method That Has Helped Families All over the World* (New York, NY: St. Martin's Press, 2015).

34 C. Gray, *My Social Stories Book* (Philadelphia: Jessica Kingsley, 2001).

35 C. Gray, *The New Social Stories Book* (Arlington, Texas: Future Horizons, 2010).

36 C. Gray, *The New Social Story Book, Revised and Expanded 15th Anniversary Edition: Over 150 Social Stories That Teach Everyday Social Skills to Children and Adults with Autism and their Peers* (Arlington, Texas: Future Horizons, 2015).

37 L. Kuypers, *The Zones of Regulation®: A Curriculum Designed to Foster Self-Regulation and Emotional Control* (Social Thinking Publishing, 2011).

38 Gray C. Jenison, *Autism Journal*, Winter 2002, 1–19.

39 B. M. Prizant et al., *The SCERTS™ Model: A Comprehensive Educational Approach for Children with Autism Spectrum Disorders* (Baltimore, MD: Brookes Pub, 2005).

40 B. M. Prizant and A. M. Wetherby, *Autism Spectrum Disorders: A Transactional Developmental Perspective* (Baltimore, MD: Brookes Publishing, 2000).

41 B. M. Prizant, *Uniquely Human: A Different Way of Seeing Autism* (New York, NY: Simon & Schuster, 2022).

42 G. F. Melson, "Child Development and the Human-companion Animal Bond," *American Behavioral Scientist*, 2003, 47:1, 31–9.

43 M. E. O'Haire et al., "Social Behaviors Increase in Children with Autism in the Presence of Animals Compared to Toys," *PLoS ONE*, 2013, 8:2.

44 L. O. Nieforth et al., "Animal-Assisted Interventions for Autism Spectrum Disorder: A Systematic Review of the Literature from 2016 to 2020," *Rev Journal of Autism and Developmental Disorders*, June 2023;10(2):255-280.

45 J. J. Diehl et al., "The Clinical Use of Robots for Individuals with Autism Spectrum Disorders: A Clinical Review," *Research in Autism Spectrum Disorders*, Jan-Mar 2012: 6:1, 249-62.

46 J. Mower, "VR for Autism: Enhancing Social and Emotional Learning," *Autism Parenting Magazine*, Feb. 13, 2025, https://www.thephoenixcensternj.org/app/uploads/2024/11/Autism-Parenting-Magazine-Article_2024_VR.pdf.

47 A. C. M. Rêgo and I. Araújo-Filho, "Artificial Intelligence in Autism Spectrum Disorder: Technological Innovations to Enhance Quality of Life: A Holistic Review of Current and Future Applications," *International Journal of Innovative Research in Medical Science*, Sept. 2024, 9(09):539-552.

48 A. Iannone and D. Giansanti, "Breaking Barriers—The Intersection of AI and Assistive Technology in Autism Care: A Narrative Review," *Journal of Personalized Medicine*, Dec 28, 2023;14(1):41.

49 D. Goleman, *Emotional Intelligence* (New York: Bantam, 2005).

Chapter 20

1 R. Allington, "The Five Pillars of Reading Instruction," *Reading Today*, Jun/Jul 2005.

2 M. B. Bowan, "Learning Disabilities, Dyslexia and Vision: A Subject Review," *Journal of the American Optometric Assoc*, Sept. 2002, 73:9, 553–70.

3 P. Suppes, "Eye-movement Models for Arithmetic and Reading," in Kowler, ed., *Eye Movements and Their Role in Visual and Cognitive Processes* (Elsevier, 1990).

4 P. A. Carpenter and M. A. Just, "What Your Eyes Do While Your Mind Is Reading," in Rayner, ed., *Eye Movements in Reading: Perceptual and Language Processes* (New York: Academic Press, 1983), 275–307.

5 H. G. Furth and H. Wachs, *Thinking Goes to School* (New York: Oxford Univ Press, 1975).

6   M. L. Bender, *Bender-Perdue Reflex Test* (San Rafael, CA: Academic Therapy Publications, 1976).

7   O. Pavlides and T. Miles, *Dyslexia Research and its Application to Education* (Chichester, England: Wiley Publications, 1987).

8   B. O'Hara, *Movement and Learning: Wombat and his Mates Songbook* (Victoria, Australia: The F# Music Company, 2003), 34.

9   N. O'Dell and P. Cook, *Stopping ADHD: A Unique and Proven Drug-free Program for Treating ADHD in Children and Adults* (New York, NY: Avery Publishing Co. 2004).

10   D. Wilson, *The Polyvagal Path to Joyful Learning: Transforming Classrooms One Nervous System at a Time* (New York, NY: Norton Professional Books, 2023).

11   D. Wilson, *The Resilient Learner's Backpack: Mind-Body Activities for Focus, Regulation, and Engagement* (Oro Valley, AZ: Integrated Learner Press, 2024).

12   M. S. Williams and S. Shellenberger, *How Does Your Engine Run? A Leaders' Guide to the Alert Program for Sensory Regulation* (Albuquerque, NM: Therapy Works, 1996).

13   M. S. Williams and S. Shellenberger, *Take Five! Staying Alert at Home and School* (Albuquerque, NM: Therapy Works, 2001).

14   D. Henry, *Tool Chest for Teachers, Parents and Students* (Youngtown, AZ: Henry OT Services, 1999).

15   J. Bluestone, *The Fabric of Autism: Weaving the Threads into a Cogent Theory* (Seattle, WA: The HANDLE Institute, 2004).

16   I. Cohen and M. Goldsmith, *Hands On: How to Use Brain Gym in the Classroom* (Ventura, CA: Edu-Kinesthetics, 2003).

17   P. E. Dennsion and G. E. Dennison, *Brain Gym Teacher's Edition* (Ventura, CA: Edu-Kinesthetics, 2010).

18   C. Hannaford, *Smart Moves: Why Learning Is Not All in Your Head* (Salt Lake City, UT: Great River Books, 2007).

19   D. E. Wilson, *School Moves Poster PE DVD* (Shasta, CA: School Moves, 2002).

20   B. O'Hara, *Beanbag Ditties CD* (Victoria, Australia: The F# Music Company, 2003).

21   P. A. Mueller and D. M. Oppenheimer, "The Pen Is Mightier Than the Keyboard: Advantages of Longhand Over Laptop Note Taking," *Psychological Science*, 25(6), 1159–1168.

22   K. M. Colby and C. Parkison, "Handedness in Autistic Children," *Journal of Autism and Childhood Schizophrenia*, 1977, 7:1.

23   F. Flam, "What's So Special about Left Handers?," The Seattle Times, Nov. 3, 2007, https://www.seattletimes.com/nation-world/whats-so-special-about-left-handers/.

24   N. Geschwind and A. M. Galaburda, "Cerebral Lateralization: Biological Mechanisms, Associations, and Pathology: A Hypothesis and a Program for Research," *Archives of Neurology*, May 1985, 42.

25   N. Bell, *Visualizing and Verbalizing for Language Comprehension and Thinking* (San Luis Obispo, CA: Gander Publications, 2007).

26   S. Wieder and H. Wachs, *Visual/Spatial Portals to Thinking/Feeling and Movement: Advancing Competencies and Emotional Development in Children with Learning and Autism Spectrum Disorders* (New York: Profectum, 2012).

27   R. B. Mumford, "The Role of Lighting," *Visual Impairment Research*, 2004, 6:1, 29–33.

Chapter 21

1   *Ocean Heaven* (Edko Films Limited, 2010).

2   M. Sarris, "Coming of Age: What Awaits Young Adults with Autism?" Simons Powering Autism Research (SPARK), Sept. 30, 2024, https://sparkforautism.org/discover_article/coming-of-age-autism-and-the-transition-to-adulthood/.

3   P. T. Shattuck et al., National Autism Indicators Report: High School Students on the Autism Spectrum. (Philadelphia, PA: Life Course Outcomes Program, A.J. Drexel Autism Institute, Drexel University, 2018) https://drexel.edu/~/media/Files/autismout comes/publications/2018%20NAIR_final_digital_onepage.ashx.

4   A. S. Halpern, "The Transition of Youth with Disabilities to Adult Life: A Position Statement of the Division on Career Development and Transition," *The Council of Exceptional Children*, Fall 1994, 17:2, 117.

5   Individuals with Disabilities Education Improvement Act (IDEIA). 20 U.S.C. § 1401 (3) (26) §300, 2004.

6   "Guardianship Capacity Questionnaire," North Carolina Administrative Office of the Courts, https://www.nccourts.gov/assets/documents/forms/sp208-en.pdf, accessed Apr. 26, 2025.

7   The Center for Special Needs Trust Administration Inc., www.CenterTrust.org, accessed Apr. 25, 2025.

8   F. Nietzsche, https://www.brainyquote.com/quotes/friedrich_nietzsche_101616, accessed Apr. 25, 2025.

9   T. W. Welch, *The Breakaway: A Parent's Guide to Transitioning the Autistic and Twice Exceptional Adolescent into Young Adulthood* (2021).

Chapter 22

1   J. Denler et al., *Creating Quality of Life for Adults on the Autism Spectrum*: The Story of Bittersweet Farms (Routledge, 2022).

2   T. Grandin and K. Dufy, *Developing Talents: Career Planning, Including Higher Education, for Students with Autism and Asperger Syndrome* (Shawnee Mission, KS: Autism Asperger Publishing Company, 2004).

3   A. Schoettle, "Hotel to Pioneer Unique Training Program for People with Disabilities," Indianapolis Business Journal, July 11, 2015. https://www.ibj.com/articles/53959.

4   C. Zillman, "Autistic? More Companies Say Add It to Your Resume," *Fortune Magazine*, Oct .26, 2016. http://fortune.com/2016/10/26/autism-jobs-employment-ey/.

5   J. Chu, "Why SAP Wants to Train and Hire Nearly 700 Adults with Autism," Inc., June 12, 2015, https://www.inc.com/jeff-chu/sap-autism-india.html. Accessed May 15, 2025.

6   Smith, ME. "Microsoft Announces Pilot Program to Hire People with Autism," Microsoft Corporate Blogs, Apr. 3, 2015, https://blogs.microsoft.com/on-the-issues/2015/04/03 /microsoft-announces-pilot-program-to-hire-people-with-autism/#sm.0001tg7031uraei iwl71xkrh32vy6.

7   T. D'Eri, *The Power of Potential: How a Nontraditional Workforce Can Lead You to Run Your Business Better* (New York, NY: Harper Collins Leadership, 2023).

8   L. J. Smethurst et al., "'I've Absolutely Reached Rock Bottom and Have No Energy': The Lived Experience of Unemployed and Underemployed Autistic Adults," *Autism in Adulthood*, May 29, 2024.

9   T. Grandin, *The Loving Push: How Parents and Professionals Can Help Spectrum Kids become Successful Adults* (Arlington, TX: Future Horizons, 2016).

10   J. Endow and M. Mayfield, *The Hidden Curriculum of Getting and Keeping a Job: Navigating the Social Landscape of Employment A Guide for Individuals with Autism Spectrum and Other Social-Cognitive Challenges* (Shawnee Mission, KS: AAPC Publishing, 2012).

11   C. Filler and M. Rosenshein, *Transition to Adulthood, Guidelines for Individuals with Autism Spectrum Disorders, Second Edition* (OCALI, 2012).

12   M. Bernick and L. A. Vismara, *The Autism Full Employment Act: The Next Stage of Jobs for Adults with Autism, ADHD, and Other Learning and Mental Differences* (New York, NY: Skyhorse Publishing, 2021).

13   M. Bernick and R. Holden, *The Autism Job Club: The Neurodiverse Workforce in the New Normal of Employment* (New York, NY: Skyhorse Publishing, 2018).

14   J. E. Robison, *Look Me in the Eye* (New York, NY: Random House, 2007).

15   R. Steward, *The Autism-Friendly Guide to Self-Employment* (London, UK: Jessica Kingsley Publishers, 2021).

16   G. Edmonds and D. Worton, *The Asperger Love Guide: A Practical Guide for Adults with Asperger's Syndrome to Seeking, Establishing and Maintaining Successful Relationships* (Sage Publications, UK, 2005).

17   K. Kerr, "Information Sheet: Romantic Relationships and Autism Spectrum Disorder," Amaze, https://theautismlifespan.com/wp-content/uploads/2024/05/Romantic-Relationships -and-Autism.pdf, accessed May 6, 2025.

18   American Psychiatric Association, *Diagnostic and Statistical Manual of Mental Disorders (5th ed.)* (Arlington, VA: American Psychiatric Publishing, 2013).

19   S. Bererot et al., "Endocrine Disruptors, the Increase of Autism Spectrum Disorder and Its Cormorbidity with Gender Identity Disorder—a Hypothetical Association," *International Journal of Andrology*, 2011, 34: e350.

20   S. H. Swan et al., "Prenatal Phthalate Exposure and Reduced Masculine Play in Boys," *International Journal of Andrology*, 2010: 33: 259–269.

21   H. Hellemans et al., "Sexual Behavior in High-Functioning Male Adolescents and Young Adults with Autism Spectrum Disorder," *Journal of Autism and Developmental Disorders*, 37, 260–269.

22   D. A. Geier and M. R. Geier, "A Prospective Assessment of Androgen Levels in Patients with Autistic Spectrum Disorders: Biochemical Under-pinnings and Suggested Therapies, *Neuro Endrocrinal Lett*ers, 28: 565–573.

23   D. Steinman, *Raising Healthy Kids: Protecting Your Children from Hidden Chemical Toxins* (New York, NY: Skyhorse Publishing, 2024), 183–186.

24   D. F. Swaab and A. Garcia-Falgueras, "Sexual Differentiation of the Human Brain in Relation to Gender Identity and Sexual Orientation," *Functional Neurology*, 24:1, 17–28.

25   M. Lemaire et al., "Gender Identity Disorder and Autism Spectrum Disorder in a 23-Year-Old Female," *Archives of Sexual Behavior*, July 9, 2013, 1573–2800.

26   A. L. C. deVries et al., "Autism Spectrum Disorders in Gender Dysphoric Children and Adolescents," *Journal of Autism and Developmental Disorders*, Aug. 2010: 40: 8, 930–36.

27   M. Tateno et al., "Gender Dysphoria in Pervasive Developmental Disorders, *Seishin Shinkeigaku Zasshi*. 2011;113(12):1173–83.

28   R. Rubio, "Mind/Body Techniques for Asperger's Syndrome (Philadelphia, PA: Jessica Kingsley Publishers, 2008).

29   C. Mitchell, *Asperger's Syndrome and Mindfulness* (Philadelphia, PA: Jessica Kingsley Publishers, 2009).

30   K. P. Koenig et al., "Efficacy of the Get Ready to Learn Yoga Program among Children with Autism Spectrum Disorders: A Pretest–Posttest Control Group Design," *American Journal of Occupational Therapy*, Sept./Oct. 2012: 66 :5 538–546.

31   M. Jones, *The Mindful Warrior: Martial Arts as a Path to Mental Wellbeing* (self-published, 2025).

Chapter 23

1 M. McDonnell, "What Can be Done to Prevent Autism Now?" Age of Autism, Mar. 28, 2010, https://www.ageofautism.com/2010/03/what-can-be-done-to-prevent-autism-now.html.

2 "U.S. Fertility Rate Drops to Another Historic Low," press release from the CDC National Center for Health Statistics, Apr. 25, 2024, https://www.cdc.gov/nchs/pressroom/nchs_press_releases/2024/20240525.htm.

3 "Infant Mortality in the United States," World Population Review, https://worldpopulationreview.com/country-rankings/infant-mortality-rate-by-country#infant-mortality-in-the-united-states, accessed May 21, 2025.

4 J. Margulis, *The Business of Baby: What Doctors Don't Tell You, What Corporations Try to Sell You, and How to Put Your Pregnancy, Childbirth, and Baby BEFORE Their Bottom Line* (New York: Scribner, 2013).

5 S. Kierkegaard, https://www.brainyquote.com/quotes/soren_kierkegaard_105030, accessed May 21, 2025.

6 C. B. Pearson, "How I Gave My Son Autism," Aug. 29, 2014, https://thinkingmomsrevolution.com/how-i-gave-my-son-autism/.

7 P. Lemer, *Envisioning a Bright Future: Interventions That Work for Children and Adults with Autism Spectrum Disorders* (Santa Ana, CA: OEPF, 2008).

8 B. Lambert, *A Compromised Generation* (Boulder, CO: Sentient Publications, 2010).

9 J. Valley, *Healthy From Day 1* (Amazon, 2011).

10 S. Ozonoff et al., "Recurrence Risk for Autism Spectrum Disorders: A Baby Siblings Research Consortium Study," *Pediatrics*, Sept. 2011, 128:3.

11 D. Berger, "From Preconception to Infancy: Environmental and Nutritional Strategies for Lowering the Risk of Autism," *Autism Science Digest*, Issue #4, 2012.

12 R. Hauser, "The Environment and Male Fertility: Recent Research on Emerging Chemicals and Semen Quality," *Semin Reprod Med*, July 2006, 24:3, 156–67.

13 P. Grandjean, *Only One Chance: How Environmental Pollution Impairs Brain Development—and How to Protect the Brains of the Next Generation* (New York: Oxford University Press, 2013).

14 J. M. Skowron, "Autism and the Endocrine Response," https://www.scribd.com/document/379546417/Autism-and-the-Endocrine-Response-docx, accessed July 29, 2025.

15 C. Dosiou, "Thyroid and Fertility: Recent Advances," *Thyroid*, 2020.

16 "Hypothyroidism and Pregnancy," Johns Hopkins Medicine, https://www.hopkinsmedicine.org/health/conditions-and-diseases/staying-healthy-during-pregnancy/hypothyroidism-and-pregnancy, accessed May 22, 2025.

17 R. Zoeller et al., "Thyroid Disruption and Brain Development: What Is It That We Don't Know?," *Neurotoxicology and Teratology*, 2008, 30:3, 248.

18 G. Román et al., "Association of Gestational Maternal Hypothyroxinemia and Increased Autism Risk," *Annals of Neurology*, Aug. 13, 2013.

19 T. Eldar-Geva et al., "Subclinical Hypothyroidism in Infertile Women: The Importance of Continuous Monitoring and the Role of the Thyrotropin-releasing Hormone Stimulation Test," *Gynecological Endcrinology*, 2007, 23:6, 332–37.

20 W. B. Grant and C. M. Soles, "Epidemiologic Evidence Supporting the Role of Maternal Vitamin D Deficiency as a Risk Factor for the Development of Infantile Autism," *Dermatoendocrinology*, July–Aug. 2009, 1:4, 223–28.

21 F. Williams, "Toxic Breast Milk," *New York Times Magazine*, Jan. 9, 2005.

22   J. Mercola, "Still Carrying Around Mercury Next to Your Brain?," Mercola, Sept. 4, 2011, https://articles.mercola.com/sites/articles/archive/2011/09/04/mercury-poisoning-from-silver-fillings.aspx.

23   "Body Burden: The Pollution in Newborns," Environmental Working Group, July 14, 2005, https://www.ewg.org/research/body-burden-pollution-newborns.

24   C. Maren, "Why You Should Detox before Pregnancy," Dr. Christine Maren, June 12, 2025, https://drchristinemaren.com/the-importance-of-preconception-detox/.

25   B. Boyers, "New York Rescue Workers Detoxification Project.," Calm Full Living, https://web.archive.org/web/20180207011736/http://calmfulliving.com/project/the-new-york-rescue-workers-detoxification-project/.

26   "EWG's Good Seafood Guide," Environmental Working Group, https://www.ewg.org/consumer-guides/ewgs-consumer-guide-seafood, accessed May 23, 2025.

27   "BioInitiative Working Group Comments on 2014 SCENIHR Preliminary Opinion on Potential Health Effects of EMF," BioInitiative 2012, Apr. 16, 2014. https://bioinitiative.org/potential-health-effects-emf/.

28   A. J. Romm, *The Natural Pregnancy Book, Third Edition: Your Complete Guide to a Safe, Organic Pregnancy and Childbirth with Herbs, Nutrition, and Other Holistic Choices* (Berkeley, CA: Ten Speed Press, 2014).

29   J. Margulis, *Your Baby, Your Way: Taking Charge of your Pregnancy, Childbirth, and Parenting Decisions for a Happier, Healthier Family* (New York: Scribner, 2015).

30   Q. Jiang et al., "Effects of Yoga Intervention During Pregnancy: A Review for Current Status," *American Journal of Perinatology*, May 2015; 32:6, 503–14.

31   C. L. Borggren, "Pregnancy and Chiropractic: A Narrative Review of the Literature," *Journal of Chiropractic Medicine*, Spring 2007; 6:2, 70–74.

32   S. Buckley, "Ultrasound Scans: Cause for Concern," *Nexus*, Oct–Nov 2002, 9:6.

33   E. L. Williams and M. F. Casanova, "Ultrasound and Autism: How Disrupted Redox Homeostasis and Transient Membrane Porosity Confer Risk," in Dietrich-Muszalska, Gagnon, Chauhan, eds. *Studies on Psychiatric Disorders* (New York: Humana Press, 2013).

34   N. Lund et al., "Selective Serotonin Reuptake Inhibitor Exposure in Utero and Pregnancy Outcomes," *Archives of Pediatrics & Adolescent Medicine*, 2009, 163, 949–54.

35   "Vaccines during Pregnancy," American College of Obstetricians and Gynecologists (ACOG), March 2025, https://www.acog.org/womens-health/infographics/vaccines-during-pregnancy.

36   M. Nevradakis, "Another Study Shows Higher Miscarriage Rate among Women Who Received COVID Vaccines," *The Defender*, May 5, 2025. https://childrenshealthdefense.org/defender/spanish-study-high-miscarriage-rate-covid-vaccine-women/.

37   "Cocooning Protects Babies," Immunization Action Coalition, https://www.immunize.org/wp-content/uploads/va/va37_cocooning.pdf, accessed May 24, 2025.

38   A. E. Blain et al., "An Assessment of the Cocooning Strategy for Preventing Infant Pertussis—United States, 2011," *Clinical Infectious Diseases*, Dec. 1, 2016; 63 (Suppl 4): S221–S226.

39   K. Johnson et al., "Outcomes of Planned Home Births with Certified Professional Midwives: Large Prospective Study in North America," *BMJ*, June 2005, 330, 1416.

40   H. Gardener et al., "Prenatal Risk Factors for Autism: A Comprehensive Meta-analysis," *British Journal of Psychiatry*, July 2009, 195:1, 7–14.

41   R. Mendelsohn, *How to Raise a Healthy Child in Spite of Your Doctor* (New York: Ballantine Books, 1987).

42   J. Zand et al., *Smart Medicine for a Healthier Child* (Garden City Park, NY: Avery, 2003).

43 R. Neustaedter, *The Holistic Baby Guide: Alternative Care for Common Health Problems* (Oakland, CA: New Harbinger Publications, 2010).

44 "Safe Baby Monitors," Dr. Magda Havas, PhD, Nov. 24, 2015, https://magdahavas.com/health-issues/safe_baby_monitors/.

45 E. Seli and T. L. Horvath, "Natural Birth-Induced UCP2 in Brain Development," *Rev Endocr Metab Disord*, Dec. 2013; 14:4, 347–50

46 N. T. Mueller et al., "The Infant Microbiome Development: Mom Matters," *Trends in Molecular Medicine*. Feb. 2015; 21:2, 109–117.

47 M. G. Dominguez-Bello et al., "Partial Restoration of the Microbiota of Cesarean-Born Infants Via Vaginal Microbial Transfer," *Nature Medicine*, 2016, Vol. 22, 250–3.

48 G. Dick-Read, *Childbirth without Fear: The Principles and Practice of Natural Childbirth* (London: Pinter & Martin Ltd, 2013).

49 R. Dekker, "Placenta Encapsulation," Evidence Based Birth, Feb. 22, 2017, https://evidencebasedbirth.com/evidence-on-placenta-encapsulation/.

50 C. Bradburn, "Why Infants Need Chiropractic Care," *Pathways to Wellness*, Issue #50, Summer, 2016.

51 O. Ressel, *Kids First: Health with No Interference* (Garden City Park, NY: Square One Publishers, 2013).

52 J. Y. Meek, "Infant Benefits of Breastfeeding," Nov. 7, 2024, https://www.uptodate.com/contents/infant-benefits-of-breastfeeding.

53 "What Are the Recommendations for Breastfeeding?," National Institute of Child Health and Human Development, https://www.nichd.nih.gov/health/topics/breastfeeding/conditioninfo/recommendations, accessed May 25, 2025.

54 "Breastfeeding," World Health Organization, https://www.who.int/health-topics/breastfeeding#tab=tab_2, accessed June 6, 2025.

55 "Infant Food and Feeding," American Academy of Pediatrics, https://www.aap.org/en/patient-care/healthy-active-living-for-families/infant-food-and-feeding/, accessed May 25, 2025.

56 "Is Baby Ready for Solid Foods," Kelly Mom, https://kellymom.com/nutrition/starting-solids/solids-when/. accessed May 25, 2025.

57 "Caution with Walkers, Jumpers, Exersaucers," Children's Rehabilitation Institute, https://critusa.org/caution-with-walkers-jumpers-exersaucers/, accessed June 5, 2025.

58 J. Deardorff, "Therapists See No Developmental Benefits from Seats," *Chicago Tribune*, Mar. 15, 2012.

59 S. Newman, "How Digital Devices Affect Infants and Toddlers," *Psychology Today*, July 31, 2014.

60 "Safe to Sleep," National Institute of Child Health and Human Development, https://safetosleep.nichd.nih.gov/, accessed May 25, 2025.

61 G. S. Goldman and R. Z. Cheng, "The Immature Infant Liver: Cytochrome P450 Enzymes and their Relevance to Vaccine Safety and SIDS Research," *International Journal of Medical Sciences*, Apr. 28, 2025;22(10):2434–2445.

62 B. A. Richardson, "Sudden Infant Death Syndrome: A Possible Primary Cause," *Journal of Forensic Sci Soc* 1994; 34:199–204.

63 T. J. Sprott, *The Cot Death Cover-up* (Auckland, NZ: Penguin Environmental, 1996).

64 A. Baker, "VAERS Data Indicates Significant Number of Infant Deaths Within a Week of Vaccination," *The Vaccine Reaction,* Mar. 10, 2025, https://thevaccinereaction.org/2025/03/study-finds-sudden-infant-deaths-have-increased/#_edn5.

65 K. E. Adolph, "Developmental Continuity? Crawling, Cruising, and Walking," *Developmental Science*, Mar. 2011; 14:2, 306–18.

66   "Link between Autism and Vaccination Debunked," Mayo Clinic Health System, Mar. 24, 2022, https://www.mayoclinichealthsystem.org/hometown-health/speaking-of-health/autism-vaccine-link-debunked.

67   D. Moon, "Vaccines and the MTHFR Mutation," Genetic Lifehacks, Sept. 16, 2024, https://www.geneticlifehacks.com/mthfr-and-vaccinations/.

68   W. Shaw, "Evidence That Increased Acetaminophen Use in Genetically Vulnerable Children Appears to Be a Major Cause of the Epidemics of Autism, Attention Deficit with Hyperactivity, and Asthma," *Journal of Restorative Medicine*, 2013, 2, 9–16.

69   "Vaccine Exemptions by State 2025," World Population Review, https://worldpopulation review.com/state-rankings/vaccine-exemptions-by-state, accessed May 26, 2025.

70   "State Vaccine Laws," National Vaccine Information Center (NVIC), https://www.nvic .org/law-policy-state/vaccine-laws, accessed June 6, 2025.

71   "Top 20 Questions about Vaccination," History of Vaccines, http://www.historyofvaccines .org/content/articles/top-20-questions-about-vaccination#5, accessed May 26, 2025.

72   M. Sarris, "Does a Baby's Gaze Hold a Clue to Autism?," Simons Powering Autism Research (SPARK), May 8, 2024, https://sparkforautism.org/discover_article/gaze -clue-to-autism/.

73   D. Berger, *How to Prevent Autism: Expert Advice from Medical Professionals* (New York, NY: Skyhorse Publishing, 2017).

Afterword

1   The President's Make America Healthy Again Commission, "The MAHA Report: Making Our Children Healthy Again—Assessment," The White House, https://static01 .nyt.com/newsgraphics/documenttools/fd441e56ad4bcf36/2f18e38b-full.pdf, accessed May 27, 2025.

2   S. Ji, "The Day the Grassroots Became Government: A Celebration of Courage, Community, and Change," GreenMedInfo, May 23, 2025, https://greenmedinfo.com/content/day -grassroots-became-government-celebration-courage-community-and-change.

INDEX

norepinephrine, 212–213
nosodes, 248
Nsouli, Talal, 162
NT. *See* neurotypical (NT)
number learning, 72–73
nutrigenomics, 82
nutrigenomic testing, 79, 82–83, 439
nutrition. *See* diet; food

O

obsessive-compulsive and related disorder
  (OCD)
  diagnosis of, 7
  employment and, 425
Obukhanych, Tetyana, 178
OCD. *See* obsessive-compulsive and
  related disorder (OCD)
OCN. *See* On Cloud Nine (OCN)
O'Hara, Brendan, 384
O'Hara, Nancy, 200
oils, essential, 256–257
Olmstead, Dan, 13
Olsen, Jan, 393
OMT. *See* osteopathic manipulative
  therapy (OMT)
On Cloud Nine (OCN), 392
One-Minute Moves, 392
ophthalmologists, 321
opioid excess theory, 125–127
optometric phototherapy, 328
optometrists, 321
oral microbiome, 259
oral praxis, 291
organic food, 63
organizational skills, 397–398
osteopathic manipulative therapy (OMT),
  110–111
osteopathy, 110–111
O'Toole, Zoey, 179
overdiagnosis, 13–14
Owens, Susan, 149
oxalates, 149–150
oxidation, 226
oxidative stress, 26–27
oxygen therapies, 255

oxytocin (OXT), 210–211, 221
oxytocin treatment, 220–222
ozone therapy, 255

P

PACE, 389
Palevsky, Larry, 178
Palmer, Christopher, 23–24, 147, 209,
  214
PANDAS. *See* pediatric autoimmune
  neuropsychiatric disorder associated
  with streptococcal infections
  (PANDAS)
PANS. *See* pediatric acute-onset
  neuropsychiatric syndrome (PANS)
parasites, 201
partials, in dentistry, 268–269
particulate matter, 34
pathogens, toxins from, 39–40
pathological demand avoidance (PDA), 8
PBDEs. *See* polybrominated diphenyl
  ethers (PBDEs)
PCBs. *See* polychlorinated biphenyls
  (PCBs)
P-cresol sulfite, 120
PDA. *See* pathological demand avoidance
  (PDA)
PDD. *See* pervasive developmental
  disorder (PDD)
pediatric acute-onset neuropsychiatric
  syndrome (PANS), 163, 198–199
pediatric autoimmune neuropsychiatric
  disorder associated with streptococcal
  infections (PANDAS), 163, 198–199
pediatricians, 20–21
  developmental, 21
  as testing resource, 79–80
PEMF. *See* pulsed electromagnetic
  frequency therapy (PEMF)
peristalsis, 116
personal care products, 38, 76
person-first language (PFL), 9–11
pervasive developmental disorder
  (PDD), 6
pest control, 65

ABOUT THE AUTHOR

Photo credit: Stanley Klein

Patricia S. Lemer is a pioneer and an internationally recognized authority on applying a holistic model of care to autism, ADHD, learning and mental health conditions. For over fifty years she has demonstrated an unwavering commitment to identifying and ameliorating root causes rather than treat symptoms.

Ms. Lemer was the first to apply the "Total Load Theory" to neurodevelopmental disorders. This engineering concept postulates that each unique individual, like a bridge, accumulates an overload of stressors that eventually causes dysfunction.

Ms. Lemer cofounded one of the first organizations offering information online. She helped establish the first center for young adults with disabilities in Kuwait. She is the author of four books, including the award-winning *Outsmarting Autism* (2019) and the moderator of the podcast "The Autism Detective." Patricia's ability to weave complex concepts with practical solutions for empowering individuals of all ages has earned her the admiration of families and professionals alike.

NOTES

NOTES

NOTES

NOTES

NOTES

NOTES

NOTES

NOTES

NOTES

NOTES

NOTES